Origin and Evolution of the Elements

This comprehensive volume reviews current understanding of the origin
and evolution of elements, from stellar nucleosynthesis to the chemical
evolution of the cosmos. With chapters by leading authorities in the field,
it describes models of how the elements are produced by stars, the nuclear
processes involved, and how the quantity of elements evolved in our
Galaxy and distant galaxies. The observed chemical composition of stars
in defferent locations within our Galaxy and nearby galaxies is discussed,
as are the compositions of hot and cold gases, of dust grains found
between stars and in meteorites, and of the integrated light from distant
galaxies and quasars. This authoritative volume is a valuable resource for
graduate students and professional research astronomers.

ANDREW MCWILLIAM obtained his Ph.D. in astronomy from the
University of Texas at Austin. He has been a staff astronomer at the
Carnegie Observatories since 1997.

MICHAEL RAUCH obtained his Ph.D. in astronomy from Cambridge
University. He has been a staff astronomer at the Carnegie Observatories
since 2000.

This series of four books celebrates the Centennial of the Carnegie
Institution of Washington, and is based on a set of four special symposia
held by the Observatories in Pasadena. Each symposium explored an
astronomical topic of major historical and current interest at the
Observatories, and each resulting book contains a set of comprehensive,
authoritative review articles by leading experts in the field.

Series Editor: Luis C. Ho.
Luis Ho received his undergraduate education at Harvard University and
his Ph.D. in astronomy from the University of California at Berkeley. He is
currently a staff astronomer at the Carnegie Observatories, where he
conducts research on black holes, accretion physics in galactic nuclei, and
star formation processes.

Carnegie Observatories Astrophysics Series
Volume 4

ORIGIN AND EVOLUTION OF THE ELEMENTS

Edited by

ANDREW McWILLIAM

and

MICHAEL RAUCH

CAMBRIDGE
UNIVERSITY PRESS

CAMBRIDGE UNIVERSITY PRESS
Cambridge, New York, Melbourne, Madrid, Cape Town, Singapore,
São Paulo, Delhi, Dubai, Tokyo, Mexico City

Cambridge University Press
The Edinburgh Building, Cambridge CB2 8RU, UK

Published in the United States of America by Cambridge University Press, New York

www.cambridge.org
Information on this title: www.cambridge.org/9780521143950

First published 2004
First paperback printing 2010

A catalogue record for this publication is available from the British Library

ISBN 978-0-521-75578-8 Hardback
ISBN 978-0-521-14395-0 Paperback

Contents

Introduction

This volume contains the review articles presented at the Carnegie Symposium on *Origin and Evolution of the Elements*, held in Pasadena, February 2003; contributed papers and poster papers presented at the meeting can be obtained online from the Carnegie web site.

Given the pivotal role played by Carnegie astronomers in our understanding and interpretation of spectra of the elements, it was altogether fitting that we celebrate the Carnegie 100 year anniversary with a conference on the origin and evolution of the elements.

In planning the meeting one of our objectives was to connect observers from disparate sub-fields: from those studying microscopic pre-solar grains embedded in meteorites, under laboratory investigation, to observers measuring element abundances in high-redshift clouds, at the largest distance scale, using the world's largest telescopes. Our hope was that by combining groups of researchers with such diverse, but related, interests in element abundances we might foster unexpected connections of interest and benefit to all. We feel that this is particularly relevant at a time when large telescopes are now making it possible to study individual stars in distant galaxies, with detail previously only available for the solar neighborhood, thus blurring the distinction between the study of Galactic evolution and the evolution of galaxies. At the very least we hoped to educate ourselves and to foster an appreciation by our colleagues of the different sub-disciplines of chemical evolution. We also wished to combine the results of observations with theoretical predictions of element yields from stellar sources and galactic chemical evolution models.

Given the great breadth of our field, and the limited time for the conference, it was not possible to include all aspects of the origin and evolution of the elements; for this reason we elected not to cover some important sub-disciplines, such as Big Bang nucleosynthesis.

Andrew McWilliam and Michael Rauch
Carnegie Observatories
December 2003

List of Participants

Adelberger, Kurt	Harvard University, USA
Allende Prieto, Carlos	University of Texas, Austin, USA
Andersen, Johannes	Nordic Optical Telescope, Denmark
Arnett, Dave	University of Arizona, USA
Balachandran, Suchitra	University of Maryland, USA
Barbieri, Cesare	University of Padova, Italy
Becker, Stephen A.	Los Alamos National Lab, USA
Bensby, Thomas	Lund Observatory, Sweden
Boesgaard, Ann	IfA, University of Hawaii, USA
Boissier, S.	Carnegie Observatories, USA
Bolte, Michael	U. C. Santa Cruz, USA
Bosler, Tammy L.	U. C. Irvine, USA
Bresolin, Fabio	IfA, University of Hawaii, USA
Brinchmann, Jarle	MPA, Garching, Germany
Burbidge, Margaret	U. C. San Diego, USA
Busso, Maurizio	Univ. of Perugia, Italy
Calura, Francesco	Univ. of Trieste, Italy
Campbell, Simon	Monash Univeristy, Australia
Carrera, Ricardo	Instituto de Astrofisica de Canarias, Spain
Carswell, Bob	Institute of Astronomy, Cambridge, UK
Cayrel, Roger	Observatoire de Paris, France
Charbonnel, Corinne	Lab. D'Astrphysique de Toulouse, France
Chiappini, Cristina	Osservatorio Astronomico di Trieste, Italy
Christlieb, Norbert	Univ. of Hamburg, Germany
Clayton, Donald	Clemson University, USA
Cohen, Judy	Caltech, USA
Cole, Andrew	Kapteyn Institute, Netherlands
Cowan, John	Univ. of Oklahoma, USA
Cunha, Katia	Observatorio Nacional, Brazil
Datta, Srabani	Univ. of Calcutta, India
Dietrich, Matthias	Georgia State University, USA
Dinerstein, Harriet	University of Texas, Austin, USA

Draine, Bruce	Princeton University, USA
Duncan, Doug	University of Colorado, USA
Elliott, Lisa	Monash Univeristy, Australia
Ellison, Sara	Universidad Catolica. Chile
Feltzing, Sofia	Lund Observatory, Sweden
Fenner, Yeshe	Swinburne University, Australia
Ford, Allie	Monash Univeristy, Australia
Franchini, Mariagrazia	NAF-Osservatorio Astronomico di Trieste, Italy
Friel, Eileen	National Science Foundation, USA
Fulbright, Jon	Carnegie Observatories, USA
Gallagher, John S.	University of Wisconsin, USA
Gallart, Carme	Instituto de Astrofisica de Canarias, Spain
Gallino, Roberto	Univ. of Torino, Italy
Garnett, Don	University of Arizona, USA
Gawiser, Eric	U. C. San Diego, USA
Gibson, Brad	Swinburne University, Australia
Grebel, Eva	MPIA, Heidelberg, Germany
Gustafsson, Bengt	Univ. of Uppsala, Sweden
Hamann, Fred	Univ. of Florida, USA
Henry, Dick	Univ. of Oklahoma, USA
Hill, Vanessa	Observatoire de Paris, France
Ho, Luis	Carnegie Observatories, USA
Hughes, John	Rutgers University, USA
Israelian, Garik	Instituto de Astrofisica de Canarias, Spain
Ivans, Inese	Caltech, USA
Jenkins, Edward	Princeton University, USA
Johnson, Jennifer	Carnegie Observatories, USA
Jonsell, Karin	Univ. of Uppsala, Sweden
Jugaku, Jun	Research Inst. of Civilization, Japan
Karakas, Amanda	Monash Univeristy, Australia
Korn, Andreas	MPE, Garching, Germany
Kraft, Robert	UCO Lick Observatory, USA
Lambert, David	University of Texas, Austin, USA
Lattanzio, John	Monash Univeristy, Australia
Lee, Henry	MPIA, Heidelberg, Germany
Linsky, Jeffrey	Univ. Colorado, JILA, USA
Loewenstein, Michael	NASA/GSFC, USA
Lopez, Sebastian	Universidad de Chile, Chile
Lubowich, Donald	American Inst. Phys. and Hofstra Univ., USA
Maiolino, Roberto	Osservatorio Astro. di Arcetri, Italy
Martin, Crystal	U. C. Santa Barbara, USA
Mathews, Grant J.	Univ. of Notre Dame, USA
Matteucci, Francesca	Trieste University, Italy
McWilliam, Andrew	Carnegie Observatories, USA
Molaro, Paolo	Trieste University, Italy
Molla, Mercedes	Univ. Autonoma de Madrid, Spain
Nissen, Poul	Univ. of Aarhus, USA
Nittler, Larry	DTM/Carnegie, USA

Nomoto, Ken'ichi	University of Tokyo, Japan
Nordstrom, Birgitta	Nils Bohr Institute, Denmark
Norris, John	Australian National Univ., Australia
Oey, Sally	Lowell Observatory, USA
Origlia, Livia	Bologna Observatory, Italy
Otsuki, Kaori	Univ. of Notre Dame, USA
Pagel, Bernard	Univ. of Sussex, UK
Pancino, Elena	Osservatorio Astronomico di Bologna
Peroux, Celine	Osservatorio Astronomico di Trieste
Peterson, Ruth	UCO Lick Observatory, USA
Philip, Davis	ISO
Preston, George	Carnegie Observatories, USA
Prochaska, Jason	UCO Lick Observatory, USA
Przybilla, Norbert	Univ. Obs. Munich, Germany
Rauch, Michael	Carnegie Observatories, USA
Reddy, Bacham	University of Texas, Austin, USA
Reyniers, Maarten	Inst. of Astronomy, K.U. Leuven, Belgium
Rich, R. Michael	UCLA, USA
Sargent, Wallace	Caltech, USA
Scalo, John	University of Texas, Austin, USA
Schuster, William J.	UNAM, Mexico
Shetrone, Matthew	University of Texas, Austin, USA
Simcoe, Rob	Caltech, USA
Simmerer, Jennifer	University of Texas, Austin, USA
Smecker-Hane, Tammy	U. C. Irvine, USA
Smith, Michael	Oak Ridge National Lab., USA
Smith, Verne	University of Texas at El Paso, USA
Sneden, Chris	University of Texas, Austin, USA
Spite, Monique	Observatoire de Paris, France
Stanford, Laura	Mt. Stromlo Obs., Australia
Starrfield, Sumner	Arizona State University, USA
Straniero, Oscar	Osservatorio Astronomico Collurania, Italy
Trager, Scott	Kapteyn Institute, Netherlands
Travaglio, Claudia	MPA, Garching, Germany
Trimble, Virginia	U. C. Irvine, USA
Tsujimoto, Takuji	National Astronomical Obs., Japan
Venn, Kim	Macalester College, USA
Verner, Ekaterina	GSFC/NASA, USA
Wallerstein, George	Univ. of Washington, USA
Wasserburg, Jerry	Caltech, USA
Wiklind, Tommy	Space Telescope Science Institute, USA
Wolfe, Art	U. C. San Diego, USA
Woolf, Vincent	Univ. of Washington, USA
Worthey, Guy	Washington State University, USA
Yong, David	University of Texas, Austin, USA
Zhao, Gang	National Astro. Obs. of China, China
Zickgraf, Franz-Josef	Univ. of Hamburg, Germany
Zucker, Daniel	MPIA, Heidelberg, Germany

1

Mount Wilson Observatory
contributions to the study of cosmic
abundances of the chemical elements

GEORGE W. PRESTON
The Observatories of the Carnegie Institution of Washington

This gathering is a centennial celebration of Carnegie Astronomy in Pasadena as well as a stand-alone scientific symposium. My job is to present the celebratory stuff and leave current science to experts, who abound in this audience. I focus my attention on the contributions of the six scientists in Figure 1.1 who, in their own ways, made significant contributions to the investigation of cosmic abundances during their researches at the Mount Wilson Observatory. They serve as archetypes of scientists who worked at Santa Barbara Street in the first half of the 20th century.

George Ellery Hale, our first Director, is hardly representative of a class. He is a class unto himself. I make two observations about Hale's early years at Carnegie.

First, he had to earn his niche. Hale was not simply given the Directorship of a big Carnegie-sponsored observatory in California. In 1904 the final configuration of the Carnegie Institution of Washington was still up for grabs. President Gilman favored NSF-style grants to individuals who would work in Universities. Secretary Charles Walcott, Andrew Carnegie's trusted advisor, envisioned a Carnegie campus with structure like that of his own USGS or NBS. It was in this period, when the Trustees themselves were responding to requests to study, variously, volcanic explosions on Martinique and snake venoms (one of the trustees had been bitten by a snake!), that Hale applied for money to investigate Mount Wilson as a site to pursue his new "astrophysics" (Yochelson 1994). On reading in the Chicago Tribune about Andrew Carnegie's venture into philanthropy, Hale fired off a proposal, and soon discovered that he was in direct competition with "classical" astronomers who were also seeking and receiving Carnegie support, as shown in Table 1.1. Thus, in 1904 Hale's Mount Wilson proposal garnered less than half of the Carnegie allocation for astronomy. However, by 1910 Carnegie's program of individual grants had all but disappeared, three solar telescopes and the 60-inch reflector were operating on Mount Wilson, and Hale had won the battle for Carnegie's astronomy money.

Second, Hale surrounded himself with builders and experimenters, not scholars. In 2003 it is no big deal to execute an ambitious observational abundance program, because there is a high-resolution spectrograph on virtually every major telescope in the world. In contrast there were none in 1903, and Hale set out to rectify that situation. Inspired by the pioneering identifications of chemical elements in the Sun and stars by Kirchoff, Bunsen, and Huggins, Hale set out to put spectrographs at the foci of large telescopes. To accomplish this, he hired engineers, physicists, an optician, and a photographer. The reader may be surprised to learn that during the first five years of operation, Walter Adams was the only astronomer on the

G. E. HALE
(c. 1900)

A. S. KING
(c. 1925)

H. N. RUSSELL
(c. 1915)

H. W. BABCOCK
(c. 1950)

L. H. ALLER
(c. 1950)

P. W. MERRILL
(c. 1950)

Fig. 1.1. Photographs of six scientists who worked at Santa Barbara Street in the first half of the 20th century.

Mt. Wilson staff, and, as shown in Table 1.2, after a decade there were but three astronomers among a staff of 12.

Hale also recognized that he must understand the spectra emitted by atoms before he could interpret the spectra of stars, and to this end an atomic physicist was among his early appointments.

Arthur Scott King joined Hale in Pasadena in 1909, six years after earning his PhD in physics at UC Berkeley, and five years before Niels Bohr published his quantum theory of the atom. In his dissertation King had investigated how atomic emission lines are affected by the conditions under which they are produced—just what Hale was looking for. King was immediately put to work supervising the construction of a laboratory on Mount Wilson, later moved to Pasadena where there was sufficient commercial power to operate an electric furnace. The core of the laboratory was a 30-foot pit spectrograph, surrounded by the gadgets of atomic spectroscopy.

In this laboratory King undertook a remarkable series of investigations of atomic and

Table 1.1. *Carnegie Grants For Astronomy in 1904*

Investigator	Topic		Award
L. Boss	star position catalog	$	5,000
W. Campbell	star positions, radial velocities	$	4,000
H. Davis	meridian circle data	$	˙ 1,500
G. E. Hale	parallaxes with Yerkes 40-inch	$	4,000
S. Newcomb	tests of the law of gravity	$	2,500
W. Reed	variable star observers	$	1,000
H. N. Russell	photographic parallax	$	1,000
G. E. Hale	Mt. Wilson site testing et al.	$	15,000

Table 1.2. *Hale's Mt. Wilson Appointments 1904–1913*

Year	Name	Expertise	From
1904	Walter Adams	Astronomy	Yerkes Obs.
1904	Ferdinand Ellerman	Photography	Yerkes Obs.
1904	Francis Pease	Mechanical Eng.	Yerkes Obs.
1904	George Ritchey	Optics	Yerkes Obs.
1906	Henry Gale	Physics	U. Chicago
1908	Charles St. John	Physics	Oberlin College
1909	Harold Babcock	Electrical Eng.	U. C. Berkeley
1909	Arthur King	Physics	U. C. Berkeley
1909	Frederick Seares	Astronomy	U. Missouri
1912	John Anderson	Physics	Johns Hopkins U.
1912	Adrian van Maanen	Astronomy	U. Utrecht
1913	No appointments	—	—

molecular spectra during the next three decades. He obtained high-resolution spectra, measured accurate wavelengths, and made intensity estimates for lines produced in an electric arc, a spark, between the poles of an electromagnet, and in a furnace that operated at temperatures up to 3000 K. He and his son Robert used absorption tube furnace spectra to calculate relative f-values for lines of Fe I (King & King 1935, 1938). Curious about the quality of their results, I plotted them against the data tabulated by Fuhr, Martin, & Weiss (1988) to obtain Figure 28.2. Considering the checkered history of Fe I f-values, the King data aren't half bad. In the course of these investigations King discovered the doublet structure of the Li I resonance lines (King 1916), and Moore & King (1943) identified the ultimate line of Th II, now used to estimate stellar ages. Analysis of the spectrum the C_2 molecule with Raymond Birge (King & Birge 1930) led to the discovery of a new isotope of carbon, ^{13}C, the abundance of which, relative to ^{12}C, provides important clues about stellar structure and nucleosynthesis in stars. Russell (1935) summed up King's work in his Darwin Lecture thus: "Thirty years of King's assiduous work have resulted in temperature classification, on a uniform basis, for a majority of all the elements, including some of the rarest, such as europium and rhenium. His results have been of inestimable value in the analysis of complex spectra." We all use King's work every time we select atomic lines for abundance analyses. A measure of its importance is that we take it for granted.

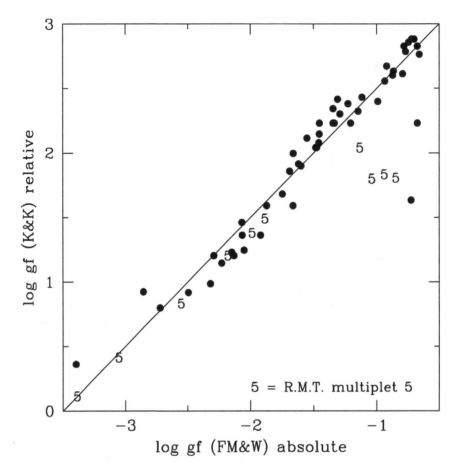

Fig. 1.2. A plot of relative log *gf* values from King & King (1935, 1938) against absolute log *gf* values tabulated by Fuhr et al. (1988). The solid line indicates a 1:1 correlation. Note that not all R.M.T. multiplet 5 lines follow the trend.

Henry Norris Russell is my archetype for the Research Associates that Hale and Adams invited to Pasadena to exploit the Mount Wilson/SBS facilities (other examples: Jeans, Kapteyn, Michelson, and Stebbins). Each was offered salary during recurrent leaves of absence from a permanent position elsewhere. Their associations were long-lasting, amounting to decades in most instances.

Russell first came to Pasadena in 1921 and returned for periods of two to three months annually for the next twenty years. Already famous (with Ejnar Hertzsprung) for the diagram that followed from his work on stellar parallaxes, he came to Pasadena hunting for stellar data to test Saha's (1921) new ionization theory that had proven so successful in predicting Annie Cannon's spectral sequence with an apparently universal set of cosmic chemical abundances. The Mount Wilson Observatory possessed three archives that contained such data: (1) a large database of solar spectrum measurements, (2) a growing archive of high-resolution stellar spectra gathered by Adams, Joy, Merrill, and Sanford, and (3) Arthur King's collection of atomic and molecular spectra.

What the Observatory got was an education: at Hale's request Russell gave lectures on atomic physics, and attendance by the Observatory staff was mandatory. What Russell got is now history. Russell began to solve the many puzzles of atomic structure presented by King's data, searching for series amid forests of lines, using King's temperature classes and Harold Babcock's Zeeman patterns to classify levels, and finally deducing electronic configurations and their associated energy-level diagrams. Russell likened these analyses to cross-word puzzles in no fewer than 16 *Scientific American* articles in the 1920's (DeVorkin 2000). He even devoted a portion of his George Darwin Lecture (Russell 1935) to this analogy. In parallel with his atomic analyses Russell, Adams, & Moore (1928) calibrated the Roland intensities of solar lines, and Adams & Russell (1928) outlined procedures for the analysis of stellar spectra. This effort culminated in Russell' s (1929) abundance analysis of the Sun. His results have stood the test of time: the steady decline of abundances with increasing atomic number, the well-known prominence of even atomic numbers over odd, the exceptionally large abundance of iron, and the extreme rarity of very light lithium and beryllium. In 50 papers about the structures of atoms and stellar atmospheres published in *The Astrophysical Journal* and *Physical Review* from 1921 to 1941 Russell had used his association with the staff of the Mount Wilson Observatory to lay the foundations of quantitative spectrum analysis of the stars. During this time Russell & Saunders (1925) devised their LS coupling scheme for atoms with two valence electrons, and the results were published in Hale's *Astrophysical Journal* under a disarming title: *New Regularities in the Spectra of the Alkaline Earths*. The Abstract of this elegant paper is worth a read.

Horace Welcome Babcock, last manager of the Mount Wilson Grating Laboratory, is representative of the Instrumentalists who have provided Carnegie astronomers over the years with the wherewithal to pursue spectroscopic investigations of the stars. One major effort was devoted to the production of diffraction gratings. Under the supervision of John Anderson (1912–1928) and Harold Babcock (1929–1947), two ruling engines were developed. The second of these was finally brought to a state that satisfied the requirements for high-resolution stellar spectroscopy (Babcock & Babcock 1951). In 1948 Horace Babcock assumed responsibility for the grating laboratory and began to generate useful gratings on a routine basis. "Routine" is perhaps an inappropriate descriptor: the sub-basement at Santa Barbara Street is filled with carefully labeled "factory seconds" set aside because of minor flaws. Horace distributed successful ones to 30 institutions around the world (Babcock 1986). Charity begins at home: three interchangeable gratings were installed at the Mount Wilson 100-inch coudé spectrograph, and a mosaic of four were provided for the Palomar 200-inch coudé spectrograph. The competition was not neglected. I recall, as a graduate student at Lick Observatory in 1958, the arrival of Olin Wilson from Pasadena carrying two of Horace's gratings for the newly completed 120-inch coudé spectrograph. George Herbig would later use one of these gratings to conduct his pioneering survey of lithium in G-dwarfs.

Lawrence H. Aller, one of the many distinguished Guest Investigators at Mount Wilson, co-authored a celebrated paper (Chamberlain & Aller 1951) that established convincingly, for the first time, that so-called A-type "subdwarfs" were actually much cooler than A-type stars, and possessed startlingly low abundances of the metals. Curious about the origins of this paper and the reception it received, I visited Aller at his home and talked to Chamberlain by telephone last December.* I learned that circa 1950 Aller spent summers in Pasadena, "doing some observing and hanging around (sic) Santa Barbara Street the rest of the time." Aller didn't make the observations of HD 19445 and HD 140283. Roscoe Sanford, a "nice guy" (Aller's words), gave him the spectra, suggesting that an investigation of these hydrogen deficient (sic) "intermediate white dwarfs" (Adams & Joy 1922) might be rewarding. Aller developed his new ideas about these stars while discussing them in his graduate course at the University of Michigan. He gave the problem to Joseph Chamberlain, a graduate student, who "did all the dirty work" (Aller's words).

Acceptance of their result was neither universal nor immediate. Chamberlain told me that before he read the paper at the Bloomington AAS meeting, Aller warned him "We're gonna catch hell from Jesse (Greenstein)," but, Aller recalled, "Jesse never said a word." So far so good, because Greenstein had become the Local Guru following publication of his famous 1948 paper on the analysis of F-type stars (Greenstein 1948). However, when I asked Aller about reactions he didn't mention Jesse. He just said "Unsold didn't believe it" (Unsold was The Guru of Physik der Sternatmosphaeren to whom all looked for approval). Nor, I might add, did Unsold believe it seven years later in 1958 during a visit to Lick Observatory, when I tried to tell him about the variety of metal-poor RR Lyrae stars I was finding in my thesis study of their K-line strengths. Unsold dismissed my results, patiently reminding me that weak K lines also had been found in metallic line stars and that explanations undoubtedly lay, not in low abundances, but in peculiarities of their reversing layers, perhaps unusual opacities or abnormal temperature structures. And Unsold was not alone. As late as 1962 Otto Struve, still a towering figure in stellar astrophysics, expressed reservations (Struve & Zeebergs 1962) about the concept of chemical evolution that had been discussed at length five years earlier by B^2FH (Burbidge et al 1957). Long-held dogma about universal cosmic abundances died slowly.

Paul Willard Merrill, a Mount Wilson Observatory Staff Member, earns a place in my sextet because of his discovery of purely radioactive technetium in a subset of long-period variable stars (Merrill 1952) we now know to be s-process rich. His discovery provided the first direct evidence for recent nucleosynthesis of heavy elements within a star. The title of Merrill's discovery paper bears his characteristically conservative stamp: *Spectroscopic Observations of Stars of Class S*. He was generally skeptical of theoretical arguments, and preferred a Baconian process of seeking truth through exhaustive observation and measurement. It was his penchant for careful identification that enabled Merrill to recognize feeble technetium lines in the extraordinarily complicated spectra of cool stars.

I cannot resist the temptation to conclude my remarks with a few recollections about my Mount Wilson colleagues that you probably will never hear about from anyone else. For example, you probably will never hear:

(1) That Roscoe Sanford spent most and perhaps all Christmas and/or New Year's days from 1922 through 1931 observing on Mount Wilson. I learned this by accident from his table of radial velocities for U Monocerotis (Sanford 1933).

(2) That Paul Merrill ate a peanut butter sandwich at the Reyn restaurant every weekday during odd-numbered years. In even-numbered years he ate a fried egg sandwich and Alfred Joy ate the peanut butter sandwich. I discovered only the odd-numbered part as an

---

* I am sorry to report that Lawrence Aller died on March 16, 2003, before I could show him this manuscript.

undergraduate research assistant to Guest Investigator Martin Schwarzschild in 1951. Years later I learned about the even-numbered years from Olin Wilson.

(3) That Paul Merrill firmly believed that 1-N emulsions were best sensitized by soaking them in lemon juice.

(4) That Ira Bowen always inspected change left as tips by his colleagues at the Reyn restaurant with his pocket ocular in search of rare coins. And finally,

(5) That Horace Babcock harrumphed as "inappropriate" Robert Howard's proposal, made during a 1970s Observatories Committee meeting, that Carnegie follow Caltech's lead and refer to itself as Carwash.

References

Adams, W., & Joy, A. 1922, PASP, 56, 242
Adams, W., & Russell, H. N. 1928, ApJ, 68, 9
Babcock, H. D., & Babcock, H. W. 1951, AJ, 56, 120
Babcock, H. W. 1986, Vistas in Astronomy, 29, 153
Burbidge, E. M., Burbidge, G. R., Fowler W. A., & Hoyle, F. 1957, Rev. Mod. Phys., 29, 547
Chamberlain, J. W., & Aller, L. H. 1951, ApJ, 114, 52
DeVorkin, D. H. 2000, Henry Norris Russell, Dean of American Astronomers (Priceton: Princeton Univ. Press)
Fuhr, J. R., Martin, G. A., & Wiese, W. L. 1988, Phys. Chem. Ref. Dat., 17, No. 4
Greenstein, J. L. 1948, ApJ, 107, 151
King, A. S. 1916, ApJ, 44, 169
King, A. S., & Birge, R. T. 1930, ApJ, 72, 19
King, R. B., & King A. S. 1935, ApJ, 82, 377
——. 1938, ApJ, 87, 24
Merrill, P. W. 1952, ApJ, 116, 21
Moore, C. E., & King, A. S. 1943, PASP, 55, 36
Russell, H. N. 1929, ApJ, 70, 11
——. 1935, MNRAS, 95, 610
Russell, H. N., Adams, W., & Moore C. 1928, ApJ, 68, 1
Russell, H. N., & Saunders, F. A. 1925, ApJ, 61, 38
Saha, M. 1921, Proc. Roy. Soc. London Series A, 99, 135
Sanford, R. 1933, ApJ, 77,120
Struve O., & Zeebergs V. 1962, Astronomy of the 20th Century (New York: The Macmillan Company), 238
Yochelson, E. L. 1994 in The Earth, the Heavens and the Carnegie Institution of Washington, ed. G. Good
   (Washington, DC: American Geophysical Union), 1

2

Synthesis of the elements in stars: B²FH and beyond

E. MARGARET BURBIDGE
University of California, San Diego

2.1 Introduction

Fred Hoyle was the H in the alphabetically listed authors of the long paper (Burbidge et al. 1957) on building the chemical elements through nuclear reactions occurring during the evolution of stars in our Galaxy. But this work had its origin 11 years earlier, in the seminal 40-page paper by Fred Hoyle, which was presented at a meeting of the Royal Astronomical Society in London on November 8, 1946. That paper was entitled "The Synthesis of the Elements from Hydrogen" (Hoyle 1946). I have never forgotten the experience of listening to Fred giving this exciting account of his work on building the elements in the abundance peak around iron, where statistical equilibrium would prevail in high-temperature, high-density interiors of evolved stars.

This was indeed a seminal paper. Work on the composition of the Sun's atmosphere had provided data on the relative abundances of the elements, available before its publication by Hans Suess and Harold Urey (see, e.g., Suess & Urey 1956), and Fred had noticed that there was a peak in the abundances of the elements centered on iron, where he knew the packing fraction, or binding energy per nucleon, reached a maximum, making such elements stable at high temperatures and pressures. Color-magnitude diagrams of stars in various clusters, from the work of Allan Sandage, had given information on the evolutionary tracks of the stars of different masses when they leave the main sequence, so Fred was free to calculate what would happen when stars reached the end of the observed tracks.

Fred Hoyle's 1946 paper came at a time when the current theory was that the elements were created primordially by the coagulation of neutrons just after the birth of the Universe. It was something of a joke that such a process would meet with trouble at masses 5 and 8, there being no stable elements here. Gamow, always ready with diagrams and drawings to illustrate his work, drew pictures of two ditches at masses 5 and 8, with himself jumping over them. George Gamow, Maria Goeppert-Mayer, and Edward Teller were working at that time on this primordial agglomeration of neutrons, in spite of the difficulty at masses 5 and 8.

2.2 Iron-peak Elements and Supernovae

But it was Hoyle who realized the significance of the peak in the abundances of the elements centered on iron, and stretching from nuclear mass number $A \simeq 40$ to $A \simeq 60$, containing the isotopes of elements from Ca to Zn. A modern discussion of what B²FH named the *e*-process (*e* for equilibrium), is given by Bradley Meyer in the publication by Wallerstein et al. (1997). There is a section entitled "The e-Process and the Iron-Group

Nuclei" by Meyer which gives many references both to the physics involved and to work on supernovae, relating this to the decays of the various isotopes involved in the physics of supernova explosions. Supernovae were the locations described in B²FH as the sites of the *r*-process. That section, in Meyer's account, referring to the excess abundance of ^{44}Ca found in graphite and silicon carbide grains in primitive meteorites, ends with the words: "In 1957, B²FH could only have dreamed of having such tangible evidence of element formation in stars!"

2.3   Paul Merrill and Technetium

Actually, we had astrophysical evidence—not "tangible," but very clear-cut—for the location of the *s*-process, in which heavier elements are built from lighter ones by the successive captures of neutrons. We named this the *s*-process, "*s*" for "slow," because after each neutron capture that produces an unstable element, there is time for that element to β-decay before capturing another neutron. The clue for the location of the *s*-process came in a landmark Mt. Wilson 1952 paper by Paul W. Merrill—he identified spectroscopic evidence for the unstable element technetium in certain red giant stars (Merrill 1952), stars of class S. With a half-life of $\sim10^5$ yr*, technetium obviously had to be produced actually in these red giant stars, and then mixed to the surface. We could therefore see such stars as the astrophysical locations where free neutrons must be produced and made available in an interior region, and, after their capture by Fe-peak elements, the products mixed to the stellar surface.

Stars in the relatively stable phase of stellar evolution along the asymptotic giant branch provide a location for the occurrence of the *s*-process. In B²FH we had as an analogy the flow of water over a riverbed with deep holes in it. Water flowing along and encountering such holes would accumulate there until the hole was filled, when it would flow on to the next hole. The relevant time scale between neutron captures was $\sim10^5$ yr. The flux of neutrons was represented by the flow of water, and the holes were represented by nuclei with closed shells, at the "magic numbers" of Maria Goeppert Mayer's shell model of nuclei.

During the work of the four of us, we realized that it would be a good idea to get high-dispersion spectra of a star showing evidence for the occurrence of the *s*-process, and determine the abundances of the elements in this and in a standard star for comparison. Merrill's stars of type S would be good, but we knew from our recent work on abundance determination in stars that the analysis of such cool stars would be a real problem. We knew, however, from work by W. Bidelman, that a class of stars known as Ba II stars, seeming to have an overabundance of the element barium, formed in the *s*-process could provide the evidence we needed.

Looking through the literature, we picked the star HD 46407. Geoff, as a Carnegie Fellow, was entitled to apply for observing time on Mt. Wilson. Women, in those far-off days, were not allowed on Mt. Wilson, but, through the persuasive efforts of Willy Fowler and Allan Sandage, Director Ira S. Bowen gave permission for me to accompany Geoff who was awarded time on both the 60-inch and the 100-inch telescopes, as long as we used our own transportation, lived in the summer cottage—the Kapteyn Cottage—instead of in the dormitory called the Monastery, and brought our own food.

The Table shows the results that Geoff and I published (Burbidge & Burbidge 1957). The

---

* The longest lived isotope of technetium, ^{98}Tc, has a half-life of 4.2×10^6 yr, whereas the isotope formed in the *s*-process is ^{99}Tc, which has a half-life of 2×10^5 yr.

Table 2.1. *The s-Process Ba II Star HD 46407*[†]

Element	Atomic Weights, s-Isotopes	y'	$\langle\sigma N\rangle y'$
Strontium	86, 87, 88	4.7	381
Yttrium	89	7.8	1326
Zirconium	90, 91, 92, 94	4.9	884
Niobium	93	5.2	832
Molybdenum	95, 96, 97	4.8	343
Barium	134, 136, 138	14.9	281
Lanthanum	139	9.8	647
Cerium	140	10.7	407
Praseodymium	141	28.0	672
Neodymium	142, 143, 144, 145, 146	15.7	443
Tungsten	182, 184	13.0	914

[†]Burbidge & Burbidge (1957)
σ = neutron capture cross-section
N = solar system abundance
y = observed overabundance
y' = overabundance ratio of isotopes built by *s*-process alone
Within a factor~2, $\langle\sigma N\rangle y'$ is constant: mean $\langle\sigma N\rangle y' = 648$

elements listed, strontium to tungsten, are elements with one or more isotopes made in the *s*-process. From our observed overabundance, *y*, for these elements, we used the emerging calculations from B^2FH to calculate y', the observed overabundance of isotopes built by the *s*-process alone. We used recently available neutron capture cross-sections, σ, to calculate the product $\langle\sigma N\rangle y'$, and, despite the fact that the cross-sections were not very accurate (they were substantially improved soon afterwards), the product $\langle\sigma N\rangle y'$ was found to be constant to within a factor $\sim$2.

2.4   Recent Work

With the large telescopes and modern instruments available today, the progress of unraveling the details of nucleosynthesis after stars leave the main sequence is one of the most exciting branches of astrophysics. Stars of very low Fe abundance can now be picked out in the halo of the Galaxy, and the study of *r*-process elements there are giving information on the early stellar population and the seeding of the young Galaxy with the products of early supernovae (e.g., Sneden et al. 1998).

But the *s*-process is also yielding very interesting data on stellar evolution. The astronomers working with the 3.6 m telescope of the European Southern Observatory observed a group of same-temperature CH (carbon rich) evolved stars, and found that the line Pb I λ4057.31 is clearly visible and quite strong in three but absent in one. Pb is the last stable element produced in the slow neutron capture chain of nucleosynthesis (Van Eck et al. 2001). Such observations will surely go hand-in-hand with the theoretical work on the late evolutionary stages of stars, and build our knowledge of the history of the Galaxy.

References

Burbidge, E. M., & Burbidge, G. R. 1957, ApJ, 126, 357

Burbidge, E. M., Burbidge, G. R., Fowler, W. A., & Hoyle, F. 1957, Rev. Mod. Phys., 29, 547

Hoyle, F. 1946, MNRAS, 106, 343

Merrill, P. W. 1952, Science, 115, 484

Sneden, C., Cowan, J. J., Burris, D. L., & Truran, J. W. 1998, ApJ, 496, 235

Suess, H. E., & Urey, H. C. 1956, Rev. Mod. Phys., 28, 53

Van Eck, S., Goriely, S., Jorissen, A., & Plez, B. 2001, The Messenger, 106, 37

Wallerstein, G., et al. 1997, Rev. Mod. Phys., 69, 995

3

Stellar nucleosynthesis: a status report 2003

DAVID ARNETT
Steward Observatory, University of Arizona

3.1 Introduction

How well can we predict the production of nuclei during the evolution and death of a generation of stars? Although the basic framework for understanding the synthesis of nuclei by thermonuclear reactions in stars was established almost half a century ago (Burbidge et al. 1957, B^{2}FH; Cameron 1957), quantitative simulation of nucleosynthesis in a stellar model had to await for access to faster computers. The first nucleosynthesis computation in a stellar evolutionary sequence (using a reaction network of 23 nuclei including the iron peak), was done at Sir Fred Hoyle's Institute of Theoretical Astronomy (Arnett 1968). The computer resourses required were such as to insure that such computations were rare! Nonetheless, it uncovered the characteristic SiSArCa production in oxygen burning and the beginnings of core convergence due to neutrino cooling. We have now had 23 doublings (a factor of $\sim 10^7$) in computer power from Moore's Law, so that it is now possible to use larger networks and explore parameter space. In this review we will place recent computations in a historical context, with emphasis upon the challenges for the future as well as the successes of the present.

The pioneering work of the Burbidges, Fowler, Hoyle, and Cameron was stimulated by precise abundance data for the solar system just made available by Seuss & Urey (1956); similarly, modern nucleosynthesis theory is driven by increasingly precise data from cosmic rays, meteorites, and astronomical objects. With the advent of microanalysis of pre-solar grains, high-energy density lasers, 8-meter class ground-based telescopes, space-based multiple wavelength detectors, faster computers, and a wealth of nuclear physics data, theory is still driven by a strong and well-based empiricism.

3.2 Fundamentals

The solar system abundance pattern indicates that the values of several critical parameters are reproduced under stellar conditons. Theorists know that these constraints must be satisfied by any acceptable theory. The challenge presently confronting us is not whether or not we can reproduce the solar system pattern of isotopic abundances, but whether we can determine—uniquely—how nature did it.

3.2.1 Radiation Entropy

The vast difference between stellar nucleosynthesis and Big Bang nucleosynthesis is due to the very high entropy of the Big Bang, and the modest entropy of stars. The thermal energy of the Big Bang at the time of nucleosynthesis is dominated by black body photons

(Kolb & Turner 1990); the entropy of this radiation (per nucleon) is $S_\gamma/\Re = 4aT^3/3\rho\Re$ in Boltzmann units. Here T is the temperature, ρ the mass density, a the radiation constant, and $\Re$ the gas constant. This is simply related to the ratio of photon to baryon number by $n_\gamma/n_b = 0.2776 S_\gamma/\Re$. The ratio T^3/ρ also appears in a fundamental way in the theory of hydrostatic gravitating structures; Eddington (1926) introduced a related quantity β in his Standard model, $S_\gamma/\Re = 4Y(1-\beta)/\beta$, which appears in the famous quartic equation, $1-\beta = 0.00298\,(m/Y^2)^2\,\beta^4$, where m is the stellar mass in solar units and Y is the number of pressure-contributing particles per baryon (an ideal gas of matter plus black body radiation is assumed). For most stars of interest to us for nucleosynthesis, $S_\gamma/\Re$ lies in the range of 0.3 to 30, while cosmological values range from 10^8 to 10^{10}. Values for stellar quasistatic burning range about unity, while explosive values are usually around 10, an interestingly restricted range.

The radiation entropy is one parameter that affects the production ratio of carbon to oxygen. These elements, the fourth and third most abundant, are predominantly in the form of the isotopes ^{12}C and ^{16}O, which are made in helium burning. Carbon is made by the triple-alpha reaction, which is proportional to the number of ^{4}He triples (proportional to ρ^3 in a unit volume), while destruction (to form ^{16}O) is proportional to the number of ^{4}He + ^{12}C pairs (proportional to ρ^2). Thus, carbon production beats carbon destruction at higher density. Massive stars have higher $S_\gamma/\Re$, hence lower density for a given burning temperature. Massive stars make almost all of the ^{16}O, while intermediate-mass stars make significant amounts ^{12}C.

A complication to this simple picture of mass dependence is that ^{12}C depletion occurs rapidly as ^{4}He is exhausted, so that incomplete helium burning favors ^{12}C production (this case applies to yields from intermediate-mass stars). Late mixing of new ^{4}He into a helium burning region enhances destruction over ^{12}C production. The theorist may adjust the predicted carbon and oxygen yields by changing the reaction rates, the ratio of massive to intermediate-mass stars, or the nature of the convective mixing process. Plausible choices do give plausible results, but this is still a thorny issue.

3.2.2   *The Bulk Yields*

The *bulk yield* specifies how much of a given star is processed by each burning stage, for the most abundant elements. The high abundance of carbon (fourth) and oxygen (third) in the solar pattern is due to neutrino cooling; without it these fuels would themselves be consumed by subsequent burning stages. Helium burning, which is cooled by photon diffusion, produces C and O. The next stage, carbon burning, is cooled by neutrino emission; because this process is local and is more vigorous at higher temperature, the stellar structure changes. The carbon burning region tends to be smaller (in mass) so that more C and O survive. Subsequent burning stages (neon, oxygen, silicon) increase the effect.

The burning separates, usually, into the following stages: hydrogen, helium, carbon, neon, oxygen, and silicon burning. Lower-mass stars may not proceed through all these stages if they can loose mass and become a white dwarf. Table 3.1 summarizes the properties of each stage. Together these produce all of the abundant elements except for the H and He from the Big Bang.

The s-process, r-process, and p-process are important diagnostics (Meyer 1994), as they make many rare nuclei but involve less than 10^{-6} of the solar abundance pattern. Only the

Table 3.1. *Bulk Nucleosynthesis Processes*

Process	Reaction	Temperature (K)	Cooling	Products
Hydrogen	pp	2×10^7	photons	He
	CNO	3×10^7	photons	He; N, Na
Helium	$3\alpha \rightarrow {}^{12}C$	2×10^8	photons	C
	${}^{12}C(\alpha, \gamma){}^{16}O$			O
Carbon	${}^{12}C + {}^{12}C$	9×10^8	neutrinos	Ne, Na, Mg, Al
Neon	${}^{20}Ne(\gamma, \alpha){}^{16}O$	1.5×10^9	neutrinos	O, Mg, Al
Oxygen	${}^{16}O + {}^{16}O$	2×10^9	neutrinos	Si, S, Ar, Ca
Silicon	${}^{28}Si(\gamma, \alpha)$	3.5×10^9	neutrinos	Fe,

s-process is thought to be produced in a photon-cooled environment (double shell burning of H and He in intermediate-mass stars, predominantly).

3.2.3   *Neutron Excess*

The yields of nuclei depend upon the nuclei being produced, but also upon their surviving. Nuclear stability reflects nuclear binding, which exhibits strong systematic properties. These systematics were observed in the Seuss & Urey (1956) abundances by B^{2}FH and Cameron (1957), and still have relevance. Nuclei tend to be more stable if they have equal numbers of neutrons and protons (Z = N), even numbers of neutrons (even N), and even numbers of protons (even Z). However, protons repel by Coulomb forces, so that nuclei with larger Z are more tightly bound if they are neutron rich (this is why supernovae shine: explosive burning of Z = N matter proceeds up to ^{56}Ni, the most bound nucleus with Z = N, and this decays by the weak interaction to ^{56}Fe, sufficiently slowly that it illuminates the expanded gas efficiently).

Below the iron peak, the abundance of odd-Z elements is sensitive to the relative abundance of neutrons and protons in the environment in which they are formed. The most bound isotopes will have even N. For example, ^{23}Na has Z = 11 and N = 12. There is one excess neutron. Similarly, ^{27}Al has Z = 13 and N = 14, with one excess neutron. If there is a paucity of excess neutrons, Na and Al will be rare.

The *neutron excess* is defined as $\eta = \Sigma_i Y_i (N_i - Z_i)$, the fraction of extra neutrons per baryon. Charge neutrality of the plasma relates this to the ratio of electrons to baryons, $Y_e = \Sigma_i Y_i Z_i = (1 - \eta)/2$.

The neutron excess is modified in stars by several processes. Ordinary hydrogen burning is equivalent to $4p + 2e^- \rightarrow {}^4He + 2\nu$. If we start with pure H and He, then the original proton-rich material changes to $\eta \rightarrow 0$. After hydrogen burning we are interested in small deviations of η above zero.

The most important processes are:

- CNO-produced nitrogen gives $^{14}N(\alpha,\gamma)^{18}F(\beta^+,\nu)^{18}O$, with and excess of two neutrons per ^{18}O. For solar system initial abundances, $\eta \approx 2 \times 10^{-3}$, which is about right for nucleosynthesis.
- Auxiliary reactions during carbon burning are $^{12}C(p,\gamma)^{13}N(\beta^+,\nu)^{13}C$, and $^{12}C+^{12}C \rightarrow ^{23}Mg+n$, followed by $^{23}Mg \rightarrow ^{23}Na+e^++\nu$ and related reactions (Arnett & Truran 1967). This gives a neutron excess that is less than, but approaching, $\eta = 2 \times 10^{-3}$.
- Electron captures during hydrostatic oxygen burning and silicon burning (Thielemann & Arnett 1985). This can greatly exceed $\eta = 2 \times 10^{-3}$ and be too large for nucleosynthesis of the solar system abundance pattern. Such matter is presumed to be doomed for incorporation in a neutron star or black hole, and theorists often enforce this result.

The N abundance from hydrogen burning derives from the CNO isotopes initially present in the star, which come from an earlier epoch of nucleosynthesis. This is an example of *seconday* nucleosynthesis (Arnett 1971). The carbon burning example does not depend upon any previous heavy elements ($Z > 2$), and is a *primary* nucleosynthesis process because the ^{12}C was formed by helium burning in the same star.

In this simple picture we would expect the abundance of odd-Z elements below the iron peak to be even more underabundant in metal-poor stars (*secondary*). This effect is mitigated by the *primary* production of neutron excess in carbon burning, which also produces Na and Al. Neon and oxygen burning further damp out the effect, with a resulting trend toward *primary* behavior.

The r-process and the s-process are dominated by the capture of free neutrons, with the captures being rapid compared to beta decay for the r-process and slow for the s-process. The source of the neutrons is closely related to the question of neutron excess.

For the s-process in intermediate-mass stars, the sources are $^{13}C(\alpha,n)^{16}O$ and marginally $^{22}Ne(\alpha,n)^{25}Mg$ for intermediate-mass stars (Busso, Gallino, & Wasserburg 1999; Busso et al. 2001; Lugaro et al. 2003). The predictions involve parameterized estimates of mixing during the inter-flash phase of the double shell evolution. A weak component of the s-process occurs in core helium burning and shell carbon burning in massive stars (Arnett 1996; The, El Eid, & Meyer 2000). Here the results depend upon models of the mixing process used in the stellar model construction.

The site of the r-process is still unclear, but is probably related to matter of high neutron excess in a core collapse supernova. Given the ambiguity of these conditions, the details of the process itself are uncertain (Freiburghaus et al. 1999; Meyer 2002). The initial enthusiasm for the neutrino wind model (Woosley & Hoffman 1992) seems to have been premature (Witti, Janke, & Takahashi 1994; Woosley et al. 1994; Freiburghaus et al. 1999). To circumvent these difficulties, Qian & Wasserburg (2003) have used a phenomenological approach. With 8-meter class telescopes, direct observations, such as those by Johnson (2002) and Johnson & Bolte (2002), are beginning to provide significant constraints.

3.2.4 Time Scales

Nucleosynthesis in stars may be divided into two qualitatively different categories: *quasistatic* or *explosive* burning. The r-process, with its time scale estimated from the needed beta decay rates to be of order of seconds (B^2FH; Cameron 1957), was seen early to be an explosive process, and the s-process to be quasistatic. Internal evidence in the isotopic abundances above iron suggest that carbon and neon burning were quasistatic (Arnett 1996, § 9.6; see also The et al. 2000). Oxygen burning, on the other hand, produces too large a neutron excess during quasistatic burning to match the solar system isotopic pattern, and is likely to be an explosive process for the matter that contributes to the yield.

3.3   Problems

There are several open questions that reduce the reliability of predictions of stellar yields.

3.3.1   *Initial Mass Function*

Any prediction of stellar nucleosynthesis must involve the question of how many stars, and of what kind, compose a stellar generation? The initial distribution of stars in mass, for a given generation of star formation, is called the initial mass function. Figure 3.1 illustrates the problem in a beautiful way; it is a galaxy in starburst mode. The clumps of massive new stars are eye-catching, but emphasize that massive stars are not uniformly distributed in space. Further, this galaxy is in an active time for star formation, so that there may be a nonuniform distribution in time as well.

Figure 3.2 illustrates the problem quantitatively. The logarithm of the return mass f, in solar units, is plotted versus the stellar mass, for an idealized cluster having $10^4 M_\odot$ total mass in stars of stellar mass $m > 1 M_\odot$. The initial mass function has a Salpeter slope. The binning is logarithmic in stellar mass (horizontal axis); the error bars represent Gaussian fluctuation due to a finite number of stars in each bin. Some interesting mass values are indicated on the horizontal axis: the mass at which the helium flash occurs (He flash), M_{up} is the mass above which nondegenerate carbon ignition follows helium exhaustion directly, O+O is the mass above which oxygen burning occurs, SN 1987A denotes the putative mass of that supernova progenitor, and WR denotes a typical mass of a Wolf-Rayet star.

Modern estimates of the mass of white dwarfs from intermediate-mass stars imply that of order half of the mass in stars above $1 M_\odot$ is returned to the interstellar medium. This is considerably larger than old estimates, and dilutes the yield of heavy elements from more massive stars (and modifies old arguments: Arnett 1978; Wheeler, Miller, & Scalo 1980; see also Chiosi 1979; Maeder 1981). The limitations on the brown dwarf mass function (Chabrier 2002) further reduce the amount of matter removed per generation from the stellar birth-death cycle.

For stars above $10 M_\odot$ the fluctuations in stellar number become an increasing worry. This idealized cluster is more massive than usual for a typical open cluster, but is consistent with what Slesnick, Hillenbrand, & Massey (2002) find for h and χ Persei (compare their Fig. 8 with Fig. 3.2).

If we take M_{up} as the boundary between intermediate-mass and massive stars, we see that the dominant mass return is from the intermediate-mass stars. Relatively modest nucleosynthesis in this mass range may have significant consequences, even beyond the ususal discussion of the s-process.

3.3.2   *Nonexplosive Mass Loss*

The radiation-driven wind theory of mass loss from OB stars now gives a reliable and understandable basis for stellar evolution with mass loss for massive stars in this part of the Hertzsprung-Russell diagram (Kudridzki & Puls 2000; Paldrach, Hoffman, & Lennon

Fig. 3.1. A galaxy (NGC 3310) in starburst mode, indicating the inhomegeity of massive stars in both time and space (Hubble Heritage, NASA, STScI).

2001; Kudridzki 2002). The situation for cool stars is less settled (Wallerstein & Knapp 1998; Willson 2000). ISOCAM observations of massive Galactic globular clusters suggest that significant mass loss occurs only at the very end of the red giant branch, is episodic, and does not show a crucial dependence upon cluster metallicity (Origlia et al. 2002). These stars are of low mass, but do contribute to the mass budget for matter returned to the interstellar medium, and may give clues for the mechanism for mass loss by more massive cool stars. For the mass range $1 \leq m/M_\odot \leq 10$, Willson (2000) suggests that empirical relationships between mass loss rates and stellar parameters, so beloved by stellar theorists, are determined mostly by selection effects, and that mass loss rates increase precipitously

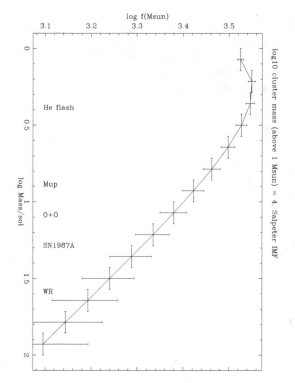

Fig. 3.2. The massive and the intermediate-mass stars dominate the return of matter to the interstellar medium. The amount of returned matter is significantly higher with modern estimates of the mass of the white dwarf remnant and the brown dwarf mass function.

as stars become brighter, larger, and cooler. Unfortunately, these are the stars that ascend the asymptotic giant branch and are thought to produce most of the *s*-process, and the mass loss affects the predictions of yields (Frost & Lattanzio 1996; Charbonnel 2002; Lattanzio 2002). There are a variety of unexplored aspects of the physics of such mass loss (e.g., Van Horn et al. 2003). The situation for massive cool stars is more obscure, both figuratively and literally; they tend to be shrouded in dust formed in ejected material. Their evolutionary status is a subject of debate (Langer et al. 1994; Crowther, Hiller, & Smith 1995a, b, c; Crowther et al. 1995), and the detailed mechanism of their mass loss seems unclear.

3.3.3   *Rotation*

While rotation has been an issue for star formation theory for centuries, it is also an issue for stellar death, as Figure 3.3 vividly shows. What happens between these extremes?

Most discussions of rotation and stellar evolution follow the approach of Endal & Sofia (1976), in which the Lagrangian shells of a stellar model are presumed to be rotationally deformed from spheres into oblate figures. Because only one dimension is computed, state variables must be uniform along these shells. This approximation can only be valid if there is some physical mechanism that insures it. If mixing along horizontal directions is fast, as would be the case if g-modes are a dominant mechanism for mixing (Talon & Zahn 1997; Maeder & Zahn 1998), this could be valid. The implication is that the evolutionary time scale is long compared to the horizontal mixing time scale. Maeder and Meynet have

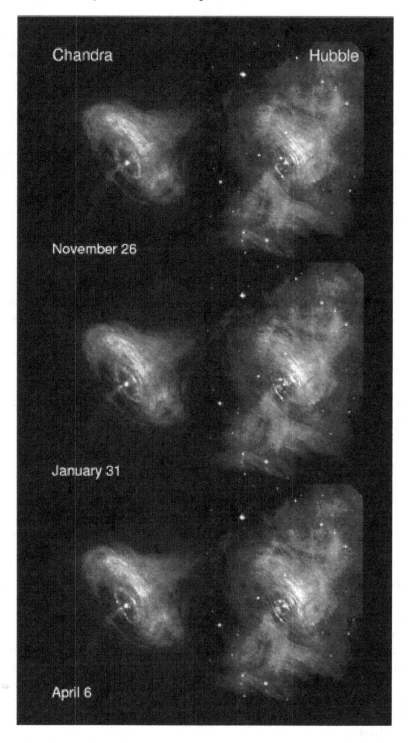

Fig. 3.3. The Crab nebula shows dramatic evidence of rotation in its collapsed core. The images are in X-rays (*Chandra*) and visible light (*HST*) (NASA/CXC/ASUIJ, Hester et al. 2002; NASA/HST/ASUIJ, Hester et al. 2002).

explored stellar evolution with rotation from this perspective in a extensive series of papers (Maeder 1997, 1999; Meynet & Maeder 1997, 2000, 2001, 2002; Maeder & Meynet 2000a, b, 2001), which set the standard for this topic. Heger, Woosley, & Langer (2000) have run a grid of models of rotating and nonrotating massive stars of solar metallicity from pre-main sequence to core collapse. Langer et al. (1999) have extended this to intermediate-mass stars.

Even with the most optimism that can be gathered, it appears that such a one-dimensional approximation breaks down badly for the oxygen burning state (Arnett 1994; Bazán & Arnett 1998), because the mixing time scale and the evolutionary time scale are becoming similar. Whatever the outcome, it is clear that we must explore the nonspherical models of late stellar evolution before we understand these stages. They are crucial for yields as they involve major shell burning stages (carbon, neon, and oxygen), which provide much of the pre-explosion nucleosynthesis. The oxygen shell provides most of the explosive burning, including the formation of ^{56}Ni. The lack of symmetry of these stages will be imprinted upon the core collapse, and may affect the explosion mechanism.

3.3.4   *Binarity*

Perhaps half of the massive stars have companions (i.e. are in binaries). The nature of their joint evolution is affected by mass loss, which, as we have seen, is a problem for the cooler side of the Hertzsprung-Russell diagram. Angular momentum loss affects the orbital properties of the binary, complicating the picture. Mass stripping, tidal interactions, and merger are fundamentally three-dimensional phenomena, and to make things more difficult, may also involve detailed radiative transfer (not just diffusion) and magnetohydrodynamics. It is widely stated that Type Ia supernova are produced in binary systems, but quantitative treatment of the evolution to explosion, in a realistically multidimensional way, is only beginning (e.g., Piersanti et al. 2003).

3.3.5   *Mix*

Figure 3.4 shows the inhomogeneity in the products of explosive nucleosynthesis in the supernova remnant Cassiopea A. Silicon is produced by explosive or quasistatic oxygen burning. Calcium is produced by explosive oxygen burning, most likely. Iron can be produced by oxygen burning that continues into silicon burning. All have different shapes in the image. This could be due to different degrees of explosive burning, with large burning making Ca, little burning leaving the Si from quasistatic burning intact, and extreme burning making Fe. It could also be due to inhomogeneity in the progenitor.

Optical data (Fesen 2001), infrared data (Douvion, Lagage, & Cesarsky 1999) from ISO-CAM, X-ray data (Hughes et al. 2000) from *Chandra* and (Willingale et al. 2003) from *XMM-Newton*, and gamma-ray lines (Vink et al. 2001) are giving a much improved picture of Cass A. In addition to Si, Ca, and Fe, the optical data include O and Ar (incomplete oxygen burning) and N and H (pre-explosion mass loss?). The infrared data include a silicate layer, a S and Ar layer, and a Ne layer. The gamma-lines give ^{44}Ti. All show evidence for inhomogeneity in composition and kinematics. The best choice for the progenitor seems to be fairly massive, possibly a 10 to 30 solar mass Wolf-Rayet star, but details of the identification are sketchy.

Although there is some success in identifying isotopic anomalies in pre-solar grains with asymptotic giant branch stars (Choi, Wasserburg, & Huss 1999), there are problems with

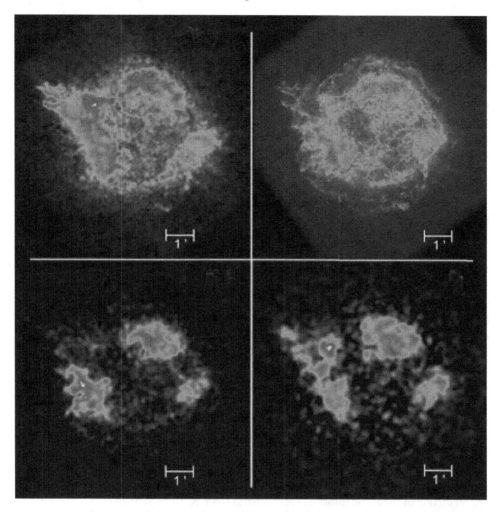

Fig. 3.4. This *Chandra* X-ray image of Cassiopea A shows inhomogeneity in the products of explosive nucleosynthesis (NASA/GSFC, Hwang et al. 2001). The upper left image is broadband, and in clockwise order are images of emission from silicon, iron, and calcium.

grains exhibiting supernova nucleosynthesis. Although aspects of explosive processes, such as extinct ^{44}Ti and ^{28}Si excesses, are found, extreme mixing models are required to match the supernova predictions (Zinner 1998). Perhaps this difficulty is related to the enforced spherical symmetry in the supernova models.

If highly evolved stars no longer maintain the symmetry of spherical shells, what will be the consequences for nucleosynthesis yields? Suppose the star to be made not of spherical shells, but of a patchwork of blobs. Neutrino cooling would lower the entropy of a blob, causing it to sink, moving past its neighbors. This would continue until the temperature rose enough to ignite a fuel. This would increase the entropy, making the blob buoyant, and causing it to rise. We might imagine a cycle of repeated rising and falling, with some fuel consumed each time. The temperature at which falling was reversed would depend upon the composition of the blob, so that on average there *would* be a layering according to composition, but it would be broken by fluctuations. Some burning flashes would be more vigorous, giving larger excursions, spreading ashes in ways not seen in one-dimensional simulations. Could a few vigorous oxygen flashes make ^{56}Ni and carry it out far enough to explain the early appearance of ^{56}Co gamma-ray lines in SN 1987A? What would be the effect of a supernova shock on the inhomogeneous oxygen-rich layer? Are nuclear-energized pulsations driven by the accelerating contraction of the core during silicon burning? How does rotation couple to the burning, and does this qualitatively change the nature of the inital core collapse model? The range of possiblities is larger than yet considered.

3.3.6   *Progenitors*

Mathematically, explosions are exponentiating solutions, which are sensitive to initial conditions. Given that we have no agreed-upon mechanisms for either thermonuclear or core collapse supernovae, it would be good to know what the progenitors are. For SN 1987A we know the progenitor had a helium core of about $6M_\odot$ from the luminosity of the B3 star, but beyond that disagreement sets in (the main sequence mass for a single star would have been $20M_\odot$). For SN 1993J, Smart et al. (2002) suggest a similar luminosity, implying a similar mass, but a K0 spectral type. These authors estimate the mass of the progenitor of SNe II-P 1999em to be $12^{+1}_{-1}M_\odot$. Smartt et al. (2001) estimated the mass of the progenitor of SN 1999gi to be $9^{+3}_{-2}M_\odot$. For SN II-P 2001du, Smartt et al. (2003) estimate the mass to be less than $15M_\odot$, and with new data, revise that of SN 1999gi to $12M_\odot$, and that of SN 1998em to less than $15M_\odot$.

From a different perspective, Hamuy (2003) has used photometry and spectroscopy of 24 SNe II-P to derive properties of their explosions and their progenitors. The nickel masses range from 0.00166 to $0.26M_\odot$. The explosion energies ranged between 0.6 to 5.5×10^{51} erg, ejected masses from 14 to $56M_\odot$, and radii from 80 to $600R_\odot$. He infers a continuous distribution of energies below 8×10^{51} erg.

Howell (2001) finds that subluminous SNe Ia come from an old population. The fact that subluminous SNe Ia and overluminous SNe Ia come from different progenitor populations is a prediction of the white dwarf merger scenario (Iben & Tutukov 1984).

While still incomplete, these new data provide interesting challenges: can we connect observed supernovae to stellar evolutionary sequences?

3.3.7   *Ejection and Explosion Mechanisms*

The question of explosion mechanisms is still open. Core collapse clearly happened in SN 1987A, but the theoretical understanding is still incomplete.

Travaglio, Kifonidis, & Müller (2004) discuss the nucleosynthesis in core collapse supernovae, as seen in 1-D and 2-D hydrodynamic simulations; see Kifonidis et al. (2003) for a discussion of the hydrodynamic behavior. A satisfactory explanation for the early escape of gamma-lines from ^{56}Co from SN 1987A still eludes us. These simulations basically assume an explosion from core collapse and examine its consequences.

The problem of that explosion is unsolved (see, e.g., Janka 2001; Liebendörfer et al. 2001; Rampp et al. 2002; Buras et al. 2003). Fully 3-D simulations of the core collapse are beginning to appear (Fryer & Warren 2002), which may clarify the effect of 2-D versus 3-D.

For thermonuclear explosions the question remains: detonation or deflagration? One of the most interesting developments are the large-scale 3-D simulations of Gamezo et al.

Table 3.2. *Stellar Evolution with Detailed Yields*

Reference	Masses/$M_\odot$	Metallicity	Networks	Comment
WZW78	15, 25	solar	19 NSE(121)	pre-MS to collapse
WW95	11–40	$Z = 0,10^{-4},0.01,$ 0.1,1	19 200	pre-MS to collapse
R02	15, 19, 20, 21, 25	solar	19 700–2200	NON-SMOKER rates
HW02	65–130, steps of $\Delta M = 5$	solar	19 304,477(exp)	He cores
TNH96	13, 15, 20, 25	solar	299	
I99	1.45	solar	299	SNe Ia
MM	0.8–120	$Z = 0.001$ to 0.08	19	mass loss rotation
CLS98	25	solar	149	pre-MS to collapse
LSC00	13–25	solar	149	pre-MS to collapse
CL02	15, 20, 25, 35, 50, 80	$Z = 0$	179	pre-MS to collapse
LC98	13–35	solar	267	pre-MS to collapse
TEM00	15, 20, 25, 30	solar	659	Ne ignition
APB96	20	solar	96	Ne-exhaustion Göttigen code

References: WZW78, Weaver, Zimmermann, & Woosley (1978); WW95, Woosley & Weaver (1995); R02, Rauscher et al. (2002); HW02, Heger & Woosley (2002); TNH96, Thielemann, Nomoto, & Hashimoto (1996); I99, Iwamoto et al. (1999); MM, Maeder & Meynet (1997-2003); CLS98, Chieffi, Limongi, & Straniero (1998); LSC00, Limongi, Straniero, & Chieffi (2000); CL02, Chieffi & Limongi (2002); LC98, Limongi & Chieffi (1998); TEM00, The, El Eid, & Meyer (2000); APB96, Aubert, Prantzos, & Baraffe (1996).

(2003). They numerically simulate the deflagration stage of a thermonuclear supernova (SN Ia), and find that while the turbulent flame does provide a healthy explosion, it is subject to gravity-induced Rayleigh-Taylor instabilities, and leaves large amounts of unburnt and partially burnt material, whereas observations imply that such material only exists in the outer (high-velocity) layers. They suggest that a transition of deflagration to detonation may be the solution.

3.4 Progress

The current state-of-the-art for supernova nucleosynthesis simulations is illustrated by the work of several groups, in which extensive networks have been coupled to the stellar evolutionary equations. Several are summarized below in Table 3.2; the list is quite incomplete, but fairly representative. Woosley, Heger, & Weaver (2002) have recently produced an extensive review. In addition, the intermediate-mass stars make an important contribution (see above).

3.5   Conclusions

The numerical simulations of stellar evolution and nucleosynthesis in supernova explosions and double shell burning now produce predictions relevant to observation and experiment. There are still significant disagreements between groups about detail. While we are making progress, we still have a lot of work to do to place these predictions upon a reliable basis, as the discussion above suggests.

References

Arnett, D. 1968, in Stellar Evolution, ed. R. Stein & A. G. W. Cameron (New York: Plenum Press)
——. 1971, ApJ, 166, 153
——. 1978, ApJ, 219, 1008
——. 1994, ApJ, 427, 932
——. 1996, Supernovae and Nucleosynthesis (Princeton: Princeton Univ. Press)
Arnett, D., & Truran, J. W. 1967, ApJ, 157, 339
Aubert, O., Prantzos, N., & Baraffe, I. 1996, A&AS, 312, 845
Bazán, G., & Arnett, D. 1998, ApJ, 496, 316
Buras, R., Rampp, M., Janka, H.-Th., & Kifonidis, K. 2003, Phys. Rev. Lett., 90, 1101
Burbidge, E. M., Burbidge, G., Fowler, W. A., & Hoyle, F. 1957, Rev. Mod. Phys., 29, 547
Busso, M., Gallino, R., Lambert, D., Travaglio, C., & Smith, V. 2001, ApJ, 557, 802
Busso, M., Gallino, R., & Wasserburg, G. J. 1999, ARA&A, 37, 239
Cameron, A. G. W. 1957, Chalk River Report CRL-41
Charbonnel, C. 2002, Ap&SS, 281, 161
Charbrier, G. 2002, ApJ, 567, 304
Chieffi, A., & Limongi, M. 2002, ApJ, 577, 281
Chieffi, A., Limongi, M., & Straniero, O. 1998, ApJ, 502, 737
Chiosi, C. 1979, A&AS, 80, 252
Choi, B.-G., Wasserburg, G. J., & Huss, G. R. 1999, ApJ, 522, L133
Crowther, P. A., Hillier, D. J., & Smith, L. J. 1995a, A&AS, 293, 172
——. 1995b, A&AS, 293, 403
——. 1995c, A&AS, 302, 457
Crowther, P. A., Smith, L. J., Hillier, D. J., & Schmutz, W. 1995, A&AS, 293, 427
Douvion, T., Lagange, P. O., & Cesarsky, C. J. 1999, A&AS, 532, L111
Eddington, A. S. 1926, The Internal Constitution of the Stars (New York: Dover), reprinted 1959
Endal, A. S., & Sofia, S. 1976, ApJ, 210, 184
Fesen, R. A. 2001, ApJS, 133, 161
Freiburghaus, C., Rembges, J.-F., Rauscher, T., Kolbe, E., Thielemann, F. K., Kratz, K.-L., Preiffer, B., & Cowan, J. J. 1999, ApJ, 516, 381
Frost, C. A., & Lattanzio, J. C. 1996, ApJ, 473, 383
Fryer, C., & Warren, M. 2002, ApJ, 574, L65
Gamezo, V. N., Khokhlov, A. M., Oran, E. S., Chtchelkanova, A. Y., & Rosenberg, R. O. 2003, Science, 299, 77
Hamuy, M. 2003, ApJ, 582, 905
Heger, A., & Woosley, S. E. 2002, ApJ, 567, 532
Heger, A., Woosley, S. E., & Langer, N. 2000, NewAR, 44, 297
Hester, J., et al. 2002, ApJ, 577, L49
Howell, D. A. 2001, ApJ, 554, L193
Hughes, J. P., Rakowski, C. E., Burrows, D. N., & Slane, P. O. 2000, ApJ, 528, 109
Hwang, U., Szymkowiak, A. E., Petre, R., & Holt, S. S. 2001, ApJ, 560, L175
Iben, I., Jr., & Tutukov, A. V. 1984, ApJS, 54, 335
Iwamoto, K., Brachwitz, F., Nomoto, K., Kishimoto, N., Umeda, H., Hix, W. R., & Thielemann, F.-K. 1999, ApJS, 125, 439
Janka, H.-Th. 2001, A&AS, 368, 527
Johnson, J. A. 2002, ApJS, 139, 219

Johnson, J. A., & Bolte, M. 2002, ApJ, 579, 616

Kifonidis, K., Plewa, T., Janka, H.-Th., & Müller, E. 2003, A&AS, 408, 621

Kolb, E. W., & Turner, M. S. 1990, The Early Universe (Redwood City, California: Addison-Wesley)

Kudritzki, R. P. 2002, ApJ, 577, 389

Kudritzki, R. P., & Puls, J. 2000, ARA&A, 38, 613

Langer, N., Hamann, W.-R., Lennon, M., Najarro, F., Pauldrach, A. W. A., & Puls, J. 1994, A&AS, 290, 819

Langer, N., Heger, A., Wellstein, S., & Herwig, F. 1999, A&AS, 346, L37

Lattanzio, J. C. 2002, NewAR, 46, 469

Liebendörfer, M., Mezzacappa, A., Thielemann, F.-K., Messer, O. E. B., Hix, W. R., & Bruenn, S. W, 2001, Phys. Rev. D, 63, 103004

Limongi, M., & Chieffi, A. 2003, ApJ, 592, 404

Limongi, M., Straniero, O., & Chieffi, A. 2000, ApJS, 129, 625

Lugaro, M., Herwig, F., Lattanzio, J., Gallino, R., & Straniero, O. 2003, ApJ, 586, 1305

Maeder, A. 1981, A&AS, 101, 385

———. 1997, A&AS, 321, 134

———. 1999, A&AS, 347, 185

Maeder, A., & Meynet, G. 2000a, A&AS, 361, 101

———. 2000b, A&AS, 361, 159

———. 2001, A&AS, 373, 555

Maeder, A., & Zahn, J. P. 1998, A&AS, 334, 1000

Meyer, B. 1994, ARA&A, 32, 153

———. 2002, Phys. Rev. Lett., 89, 1101

Meynet, G., & Maeder, A. 1997, A&AS, 321, 465

———. 2000, ARA&A, 38, 143

———. 2001, A&A, 373, 555

———. 2002, A&AS, 390, 561

Origlia, L., Ferraro, F. R., Fusi Pecci, F, & Rood, R. T. 2002, ApJ, 571, 458

Paldrach, A. W. A., Hoffman, T. L., & Lennon, M. 2001, A&AS, 375, 161

Piersanti, L., Gagliardi, S., Iben, I., Jr., & Tornambé, A. 2003, ApJ, 583, 885

Qian, Y.-Z., & Wasserburg, G. J. 2003, ApJ, 588, 1099

Rampp, M., Buras, R., Janka, H.-Th., & Raffelt, G. 2002. in Proc. 11th Workshop on Nuclear Astrophysics, ed. W. Hillebrandt & E. Müller (Garching: Max-Planck-Institut für Astrophysik), 119

Rauscher, T., Heger, A., Hoffman, R. D., & Woosley, S. E. 2002, ApJ, 576, 323

Seuss, H. E., & Urey, H. C. 1956, Rev. Mod. Phys., 28, 53

Slesnick, C. L., Hillenbrand, L. A., & Massey, P. 2002, ApJ, 576, 880

Smartt, S. J., Gilmore, G. E., Tout, C. A., & Hodgkin, S. T. 2002, ApJ, 565, 1087

Smartt, S. J., Gilmore, G. E., Trentham, N., Tout, C. A., & Frayn, C. M. 2001, ApJ, 556, L29

Smartt, S. J., Maund, J. R., Gilmore, G. E., Tout, C. A., Kilkenny, D., & Benetti, S. 2003, MNRAS, 343, 735

Talon, S., & Zahn, J. P. 1997, A&AS, 317, 479

The, L.-S., El Eid, M., & Meyer, B. S. 2000, ApJ, 533, 998

Thielemann, F.-K., & Arnett, D. 1985, ApJ, 295,604

Thielemann, F.-K., Nomoto, K., & Hashimoto, M. 1996, ApJ, 460, 408

Travaglio, C., Kifonidis, K., & Müller, E. 2004, in Carnegie Observatories Astrophysics Series, Vol. 4: Origin and Evolution of the Elements, ed. A. McWilliam & M. Rauch (Pasadena: Carnegie Observatories, http://www.ociw.edu/symposia/series/symposium4/proceedings.html)

Van Horn, H., Thomas, J. H., Frank, A., & Blackman, E. G. 2003, ApJ, 585, 983

Vink, F., Laming, J. M., Kaastra, J. S., Bleeker, J. A. M., Bloemen, H., & Oberlack, U. 2001, ApJ, 560, L79

Wallerstein, G., & Knapp, G. R. 1998, ARA&A, 36, 369

Weaver, T. A., Zimmermann, G. B., & Woosley, S. E. 1978, ApJ, 225, 1021

Wheeler, J. C., Miller, G. E., & Scalo, J. M. 1980, A&AS, 82, 152

Willingale, R., Bleeker, J. A. M., van der Heyden, K. J., & Kaastra, J. S. 2003, A&AS, 398, 1021

Willson, L. A. 2000, ARA&A, 38, 573

Witti, J., Janka, H.-Th., & Kakahashi, K., 1994, A&AS, 286, 841

Woosley, S. E., Heger, A., & Weaver, T. A. 2002, Rev. Mod. Phys., 74, 1015

Woosley, S. E., & Hoffman, R. D. 1992, ApJ, 395, 202

Woosley, S. E., & Weaver, T. A. 1995, ApJS, 101, 181

Woosley, S. E., Wilson, J. R., Mathews, G. J., Hoffman, R. D., & Meyer, B. S. 1994, ApJ, 433, 229

Zinner, E. 1998, Ann. Rev. Earth Planet. Sci., 26, 147

4

Advances in *r*-process nucleosynthesis

JOHN J. COWAN[1] and CHRISTOPHER SNEDEN[2]
(1) Department of Physics and Astronomy, University of Oklahoma
(2) Department of Astronomy, University of Texas at Austin

Abstract

During the last several decades, there have been a number of advances in understanding the rapid neutron-capture process (i.e., the *r*-process). These advances include large quantities of high-resolution spectroscopic abundance data of neutron-capture elements, improved astrophysical models, and increasingly more precise nuclear and atomic physics data. The elemental abundances of the heavy neutron-capture elements, from Ba through the third *r*-process peak, in low-metallicity ([Fe/H] $\lesssim$ –2.5) Galactic halo stars are consistent with the scaled (i.e., relative) solar system *r*-process abundance distribution. These abundance comparisons suggest that for elements with Z $\geq$ 56 the *r*-process is robust—appearing to operate in a relatively consistent manner over the history of the Galaxy—and place stringent constraints on *r*-process models. While not yet identified, neutron-rich ejecta outside of the core in a collapsing (Type II, Ib) supernova continues to be a promising site for the *r*-process. Neutron star binary mergers might also be a possible alternative site. Abundance comparisons of lighter *n*-capture elements in halo stars show variations with the scaled solar *r*-process curve and might suggest either multiple *r*-process sites, or, at least, different synthesis conditions in the same astrophysical site. Constraints on *r*-process models and clues to the progenitors of the halo stars—the earliest generations of Galactic stars—are also provided by the star-to-star abundance scatter of [Eu/Fe] at low metallicities in the early Galaxy. Finally, abundance observations of long-lived radioactive elements (such as Th and U) produced in the *r*-process can be used to determine the chronometric ages of the oldest stars, placing constraints on the lower limit age estimates of the Galaxy and the Universe.

4.1   Introduction

Most of the heavy elements (here, Z > 30) in the solar system are formed in neutron-capture (*n*-capture) processes, either the slow (*s*-) or rapid (*r*-) process. Our understanding of the distinction between these two processes follows from the pioneering work of Cameron (1957) and Burbidge et al. (1957). In the *s*-process the relative lifetime for neutron captures (τ_n) is much longer than for electron (β) decays (τ_β). As a result, the *n*-capture path in the *s*-process is near the so-called valley of beta stability, and the properties of nuclei involved in this nucleosynthesis are, in great part, experimentally accessible. The situation is quite different in the *r*-process where $\tau_n \ll \tau_\beta$. Thus, the *r*-process path occurs in a very neutron-rich regime far from stability, making experimental measurements of those nuclei very difficult, if not impossible.

In this review we focus on advances in our understanding—still very incomplete—of the *r*-process. We note there have been a number of earlier reviews, including those by Hille-brandt (1978), Mathews & Cowan (1990), Cowan, Thielemann, & Truran (1991a), Meyer (1994), Truran et al. (2002), and Sneden & Cowan (2003). We employ and emphasize the observed stellar *n*-capture abundances in our discussion of the *r*-process. These abundances, particularly in low-metal (i.e., low iron, [Fe/H], abundance) stars, provide direct clues to the natures of the *r*-process and *s*-process formation sites. In addition, the abundances of the *n*-capture elements and related chemical evolution studies have also provided important information concerning the *r*-process, specifically in relation to early Galactic nucleosynthesis and star formation history (see reviews of Galactic chemical evolution by, e.g., Wheeler, Sneden, & Truran 1989, McWilliam 1997, and Truran et al. 2002). We also note, and discuss briefly, the importance of certain long-lived radioactive elements, such as thorium and uranium, produced entirely in the *r*-process. The abundance levels of these nuclear chronometers in the most metal-poor halo stars can provide direct age determinations, and hence set lower limits on Galactic and cosmological age estimates (see, e.g., Cowan, Thielemann, & Truran 1991b).

4.2   Neutron-capture Abundances

In this section we examine the abundances of the elements, concentrating on those produced in neutron-capture processes. We show in Figure 4.1 the solar system abundances based upon the compilation of Grevesse & Sauval (1998). Earlier compilations include those of Anders & Ebihara (1982), Cameron (1982a), and Anders & Grevesse (1989). We also note the very recent solar system abundance determinations of Lodders (2003). These solar system abundances can in many ways be treated as "cosmic" and are frequently employed for stellar abundance comparisons.

We highlight in Figure 4.1 the abundances of the neutron-capture elements in solar system material. As is well known, these elements above iron are synthesized predominantly by, and are the sum of individual isotopic contributions from, the *s*- and the *r*-process. The deconvolution of the solar system material into the *s*-process and *r*-process has traditionally relied upon reproducing the "*σ* N" curve (i.e., the product of the *n*-capture cross section and *s*-process abundance). This "classical approach" to the *s*-process is empirical and, by definition, model independent. Subtracting these *s*-process isotopic contributions from the solar abundances determines the residual *r*-process contributions. Early deconvolutions of solar system material into respective *s*- and *r*-process contributions were performed by Clayton et al. (1961), Seeger, Fowler, & Clayton (1965), and Cameron (1982b). We show in Figure 4.2 a more recent such deconvolution based upon *n*-capture cross section measurements (Käppeler, Beer, & Wisshak 1989; Wisshak, Voss, & Käppeler 1996). In addition to this classical approach, more sophisticated models, based upon *s*-process nucleosynthesis in low-mass AGB stars, have been developed recently (Arlandini et al. 1999). Comparing the solar system elemental *r*-process abundance predictions obtained from these model calculations (Arlandini et al. 1999) with the classical approach (Burris et al. 2000) indicates good agreement between each other and with observed stellar abundances (Cowan et al. 2002; Sneden et al. 2003). We also note in Figure 4.2 that the *s*- and *r*-process solar system abundance distributions indicate that not just individual isotopes but entire elements were synthesized primarily in the *s*-process (e.g., Sr, Ba) or the *r*-process (e.g., Eu, Pt) in solar system material. These solar *s*-process, and corresponding *r*-process, elemental abundance distributions have been tabulated in, for example, Sneden et al. (1996) and Burris et al. (2000).

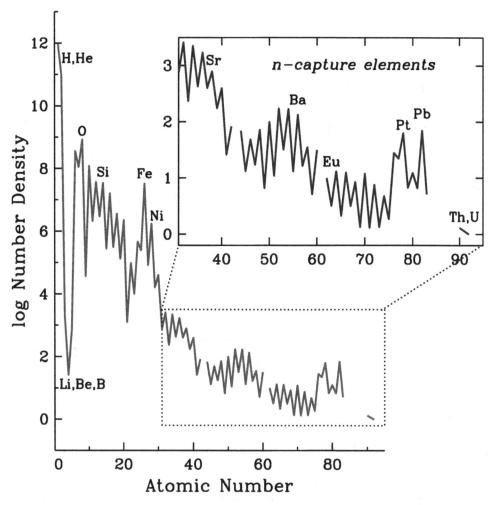

Fig. 4.1. Abundances of elements in the Sun and in solar system material. This abundance set is normalized by convention to log N(H) = 12. The main figure shows the entire set of stable and long-lived radioactive elements, while the inset is restricted to only those (neutron-capture) elements with Z > 30.

4.3 Stellar Abundance Observations

Many of the new advances in understanding the *r*-process have come from stellar abundance observations, and we highlight some of those critical new results in this section.

4.3.1 Metal-poor Stars

Various research groups over the last several decades have been employing the observed abundance distributions in metal-poor ([Fe/H] < −1) Galactic halo stars—bright giants with relatively "uncrowded" spectra—to try to identify, and to understand, the signatures of the *r*- and the *s*-process (see, e.g., Spite & Spite 1978; Sneden & Parthasarathy 1983; Sneden & Pilachowski 1985; Gilroy et al. 1988; Gratton & Sneden 1994; Cowan et al. 1995; McWilliam et al. 1995; Ryan, Norris, & Beers 1996; Sneden et al. 1996; Burris et al. 2000;

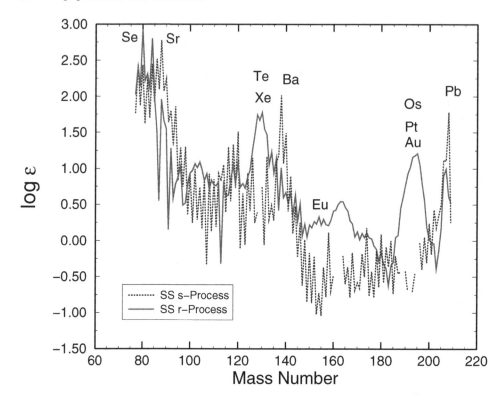

Fig. 4.2. The *s*-process (dotted line) and *r*-process (solid line) abundances in solar system matter, based upon the work by Käppeler et al. (1989). The total solar system abundances for the heavy elements are from Anders & Grevesse (1989).

Johnson & Bolte 2001; Hill et al. 2002). These studies have all suggested the dominance of the *r*-process in the oldest and most metal-poor Galactic halo stars. We show in Figure 4.3 the abundances of *n*-capture elements in CS 22892–052 ([Fe/H] = –3.1) compared with scaled solar system *r*-process (Burris et al. 2000, solid line) and *s*-process (Burris et al. 2000, dashed line) abundance distributions (Sneden et al. 2003). It is very clear that the *n*-capture element abundances in this star are entirely consistent with the relative solar system *r*-process abundance distribution. (The stellar abundances are also well fit with the solar system *r*-process predictions of Arlandini et al. 1999.) It is also clear that *s*-process nucleosynthesis was not responsible for forming the elements observed in this star, at least in anything resembling solar proportions.

Detailed abundance distributions, with more than a few *n*-capture elements, have been obtained for relatively few cases—probably fewer than 20 of the metal-poor halo stars. This picture has been changing in the last few years, however, with new comprehensive abundance studies of the stars CS 22892–052 (Sneden et al. 1996, 2000a, 2003), HD 115444 (Westin et al. 2000), BD +17°3248 (Cowan et al. 2002), and CS 31082–001 (Hill et al. 2002). We show in Figure 4.4 relative abundance distributions in those four stars, again compared with a scaled solar system *r*-process distribution (solid lines). Particularly noteworthy has been: (1) the increasing accessibility of elements in the third *r*-process peak,

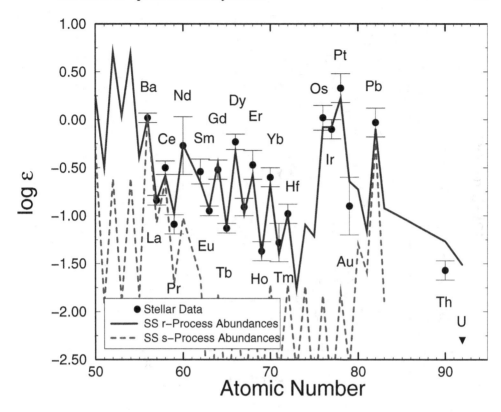

Fig. 4.3. The heavy element abundance pattern for CS 22892–052, normalized to Ba, is compared with the scaled solar system *r*-process (solid line) and *s*-process (dashed line) abundance distributions.

typically only available with observations in the UV with the *Hubble Space Telescope*; and (2) the increasingly precise abundance determinations, resulting in great part from marked improvements in the atomic physics data (see, e.g., Lawler, Bonvallet, & Sneden 2001; Den Hartog et al. 2003, and references therein). In fact, advancements in the atomic physics input have eliminated more and more discrepancies between observed metal-poor stellar abundances and solar *r*-process abundance predictions. Further improvements in individual elemental abundance determinations might be employed to constrain the various theoretical predictions of the actual solar system *r*-process abundances.

Figure 4.4 makes clear that for all four of these *r*-process-rich stars, the elemental abundances, from Ba through the third *r*-process peak, are consistent with relative solar *r*-process proportions. This suggests that for elements with $Z \geq 56$ the *r*-process is very robust, appearing to operate in a relatively consistent manner over the history of the Galaxy. This might imply a similar range of conditions (both astrophysical and nuclear) for the operation of the *r*-process (Freiburghaus et al. 1999a), and perhaps even a narrow range of masses for supernovae sites of the *r*-process (e.g., Mathews, Bazan, & Cowan 1992; Wheeler, Cowan, & Hillenbrandt 1998; Ishimaru & Wanajo 1999).

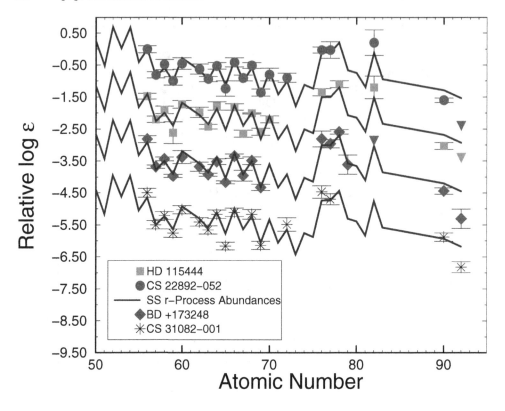

Fig. 4.4. The heavy element abundance patterns for the four stars CS 22892–052, HD 155444, BD +17°3248, and CS 31082–001 are compared with the scaled solar system *r*-process abundance distribution (solid line) (see Westin et al. 2000; Cowan et al. 2002; Hill et al. 2002; Sneden et al. 2003). The absolute abundances have been shifted for all stars except CS 22892–052 for display purposes. Upper limits are indicated by inverted triangles.

4.3.2   *Isotopic Abundances*

While there have been a continuing number of observations of elements in the metal-poor stars, isotopic abundance studies have been relatively rare. This has been predominantly due to observational difficulties—isotopic wavelength shifts for transitions of most all *n*-capture elements are small compared to the thermal and turbulent line widths in stellar spectra. Recently, Sneden et al. (2002) and Aoki et al. (2003) have determined europium isotopic abundance fractions in four very metal-poor, *r*-process-rich stars: CS 22892–052, HD 115444, BD +17°3248, and CS 31082–001. The abundance fractions for Eu in these stars are in excellent agreement with each other and with their values in the solar system: $\mathrm{fr}(^{151}\mathrm{Eu}) \simeq \mathrm{fr}(^{153}\mathrm{Eu}) \simeq 0.5$. Additional Eu stellar abundance studies that demonstrate this same solar system *r*-process agreement have been reported (Ivans 2003, private communication). These isotopic abundance observations support earlier studies that indicate that stellar elemental abundances for $Z \geq 56$ match very closely those of a scaled solar system *r*-process abundance distribution. With only two isotopes, perhaps it is not totally surprising that the Eu abundance fractions in the metal-poor halo stars are the same as in the solar system. It does suggest, though, that the solar abundances are cosmic, in some sense, and that the *r*-process (for the heavier *n*-capture elements) is robust over the lifetime of the Galaxy. Nevertheless, a more definitive test would be an isotopic analysis of the element Ba with many more isotopes than Eu. Lambert & Allende-Prieto (2002) have done such an analysis in another metal-poor star, HD 140283, and find that the isotopic fractions of Ba are also consistent with the solar *r*-process values. Additional stellar isotopic abundance studies will be necessary to strengthen and extend these findings.

4.4   The *r*-Process

Although the *r*-process has been studied for many years, the actual site for this nucleosynthesis has not been identified. Further complicating this search is the possibility of more than one such site. Nevertheless, there has been much progress in our understanding of the astrophysical models and the related nuclear physics of the *r*-process.

4.4.1   *Astrophysical Sites and Models*

The nature of the *r*-process requires high neutron number densities on short time scales, indicative of explosive environments. The early work of Burbidge et al. (1957) suggested that the neutron-rich ejecta outside of the core in a collapsing (Type II, Ib) supernova was the likely site for the *r*-process. Nevertheless, the detailed physics of core-collapse supernovae were poorly known at that time, to say nothing of the lack of computational tools. These hindrances prevented definitive identifications on the nature of the *r*-process and led to the consideration of other possible sites, including the shocked helium and carbon zones of exploding supernovae and jets and bubbles of neutron-rich material ejected from the collapsing core (see Cowan et al. 1991a, and references therein). Inhomogeneous Big Bang cosmological models were even studied as possible sites (Rauscher et al. 1994).

Advances in understanding supernova physics, particularly neutrino interactions, led to new promising *r*-process scenarios, such as the high-entropy neutrino wind in supernovae (Takahashi, Witti, & Janka 1994; Woosley et al. 1994; Qian & Woosley 1996; Wanajo et al. 2001, 2002; Terasawa et al. 2002). There have been some problems, however, with these models in obtaining the required entropies and in some inadequate abundance predictions (see, e.g., Meyer, McLaughlin, & Fuller 1998; Freiburghaus, Rosswog, & Thielemann 1999b; Thompson, Burrows, & Myer 2001; but see also Thompson 2003). (See Thielemann et al. 2002 for a general review of nucleosynthesis in supernovae and the related model uncertainties.) While much emphasis has been placed on determining the physics in "delayed" models, "prompt" supernova explosion scenarios have not been abandoned as a possible site for the *r*-process (see, e.g., Wheeler et al. 1998; Sumiyoshi et al. 2001; Wanajo et al. 2003). It has also been argued that not all core-collapse supernovae are responsible for *r*-process synthesis. In particular there have been a number of studies that suggest only low-mass ($\lesssim$) 11 $M_\odot$ supernovae are likely sites (Mathews & Cowan 1990; Mathews et al. 1992; Wheeler et al. 1998; Ishimaru & Wanajo 1999; Wanajo et al. 2003; but see also Wasserburg & Qian 2000 or Cameron 2001).

While most of the attention in studying the *r*-process has focused on supernovae, there has been some consideration of neutron star binaries, which have an abundance of neutron-rich material. Early studies suggested that the tidal interaction between a neutron star and a black hole, or a second neutron star, might be a possible astrophysical site for this nucleosynthesis (see, e.g., Lattimer et al. 1977). Despite encouraging recent studies by Rosswog et al. (1999)

and Freiburghaus et al. (1999a), however, there are questions about whether the frequency of these events and the amount of r-process ejecta per merger are consistent with observational constraints (Qian 2000).

Accompanying the advances in these more sophisticated astrophysical models has been a concomitant improvement in our understanding of the nuclear physics involved in the r-process - particularly more reliable nuclear information about the very neutron-rich nuclei. The r-process occurs far from stability, and, thus, in the past there has been little reliable nuclear data available. Recently, however, there have been an increasing amount of experimental determinations of critical nuclear data, including half-lives and neutron-pairing energies (see, e.g., Pfeiffer, Kratz, & Möller 2002; Möller, Pfeiffer, & Kratz 2003). In addition to these new nuclear data, there have been recent advances in theoretical prescriptions for very neutron-rich nuclear data (see, e.g., Chen et al. 1995; Pearson, Nayak, & Goriely 1996; Möller, Nix, & Kratz 1997). In particular, these developments include nuclear mass formulae that are more reliable and physically predictive for nuclei far from stability—especially crucial for chronometer studies—for example, such mass models as ETFSI-Q and HFBCS-1 (see Schatz et al. 2002 for discussion and additional references therein). The combination of more nuclear data and advances in theoretical mass models has led to increasingly more reliable descriptions for very neutron-rich nuclei, necessary for a better understanding of the r-process (see also Pfeiffer et al. 2001 for further discussion)

4.4.2   Two r-Processes?

The observations (discussed above) demonstrate that the heavier (Ba and above, Z $\geq$ 56, or A $\gtrsim$ 130–140) neutron-capture elements, particularly in r-process-rich stars, are consistent with a scaled solar system r-process curve. Until very recently, however, there has been relatively little data for elements between Zr and Ba. We show in Figure 4.5 the total abundance summary of the elements in CS 22892–052. A total of 58 elements (53 detections and 5 upper limits) have been observed in this star, which appears to be the most of any other star except the Sun at the time (Sneden et al. 2003). The dashed line in the figure indicates the iron abundance ([Fe/H] = −3.1) in this star. It is clear from the abundances of the heavy n-capture elements why this star has been so well studied—[Eu/Fe], for example, is enhanced by approximately a factor of 45 above the iron abundance level. It is also seen in this figure that the abundances of the light n-capture elements in the little-explored element regime of Z = 40–50 mostly lie below those of the heavy n-capture elements.

This difference is seen in more detail in Figure 4.6, where the abundances in CS 22892–052 (Sneden et al. 2003) are compared with two predictions for solar system r-process abundances, by Burris et al. (2000; top panel) and Arlandini et al. (1999; bottom panel). The dotted line in each panel indicates the unweighted mean difference for elements in the range 56 $\leq$ Z $\leq$ 79. It is clear that the stellar abundance data are well fit by both of these distributions for Z $\geq$ 56, confirming earlier such results (as discussed above and shown in Figs. 4.3 and 4.4). The data, however, seem to indicate that some of the lighter n-capture elements from Z = 40–50 (for example Ag and Mo) are not consistent with (i.e., in general fall below) those same scaled r-process curves that fit the heavy n-capture elements. There are exceptions, with the abundances of Nb and Rh seemingly consistent with the scaled solar system r-process curve, but on average these lighter elements do seem to have been synthesized at a lower abundance level than the heavier n-capture elements.

There are several possible explanations for the differences in the abundance data for the

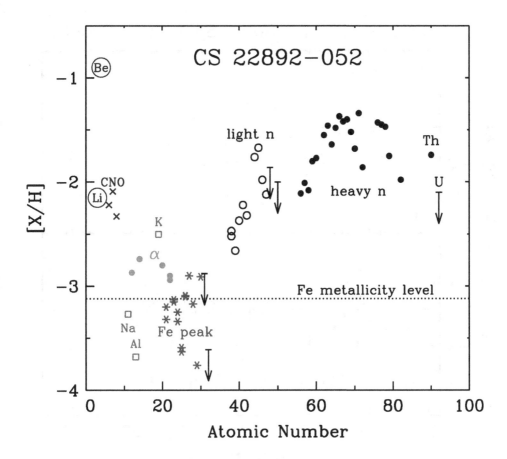

Fig. 4.5. Total abundance pattern in CS 22892–052 with respect to solar system values. The dashed line represents the iron abundance (i.e., the metallicity of the star). Upper limits are denoted by downward-pointing arrows. Li and Be are displaced from their actual abundance values for display purposes.

lighter and heavier *n*-capture elements. These observations might support earlier suggestions of two *r*-processes based upon solar system meteoritic (isotopic) data (Wasserburg, Busso & Gallino 1996). It has been suggested, for example, that perhaps, analogously to the *s*-process, the lighter elements might be synthesized in a "weak" *r*-process with the heavier elements synthesized in a more robust "strong" (or "main") *r*-process (Truran et al. 2002). Thus, the helium zones of exploding supernovae, have been suggested as possible second *r*-process sites that might be responsible for the synthesis of nuclei with $A \lesssim 130$–140 (Truran & Cowan 2000). Or the two sites might come from supernovae of a different mass range or frequency (Wasserburg & Qian 2000), or perhaps a combination of supernova and neutron star binaries (Truran et al. 2002). Alternative interpretations have suggested that the entire abundance distribution could be synthesized in a single core-collapse supernova (Sneden et al. 2000a; Cameron 2001). We note, however, that only a few (three) stars have detailed abundance distributions that include data in this $Z = 40$–50 element domain. Crawford et

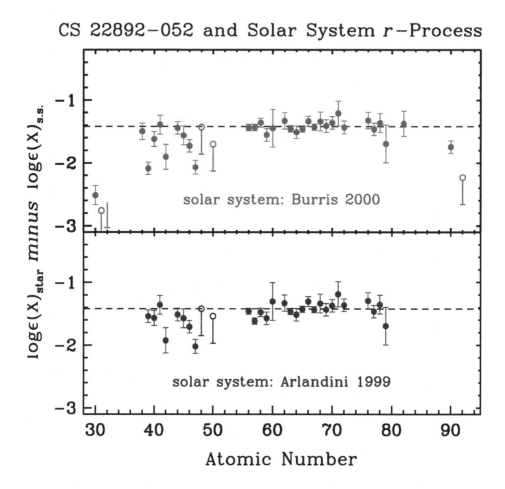

Fig. 4.6. Differences between CS 22892–052 abundances and two scaled solar system abundance distributions, after Sneden et al. (2003). The dotted line in each panel indicates the unweighted mean difference for elements in the range $56 \leq Z \leq 79$. In the top panel the abundance differences relative to those of Burris et al. (2000; their Table 5) are shown. Upper limits are indicated by open circles, except for Ge ($Z = 32$), because its upper limit lies below the lower limit boundary of the plot. In the bottom panel the abundance differences are relative to those of Arlandini et al. (1999), who tabulated *n*-capture abundances only for the atomic number range $39 \leq Z \leq 81$.

al. (1998), however, detected silver in four halo stars, and Sr, Pd, and Ag abundances for a sample of metal-poor stars have been reported by Johnson & Bolte (2002). Further detailed spectroscopic studies, in conjunction with additional theoretical efforts, will be necessary to determine any differences in the nature and history of the synthesis of the lighter and heavier *r*-process elements.

4.5   r-Process Abundance Scatter in the Galaxy

A number of observational and theoretical studies have demonstrated that at the earliest times in the Galaxy the *r*-process was primarily responsible for *n*-capture element formation, even for elements (such as Ba) that are formed primarily in the *s*-process in solar system material (Spite & Spite 1978; Truran 1981; Sneden & Parthasarathy 1983; Sneden & Pilachowski 1985; Gilroy et al. 1988; Gratton & Sneden 1994; McWilliam et al. 1995; Cowan et al. 1995; Sneden et al. 1996; Ryan et al. 1996). The presence of these *r*-process elements in the very oldest stars in our Galaxy strongly suggests the astrophysical *r*-process site is short-lived. Thus, the first stars, the progenitors of the halo stars, were likely massive and evolved quickly, synthesized the *r*-process elements and ejected them into the interstellar medium before the formation of the currently observed stars. In contrast, the primary site for *s*-process nucleosynthesis is low- or intermediate-mass stars (i.e., $M \simeq 0.8 - 8\ M_\odot$) with long evolutionary time scales (Busso, Gallino, & Wasserburg 1999). Thus, these stars would not have had time to have synthesized the first elements in the Galaxy.

Further clues about the nature of the *r*-process are found in examining the abundance scatter of *n*-capture elements in the early Galaxy. This trend was first noted by Gilroy et al. (1988) and then studied in more detail by Burris et al. (2000). We show in Figure 4.7 [Eu/Fe] as a function of metallicity for a number of halo and disk stars from Sneden & Cowan (2003) (see also Truran et al. 2002 and references therein). The increasing level of star-to-star scatter of [Eu/Fe] with decreasing metallicity, particularly at values below [Fe/H] $\approx$ –2.0, suggests an early, chemically unmixed and inhomogeneous Galaxy. These data also suggest that not all early stars are sites for the formation of both *r*-process nuclei and iron. Instead, this scatter is consistent with the view that only a small fraction (2%– 10%) of the massive stars that produce iron also yield *r*-process elements (Truran et al. 2002). Various theoretical models to explain this abundance scatter have been proposed by, for example, Qian & Wasserburg (2001) and Fields, Truran, & Cowan (2002). Observations to help differentiate between these models and provide new insight into the nature and site of *r*-process nucleosynthesis in the early Galaxy are ongoing (Norris 2004).

4.6   r-Process Chronometers

The abundances of certain radioactive *n*-capture elements, known as chronometers, can be utilized to obtain age determinations for the oldest stars, which in turn put lower limits on age estimates for the Galaxy and the Universe. Thorium, with a half-life of 14 Gy, in ratio to Nd (Butcher 1987) and to Eu (Pagel 1989) was suggested as such a chronometer. Th/Eu is a preferred ratio—both are *r*-process elements and any possible evolutionary effects of the predominantly *s*-process Nd are avoided. The detection of thorium in very metal-poor stars was pioneered by François, Spite, & Spite (1993), and since then has been observed in a number of these stars. Chronometric ages, based upon the Th/Eu ratios, have typically fallen in the range of 11–15 Gyr for the observed stars (e.g., Sneden et al. 1996; Cowan et al. 1997, 1999; Pfeiffer, Kratz, & Thielemann 1997; Sneden et al. 2000a, 2003; Westin et al. 2000; Johnson & Bolte 2001; Cowan et al. 2002). Th/Eu ratios have also been determined for several giants in the globular cluster M 15 (Sneden et al. 2000b), who estimated their average, and hence the cluster, age at 14 ± 4 Gyr.

These age estimates all typically have errors $\sim \pm 3 - 4$ Gyr resulting from both observational and nuclear uncertainties. In particular the chronometric age estimates depend sensitively upon the initial predicted values of Th/Eu and hence on the nuclear mass formulae

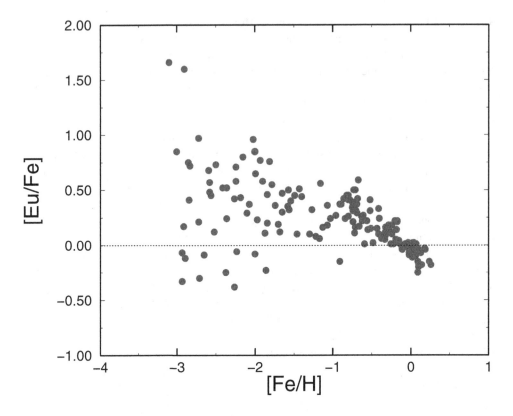

Fig. 4.7. Abundance scatter of [Eu/Fe] versus metallicity for samples of halo and disk stars. (Reprinted with permission from Sneden, C., & Cowan, J. J. 2003, Science, 299, 70. Copyright 2003 AAAS.)

and *r*-process models employed in making those determinations. We show, for example, in Figure 4.8 theoretical predictions for these ratios in comparison with recent abundance determinations in CS 22892–052 (Sneden et al. 2003). Utilizing the ETFSI-Q mass formula, the top panel shows predictions from Cowan et al. (1999), while the bottom panel shows a newer prediction, constrained by some recent experimental data (Sneden et al. 2003). These differences lead to age uncertainties of ~2 Gyr, while very different mass formulae lead to a wider range of initial abundance ratios and correspondingly wider range in age estimates (see Cowan et al. 1999; Truran et al. 2002). The large separation in nuclear mass number between Th and Eu might also exacerbate uncertainties in these initial predictions (see, e.g., Goriely & Arnould 2001). Thus, it would be preferable to obtain abundances of stable elements nearer in mass number to thorium (third *r*-process peak elements, for example), or, even better, to obtain two long-lived chronometers such as Th and U.

Uranium was first detected in any halo star by Cayrel et al. (2001), who initially estimated the age of CS 31082–001 at 12.5 ± 3 Gyr. More refined abundance determinations (Hill et al. 2002) and recent theoretical studies have suggested an age of 15.5 ± 3.2 Gyr (Schatz et al. 2002). We note a fundamental difference in the abundance pattern of this star in comparison with several other metal-poor halo stars shown in Figure 4.4. Both Th and U are enhanced

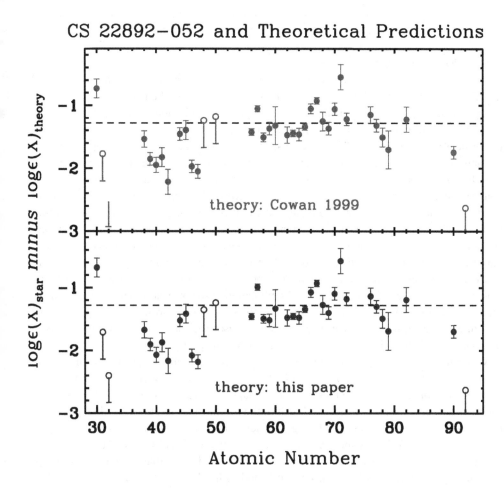

Fig. 4.8. Differences between CS 22892–052 abundances and two scaled *r*-process theoretical predictions, after Sneden et al. (2003). The top panel shows the abundance differences relative to those of Cowan et al. (1999), while the bottom panel differences are relative to Sneden et al. (2003).

greatly with respect to the other stable elements, including Eu, in CS 31082–001. Thus, employing Th/Eu as a chronometer gives an unrealistic, even a negative, age in this case (Hill et al. 2002; Schatz et al. 2002; see also Cayrel 2004 for a more complete discussion of chronometers in this star). Interestingly, another tentative U detection has been recently reported in the metal-poor halo star BD +17°3248 (see Fig. 4.4), and in this case, Th/U and Th/Eu give comparable age values. It is not clear yet why CS 31082–001, with its very large overabundances of Th and U, is so different, but it clearly suggests that Th/U is a more reliable chronometer than Th/Eu—certainly it is in this star. Unfortunately, it may be difficult to detect uranium in many halo stars; note the nondetection of this element in CS 22892–052 (Sneden et al. 2003). Additional such detections of U, as well as continually improving nuclear descriptions of the very neutron-rich nuclei participating in the *r*-process, will be necessary to strengthen this technique and reduce the age uncertainties.

4.7   Summary and Conclusions

A wealth of stellar abundance data has been assembled during the last several decades. These high-resolution spectroscopic studies of low-metallicity halo stars—accompanied by significant new experimental atomic physics data—have provided significant clues about the nature of the *r*-process and, at the same time, have imposed strong constraints on astrophysical and nuclear model calculations. The heavy *n*-capture abundances, from Z ≥ 56, appear to be consistent with the relative solar system *r*-process abundance fractions, at least for the *r*-process-rich stars. This consistency suggests a similar mechanism, or well-constrained astrophysical conditions, for the operation of the *r*-process over many billions of years. While there are less data available of the lighter *n*-capture elements, the abundances of those elements appear to be not consistent—on average lower—with the same scaled solar system *r*-process distribution that fits the heavier *n*–capture elements. Various explanations have been offered to explain this difference between the lighter and heavier *n*-capture abundance distributions: the possibility of two astrophysical sites for the *r*-process (e.g., different masses or frequencies of supernovae, or a combination of supernovae and neutron star binary mergers), or models with different conditions in the same single core-collapse supernova. At this time, however, it is not clear what the exact causes are for the apparent differences in the abundance distributions of the lighter and heavier *n*-capture elements.

While the actual site for the *r*-process has not been definitively identified, there have been many advances in our understanding of the astrophysical models and the related nuclear physics of this nucleosynthesis. Core-collapse supernovae remain a promising site for the origin of the *r*-process nuclei. Much of the recent focus has been on obtaining more physically reliable supernova models (e.g., including an improved treatment of neutrino processes). Neutron star binary mergers, with improved treatments of their evolution and coalescence, have also been studied and can still be considered a possible site for the *r*-process. Accompanying these improved astrophysical models has been more experimental nuclear data and more reliable theoretical prescriptions for neutron-rich nuclei far from stability.

The abundance patterns in the oldest Galactic halo stars have also provided additional new insights into the origins and sites of the *r*-process. First, the elemental and isotopic abundances in the oldest halo stars are consistent with an *r*-process-only origin at the earliest times in the history of the Galaxy. These results suggest that the *r*-process sites in the earliest stellar generations, the progenitors of the halo stars, were rapidly evolving—ejecting *r*-process-rich material into the interstellar medium long before the major onset of Galactic *s*-process nucleosynthesis from low- and intermediate-mass stars. In addition, the star-to-star abundance scatter (e.g., [Eu/Fe]) observed in the lowest metallicity (i.e., oldest) Galactic halo stars places strong constraints on models of nucleosynthesis and suggests that not all early stars were sites for the formation of both *r*-process nuclei and iron. Further, this abundance scatter suggests an early chemically unmixed Galaxy.

The detection of the long-lived chronometers thorium, and now uranium, in some of the metal-poor halo stars has allowed for the radioactive dating of the oldest stars. This technique depends sensitively upon the observed stellar values and the theoretical predictions of the initial abundance ratios (Th/Eu, Th/U, etc.) of elements synthesized in the *r*-process. While there have been significant advances in nuclear physics, both in experiment and theory, we still need to better define the properties of very *n*-rich (radioactive) nuclei. These improvements will be necessary to better understand the origin and nature of the *r*-process and to reduce chronometric age uncertainties—strengthening the radioactive dating technique, and providing more precise age estimates for the Galaxy and the Universe.

Acknowledgements. We thank all of our colleagues for their contributions to the work described here. We also thank the referee, Friedel Thielemann, for helpful comments. This research has been supported in part by NSF grants AST-9986974 and AST-0307279 (JJC), AST-9987162 and AST-0307495 (CS), and by STScI grants GO-8111 and GO-08342.

References

Anders, E., & Ebihara, M. 1982, Geochim. Cosmo. Acta, 46, 2363
Anders, E., & Grevesse, N. 1989, Geochem. Cosmo. Chem., 53, 197
Aoki, W., Honda, S., Beers, T. C., & Sneden, C. 2003, ApJ, 586, 506
Arlandini, C., Käppeler, F., Wisshak, K., Gallino, R., Lugaro, M., Busso, M., & Staniero, O. 1999, ApJ, 525, 886
Burbidge, E. M., Burbidge, G. R., Fowler, W. A., & Hoyle, F. 1957, Rev. Mod. Phys., 29, 547
Burris, D. L., Pilachowski, C. A., Armandroff, T. A., Sneden, C., Cowan, J. J., & Roe, H. 2000, ApJ, 544, 302
Busso, M., Gallino, R., & Wasserburg, G. J. 1999, ARA&A, 37, 239
Butcher, H. R. 1987, Nature, 328, 127
Cameron, A. G. W. 1957, Chalk River Report CRL-41
——. 1982a, in Essays in Nuclear Astrophysics, ed. C. A. Barnes, D. D. Clayton, & D. N. Schramm (Cambridge: Cambridge Univ. Press), 23
——. 1982b, Astrophys. Space Sci., 82, 123
——. 2001, ApJ, 562, 456
Cayrel, R. 2004, in Carnegie Observatories Astrophysics Series, Vol. 4: Origin and Evolution of the Elements, ed. A. McWilliam & M. Rauch (Pasadena: Carnegie Observatories, http://www.ociw.edu/ociw/symposia/series/symposium4/proceedings.html)
Cayrel, R., et al. 2001, Nature, 409, 691
Chen, B., Dobaczewski, J., Kratz, K.-L., Langanke, K., Pfeiffer, B., Thielemann, F.-K., & Vogel, P. 1995, Phys. Lett., B355, 37
Clayton, D. D., Fowler, W. A., Hull, T. E., & Zimmerman, B. A. 1961, Ann. Phys., 12, 331
Cowan, J. J., et al. 2002, ApJ, 572, 861
Cowan, J. J., Burris, D. L., Sneden, C., McWilliam, A., & Preston, G. W. 1995, ApJ, 439, L51
Cowan, J. J., McWilliam, A., Sneden, C., & Burris, D. L. 1997, ApJ, 480, 246
Cowan, J. J., Pfeiffer, B., Kratz, K.-L., Thielemann, F.-K., Sneden, C., Burles, S., Tytler, D., & Beers, T. C. 1999, ApJ, 521, 194
Cowan, J. J., Thielemann, F.-K., & Truran, J. W. 1991a, Phys. Rep., 208, 267
——. 1991b, ARA&A, 29, 447
Crawford, J. L., Sneden, C., King, J. R., Boesgaard, A. M., & Deliyannis, C. P. 1998, AJ, 116, 2489
Den Hartog, E. A., Lawler, J. E., Sneden, C., & Cowan, J. J. 2003, ApJS, 148, 543
Fields, B. D., Truran, J. W., & Cowan, J. J. 2002, ApJ, 575, 845
François, P., Spite, M., & Spite, F. 1993, A&A, 274, 821
Freiburghaus, C., Rembges, J.-F., Rauscher, T., Thielemann, F.-K., Kratz, K.-L., Pfeiffer, B., & Cowan, J. J. 1999a, ApJ, 516, 381
Freiburghaus, C., Rosswog, S., & Thielemann, F.-K. 1999b, ApJ, 525, L121
Gilroy, K. K., Sneden, C., Pilachowski, C. A., & Cowan, J. J. 1988, ApJ, 327, 298
Goriely, S., & Arnould, M. 2001, A&A, 379, 1113
Gratton, R., & Sneden, C. 1994, A&A, 287, 927
Grevesse, N., & Sauval, A. J. 1998, Sp. Sci. Rev., 85, 161
Hill, V., et al. 2002, A&A, 387, 560
Hillebrandt, W. 1978, Sp. Sci. Rev., 21, 639
Ishimaru, Y., & Wanajo, S. 1999, ApJ, 511, L33
Johnson, J. A., & Bolte, M. 2001, ApJ, 554, 888
——. 2002, ApJ, 579, 616
Käppeler, F., Beer, H., & Wisshak, K. 1989, Rep. Prog. Phys., 52, 945

Lambert, D. L., & Allende Prieto, C. 2002, MNRAS, 335, 325

Lattimer, J. M., Mackie, F., Ravenhall, D. G., & Schramm, D. N. 1977, ApJ, 213, 225

Lawler, J. E., Bonvallet, G., & Sneden, C. 2001, ApJ, 556, 452

Lodders, K. 2003, ApJ, 591, 1220

Mathews, G. J., Bazan, G., & Cowan, J. J. 1992, ApJ, 391, 719

Mathews, G. J., & Cowan, J. J. 1990, Nature, 345, 491

McWilliam, A. 1997, ARA&A, 35, 503

McWilliam, A., Preston, G. W., Sneden, C., & Searle, L. 1995, AJ, 109, 2757

Meyer, B. S. 1994, ARA&A, 32, 153

Meyer, B. S., McLaughlin, G. C., & Fuller, G. M. 1998, Phys. Rev. C, 58, 3696

Möller, P., Nix, J. R., & Kratz, K.-L. 1997, At. Data Nucl. Data Tables, 66, 131

Moïler, P., Pfeiffer, B., & Kratz, K.-L. 2003, Phys. Rev., C67, 055802

Norris, J. E. 2004, in Carnegie Observatories Astrophysics Series, Vol. 4: Origin and Evolution of the Elements, ed. A. McWilliam & M. Rauch (Cambridge: Cambridge Univ. Press), in press

Pagel, B. E. J. 1989, in Evolutionary Phenomena in Galaxies, ed. J. E. Beckman & B. E. J. Pagel (Cambridge: Cambridge Univ. Press), 201

Pearson, J. M., Nayak, R. C., & Goriely, S. 1996, Phys. Lett., B387, 455

Pfeiffer, B., Kratz, K.-L., & Mïler, P. 2002, Progr. Nucl. Energ., 41, 39

Pfeiffer, B., Kratz, K.-L., & Thielemann, F.-K. 1997, Z. Phys. A, 357, 235

Pfeiffer, B., Kratz, K.-L., Thielemann, F.-K., & Walters, W. B. 2001, Nucl. Phys., A693, 282

Qian, Y.-Z. 2000, ApJ, 534, L67

Qian, Y.-Z., & Woosley, S. E. 1996, ApJ, 471, 331

Qian, Y.-Z., & Wasserburg, G.J. 2001, ApJ, 559, 925

Rauscher, T., Applegate, J. H., Cowan, J. J., Thielemann, F.-K., & Wiescher, M. 1994, ApJ, 429, 499

Rosswog, S., Liebendorfer, M., Thielemann, F.-K., Davies, M. B., Benz, W., & Piran, T., 1999, A&A, 341, 499

Ryan, S. G., Norris, J. E., & Beers, T. C. 1996, ApJ, 471, 254

Schatz, H., Toenjes, R., Pfeiffer, B., Beers, T. C., Cowan, J. J.; Hill, V., & Kratz, K.-L. 2002, ApJ, 579, 626

Seeger, P. A., Fowler, W. A., & Clayton, D. D. 1965, ApJS, 11, 121

Sneden, C., et al. 2003, ApJ, 591, 936

Sneden, C., & Cowan, J. J. 2003, Science, 299, 70

Sneden, C., Cowan, J. J., Ivans, I. I., Fuller, G. M., Burles, S., Beers, T. C., & Lawler, J. E. 2000a, ApJ, 533, L139

Sneden, C., Cowan, J. J., Lawler, J. E., Burles, S., Beers, T. C., & Fuller, G. M. 2002, ApJ, 566, L25

Sneden, C., Johnson, J., Kraft, R. P., Smith, G. H., Cowan, J. J., & Bolte, M. S. 2000b, ApJ, 536, L85

Sneden, C., McWilliam, A., Preston, G. W., Cowan, J. J., Burris, D. L., & Armosky, B. J. 1996, ApJ, 467, 819

Sneden, C., & Parthasarathy, M. 1983, ApJ, 267, 757

Sneden, C., & Pilachowski, C. A. 1985, ApJ, 288, L55

Spite, M., & Spite, F. 1978, A&A, 67, 23

Sumiyoshi, K., Terasawa, M., Mathews, G. J., Kajino, T., Yamada, S., & Suzuki, H. 2001, ApJ, 562, 880

Takahashi, K., Witti, J., & Janka, H.-T. 1994, A&A, 286, 857

Terasawa, M., Sumiyoshi, K., Yamada, S., Suzuki, H., & Kajino, T. 2002, ApJ, 578, L137

Thielemann, F.-K., et al. 2002, Astrophys. Space Sci., 281, 25

Thompson, T. A. 2003, ApJ, 585, L33

Thompson, T. A., Burrows, A., & Meyer, B. 2001, ApJ, 562, 887

Truran, J. W. 1981, A&A, 97, 391

Truran, J. W., & Cowan, J. J. 2000, in Nuclear Astrophysics, ed. W. Hillebrandt & E. Müller (Munich: MPI), 64

Truran, J. W., Cowan, J. J., Pilachowski, C. A., & Sneden, C. 2002, PASP, 114, 1293

Wanajo, S., Itoh, N., Ishimaru, Y., Nozawa, S., & Beers, T. C. 2002, ApJ, 577, 853

Wanajo, S., Kajino, T., Mathews, G. J., & Otsuki, K. 2001, ApJ, 554, 578

Wanajo, S., Tamamura, M., Itoh, N., Nomoto, K., Ishimaru, Y., Beers, T. C., & Nozawa, S. 2003, ApJ, 593, 968

Wasserburg, G. J., Busso, M., & Gallino, R. 1996, ApJ, 466, L109

Wasserburg, G. J., & Qian, Y.-Z. 2000, ApJ, 529, L21

Westin, J., Sneden, C., Gustafsson, B., & Cowan, J. J. 2000, ApJ, 530, 783

Wheeler, C., Cowan, J. J., & Hillebrandt, W. 1998, ApJ, 493, L101

Wheeler, J. C., Sneden, C., & Truran, J. W. 1989, ARA&A, 27, 279

Wisshak, K., Voss, F., & Käppeler, F. 1996, in Proceedings of the 8th Workshop on Nuclear Astrophysics, ed. W. Hillebrandt & E. Müller (Munich: MPI), 16

Woosley, S. E., Wilson, J. R., Mathews, G. J., Hoffman, R. D., & Meyer, B. S. 1994, ApJ, 433, 229

5

Element yields of intermediate-mass stars

RICHARD B. C. HENRY
University of Oklahoma

Abstract

Intermediate-mass stars occupy the mass range between 0.8 and 8 $M_\odot$. In this contribution, evolutionary models of these stars from numerous sources are compared in terms of their input physics and predicted yields. In particular, the results of Renzini & Voli, van den Hoek & Groenewegen, and Marigo are discussed. Generally speaking, it is shown that yields of ^{4}He, ^{12}C, and ^{14}N decrease with increasing metallicity, reduced mass loss rate, and increased rotation rate. Integrated yields and recently published chemical evolution model studies are used to assess the relative importance of intermediate-mass and massive stars in terms of their contributions to universal element build-up. Intermediate-mass stars appear to play a major role in the chemical evolution of ^{14}N, a modest role in the case of ^{12}C, and a small role for ^{4}He. Furthermore, the time delay in their release of nuclear products appears to play an important part in explaining the apparent bimodality in the distribution of damped Lyα systems in the N/α–α/H plane.

5.1   The Nature of Intermediate-mass Stars

Intermediate-mass stars (IMS) comprise objects with zero-age main sequence masses between 0.8 and 8 $M_\odot$*, corresponding to spectral types between G2 and B2. The lower mass limit is the minimum value required for double-shell (H and He) fusion to occur, resulting in thermal pulsations during the asymptotic giant branch (AGB) phase and eventually planetary nebula formation. Above the upper mass limit stars are capable of additional core-burning stages, and it is generally assumed that these stars become supernovae. A Salpeter (1955) initial mass function (IMF) can be used to show that IMS represent about 4% of all stars above 0.08 $M_\odot$, but this may be a lower limit if the IMF is flat at low stellar masses (Scalo 1998).

IMS evolution is an interesting and complex subject, and the literature is extensive. A good, complete, generally accessible review of the subject is given by Iben (1995). Shorter reviews focusing on the AGB stage can be found in Charbonnel (2002) and Lattanzio (2002). I will simply summarize here.

IMS spend about 10%–20% of their nuclear lives in post-main sequence stages (Schaller et al. 1992). Fresh off the main sequence, a star's core is replete with H-burning products

---

* Stars within this range but below 2–2.2 $M_\odot$ form degenerate He cores at the tip of the red giant branch, while those above this threshold form nondegenerate He cores. Some authors refer to the former group as low-mass stars and reserve the intermediate-mass classification for the latter group. For the purposes of this review, however, this distinction is not emphasized.

such as ^{4}He and ^{14}N. The shrinking core's temperature rises, a H-burning shell forms outward from the core, and shortly afterwards the base of the outer convective envelope moves inward and encounters these H-burning products, which are then mixed outward into the envelope during what is called the *first dredge-up*. As a result, envelope levels of ^{4}He, ^{14}N, and ^{13}C rise. Externally, the star is observed to be a red giant.

As the shrinking He core ignites, the star enters a relatively stable and quiescent time during which it synthesizes ^{12}C and ^{16}O. Once core He is exhausted, the star enters the AGB phase, characterized by a CO core along with shells of H and He-fusing material above it. Early in this phase, for masses in excess of 4 $M_\odot$, *second dredge-up* occurs, during which the base of the convective envelope again extends inward, this time well into the intershell region, and dredges up H-burning products, increasing the envelope inventory of ^{4}He, ^{14}N, and ^{13}C as before.

Later in the AGB phase, however, the He shell becomes unstable to runaway fusion reactions, due to its thin nature and the extreme temperature sensitivity of He burning. The resulting He-shell flash drives an intershell convective pocket that mixes fresh ^{12}C outward toward the H shell. But as the intershell expands, H-shell burning is momentarily quenched, and once again the outer convective envelope extends down into the intershell region and dredges up the fresh ^{12}C into the envelope, an event called *third dredge-up*. Subsequently, the intershell region contracts, the H shell reignites, and the cycle repeats during a succession of thermal pulses. Observational consequences of thermal pulsing and third dredge-up include the formation of carbon stars, Mira variables, and barium stars.

Now, in IMS more massive than about 3–4 $M_\odot$, the base of the convective envelope may reach temperatures that are high enough ($\sim$60 million K) to cause further H burning via the CN cycle during third dredge-up. As a result, substantial amounts of ^{12}C are converted to ^{14}N in a process referred to as hot-bottom burning (HBB; Renzini & Voli 1981). HBB not only produces large amounts of ^{14}N but also results in additional neutron production through the ^{13}C(α,n)^{16}O reaction, where extra mixing is required to produce the necessary ^{13}C. These additional neutrons spawn the production of *s*-process elements that are often observed in the atmospheres of AGB stars. Note that carbon star formation is precluded by HBB in those stars where it occurs. Other nuclei that are synthesized during thermal pulsing and HBB include ^{22}Ne, ^{25}Mg, ^{26}Al, ^{23}Na, and ^{7}Li (Karakas & Lattanzio 2004).

The thermal pulsing phase ends when the star loses most of its outer envelope through winds and planetary nebula (PN) formation, and thus the main fuel source for the H shell (and for the star) is removed, and evolution is all but over. Note that the PN contains much of the new material synthesized and dredged up into the atmosphere of the progenitor star during its evolution. As this material becomes heated by photoionization, it produces numerous emission lines whose strengths can be measured and used to infer physical and chemical properties of the nebula.

5.2   Stellar Models and Yields

5.2.1   The Predictions

Models of IMS evolution are typically synthetic in nature. A coarse grid of models, in which values for variable quantities are computed directly from fundamental physics, is first produced. Then interpolation formulas are inferred from this grid, which are subse-

Table 5.1. *Model Calculations of IMS Stellar Yields*

Authors[1]	M_l/M_u ($M_\odot$)	Z_l/Z_h	HBB/CBP	$\dot{M}$/Rot	Isotopes
IT	1/8	0.02	n/n	R/n	$^{12}C,^{13}C,^{14}N,^{22}Ne$
RV	1/8	0.004/0.02	y/n	R/n	$^{12}C,^{13}C,^{14}N,^{16}O$
HG	0.8/8	0.001/0.04	y/n	R/n	$^{12}C,^{13}C,^{14}N,^{16}O$
FC	3/7	0.005/0.02	y/n	R/n	$^{7}Li,^{12}C,^{14}N$, and more
BS	0.8/9	0.0001/0.02	n/y	R/n	$^{12}C,^{13}C,^{14}N,^{16,17,18}O$
BU	1/8	0.002/0.03	y/n	VW/n	$^{12,13}C,^{14}N,^{16}O$
M01	0.8/6	0.004/0.019	y/n	VW/n	$^{12,13}C,^{14,15}N,^{16,17,18}O$
MM	2/7	10^{-5}	n/n	n/y	$^{12}C,^{14}N,^{16}O$

[1]References: IT = Iben & Truran (1978); RV = Renzini & Voli (1981); HG = van den Hoek & Groenewegen (1997); FC = Forestini & Charbonnel (1997); BS = Boothroyd & Sackmann (1999); BU = Buell (1997); M01 = Marigo (2001); MM = Meynet & Maeder (2002).

quently used in a much larger run of models, thus reducing the computation time requirements. The models described below are of this type.

The major parameters that serve as input for IMS models include stellar mass and metallicity, the value of the mixing length parameter, the minimum core mass required for HBB, the formulation for mass loss, and third dredge-up efficiency.

The first substantial study of IMS surface abundances using theoretical models was carried out by Iben & Truran (1978), whose calculations accounted for three dredge-up stages including thermal pulsing. Renzini & Voli (1981; hereafter RV) introduced HBB and the Reimers (1975) mass loss rate to their models and explicitly predicted PN composition and total stellar yields. Van den Hoek & Groenewegen (1997; hereafter HG) introduced a metallicity dependence, heretofore ignored, into their evolutionary algorithms along with an adjustment upwards in the mass loss rate, the latter being a change driven by constraints imposed by the carbon star luminosity function (see below). Finally, Boothroyd & Sackmann (1999) demonstrated effects of cool-bottom processing (CBP) on the $^{12}C/^{13}C$ ratio; Marigo, Bressan, & Chiosi (1996), Buell (1997), and Marigo (2001; hereafter M01) employed the mass loss formalism of Vassiliadis & Wood (1993), which links the mass loss rate to the star's pulsation period, to predict yields of important CNO isotopes; and Langer et al. (1999) and Meynet & Maeder (2002) studied the effects of stellar rotation on CNO yields.

Table 5.1 provides a representative sample of yield calculations carried out over the past two decades. To the right of the author column are columns that indicate the lower and upper limits of the mass and metallicity ranges considered, an indication of whether HBB or CBP was included in the calculations (yes or no), the type of mass loss used [R = Reimers (1975), VW = Vassiliadis & Wood (1993)], an indication of whether the calculations included stellar rotation (yes or no), and some important nuclei whose abundances were followed during the calculations.

It should be noted that the references listed in Table 5.1 represent a much larger collection of papers, which, while uncited here, nevertheless form an indispensable body of theory. The IMS enthusiast is hereby urged to consult the references in the table and explore this extensive literature.

For the purposes of a detailed discussion and comparison, I have singled out three yield sets in Table 5.1, which are used most frequently to compute chemical evolution models. These are the yields published by RV, HG, and M01. In the remainder of this subsection, each set will be considered by itself, after which a comparison of their results will be made.

- *Renzini & Voli (1981)*: The calculations published by RV were the first to fully develop and include the process of HBB. Using parameterized equations in the spirit of Iben & Truran (1978), RV followed the surface evolution of ^{4}He, ^{12}C, ^{13}C, ^{14}N, and ^{16}O for stars of birth mass between 1 and 8 $M_\odot$ and metallicities 0.004 and 0.02. They adopted the empirical mass loss scheme of Reimers (1975), which at the time was uncalibrated with respect to observables such as the carbon star luminosity function. The RV yields are still extensively used, although the models suffer from being unable to reproduce the carbon star luminosity function, probably because their adopted mass loss rate was too low (thus allowing more time for HBB to turn ^{12}C into ^{14}N).

- *van den Hoek & Groenewegen (1997)*: The models of HG covered roughly the same stellar mass range and isotope group as those of RV. In addition, HG used synthetic models with parameterized equations as did RV. HG also assumed a metallicity dependence of the algorithms, something RV had not done. Although both RV and HG employed the Reimers mass loss scheme, HG's η parameter was over 4 times larger than the one used by RV, thus elevating the mass loss rate and reducing stellar life times. As we will see below, this has a big effect on the resulting yields. HG also used fundamental parameters for mass loss and dredge-up that were actually tuned to reproduce observables such as the carbon star luminosity function and the initial-final mass relation.

- *Marigo (2001)*: M01's calculations covered a stellar mass range between 0.8 and 5 $M_\odot$[*] and a metallicity range between 0.004 and 0.019. M01's thermally pulsing AGB models were built upon the precursor models of Girardi et al. (2000), which stopped at the onset of the thermally pulsing AGB stage. A major difference in M01's calculations compared with earlier ones was the use of the mass loss scheme developed by Vassiliadis & Wood (1993), in which stellar pulsation theory and variable AGB star observations play roles. In addition, an apparent advantage of the M01 yields for chemical evolution models is that the Girardi-Marigo IMS models are based upon the same Padua library as are the massive star models and predicted yields of Portinari, Chiosi, & Bressan (1998). Thus, the two sets can be combined to form a seamless yield set over a large stellar mass range (0.8–120 $M_\odot$).

Now turning to the results of these three papers, Figures 5.1, 5.2, and 5.3 show comparisons of the yield predictions specifically for the isotopes of ^{4}He, ^{12}C, and ^{14}N, respectively. Because of space limitations, only these entities will be emphasized in the remainder of the paper.[†] Yields are plotted as a function of stellar birth mass, both in solar masses; metallicities are indicated by line type. Note that for ease of comparison, the three panels in each figure are identically scaled.

Generally speaking, IMS yields vary directly with the time between thermal pulses, the efficiency of third dredge-up, and the total number of thermal pulses that occur while the

---

[*] Marigo considered effects of convective overshooting, which lowered the threshold for nondegenerate carbon burning as well as the upper mass limit for IMS.

[†] IMS contributions to the production of many other isotopes, including those produced in the *s*-process, are discussed elsewhere in this conference by experts in the field. Interested readers should see the papers by Charbonnel, Karakas & Lattanzio, Lambert, Busso, Straniero, and Reyniers & Van Winckel for further updates and references.

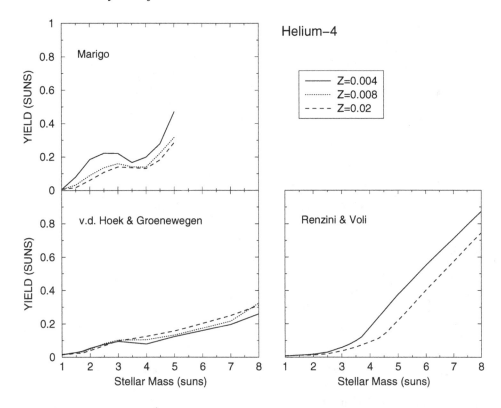

Fig. 5.1. Stellar yield for ^{4}He as a function of stellar mass, both in solar units, from M01 (top left panel), HG (bottom left panel), and RV (bottom right panel). Metallicity is shown in the legend and corresponds to line type in each figure.

star is on the AGB. These factors are, in turn, fundamentally affected by stellar mass and metallicity, as well as the values of the mixing length, dredge-up, and mass loss parameters. A full discussion of the relevant stellar theory is beyond the scope of this review, and the reader is referred to the papers listed in Table 5.1 for more details and further references. Nevertheless, general statements can be made here concerning these relations.

M01 points out that the number and duration of thermal pulses increases with stellar mass. Hence yields likewise tend to be related directly to this parameter, as can be seen clearly especially in the cases of ^{4}He and ^{14}N (Figs. 5.1 and 5.3), although the effect is overridden by HBB in the case of ^{12}C (Fig. 5.2). In addition, the mass loss rate is directly related to metallicity, and so yields tend to be greater at low metallicity, a trend that is also visible in Figures 5.1, 5.2, and 5.3. The stellar core mass tends to be larger at lower metallicity (Groenewegen & de Jong 1993), and so the first thermal pulse dredges up more core material, adding to the other effects of metallicity. Also shown by M01 but not included here is the upward trend in yields with increasing values of the mixing length parameter (see Figs. 1 and 2 in M01).

Many of the secondary features appearing in the figures relate to the details of HBB, which is effective in stars of mass greater than 3–4 $M_\odot$. Its efficiency increases directly with mass and inversely with metallicity (M01), and the effect is particularly apparent in the

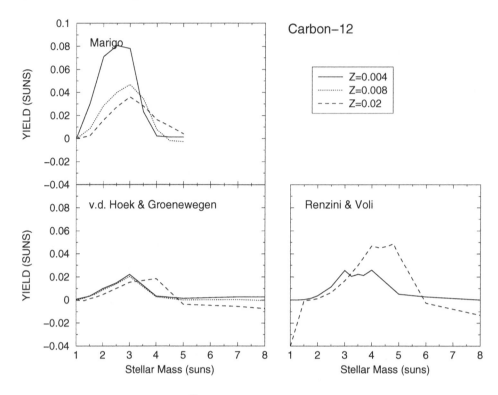

Fig. 5.2. Same as Fig. 5.1 but for ^{12}C.

behavior of ^{12}C and ^{14}N yields in Figures 5.2 and 5.3. For example, below the HBB mass threshold the process does not operate, and the increase in the ^{12}C fraction in the envelope with third dredge-up is a major result of this process. However, above the threshold ^{12}C is processed into ^{14}N.

The effect of mass loss rate is shown explicitly in Figure 5.4, which compares ^{12}C and ^{14}N yields predicted by HG, who employed the Reimers mass loss scheme, for two different values of the mass loss parameter. For $\eta = 1$ (i.e. relatively low mass low rate), yields in both cases are significantly higher than they are for the higher rate associated with the $\eta = 4$ case. The relatively low mass loss parameter employed by RV explains why their ^{4}He and ^{14}N yields, in particular, tend to be greater than those of HG and M01.

The effects of stellar rotation on IMS evolution and yields have been receiving much attention lately. The topic is nicely explored by Langer et al. (1999), Charbonnel & Palacios (2003), and by Charbonnel (2004) at this conference. In terms of rotational effects on stellar yields, Meynet & Maeder (2002) have recently computed an extensive model grid of stellar models for $Z = 10^{-5}$ spanning the mass range of 2–60 $M_{\odot}$, and their results for the IMS mass range are shown in Figure 5.5, where the predicted yields for ^{4}He, ^{12}C, and ^{14}N for the cases with and without rotation are compared*. While rotation makes little difference in the case of ^{4}He, it has a substantial effect on yields of ^{12}C (particularly at the low-mass end) and ^{14}N.

---

* Meynet & Maeder did not carry their IMS calculations past the first few thermal pulses, and therefore the effects of HBB are not included in their results.

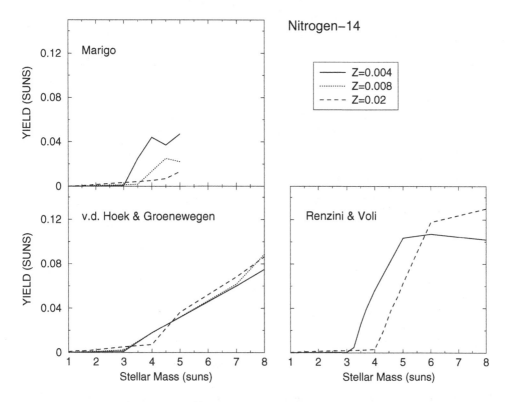

Fig. 5.3. Same as Fig. 5.1 but for ^{14}N.

According to Meynet & Maeder, rotation increases the size of the CO core through more efficient rotational mixing; the effect is particularly pronounced at low Z, where the angular velocity gradient is much steeper.

Finally, Population III nucleosynthesis for IMS has been studied by Chieffi et al. (2001), Marigo et al. (2001), and Siess, Livio, & Lattanzio (2002). The Chieffi and Siess teams conclude that IMS are likely to be major contributors to ^{12}C and ^{14}N evolution in the early Universe, although a great deal of work has yet to be done on this subject.

5.2.2   *Observational Constraints*

How well do the stellar models discussed above agree with observational constraints? At least four tests are available for comparison: the initial-final mass relation for IMS, the carbon star luminosity function, abundances in planetary nebulae, and ^{12}C/^{13}C abundance ratios in red giants.

First, realistic evolutionary models of IMS should predict the form of the initial-final mass relation that is observed. Typically, the empirical relation is established by comparing central star masses and turn-off masses in clusters (Weidemann 1987; Herwig 1995; Jeffries 1997). M01 graphically compared these relations with theoretical results extracted from her models as well as those of HG and RV. The comparison makes clear that the higher mass loss rates used in the HG and M01 calculations are more successful in explaining the empirical relation than are the RV calculations with much lower rates.

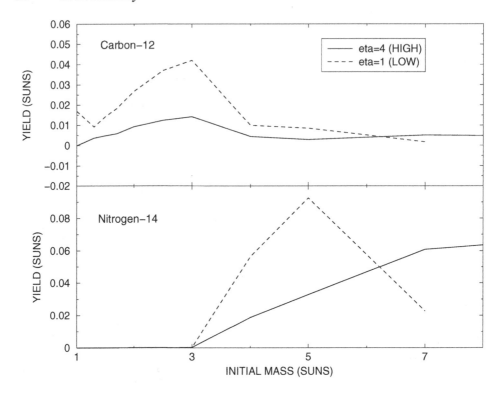

Fig. 5.4. Stellar yield versus initial stellar mass, both in solar units, for ^{12}C and ^{14}N, from HG. Results are shown for two different values of η, the mass loss parameter in the Reimers formulation.

Second, the paucity of high-luminosity (high-mass) carbon stars in the observed carbon star luminosity function compared with theoretical expectations, was first noted by Iben (1981). The finding gave rise to the introduction of HBB in order to reduce the amount of ^{12}C that is produced in the higher-mass IMS. Thus, the carbon star luminosity function provides another test for the models in general, and mass loss in particular, because the latter, as we have seen, controls the number of pulses and resultant dredge-ups that occur on the AGB. Again, M01 has compared model predictions with observations for the Magellanic Clouds and finds that the higher mass loss rates, such as those used by HG and M01, are more appropriate than the lower ones employed by RV, where the latter predict too many high-mass carbon stars.

Third, the stellar models also predict abundances in planetary nebulae (PNe) that can be compared directly with observations. Figure 5.6 shows observed logarithmic N/O, C/O, and normal He/H abundance ratios for a sample of PNe from Henry, Kwitter, & Bates (2000; open circles) and Kingsburgh & Barlow (1994; filled circles), along with model predictions by M01 (thin lines) for three different values of the mixing length parameter, as indicated in the legend. Also shown are model predictions from HG (bold lines). The models, all of which are for solar metallicity, occupy the same general area of the diagram as the observed points, and thus there appears to be consistency between theory and observation. Note that adjusting the mixing length parameter to higher values increases the extent of HBB and thus

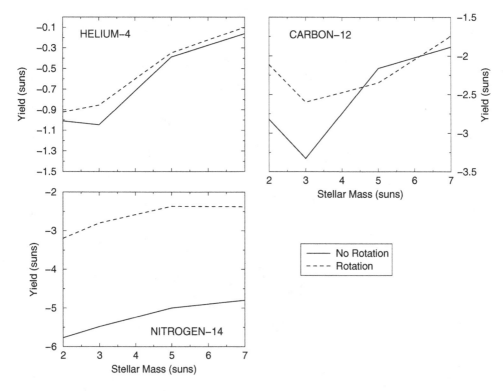

Fig. 5.5. Log of stellar yield versus stellar mass, both in solar units, for the situation of no rotation (solid line) and with rotation (v_{ini} = 300 km s^{-1}; dashed line). Results are from Meynet & Maeder (2002) for $Z = 10^{-5}$.

the amount of ^{4}He and ^{14}N predicted to be in the nebula (and in the yield). Kaler & Jacoby (1990) studied the N/O ratio and central star mass for a sample of PNe and found that when central star progenitor masses exceeded about 3 $M_\odot$, the N/O ratios in the associated PNe were several times higher than in PNe with lower progenitor masses, a finding that suggests that HBB is effective in stars more massive than around 3 $M_\odot$. Recently, Péquignot et al. (2000) and Dinerstein et al. (2003) studied abundances of O, Ne, S, and Ar in a total of four low-mass, metal-deficient PNe and found evidence for oxygen enrichment from third dredge-up in these objects. With a dozen or so Galactic halo and numerous Magellanic Cloud PNe now known, this implication of oxygen production by IMS needs to be investigated further.

Finally, in red giants the ^{12}C/^{13}C ratio is predicted to be many times lower than the solar value due to the effects of first dredge-up (Charbonnel 1994). Recently, Smith et al. (2002) have reported new measurements of red giants in the LMC having [Fe/H] values between −1.1 and −0.3 that have ^{12}C/^{13}C values consistent with expectations. Charbonnel (1994) presents models in which she predicts ratios in the observed range but claims that for stars of 2 $M_\odot$ and below an additional mixing process (CBP) is required. More recent work by Boothroyd & Sackmann (1999) supports this claim.

To summarize the picture for IMS stellar models and their predicted yields:

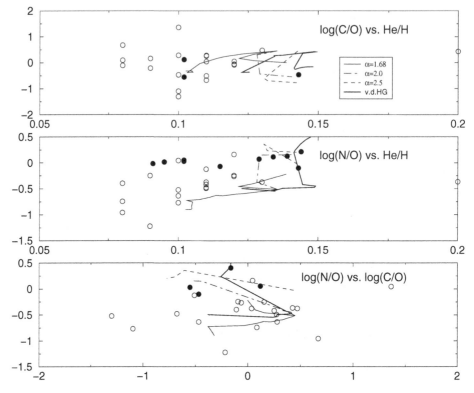

Fig. 5.6. *Top panel:* log (C/O) vs. He/H for a sample of planetary nebulae from Henry, Kwitter, & Bates (2000; open circles) and Kingsburgh & Barlow (1994; filled circles) compared with stellar evolution predictions by HG (bold line) and Marigo (2001; thin lines). Marigo's results are shown for three different values of the mixing length parameter, as indicated in the legend. *Middle panel:* Same as top panel but for log (N/O) vs. He/H. *Bottom panel:* Same as top panel but for log (N/O) vs. log (C/O).

(1) Yields generally increase with stellar mass*, reduced mass loss rate, lower metallicity, and in the case of ^{14}N, increased mixing length parameter.

(2) Mass loss rates are important: models using rates consistent with observational constraints at the same time seem more likely to reproduce several other observed trends.

(3) The threshold above which HBB occurs is somewhere between 3–4 $M_{\odot}$

(4) Rotational mixing may substantially increase the yields of ^{12}C and primary ^{14}N.

Interesting areas that especially need to be developed further are the effects of rotation and zero metallicity (Population III).

5.3   IMS and Chemical Evolution

In the grand scheme of galactic chemical evolution, do IMS matter? My discussion is divided into two parts. First, using computed values of integrated yields, I will compare predicted IMS yields with those for massive stars. Then, the question of IMS evolutionary time delay and its effects on galactic systems will be addressed.

* The ^{12}C yield increases with stellar mass up to the mass threshold for HBB, above which it declines.

Table 5.2. *Integrated Yields,* P_x

Source[1]	m_{down}	m_{up}	Z	$P_{^4\text{He}}$	$P_{^{12}\text{C}}$	$P_{^{14}\text{N}}$
RV	1	8	0.004	7.93E–3	6.05E–4	1.35E–3
RV	1	8	0.008	7.39E–3	2.14E–4	1.26E–3
RV	1	8	0.020	5.75E–3	–9.57E–4	1.00 E–3
HG	1	8	0.004	6.14E–3	7.07E–4	5.70E–4
HG	1	8	0.008	6.41E–3	6.11E–4	6.24E–4
HG	1	8	0.020	6.41E–3	4.06E–4	7.29E–4
M01	1	5	0.004	1.28E–2	3.74E–3	4.28E–4
M01	1	5	0.008	7.47E–3	1.68E–3	1.89E–4
M01	1	5	0.020	5.82E–3	1.09E–3	1.68E–4
WW	11	40	0.004	3.33E–2	9.92E–4	6.69E–5
WW	11	40	0.008	3.33E–2	9.92E–4	1.33E–4
WW	11	40	0.020	3.30E–2	9.94E–4	3.30E–4
P	6	120	0.004	4.78E–2	1.53E–3	1.72E–4
P	6	120	0.008	5.09E–2	6.56E–3	3.30E–4
P	6	120	0.020	6.28E–2	5.66E–3	8.09E–4

[1]References: RV = Renzini & Voli (1981); HG = van den Hoek & Groenewegen (1997); M01 = Marigo (2001); WW = Woosley & Weaver (1995); P = Portinari et al. (1998).

5.3.1 Integrated Yields

The relative impact of the two stellar groups, IMS and massive stars, can be assessed by integrating their yields over an IMF for the two relevant mass ranges. Henry, Edmunds, & Köppen (2000) defined the integrated yield for isotope x, P_x, as

$$P_x \equiv \int_{m_{\text{down}}}^{m_{\text{up}}} m p_x(m)\phi(m)dm, \tag{5.1}$$

where $mp_x(m)$ is the stellar yield of isotope x in solar masses of a star of mass m, $\phi(m)$ is the IMF, and the upper and lower mass limits are m_{up} and m_{down}, respectively. P_x is then the mass of isotope x that is newly produced and ejected per mass of new stars formed, ranging in mass $m_{\text{down}} \leq M \leq m_{\text{up}}$. For this specific exercise, IMS yields of RV, HG, and M01, along with massive star yields of Woosley & Weaver (1995) and Portinari et al. (1998) were used. Integrations were performed over a Salpeter (1955) IMF.

Results of these calculations are provided in Table 5.2. The source of the yields is indicated in column 1; the lower and upper mass limits for the integration are given in columns 2 and 3; column 4 gives the metallicity applicable to the yields; and columns 5, 6, and 7 give the values of the integrated yield for ^{4}He, ^{12}C, and ^{14}N, respectively.

Values for P_x in Table 5.2 are in turn plotted against metallicity in Figure 5.7. The figure legend identifies the correspondence between line type and yield source, where the abbreviations are the same as those defined in the footnote to Table 5.2. Note that massive star integrated yields are indicated with bold lines, while thin lines signify IMS integrated yields.

For ^{4}He note that the three IMS yield sets predict similar results except at low metallicity, where the M01 yields are higher. According to M01, this difference is presumably due to

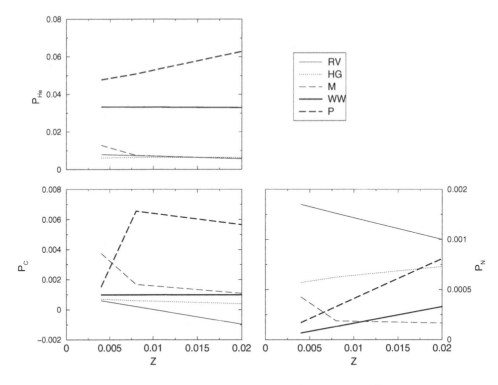

Fig. 5.7. Integrated yield, P_x, versus metallicity, Z, for ^{4}He (top left), ^{12}C (bottom left), and ^{14}N (bottom right). Plotted values are taken from Table 5.2.

the earlier activation and larger efficiency of third dredge-up in her models. It is also clear that IMS contribute to the cosmic build-up of ^{4}He at roughly the 20%–30% level.

The RV yields for ^{12}C tend to be less than those of HG, while those of M01 are greater, due to differences in onset time and average efficiency of third dredge-up. Globally, the role of IMS in ^{12}C production is therefore ambiguous, because it depends upon which set of massive star yields one uses to compare with the IMS yields. For example, IMS yields are comparable to the massive star yields of Woosley & Weaver (1995), yet significantly less than those of Portinari et al. (1998).

Finally, there is a significant difference between the three yields sets where ^{14}N is concerned. The lifetimes of the thermal pulses in the stars at the upper end of the mass range in M01's calculations are largely responsible for her ^{14}N yields being significantly less than those of others. On the other hand, RV's lower mass loss rate lengthens a star's lifetime on the late AGB and results in more ^{14}N production. When compared with massive star yields, RV and HG predict that IMS will produce several times more ^{14}N, particularly at lower Z, when compared to either the Woosley & Weaver (1995) or Portinari et al. (1998) yields. On the other hand, M01's models predict less ^{14}N. Universally speaking, then, IMS yield predictions indicate that these stars contribute significantly to ^{14}N production, moderately to ^{12}C production, and hardly at all to ^{4}He production. Remember, though, that these conclusions are heavily based upon model predictions. The strength of these conclusions is only as strong as the models are realistic.

5.3.2   IMS and Chemical Evolution Models

The respective roles of IMS and massive stars in galactic chemical evolution can be further assessed by confronting observations of abundance gradients and element ratio plots with chemical evolution models that employ the various yields to make their predictions. Because there is a time delay of at least 30 Myr between birth and release of products by IMS, these roles may be especially noticeable in young systems whose ages are roughly comparable to such delay times, or in systems that experienced a burst less that 30 Myr ago.

Henry et al. (2000) explored the C/O vs. O/H and N/O vs. O/H domains in great detail, using both analytical and numerical models to test the general trends observed in a large and diverse sample of galactic and extragalactic H II regions located in numerous spiral and dwarf irregular galaxies. Using the IMS yields of HG and the massive star yields of Maeder (1992), they were able to explain the broad trends in the data, and in the end they concluded that while massive stars produce nearly all of the ^{12}C in the Universe, IMS produce nearly all of the ^{14}N. They also illustrated the impact of the star formation rate on the age-metallicity relation and the behavior of the N/O value as metallicity increases in low-metallicity systems. In this conference, Mollá, Gavilán, & Buell (2004) report on their chemical evolution models that use the Buell (1997) IMS yields along with the Woosley & Weaver (1995) massive star yields, where the former employ the mass loss rate scheme of Vassiliadis & Wood (1993). Their model results confirm those of Henry et al. (2000) in terms of the star formation rate and the age-metallicity relation.

Recently, Pilyugin, Thuan, & Vílchez (2003) reexamined the issue of the origin of nitrogen and found that presently the stellar mass range responsible for this element cannot be clearly identified because of limitations in the available data.

Chiappini, Romano, & Matteucci (2003) have explored the CNO question using chemical evolution models and the HG and Woosley & Weaver (1995) yields to study the distribution of elements in the Milky Way disk as well as in the disk of M101 and dwarf irregular galaxies. Like Henry et al. (2000), they conclude that ^{14}N is largely produced by IMS. However, they find that by assuming that the IMS mass loss rate varies directly with metallicity, ^{12}C production in these stars is relatively enhanced at low Z. In the end, they conclude that IMS, not massive stars, control the universal evolution of ^{12}C, in disagreement with Henry et al. (2000). Figure 5.8 is similar to Figure 10 in their paper and is shown here to graphically illustrate the effects of IMS on the chemical evolution of ^{12}C and ^{14}N, according to their models. Each panel shows logarithmic abundance as a function of galactocentric distance in kpc for the Milky Way disk. Besides the data points, two model results are shown in each panel. The solid line in each case corresponds to the best-fit model in their paper, while the dashed line is for the same model but with the IMS contribution to nucleosynthesis turned off.* As can be seen, IMS make roughly a 0.5 dex and 1 dex difference in the case of C and N, respectively; their effects are sizable.

Finally, the question of IMS production of nitrogen has become entangled in the debate over the interpretation of the apparent bimodal distribution of damped Lyα systems in the N/α–α/H plane (Prochaska et al. 2002; Centurión et al. 2003).† Most damped Lyα systems fall in the region of the "primary plateau," located at a [N/α] value of $\sim$−0.7 and between metallicities of −1.5 and −2.0 on the [α/H] axis. However, a few objects are positioned noticeably below the plateau by roughly 0.8 dex in [N/α], although still within the same

---

* This altered model was kindly calculated by Cristina Chiappini upon the request of the author.

† α represents elements such as O, Mg, Si, and S, whose abundances are assumed to scale in lockstep.

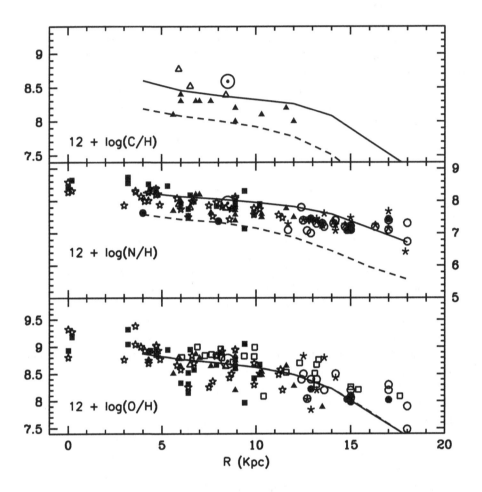

Fig. 5.8. 12 + log (X/H) vs. galactocentric distance, R (in kiloparsecs), as adapted from Fig. 10 in Chiappini et al. (2003). Data points show either H II region or stellar abundances, where the references are detailed in their paper. The solid line is their model 7, while the dashed line has been added and is model 7 but with the IMS contributions to chemical evolution turned off.

metallicity range as the plateau objects. The Prochaska group proposes that these low-N objects correspond to systems characterized by a top-heavy IMF with a paucity of IMS, or, in the same spirit, a population of massive stars truncated below some threshold mass. Either possibility works through suppressing the IMS contribution to nitrogen production by reducing the proportion of these stars in a system's stellar population. The Centurión group, on the other hand, suggests that low-N damped Lyα systems are less evolved than the plateau objects (i.e. star formation occurred within them less than 30 Myr ago), so these systems are momentarily pausing at the low-N region until their slowly evolving IMS begin to release their nitrogen. The latter picture, while not needing to invoke a nonstandard IMF (an action that causes great discomfort among astronomers), does require that the time to evolve from the low-N ledge to the plateau region be very quick; otherwise. their idea is inconsistent with the observed absence of a continuous trail of objects connecting these points. This problem is bound to be solved when the number of damped Lyα systems with measured nitrogen abundances increases, but it nevertheless illustrates an important role that IMS play in questions involving early chemical evolution in the Universe.

5.4   Summary

IMS play an important role in the chemical evolution of ^{12}C, ^{13}C, ^{14}N, and ^{7}Li, as well as s-process isotopes. Stellar models have gained in sophistication over the past two decades, so that currently they include effects of three dredge-up stages, thermal pulsing and HBB on the AGB, metallicity, and mass loss by winds and sudden ejection. Generally speaking, yield predictions from stellar evolution models indicate that yields increase as metallicity declines, as the mass loss rate is reduced, and when rotation is included. Furthermore, observational evidence supports the claim that the lower-mass limit for HBB is between 3 and 4 $M_\odot$.

Integration of yields over a Salpeter IMF shows clearly that IMS have little impact on the evolution of ^{4}He while at the same time playing a dominant role in the cosmic build-up of ^{14}N. The case of ^{12}C is a bit more confused. The issue of ^{14}N production is particularly important in the current discussion of the distribution of damped Lyα systems in the N/α–α/H plane.

Finally, what I believe is needed are grids of models that attempt to treat IMS and massive stars in a consistent and seamless manner. The role of each stellar mass range would be easier to judge if yield sets of separate origins did not have to be patched together in chemical evolution models. Otherwise, it is not clear to what extent the various assumptions that are adopted by stellar evolution theorists impact (and therefore confuse!) the analyses.

Acknowledgements. I would like to thank the organizing committee for inviting me to write this review and to present these ideas at the conference. I also want to thank Corinne Charbonnel, Georges Meynet, Francesca Matteucci, Cristina Chiappini, Jason Prochaska, John Cowan, and Paulo Molaro for clarifying my understanding on several topics addressed in this review. Finally, I am grateful to the NSF for supporting my work under grant AST 98-19123.

References

Boothroyd, A. I., & Sackmann, I.-J. 1999, ApJ, 510, 232
Buell, J. F. 1997, Ph.D. Thesis, University of Oklahoma
Centurión, M., Molaro, P., Vladilo, G., Péroux, C., Levshakov, S. A., & D'Odorico, V. 2003, A&A, 403, 55
Charbonnel, C. 1994, A&A, 282, 811
——. 2002, Ap&SS, 281, 161
——. 2004, in Carnegie Observatories Astrophysics Series, Vol. 4: Origin and Evolution of the Elements, ed. A.
    McWilliam & M. Rauch (Pasadena: Carnegie Observatories,
    http://www.ociw.edu/ociw/symposia/series/symposium4/proceedings.html)
Charbonnel, C., & Palacios, A. 2003, in IAU Symp. 215, Stellar Rotation, ed. A. Maeder & P. Eenens (Dordrecht:
    Reidel), 1
Chiappini, C., Romano, D., & Matteucci, F. 2003, MNRAS, 339, 63
Chieffi, A., Domínguez, I., Limongi, M., & Straniero, O. 2001, ApJ, 554, 1159
Dinerstein, H. L., Richter, M. J., Lacy, J. H., & Sellgren, K. 2003, AJ, 125, 265
Forestini, M., & Charbonnel, C. 1997, A&AS, 123, 241
Girardi, L., Bressan, A., Bertelli, G., & Chiosi, C. 2000, A&AS, 141, 371

Groenewegen, M. A. T., & de Jong, T. 1993, A&A, 267, 410

Henry, R. B. C., Edmunds, M. G., & Köppen, J. 2000, ApJ, 541, 660

Henry, R. B. C., Kwitter, K. B., & Bates, J. A. 2000, ApJ, 531, 928

Herwig, F. 1995, in Stellar Evolution: What Should Be Done, 32nd Liège Int. Astrophys. Colloq., ed. A. Noels et al. (Liége: Universite de Liége), 441

Iben, I. 1981, ApJ, 246, 278

——. 1995, Phys. Rep., 250, 1

Iben, I., & Truran, J. W. 1978, ApJ, 220, 980

Jeffries, R. D. 1997, MNRAS, 288, 585

Kaler, J. B., & Jacoby, G. H. 1990, ApJ, 362, 491

Karakas, A. I., & Lattanzio, J. C. 2004, in Carnegie Observatories Astrophysics Series, Vol. 4: Origin and Evolution of the Elements, ed. A. McWilliam & M. Rauch (Pasadena: Carnegie Observatories, http://www.ociw.edu/ociw/symposia/series/symposium4/proceedings.html)

Kingsburgh, R. L., & Barlow, M. J. 1994, MNRAS, 271, 257

Langer, N., Heger, A., Wellstein, S., & Herwig, F. 1999, A&A, 346, L37

Lattanzio, J. C. 2002, NewAR, 46, 469

Maeder, A. 1992, A&A, 264, 105

Marigo, P. 2001, A&A, 370, 194 (M01)

Marigo, P., Bressan, A., & Chiosi, C. 1996, A&A, 313, 564

Marigo, P., Girardi, L., Chiosi, C., & Wood, P. R. 2001, A&A, 371, 152

Meynet, G., & Maeder, A. 2002, A&A, 390, 561

Mollá, M., Gavilán, M., & Buell, J. F. 2004, in Carnegie Observatories Astrophysics Series, Vol. 4: Origin and Evolution of the Elements, ed. A. McWilliam & M. Rauch (Pasadena: Carnegie Observatories, http://www.ociw.edu/ociw/symposia/series/symposium4/proceedings.html)

Péquignot, D., Walsh, J. R., Zijlstra, A. A., & Dudziak, G. 2000, A&A, 361, L1

Pilyugin, L. S., Thuan, T. X., & Vílchez, J. M. 2003, A&A, 397, 487

Portinari, L., Chiosi, C., & Bressan, A. 1998, A&A, 334, 505

Prochaska, J. X., Henry, R. B. C., O'Meara, J. M., Tytler, D., Wolfe, A. M., Kirkman, D., Lubin, D., & Suzuki, N. 2002, PASP, 114, 933

Reimers, D. 1975, Mem. Soc. R. Sci. Liège, 6th Series, 8, 369

Renzini, A., & Voli, M. 1981, A&A, 94, 175 (RV)

Salpeter, E. E. 1955, ApJ, 121, 161

Scalo, J. 1998, in The Stellar Initial Mass Function, 38th Herstmonceux Conference, ed. G. Gilmore, I. Parry, & S. Ryan (San Francisco: ASP), 201

Schaller, G., Schaerer, D., Meynet, G., & Maeder, A. 1992, A&AS, 96, 269

Siess, L., Livio, M., & Lattanzio, J. C. 2002, ApJ, 570, 329

Smith, V. V., et al. 2002, AJ, 124, 3241

van den Hoek, L. B., & Groenewegen, M. A. T. 1997, A&AS, 123, 305 (HG)

Vassiliadis, E., & Wood, P. R. 1993, ApJ, 413, 641

Weidemann, V. 1987, A&A, 188, 74

Woosley, S. E., & Weaver, T. A. 1995, ApJS, 101, 181

6

The impact of rotation on chemical abundances in red giant branch stars

CORINNE CHARBONNEL

Geneva Observatory, 1290 Sauverny, Switzerland
LA-OMP (CNRS UMR 5572), 14, av.E.Belin, 31400 Toulouse, France

Abstract

Low-mass stars (<2–$2.5\ M_\odot$) exhibit, at all the stages of their evolution, signatures of processes that require challenging modeling beyond the standard stellar theory. In this paper we focus on their peculiarities while they climb the red giant branch (RGB). We first compare the classical predictions for abundance variations due to the first dredge-up with observational data in various environments. We show how clear spectroscopic diagnostics probe the nucleosynthesis and the internal mixing mechanisms that drive RGB stars. Coherent data reveal in particular the existence of a nonstandard and shallow mixing process that changes their surface abundances at the so-called RGB bump. We show that the occurrence of this extra-mixing process is certainly related to rotation. Finally we discuss the so-called Li-flash, which is expected to occur at the very beginning of the extra-mixing episode.

6.1   Abundance Anomalies in RGB Stars Due to *In Situ* Processes

6.1.1   *Abundance Variations in RGB Stars Due to the First Dredge-up*

According to the classical stellar evolution theory[*], the only opportunity for low-mass stars to modify their surface abundances happens on their way to the red giant branch (RGB) when they undergo the so-called first dredge-up (Iben 1965). During this event their convective envelope deepens in mass, leading to the dilution of the surface pristine material within regions that have undergone partial nuclear processing on the main sequence. Qualitatively, this leads to the decrease of the surface abundances of the fragile Li, Be, and B elements and of ^{12}C, while those of ^{3}He, ^{13}C, and ^{14}N increase. Due to low temperatures inside main sequence low-mass stars, O and heavier elements are not affected by nuclear reactions, and their surface abundances remain unchanged subsequent to the first dredge-up. Quantitatively, the abundance variations due to the first dredge-up depend on the stellar mass and metallicity (e.g., Sweigart, Greggio, & Renzini 1989; Charbonnel 1994; Boothroyd & Sackmann 1999). After the first dredge-up, the convective envelope withdraws while the hydrogen-burning shell moves outwards in mass. No more variations of the surface abundances are then expected until the star reaches the asymptotic giant branch.

---

[*] By this we refer to the modeling of nonrotating, nonmagnetic stars, in which convection is the only mixing process considered.

6.1.2   *Observational Clues of a Second Mixing Episode on the RGB*

The first dredge-up predictions agree with the observations on the lower part of the RGB†. However, observational evidences have accumulated, that we list below, of a second and distinct mixing episode that occurs in low-mass stars after the end of the first dredge-up, and more precisely at the RGB bump.

The determination of the carbon isotopic ratio ($^{12}C/^{13}C$) for RGB stars in open clusters with various turn-off masses (Gilroy 1989) provided the first pertinent clue on this process. It was indeed shown that bright RGB stars with initial masses lower than $\sim$2–2.5 $M_\odot$ exhibit carbon isotopic ratios considerably lower than predicted by the first dredge-up. Thanks to data in stars sampling the RGB of M67 (Gilroy & Brown 1991), it clearly appeared that observations deviated from classical predictions just at the so-called RGB bump (Charbonnel 1994). Field and globular cluster stars behave similarly*. In the former case this could be established thanks to the determination of the carbon isotopic ratio in large samples of stars for which *Hipparcos* parallaxes allowed the precise determination of their evolutionary status (Charbonnel, Brown, & Wallerstein 1998; Gratton et al. 2000). We knew for a long time that the brightest RGB stars in globular clusters presented carbon isotopic ratios close to the equilibrium value. But very recently the region around the bump could finally be probed for two globular clusters: In NGC 6528 and M4, the carbon isotopic ratio drops below the first dredge-up predictions just at the RGB bump (Shetrone 2003).

The extra-mixing process affects the surface abundances of other chemical elements: Li also decreases at the RGB bump, both in the field and in globular clusters (Pilachowski, Sneden, & Booth 1993; Grundahl et al. 2002). In field stars C decreases, while N increases for RGB stars brighter than the bump (Gratton et al. 2000), confirming the CN processing of the stellar envelope. The lowering of the C abundance along the RGB is also seen in globular clusters (e.g., Bellman et al. 2001, and references therein), though in this case the picture is more confused because of a probable nonnegligible dispersion of the initial [C/Fe].

An *in situ* mechanism has also frequently been invoked to explain the abundance anomalies of heavier elements in globular cluster stars, and in particular the O-Na anticorrelation. This pattern has been observed in the brightest globular cluster RGB stars for a long time (see references in Ivans et al. 1999 and Ramírez & Cohen 2002). However, it is only thanks to 8–10 m-class telescopes that O and Na abundances could be determined for less evolved stars, and in particular for turn-off stars in a couple of globular clusters (Gratton et al. 2001; Thévenin et al. 2001; Ramírez & Cohen 2002). There, the O-Na anticorrelation extends to the main sequence. This result is crucial. Indeed, the ON- and NeNa-cycles do not occur in main sequence low-mass stars, as the involved reactions require too high temperatures that are only reached when low-mass stars are on the RGB. Thus, the fact that the O-Na anticorrelation already exists on the main sequence clearly proves that it is not produced by *in situ* processes, but by external causes, the discussion of which is out of the scope of this paper (see Charbonnel 2004).

Recently we have searched for an eventual evolution of the O and Na abundances with luminosity along the RGB for field stars. We gathered from the literature a large sample (625) of field stars with *Hipparcos* parallaxes over a large range of metallicity ([Fe/H] between

---

† One has, of course, to take into account possible variations of the surface Li abundance occurring already on the main sequence. This discussion is, however, out of the scope of the present paper.

* The literature on the various abundance anomalies within globular clusters is so large that we chose to quote here only the most recent papers to illustrate our discussion.

−2.5 and 0) for which we homogeneously redetermined the abundances of a large number of elements. No evolutionary effect could be found, neither for O nor Na (Palacios et al. 2002; Charbonnel et al. 2004), in agreement with the most recent globular cluster data mentioned above [see also Yong et al. (2003, 2004) for a discussion on Mg and Al].

Last, but not least, the recent determination of the oxygen isotopic ratios in a couple of RGB stars with low carbon isotopic ratios (Balachandran & Carr 2004) confirmed this result. In these objects, indeed, both the $^{16}O/^{17}O$ and $^{16}O/^{18}O$ ratios appear to be high, in agreement with extensive CN-processing but no dredge-up of ON-cycle material in RGB stars (see the arrow in Fig. 6.1).

6.2   Origin and Nature of the Extra-mixing Process

The observations summarized previously provide definitive clues on a second and distinct mixing episode that occurs in low-mass stars well after the end of the first dredge-up, precisely at the RGB bump. This process appears to be universal: it affects more than 95% of the low-mass stars (Charbonnel & do Nascimento 1998), whether they belong to the field, to open, or to globular clusters. Its signatures in terms of abundance anomalies are clear: the Li and the ^{12}C abundances as well as the $^{12}C/^{13}C$ ratio drop, while the ^{14}N abundance and the $^{16}O/^{18}O$ ratio increase. On the other hand ^{16}O, ^{17}O, Na, and heavier elements remain unaffected. As we shall see in §6.4, this sequence may start with a very brief episode of Li enrichment.

These coherent data indicate that during the extra-mixing episode at the RGB bump, the stellar convective envelope is connected with the radiative regions where CN-burning occurs, but not with the ON- and NeNa-processed layers (see Fig. 6.1). The important increase of the molecular weight (or μ) gradients when one approaches the hydrogen-burning shell certainly shields these deeper regions from the extra-mixing.

μ-gradients are also certainly responsible for the fact that the extra-mixing starts to be efficient only at the RGB bump. Indeed, during the first dredge-up a μ-discontinuity is built at the region of the deepest penetration of the convective envelope. This μ-barrier then probably inhibits any mixing between the convective envelope and the hydrogen-burning shell. At the RGB bump, the external regions of the hydrogen-burning shell pass through the μ-discontinuity*. From the mixing point of view, after this evolutionary point the μ-gradients between the base of the convective envelope and the hydrogen-burning shell are much smoother, allowing the occurrence of some extra-mixing in this radiative region (Sweigart & Mengel 1979; Charbonnel 1994, 1995; Charbonnel et al. 1998).

All these observational facts represent strong clues on the origin of the extra-mixing process, which is certainly rotationally induced. Indeed, μ-gradients have been known for a long time to be powerful in inhibiting meridional circulation (Mestel 1953, 1957).

6.3   Rotation-induced Mixing

Sweigart & Mengel (1979) investigated the possibility that meridional circulation might lead to the mixing of CNO-processed material in RGB stars. Since then, a lot of work has been done on the physics of the mixing induced by rotation. Zahn (1992) proposed a

---

* Due to the composition changes in the burning region, the stellar structure has to readjust, and the stellar luminosity then slightly decreases before increasing again. Low-mass stars spend a nonnegligible part of their RGB lifetime in this region; this explains why the bump clearly appears as a peak in the differential luminosity function of the Galactic globular clusters or as a change in the slope of the cumulative luminosity function (e.g., Zoccali et al. 1999).

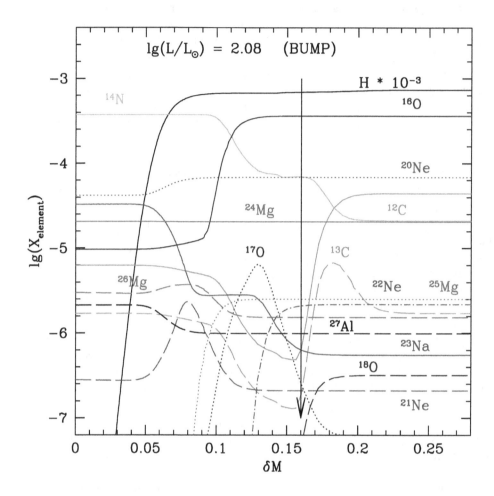

Fig. 6.1. Abundance profiles of the isotopes participating in the CNO, NeNa, and MgAl cycles inside a 0.83 $M_\odot$, [Fe/H]= −1.6 (typical of the globular cluster M13) standard model at the luminosity of the RGB bump. The abscissa is the relative mass coordinate, $\delta M = (m_r - m_{BHBS})/(m_{BCE} - m_{BHBS})$, where m_r is the mass within a given radius, m_{BHBS} is the mass at the bottom of the hydrogen-burning shell, and m_{BCE} is the mass at the base of the convective envelope; $\delta M = 0$ at the bottom of the hydrogen-burning shell and $\delta M = 1$ at the base of the convective envelope. The vertical arrow indicates the maximum depth down to which the extra-mixing process is efficient according to the observations.

description of the interaction between meridional circulation and shear turbulence, pushing forward the idea of choking of meridional circulation by μ-gradients. Following these developments, Charbonnel (1995) showed that such a process can indeed account for the observed behavior of carbon isotopic ratios and for the Li abundances in Population II low-mass giants. Simultaneously, when this extra-mixing begins to act, ^{3}He is rapidly transported down to the regions where it burns by the ^{3}He$(\alpha, \gamma)^7$Be reaction. This leads to a decrease of the surface value of ^{3}He/H, solving the long-standing problem of the Galactic evolution of ^{3}He (e.g., Charbonnel 2002).

In these exploratory computations, however, the transport of angular momentum by the hydrodynamical processes was not treated self-consistently, though it is of utmost importance in the understanding of rotation-induced mixing. Let us now discuss the results obtained for RGB stars when the transport of angular momentum and of the chemicals, as described by Zahn (1992) and latter on by Talon et al. (1997), Maeder & Zahn (1998), and Palacios et al. (2003), is taken into account.

Two independent studies have been carried out using basically the same physical description of rotational mixing: Denissenkov & Tout (2000) and Palacios, Talon, & Charbonnel (2004). They differ on some details of the input physics, such as the choice of the shear instability criteria and of the expression of the diffusion coefficient associated with vertical turbulence. But more importantly, a drastic assumption differentiates both studies: Denissenkov & Tout (2000) indeed compute the transport of angular momentum and of the chemicals in a post-processing way; in other words, they use the internal structure of a few *standard* RGB models to estimate the mixing efficiency and the corresponding changes in the surface abundances. In this approach, there is no feedback of the mixing on the stellar structure and evolution, which are computed independently.

On the other hand, Palacios et al. (2004) treat in a self-consistent way the transport of angular momentum and of the chemicals inside their stellar evolution code, thereby following the "nonstandard" evolution from the zero-age main sequence on. As a result, in this case the mixing has a feedback on the star since its early evolution. On the RGB, in particular, the mixing distorts the stationary profiles of the chemicals deep inside the star even if the surface abundances are not yet affected. In turn, the mixed star reacts to the mixing that it undergoes, thereby influencing the mixing itself (see §6.4).

This difference has a strong impact on the final predictions. In Denissenkov & Tout (2000), indeed, the mixing is applied to standard profiles, and surface variations of O and Na are predicted in RGB stars. In Palacios et al. (2004), however, the mixing is applied on self-consistently distorted profiles; as a result, hardly any variation of even the C and N isotopes is produced at the stellar surface by this process alone. One could thus question the actual role of rotation in this problem. However, something crucial happens, that we discuss below.

6.4   The Li Flash

A few ($\sim 1\%$) RGB stars present Li overabundances. The fact that these so-called super Li-rich giants are all located at the RGB bump and still exhibit "standard" carbon isotopic ratios led Charbonnel & Balachandran (2000) to conclude that the Li-rich phase was a precursor to the extra-mixing process described previously. This brought very tight constraints on the underlying physics. Indeed, the transport coefficient D_R obtained in the case of rotation-induced mixing are also too low to lead to surface lithium enrichment.

Palacios, Charbonnel, & Forestini (2001) proposed that the structural response to the mixing could actually be the cause of the increase of D_R, which is necessary to produce super Li-rich giants and to change the surface abundances of other elements. The mixing sequence is the following. At the RGB bump the external wing of the ^{7}Be peak crosses the molecular weight discontinuity before the ^{13}C peak does, and is then connected with the convective envelope by the extra-mixing process. ^{7}Be produced via ^{3}He$(\alpha, \gamma)^7$Be starts to diffuse outwards. However, due to the relatively low D_R given above, the transported ^{7}Be decays to ^{7}Li in regions where the ^{7}Li is rapidly destroyed by proton capture. A lithium-burning shell appears. ^{7}Li(p,α)α becomes the dominant reaction, leading to an increase of the local temperature and of both the local and total luminosities.

Palacios et al. suggested that the meridional circulation and the corresponding transport coefficient then increase due to the ϵ_{nuc}(^{7}Li+p) burst, allowing for lithium enrichment of the convective envelope. Indeed, in the developments for the transport of matter and angular momentum, the effective diffusivity is proportional to $[rU(r)]^2$, the vertical component of the meridional velocity. This quantity is itself proportional to terms that depend on both the velocity profile and the chemical inhomogeneities along isobars (see Maeder & Zahn 1998 for the complete expressions). But what is important for the present discussion is their dependence on the local rates of nuclear and gravothermal energies. We thus proposed that the apparition of the Li-burning shell and the significant release of ϵ_{nuc}(^{7}Li+p) do highly enhance the meridional velocity and the associated transport coefficient as required for the surface lithium abundance to increase.

A lithium flash terminates this phase. As the convective instability develops around the Li-burning shell, it erases the molecular weight gradient and the mixing is free to proceed deeper with higher D_R than initially. Once it reaches the regions where ^{12}C is converted into ^{13}C (as is observed to occur in low-mass stars after the bump) the surface material is exposed to temperatures higher than lithium can withstand. The freshly synthesized lithium is then destroyed as the surface carbon isotopic ratio decreases and the lithium-rich phase ends. This is consistent with the available data on ^{12}C/^{13}C in the lithium-rich RGB stars: those with the highest lithium abundance still have a standard (i.e. post-dredge-up) carbon isotopic value, but no star retains its peak lithium abundance once its carbon isotopic value dips (Charbonnel & Balachandran 2000). When the flash is quenched, the external convective envelope deepens again. We suggest that this creates a new μ-barrier, similar to that built up during the first dredge-up, which then inhibits again the extra-mixing during the subsequent evolution.

We thus have at hand an extra-mixing process that is induced by rotation, but that becomes highly efficient at the RGB bump due to the important release of nuclear energy inside a Li-burning shell. We can thus explain in a self-consistent way the fact that the mixing episode starts by a short period of Li enrichment, followed by a decrease of the fresh Li and of the carbon isotopic ratio. The subsequent deepening of the convective envelope creates a new μ-barrier that will inhibit further mixing after the bump episode.

In order to estimate the time scale of our process, let us define the beginning and the end of the lithium-rich phase, respectively, as the moments where the surface lithium abundance increases above its post-dredge-up value and where the carbon isotopic ratio decreases below its post-dredge-up value. While the duration of the lithium flash itself is quite short ($\sim 2 \times 10^4$ yr in a 1.5 $M_\odot$ solar-metallicity model), the lithium-rich phase so-defined lasts for $\sim 60\%$ (i.e. $\sim 6 \times 10^6$ yr) of the classical bump duration. This has to be compared with the number of stars located in the bump region that still present a post-dredge-up carbon isotopic ratio. Among the sample of stars with *Hipparcos* parallaxes and carbon isotopic ratio determinations used in Charbonnel & do Nascimento (1998), $\sim 35\%$ of the 49 objects located close to the bump do satisfy this criterion. We consider that this comparison is very encouraging in view of the limited observational sample. This preliminary estimation indicates that many lithium-rich stars should be discovered by observational programs focusing on the bump region.

During the whole sequence described above the temporary increase of the stellar luminosity causes an enhanced mass loss rate, which naturally accounts for the dust shell suggested by the far-infrared color excesses measured for some lithium-rich objects from *IRAS* fluxes (e.g., de la Reza, Drake, & da Silva 1996). Let us note that the contribution of these stars to the lithium enrichment of the Galaxy should be very modest.

6.5   Conclusions

During the last three decades, an incredible amount of work has been devoted to the understanding of the chemical anomalies exhibited by evolved low-mass stars. On the observational side, crucial breakthroughs were made possible recently thanks to the advent of 8–10 m-class telescopes. The signatures of the extra-mixing process that occurs at the RGB bump are now very clear.

On the other hand, recent theoretical models include a sophisticated description of the mixing processes induced by rotation. Those that consistently couple the transport of angular momentum and of the chemicals with the structural evolution of the star predict a Li flash and surface abundance variations on the RGB, in agreement with the observational data. However, a couple of assumptions remain to be tested, in particular the reaction of the meridional circulation and of the various instabilities to a major and local release of nuclear energy.

Rotation-induced mixing processes modify all the stellar outputs, and in particular the final yields from low-mass stars (i.e., ^{3}He is not overproduced anymore by rotating low-mass stars). Their impacts on the stellar lifetimes and on the further evolution (luminosity of the RGB tip, morphology of the horizontal branch, asymptotic giant branch phase, ...) have to be investigated now in a systematic way. This is of crucial importance if one wants to correctly understand the role of low-mass stars in the evolution (chemical and spectrophotometric) of stellar clusters and of galaxies in general.

Acknowledgements. We thank the French Programme National de Physique Stellaire and Programme National Galaxies for their support on this work.

References

Balachandran, S. C., & Carr, J. S. 2004, in CNO in the Universe, ed. C. Charbonnel, D. Schaerer, & G. Meynet (San Francisco: ASP), in press
Bellman, S., Briley, M. M., Smith, G. H., & Claver, C. F. 2001, PASP, 113, 326
Boothroyd, A. I., & Sackmann, I. J. 1999, ApJ, 510, 232
Charbonnel, C. 1994, A&A, 282, 811
——. 1995, ApJ, 543, L41
——. 2002, Nature, 415, 27
——. 2004, in CNO in the Universe, ed. C. Charbonnel, D. Schaerer, & G. Meynet (San Francisco: ASP), in press
Charbonnel, C., & Balachandran, S. 2000, A&A, 359, 563
Charbonnel, C., Brown, J. A., & Wallerstein, G. 1998, A&A, 332, 204
Charbonnel, C., & do Nascimento, J. 1998, A&A, 336, 915
Charbonnel, C., Palacios, A., Thévenin, F., & Bolmont, J. 2004, in preparation
de la Reza, R., Drake, N. A., & da Silva, L. 1996, ApJ, 456, L115
Denissenkov, P. A., & Tout, C. A. 2000, MNRAS, 316, 395
Gilroy, K. K. 1989, ApJ, 347, 835
Gilroy, K. K., & Brown, J. 1991, ApJ, 371, 578
Gratton, R. G., et al. 2001, A&A, 369, 87

Gratton, R. G., Sneden, C., Carreta, E., & Bragaglia, A. 2000, A&A, 354, 169

Grundahl, F., Briley, M., Nissen, P. E., & Feltzing, S. 2002, A&A, 336, 915

Iben, I., Jr. 1965, ApJ, 142, 1447

Ivans, I. I., Sneden, C., Kraft, R. P., Suntzeff, N. B., Smith, V. V., Langer, G. E., & Fulbright, J. P. 1999, AJ, 118, 1273

Maeder, A., & Zahn, J. P. 1998, A&A, 334, 1000

Mestel, L. 1953, MNRAS, 113, 716

——. 1957, ApJ, 126, 550

Palacios, A., Charbonnel, C., Bolmont, J., & Thévenin, F. 2002, Ap&SS, 281, 213

Palacios, A., Charbonnel, C., & Forestini, M. 2001, A&A, 375, L9

Palacios, A., Talon, S., & Charbonnel, C. 2004, in preparation

Palacios, A., Talon, S., Charbonnel, C., & Forestini, M. 2003, A&A, 399, 603

Pilachowski, C. A., Sneden, C., & Booth, J. 1993, ApJ, 407, 699

Ramírez, S. V., & Cohen, J. G. 2002, AJ, 123, 3277

Shetrone, M. 2003, ApJ, 585, L45

Sweigart, A. V., Greggio, L., & Renzini, A. 1989, ApJS, 69, 911

Sweigart, A. V., & Mengel, J. G. 1979, ApJ, 229, 624

Talon, S., Zahn, J. P., Maeder, A., & Meynet, G. 1997, A&A, 322, 209

Thévenin, F., Charbonnel, C., de Freitas Pacheco, J. A., Idiart, T. P., Jasniewicz, G., de Laverny, P., & Plez, B. 2001, A&A, 373, 905

Yong, D., Grundahl, F., Lambert, D. L., Nissen, P. E., & Shetrone, M. D. 2003, A&A, 402, 985

——. 2004, in Carnegie Observatories Astrophysics Series, Vol. 4: Origin and Evolution of the Elements, ed. A. McWilliam & M. Rauch (Pasadena: Carnegie Observatories, http://www.ociw.edu/symposia/series/symposium4/proceedings.html)

Zahn, J. P. 1992, A&A, 265, 115

Zoccali, M., Cassisi, S., Piotto, G., Bono, G., & Salaris, M. 1999, ApJ, 518, L49

7

s-processing in AGB stars and the composition of carbon stars

MAURIZIO BUSSO[1], OSCAR STRANIERO[2], ROBERTO GALLINO[3], and CARLOS ABIA[4]

(1)Dipartimiento di Fisica, Università di Perugia, Perugia, Italy
(2)Osservatorio Astronomico di Collurania, Teramo, Italy
(3)Dipartimento di Fisica Generale, Università di Torino. Torino, Italy
(4)Departamento de Física Teórica y del Cosmos, Universidad de Granada, 18071 Granada, Spain

Abstract

We briefly review the researches on *s*-process nucleosynthesis, starting from the original studies where the phenomenological approach based on solar system abundances was founded, and summarizing the improvements that subsequently led to a crisis in the traditional ideas and to a new scenario in which reliable distributions of nuclei synthesized by slow neutron captures can be provided directly (and solely) by stellar models. In analyzing these last, we concentrate on recent work for low- and intermediate-mass stars in their final evolutionary stages, when they climb for the second time along the red giant branch. We summarize our knowledge of their evolution, mixing, and neutron capture nucleosynthesis, and we discuss some of the many open problems. In particular, we underline the fact that modeling the formation of the main neutron source still requires rather free parameterizations. The results of *s*-process calculations are then illustrated and compared with spectroscopic observations at different metallicities, with special attention to carbon stars.

7.1   Introduction

Not many years after the 50th anniversary of the Carnegie Institution, the first fundamental paper on stellar nucleosynthesis appeared. Presented by Burbidge, Burbidge, Fowler, & Hoyle (1957; hereafter B[2]FH), it laid the general scenario of nuclear processes in stars, inferring their effects for the production of nuclei and the ensuing evolution of matter in galaxies. Concerning the specific problem of neutron captures for the synthesis of heavy elements beyond Fe, of comparable importance were the compilation of meteorite abundances by Suess & Urey (1956) and the works by Greenstein (1954) and by Cameron (1954, 1957) on the neutron sources that must be activated. Following B[2]FH, neutron-addition reactions have since been divided according to their slow (*s*-process) or rapid (*r*-process) time scales, as compared to those of β-decays from unstable nuclei encountered along the neutron capture path. Many improvements on the first ideas by B[2]FH were soon presented, thanks to increased precision in the measurements of isotope abundances from meteorites (Goldberg, Muller, & Aller 1960) and of neutron capture cross sections (Macklin & Gibbons 1965). Various reviews dealing with the *s*-process and with connected stellar and nuclear issues have been published over the years. For a general outline of the stellar environment and its physics, we mention those by Iben & Renzini (1983), Iben (1991), Sackmann & Boothroyd

(1991), and Wallerstein et al. (1997). Emphasis on the nucleosynthesis processes is given in Wheeler, Sneden, & Truran (1989), Meyer (1994), Gallino, Busso, & Lugaro (1997), and Busso, Gallino, & Wasserburg (1999).

The above summaries deal with the studies on stellar neutron captures in different moments of their development, and testify to an evolution in which the role of verifying the theory has gradually moved from solar system abundances to stellar spectroscopy, especially for the asymptotic giant branch (AGB) stars where neutron-rich elements are produced in the inner regions and then carried to the surface by a series of mixing phenomena known under the name of *third dredge-up* (hereafter TDU; Smith & Lambert 1990; Luck & Bond 1991; Vanture 1992). Modern upgrades of these studies have included an analysis of abundances in evolved stars of very different metallicity (Busso et al. 2001), in C-rich stars of types SC, C(N), and in post-AGB stars (Abia & Wallerstein 1998; Van Winckel & Reyniers 2000; Abia et al. 2001, 2002; see also Reyniers & Van Winckel 2004). Another source of constraints emerged when solid condensates of presolar origin began to be recovered from meteorites, many of which were ascribed to AGB star envelopes (see, e.g., Anders & Zinner 1993; Ott 1993; Zinner 1997). The high precision reached in the measurements of the isotopic composition of such grains made them an increasingly important source of data for fixing the details of the *s*-process distribution (Käppeler 2001)

In this paper we review these topics, from a summary of the classical analysis of slow neutron captures (§7.2) to the discovery of growing problems (§7.3), which ultimately led to a new generation of models. Then we present the updated view of low- and intermediate-mass star evolution provided by such models, especially for the final phases where the bulk of slow neutron captures occur (§7.4), and a comparison of model results with observational constraints at different metallicities (§7.5). In particular, a detailed analysis is given of the present understanding of carbon-rich and *s*-process-rich stars in the Galaxy (§7.6).

7.2   The Classical Analysis of the *s*-Process

A lucky circumstance favored since the beginning the understanding of the *s*-process: the distribution is controlled by a small number of parameters and can be approximated analytically. The mathematical tools were presented by Clayton et al. (1961) and by Seeger, Fowler, & Clayton (1965), who outlined the so-called *phenomenological approach* or *classical analysis*. These authors compared the experimental distribution of $\sigma N_{s,\odot}$ products (*s*-process fraction of the abundance of each nucleus times its neutron capture cross section) with a model σN_s curve, by computing analytically the *s*-process contributions N_s to each isotope. As a consequence, the ratio $(N_i^\odot - N_s)/N_i^\odot$ yielded a prediction on the more complex *r*-process distribution. Studies on the *s*-process were then shown to provide a rather deep and thorough view of heavy isotope nucleosynthesis.

Many interesting properties of slow neutron captures derive directly from the basic equations of the process. If two nuclei of atomic mass number $A-1$ and A undergo only neutron captures (i.e. the reaction chain is for them *unbranched*), then the abundance $N(A)$ varies in time as

$$\frac{dN(A)}{dt} = N(A-1)n_n\langle\sigma_{A-1}\upsilon\rangle - N(A)n_n\langle\sigma_A\upsilon\rangle, \tag{7.1}$$

where $\langle\sigma\upsilon\rangle$ indicates the Maxwellian-averaged product of cross section and relative velocity, and n_n is the neutron density. By replacing time with the time-integrated neutron flux, or *neutron exposure* τ, through the substitution

$$\tau = \int \Phi_n dt = \int n_n \upsilon_T dt, \tag{7.2}$$

(here υ_T is the thermal velocity), and indicating by $\bar{\sigma}$ the ratio $\langle \sigma \upsilon \rangle / \upsilon_T$, one has

$$\frac{dN(A)}{d\tau} = N(A-1)\bar{\sigma}(A-1) - N(A)\bar{\sigma}(A). \tag{7.3}$$

This relation, in steady-state conditions, yields $\bar{\sigma}(A)N(A) = $ constant, a rule that is rather well satisfied over large intervals of atomic mass number in the experimental σN_s curve of solar system material. The smoothness of the distribution was found to be interrupted by steep drops at nuclei with closed neutron shells, for neutron numbers $N = 50$, 82, and 126 (the so-called *magic* neutron numbers). For elements produced mainly by the *s*-process, this happens close to mass numbers $A = 88$, 138, and 208, where neutron capture cross sections become very small (a few millibarns, i.e. a few $\times 10^{-27}$ cm^2); solar abundances show correspondingly remarkable peaks for these atomic mass numbers (Anders & Grevesse 1989).

From inspection of the solar σN_s curve, two main conclusions were derived already in the early analyses. The first was that the *s*-process could not be unique. A series of different physical mechanisms, occurring in separated astrophysical environments, had to provide a multiplicity of τ values in order to bypass the bottlenecks introduced in the neutron capture path by the small cross sections of neutron magic nuclei. One of the components of the process had to account for the *s* nuclei of $A \leq 88$ (the *weak s*-component), and a second one was necessary for nuclei with $88 \leq A \leq 208$ (the *main* component). A third (*strong*) component was also assumed for producing roughly 50% of ^{208}Pb, which apparently could not be explained in other ways (Clayton & Rassbach 1967). The other remarkable conclusion was that, in order to simulate analytically the effects of the above different processes, one could assume either a limited number of intense fluxes or a continuous distribution of decreasing neutron exposures (in which many nuclei could capture a small number of neutrons, and only few nuclei could capture a large number of them): see Seeger et al. (1965) and Clayton (1968). This second alternative soon became very popular, as it was easily approximated by a power law or by an exponential distribution of neutron exposures, both accounting for the *s*-nuclei from Sr to Pb for suitable choices of only two parameters.

In particular, starting with pure Fe seeds and adopting an exponential distribution of neutron exposures,

$$\rho(\tau) = \frac{GN^\odot_{56}}{\tau_0} e^{-\frac{\tau}{\tau_0}}, \tag{7.4}$$

so that $\rho(\tau)d\tau$ represents the number of Fe seeds exposed to an integrated flux between τ and $\tau + d\tau$, one can obtain an analytical solution for the set of Equations 7.3, yielding (Clayton & Ward 1974)

$$\sigma(A)N_s(A) = GN^\odot_{56}\tau_0 \Pi^A_{i=56} \left(1 + [\sigma(i)\tau_0]^{-1}\right)^{-1} \tag{7.5}$$

A fit to the experimental σN curve then provided values for the fraction G of solar Fe that had been irradiated and the mean neutron exposure τ_0.

The success of this solution actually lead to the idea that each one of the three *s*-process components initially assumed to exist had to provide an exponential $\rho(\tau)$ (Clayton & Ward 1974). The *main* component was characterized by a value of τ_0 initially estimated around 0.2

mbarn^{-1}. This value was gradually increased over the years, reaching about 0.30 mbarn^{-1}, due to continuous improvements in the nuclear parameters (see Käppeler, Beer, & Wisshak 1989). The *weak* component was characterized by $\tau_0 = 0.06$ mbarn^{-1}, while the *strong* one was found to need $\tau_0 = 7$ mbarn^{-1}. These different values of the mean neutron exposure also explain the names given to the three components. It is here important to notice that the fortune of the exponential distribution is actually due only to its being a simple analytical formulation. In reality, any simple mathematical representation for $\rho(\tau)$ is totally insufficient (see, e.g., for the weak component Raiteri et al. 1993, and for the main component Gallino et al. 1997 and Arlandini et al. 1999).

After the basic ideas of the classical analysis were put forward, Ulrich (1973) showed how the final evolutionary phases of stars with moderate mass ($M \leq 8\ M_\odot$) could approximately provide exponential distributions of neutron exposures. He showed that this result is obtained when the neutron sources are activated in red giants during the late AGB phases, in the intermediate convective zones developing periodically as a consequence of sudden thermal instabilities in the He-burning shell (*thermal pulses*; hereafter TP). The most straightforward way to get neutrons in these He-rich layers includes the transformation of initial CNO nuclei first into ^{14}N (as accomplished by H-burning), and then into ^{22}Ne through the reaction chain ^{14}N$(\alpha,\gamma)^{18}$F$(\beta^+\nu)^{18}$O$(\alpha,\gamma)^{22}$Ne, followed at somewhat higher temperatures by the activation of the ^{22}Ne$(\alpha, n)^{25}$Mg reaction. For several years the main component was therefore ascribed to AGB stars sufficiently massive to develop, in TPs, the high temperatures needed to burn ^{22}Ne ($T \geq 3.5 \times 10^8$ K). This occurs for $8\,M_\odot \geq M \geq 4\ M_\odot$, i.e. for the so-called intermediate-mass stars. This line of research was pursued in particular by Icko Iben and coworkers (see, e.g., Iben 1975; Truran & Iben 1977; Iben & Truran 1978; Iben & Renzini 1983).

7.3   Success and Crisis of the Classical *s*-Process Analysis

7.3.1   *The Build-up of the Classical Scenario*

The "classical" analysis rapidly became a sophisticated tool, where reaction branchings, like those occurring at crucial isotopes (such as ^{79}Se, ^{85}Kr, ^{64}Cu, ^{151}Sm), also could be treated (Ward, Newman, & Clayton 1976; Ward 1977). Application of the branching analysis to specific ramifications of the process was since then used for inferring the stellar parameters (average neutron density, temperature, electron density). It was also shown by Ward & Newman (1978) that the branchings held information on the pulsed nature of the neutron flux. An account of the phenomenological approach in various stages of its development can be found in Käppeler et al. (1982, 1989), Mathews & Ward (1985), and Meyer (1994).

Over the years, the nuclear parameters became so precise, on average, that the phenomenological approach based on an exponential distribution started to show its limits. Recent studies on neutron capture cross sections in the Nd–Sm region, and on their application to the analysis of branchings, finally showed that the *s*-process, as modeled directly in stellar conditions, can reach a higher level of accuracy than the classical analysis; this is due to the fact that real neutron distributions are complex, time dependent, and not really suitable for simple analytical formulations (Arlandini et al. 1999).

The combined information coming from detailed calculations in stellar models and from the phenomenological analysis of the *s*-process also showed that the ^{22}Ne$(\alpha, n)^{25}$Mg source in intermediate-mass stars suffers for serious problems. Indeed, the neutron densities reached in TPs through ^{22}Ne-burning are rather high (above 10^{10} cm^{-3} for temperatures in excess of $T = 3.5 \times 10^8$ K). This implies a n-irradiation intermediate between the s- and the r-process (Despain 1980), in which the reaction network becomes rather complex and many branchings along the s-path are open. This inevitably leads to a nonsolar distribution of s-isotopes, and especially to excesses in the neutron-rich nuclei, like ^{86}Kr and ^{96}Zr.

Spectroscopic and photometric observations (especially on Magellanic Cloud red giants) soon confirmed that AGB stars enriched in s-elements (and in C, the main product of partial He burning) cannot be identified with intermediate-mass stars, where TPs reach a temperature sufficient to burn ^{22}Ne efficiently. Indeed, the observed luminosities are in general lower than in intermediate-mass stars (Blanco, McCarthy, & Blanco 1980). Moreover, in most AGBs Mg isotopes show a solar mix, contrary to the expectations of ^{25}Mg enhancement by the ^{22}Ne$(\alpha,$ n$)^{25}$Mg reaction (Clegg, Lambert, & Bell 1979; see also Karakas & Lattanzio 2004). On these problems, see Wallerstein & Knapp (1998) and Busso et al. (1999) for recent reviews.

7.3.2   s-Processing From ^{13}C Burning

The above problems eventually induced various authors to reconsider the conditions for the activation of the alternative source ^{13}C$(\alpha,$ n$)^{16}$O in low-mass stars. The main difficulty concerning this reaction is that of finding some mixing mechanism to inject a small amount of hydrogen from the envelope into the intershell region, where, at hydrogen re-ignition, they would be captured by the abundant ^{12}C, producing ^{13}C.

Various physical mechanisms have been proposed to solve the problem of proton ingestion (Iben 1981; Iben & Renzini 1982a,b; Hollowell & Iben 1988; Herwig, Blöcker, & Schönberner 1997; Langer et al. 1999; Cristallo et al. 2001; Denissenkov & Tout 2003; Cristallo, Gallino, & Straniero 2004). None of them, however, can prove from first principles the existence of proton fluxes in the C-rich zones. As a matter of fact, in most calculations of recent years the extent of the proton-enriched layers and/or the ^{13}C concentration in them have been assumed as free parameters to be calibrated by observational constraints.

The first models of ^{13}C burning in AGB stars assumed that the layers enriched in ^{13}C (the so-called ^{13}C-pocket) could be ingested by the subsequent pulse and burned at the typical temperature of 1.5×10^8 K (Gallino et al. 1988; Käppeler et al. 1990). In such pulses, when the intershell convection subsequently reaches its maximum extent and the temperature increases to about 3×10^8 K, the ^{22}Ne source is marginally activated, providing a second small neutron burst of relatively high peak neutron density. This neutron burst was recognized to account for several details of the solar s-process abundance distribution, mainly for nuclei depending on reaction branchings for which the phenomenological approach suggests a relatively high temperature in the production zone (Käppeler et al. 1989, 1990).

In the last decade, new stellar model calculations for the AGB phases by Straniero et al. (1995, 1997) showed instead that, whichever is the abundance of ^{13}C produced in the intershell, it burns locally in the radiative layers of the He intershell before a new convective pulse develops. The thermal conditions are rather different from those established inside the pulses; in particular, the temperature is lower [$(0.8-0.9) \times 10^8$ K], and the average neutron density never exceeds 1×10^7 n cm^{-3} (Gallino et al. 1998). Then the pocket, now highly s-enhanced, is engulfed in the convective pulse. The s-elements are therefore diffused over the whole He intershell and slightly modified by the marginal activation of the ^{22}Ne source.

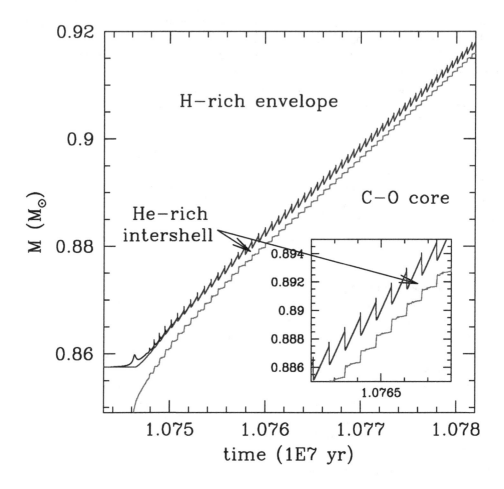

Fig. 7.1. Temporal evolution of the mass coordinates of the He-burning shell (lower line), of the H-burning shell (intermediate line), and of the inner border of envelope convection (upper line), for an AGB model with $M = 5\,M_\odot$ and $Z = 0.02$.

They are then partially brought to the surface during the following TDU episode. Such a scheme has been confirmed by all recent computations (Herwig et al. 1997; Goriely & Siess 2001; Lugaro et al. 2003). Before entering into the details of the ensuing *s*-process features, let us first outline the present scenario of the AGB evolution, as emerged in recent years.

7.4   AGB Models: Features and Uncertainties as a Function of Mass

The temporal evolution of the internal structure of a $5\,M_\odot$ with solar metallicity during the AGB phase is shown in Figure 7.1 The C-O core, whose pressure is mainly provided by degenerate electrons, is surrounded by a thin He intershell, the He-rich zone located between the He- and the H-burning shells. The convective H-rich envelope extends from the surface down to a layer located very close to the H-shell. After each TP the base of the convective envelope penetrates into the top layers of the He intershell (TDU), carrying fresh ^{12}C and *s*-process isotopes to the surface.

7.4.1   Uncertainties on Mass Loss and Convection

One of the most severe uncertainties still affecting AGB models concerns mass loss. The duration of the AGB phase and the number of TPs, the amount of mass dredged up, the impact of stellar winds on interstellar abundances, and many other important predictions depend on the assumed mass loss rate. The available data indicate that this rate ranges between 10^{-8} and 10^{-4} $M_\odot$ yr^{-1} (Kastner et al. 1993; Loup et al. 1993). Studies of Mira and semiregular variables show that stellar winds do not increase monotonically with time, and the star certainly encounters variations in its mass loss efficiency (Marengo et al. 1999; Marengo, Ivezić, & Knapp 2001), until a final violent (perhaps dynamical) envelope ejection occurs. One has to notice that a common finding of models is that immediately after a TP, during a phase of quiescent He-burning, the ratio of the gas pressure to the radiation pressure (β) decreases in a thin layer located around the H/He discontinuity. Sweigart (1999) found that, when the envelope is reduced to a sufficiently low mass (the exact value depending on the core mass and metallicity), β approaches 0. Then, the local stellar luminosity exceeds the Eddington luminosity and the whole envelope becomes dynamically unstable. Recently, this finding was independently confirmed in models of a 5 $M_\odot$ star at $Z = 0.02$ (Straniero et al. 2000). At the moment, however, one cannot say if this instability will grow indefinitely, up to the complete envelope removal, thus giving rise to the formation of a planetary nebula.

A second important uncertainty is related to the actual extension of the convective zones. Difficulties are met when the internal boundary of the convective envelope approaches regions of varying composition. In this case any perturbation, even small at the beginning, may propagate the convective instability toward the interior. In fact, as noted by Becker & Iben (1979), Castellani, Chieffi, & Straniero (1990), and Frost & Lattanzio (1996), if the H-rich convective envelope enters a region progressively enriched in He, a discontinuity of the radiative gradient forms at the interface between stable and unstable layers. This occurs because the H-rich envelope has a significantly larger opacity than the He-rich layer immediately below it. In such a case, if any mixing occurs beyond the inner border of the convective envelope, the local hydrogen abundance rises, the opacity (and the radiative gradient) grows, and the layer becomes unstable to convection. A similar situation is encountered at the outer edge of the convective core during central He-burning (Paczyński 1970; Castellani, Giannone, & Renzini 1971). Castellani et al. (1985) named this phenomenon *induced overshoot* (i.e. induced by the chemical discontinuity that forms at the boundary of a convective region), to be distinguished from *mechanical overshoot* (i.e. the mixing caused by the convective elements, which preserve a finite velocity outside the unstable region).

As far as the evolution before the TP-AGB phase is concerned (e.g., the second dredge-up, which takes place during the early AGB stages in intermediate-mass stars), the overall impact of overshoot phenomena is generally small (Castellani, Marconi, & Straniero 1998). On the contrary, Frost & Lattanzio (1996) found that the inclusion of a moderate overshoot strongly increases the efficiency of the TDU (see also Herwig et al. 1997; Herwig 2000; Cristallo et al. 2001). A change in the strength of the TDU not only modifies the surface composition of an AGB star, but greatly affects its physical properties. The comparison of model predictions with spectroscopic and photometric observations of AGB stars may help to verify the efficiency of this phenomenon and, in the long term, to calibrate mixing phenomena in stellar models. In this respect, despite the mentioned uncertainties, full stellar evolution models can now be used to provide the yields from AGB stars necessary to compute the chemical evolution of galaxies: for many years this task has been performed instead through semi-analytic or synthetic AGB models (Van der Hoek & Groenewegen 1997; Marigo 2001; see also Mollá, Gavilán, & Buell 2004).

7.4.2   Evolutionary Path as a Function of Mass

Present-generation AGB stars have masses larger than $0.8 - 0.9\,M_\odot$ (the main sequence lifetime of less massive stars is longer than the age of the Universe). The upper mass limit for AGB stars marks the inferior mass limit for "massive" stars, those which, after He exhaustion in the core, burn C, Ne, O, and Si, form a degenerate iron core, and eventually collapse. The precise value of this limit is not well defined. It has been long understood that all stars with mass below 7–$8\,M_\odot$ should skip C burning, due to the huge energy loss by plasma neutrino emission, which cools their central region (Schwarzschild & Härm 1965). These stars are the progenitors of C-O white dwarfs. However, slightly more massive stars, in the range 9–$10\,M_\odot$, can ignite carbon in degenerate conditions, but, once a degenerate O-Ne-Mg core is built, become thermally pulsing AGB stars (Ritossa, García-Berro, & Iben 1996). These stars may produce a massive O-Ne-Mg white dwarf (Domínguez, Tornambè, & Isern 1993) or may collapse before the beginning of the AGB phase (Miyaji & Nomoto 1987).

Even below $8\,M_\odot$ the AGB evolutionary scenario and related nucleosynthesis significantly change with the mass of the star. In the following we review the properties of AGB stars with different initial mass. The mass intervals reported below are appropriate for solar composition (i.e. $Z = 0.02$, $Y = 0.28$). The quantitative results have been derived from recently published AGB models computed by several authors, in particular Lattanzio et al. (1996), Forestini & Charbonnel (1997), Straniero et al. (1997, 2000), and Lattanzio & Forestini (1999).

$M < 1.3\,M_\odot$

At the beginning of the thermally pulsing AGB phase, the mass of the H-depleted region (M_H) for these low-mass stars is about $0.55\,M_\odot$. Their envelope (already eroded by mass loss during the red giant branch) is rather small. For example, taking into account a pre-AGB mass loss in agreement with the Reimers (1975) formula with $\eta = 0.4$ (Fusi Pecci & Renzini 1975), a $1\,M_\odot$ star of solar metallicity would have at the first TP a residual envelope mass of just $0.2\,M_\odot$. During the AGB this envelope is rapidly consumed, from the top by further mass loss and from the bottom by H-shell burning, so that only few TPs are possible before the envelope mass drops below the minimum threshold allowing a red giant star structure (about $0.001\,M_\odot$). For $1\,M_\odot$ models about 10 TPs are expected (with an average mass loss of $10^{-7}\,M_\odot\,\mathrm{yr}^{-1}$). However, the TDU does not take place and the envelope composition remains frozen at the composition reached after the red giant branch.

$1.3 \leq M/M_\odot < 2$

The core mass at the first TP is more or less the same as before, but the envelope mass is larger. After a few TPs (in any case fewer than 8–9) the core mass reaches about 0.6 $M_\odot$, and a moderate TDU takes place. The surface composition changes: C and s-process elements are enhanced. Note that the maximum temperature at the base of the convective zone generated by TPs is well below the threshold for the activation of the $^{22}\mathrm{Ne}(\alpha, \mathrm{n})^{25}\mathrm{Mg}$ reaction. Then, the s-process nucleosynthesis is only powered by the $^{13}\mathrm{C}(\alpha, \mathrm{n})^{16}\mathrm{O}$ reaction, which is at work during the interpulse phase. TDU continues as long as the envelope mass remains larger than about 0.4–0.5 $M_\odot$ (Straniero et al. 1997). After the envelope mass drops below this limit, the remaining few TPs occur without any TDU. In any case, when these low-mass AGB stars reach the tip of the AGB, the C/O ratio is still lower than 1. For example, a 1.5 $M_\odot$ (solar composition) star has a C/O ratio at its birth of 0.5 (Allende Prieto, Lambert, & Asplund 2002); after the first dredge-up this ratio drops to 0.35; at the end of the AGB phase TDU restores it to 0.49. Once again, we recall that the final value of the C/O ratio depends on the assumptions about mass loss (in this particular case we have used a Reimers formula with the free parameter $\eta = 0.4$) and on the efficiency of TDU. The above scenario does not consider overshoot, so that the resulting dredged-up mass should be considered as a lower limit.

$2 \leq M/M_\odot \leq 3$

These stars are probably the most efficient contributors to the main component of the *s*-process. They are also progenitors of intrinsic carbon stars observed in the Galaxy and in the Magellanic Clouds, as we shall see later. Similarly to what occurs at lower masses, TDU starts after a few TPs, when the core mass attains 0.6 $M_\odot$. However, due to the larger envelope, stronger TDU episodes take place. The amount of dredged-up material may be as large as half the (He-C-*s*)-rich region. The TDU phase continues up to the C star stage (C/O>1) and beyond. The main source of neutrons is from the $^{13}C(\alpha, n)^{16}O$ reaction. The $^{22}Ne(\alpha, n)^{25}Mg$ reaction is only marginally activated at the peak temperatures of the most advanced TPs, near 300×10^6 K. Nonetheless, it has important effects on those reaction branchings that are sensitive to the neutron density and temperature conditions.

$3 < M/M_\odot < 5$

These stars start the TP-AGB phase with a rather large core mass. A 5 $M_\odot$ model, for example, has a core mass at the first TP of about 0.86 $M_\odot$. A negligible amount of mass loss affects the pre-AGB evolution, so that the initial envelope is quite massive. TDU begins very soon (after a couple of TPs). However, due to the huge dilution of the dredged-up material (few 10^{-4} $M_\odot$ are mixed with a few $M_\odot$ of envelope), a rather slow change of the surface C/O ratio is expected. After a few TPs, the temperature at the base of the convective zones generated by the TPs reaches 3.5×10^8 K, and the $^{22}Ne(\alpha, n)^{25}Mg$ neutron source is activated efficiently. On the contrary, a less efficient $^{13}C(\alpha, n)^{16}O$ reaction is expected. In fact, the larger the core mass, the smaller the He intershell region is: for a 5 $M_\odot$ star this layer is about one order of magnitude smaller than for of a 2 $M_\odot$ model. Correspondingly, the amount of material dredged up and the extension in mass of the ^{13}C pocket are expected to be reduced by about an order of magnitude. The resulting *s*-process nucleosynthesis is therefore rather different with respect to the one of low-mass stars.

During the final part of the AGB, when most of the envelope mass has been lost, these stars may eventually become bright C stars. The most massive in this range (4–5 $M_\odot$) may also become Li-rich AGB stars. First of all, Be is produced by the $^3He(\alpha, \gamma)^7Be$ reaction. This requires a temperature of at least 30×10^6 K. Then Li may be synthesized by the Be decay $^7Be(e^-, \nu)^7Li$ ($t_{1/2} = 53$ d). Note that at high temperatures Li is quickly destroyed by the reaction $^7Li(p, \alpha)^4He$. Thus, the Li production might occur only if the mixing is fast enough to move ^7Be to a cooler external region before electron captures take place (Cameron & Fowler 1971).

$5 < M/M_\odot < 8$

The most massive AGB stars can also produce *s*-elements, mainly through the $^{22}Ne(\alpha,$

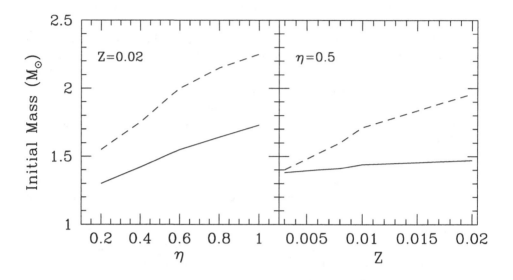

Fig. 7.2. *Left panel*: The minimum initial mass for experiencing a TDU phase (solid line) for the formation of a C star (dashed line) as a function of the η parameter in the Reimers formula for mass loss. *Right panel*: The same parameters shown in the left panel, but plotted as a function of metallicity.

n)^{25}Mg reaction activated during the TP phase. Then, s-elements and fresh He and C are mixed with the envelope by TDU. The relevant feature of these stars is that a complete CN burning occurs inside the convective envelope, so that the fresh C dredged up is almost totally converted into N. The occurrence of this *hot bottom burning* (Karakas & Lattanzio 2004) solves the longstanding problem of the lack of very bright C stars ($M_{bol} < -6$ mag; see Iben 1981). Only for a short period before the end of the AGB phase, when hot bottom burning dies out and TDU is still active, might the surface C/O ratio increase above unity, generating very luminous C stars, which, however, should be extremely rare (Lattanzio 1998).

7.4.3   The Effect of Metallicity

The limiting masses of the above ranges decrease for decreasing mass loss rate and metallicity. This is shown in Figure 7.2, where the minimum mass for the formation of a C star and the minimum initial mass for the occurrence of the TDU are shown.

7.5   *s*-Processing in the Galaxy at Different Metallicities

The ^{13}C neutron source, as operating in AGB models, is of "primary" origin, i.e. derives from processes starting directly from the original H of the star. One of the major consequences of this fact is that the ensuing s-process distribution is extremely dependent on the initial abundance of Fe-group seeds, i.e. on stellar metallicity. Indeed, the neutron exposure τ is roughly proportional to the number of available ^{13}C nuclei per Fe seed, hence inversely proportional to the metallicity. Then, starting from an AGB star of initially solar composition, the neutron exposure tends to increase with decreasing Z. This has a dramatic effect on s-process abundances of elements with different atomic mass number, as shown in

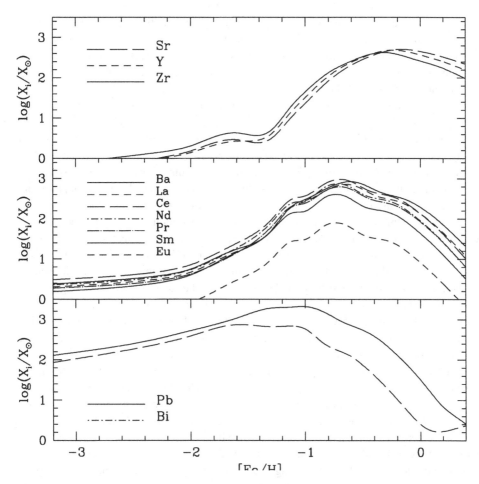

Fig. 7.3. Model predictions for the abundance of selected neutron-rich elements in the production layers of AGB stars with different metallicity. All calculations adopt the same choice for the amount of ^{13}C burnt per cycle.

Figure 7.3 as a function of [Fe/H] for Galactic disc compositions. Here [Fe/H] = log(Fe/H) − log(Fe/H)$_\odot$ is the usual logarithmic notation adopted in astronomical spectroscopy for the stellar metallicity.

With the adopted ^{13}C pocket, at solar metallicity one would produce essentially s-elements belonging to the Sr-Y-Zr peak, at the neutron magic number N = 50; decreasing Z, one would have more neutrons available per Fe seed, thus bypassing the bottleneck at N = 50 and progressively feeding elements at the second neutron magic peak, at Ba-La-Ce-Pr-Nd, with a maximum production at $Z \approx Z_\odot/4$ (Travaglio et al. 1999). For even lower metallicity one would have enough neutrons per Fe seed to feed Pb, in particular the abundant isotope ^{208}Pb at the magic neutron number N = 126. The maximum production of Pb occurs around $Z = Z_\odot/20$ (Travaglio et al. 2001). Eventually, for Galactic halo AGB stars with very low metallicities, all the initial Fe is efficiently converted directly to Pb, but the shortage of iron seeds becomes dominant, so that even the production of Pb decreases with metallicity. Due to this fact, *s*-process abundances are expected to decrease and eventually vanish at very low values of [Fe/H], where the neutron-rich elements must be dominated by the *r*-process (see, e.g., Simmerer et al. 2004, and references therein). Summing up, the *s*-process in AGB stars of different metallicity driven by the primary ^{13}C neutron source gives rise to a wide spectrum of different abundance distributions.

The mentioned dependence of the neutron source on local mixing phenomena then leads to expectations of large star-to-star scatter in the *s*-abundance distributions, even for AGB stars of the same metallicity. On the whole, the solar system distribution of *s*-elements is clearly not a "unique" process, but results from the integrated chemical evolution of the Galaxy, which mixes in the interstellar medium the outputs of many different stars, with yields changing with the initial metallicity, stellar mass, and maybe other physical properties.

The above discussion changes completely the traditional interpretation of the various *s*-process components. For example, the fact that at low metallicities Pb (^{208}Pb) becomes the dominant product of AGB nucleosynthesis reveals that an independent strong component (by some unknown astrophysical source) does not exist (Gallino et al. 1998; Goriely & Siess 2001; Travaglio et al. 2001). It also shows that at very low metallicity the neutron exposure cannot be monitored only by the elements of the Zr-peak and Ba-peak groups, as the neutron flux tends to accumulate on the neutron magic number corresponding to ^{208}Pb (Fig. 7.4). Moreover, at very low metallicities, the very low abundance of the Fe seeds is partly compensated by lighter, primary-like nuclei, from which the *s*-process nucleosynthesis directly starts, reproducing and crossing the Fe peak. Among such intermediate-mass neutron seeds one has to mention, in particular, ^{22}Ne and its progeny. In fact, CNO nuclei in the envelope become abundant due to the increased [O/Fe] ratio characterizing halo stars (see, e.g., Spite et al. 2004): this trend is actually continued through the thick and thin disk for metallicities lower than solar, as shown by Bensby, Feltzing, & Lundström (2004). In the CNO increase an important role is played also by the primary ^{12}C produced in the AGB phase. These CNO nuclei are mixed through the H-He discontinuity at TDU; here they are transformed into ^{14}N by shell H-burning and subsequently into ^{22}Ne by double α-captures in the early phases of a convective pulse, thus strongly increasing the concentration of the neutron source.

7.6   *s*-Processing in C(N) Stars

Probably the most important chemical anomaly of stars in the AGB phase is that many of them are carbon rich, i.e. they have a C/O ratio (by number) larger than unity in the envelope. For this reason, they are named carbon stars. As noted in previous sections, this carbon enrichment is the result of the simultaneous operation of TPs and TDU along the AGB phase that eventually transforms an O-rich star into a carbon star. In fact, the carbon content in the envelope is expected to increase along the spectral sequence of the AGB phase M→MS→S→SC→C(N type). Alternatively, this carbon enrichment could derive from a transfer of carbon-rich material in a binary system. In the first case the stars are usually named *intrinsic* carbon stars, while in the second case they are named *extrinsic* carbon stars. Our discussion will be concentrated on the first class. As shown before, TDU also enriches the envelope with *s*-process elements: in fact, C(N) stars are the main producers of *s*-elements in the Galaxy. They are significant contributors also to ^{12}C, ^{13}C, and rare nuclei

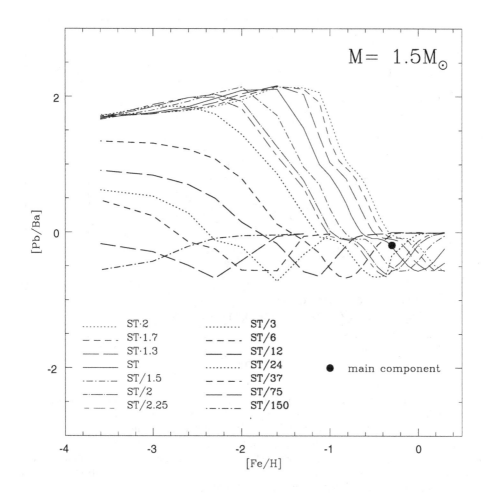

Fig. 7.4. The final envelope abundance ratio [Pb/Ba], as predicted by our AGB models for $M = 1.5\,M_\odot$, as a function of metallicity. The concentration of the ^{13}C source has been left to vary over a wide interval, as indicated by the labels. Here ST refers to the abundance of ^{13}C assumed as *standard* by Gallino et al. (1998), namely $4 \times 10^{-6}\,M_\odot$

such as ^7Li and ^{26}Al: these isotopes are affected not only by convective dredge-up, but also by more complex phenomena of extended mixing, or *cool bottom processes*, occurring in red giants (see, e.g., Charbonnel & Balachandran 2000; Nollett, Busso, & Wasserburg 2003; Busso, Gallino, & Wasserburg 2004; Charbonnel 2004).

C(N) stars, or normal carbon stars (Keenan 1993), represent a formidable challenge from the spectroscopic point of view. They show very crowded spectra due to their strong molecular-band absorption and low temperature. Thus, it is not surprising that until recently the only detailed analysis of their *s*-element abundances was that by Utsumi (1970, 1985), who found that C(N) stars are of roughly solar metallicity; he suggested that their mean

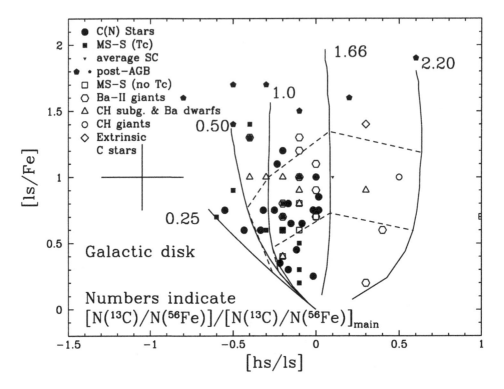

Fig. 7.5. The various symbols refer to observations of *s*-element enhancement in different classes of AGB stars at different metallicities. The curves are theoretical sequences of envelope enrichment. The number of TDU episodes increases toward the top, and dashed lines mark the occurrence of the 4th and 8th mixing episode. The curves were obtained for models with varying efficiency in the neutron captures, as monitored by the abundance ratio between the neutron source ^{13}C and Fe. C(N) stars occupy regions compatible with their being slightly more evolved than the average of S stars.

s-element enhancements ([ls/Fe] and/or [hs/Fe])* reach a factor of 10 or more above solar. This level of enrichment would be considerably larger than for the O-rich S stars: however, it was accepted as a rather extreme outcome of the synthesis of *s* nuclei during the AGB phase. More recent studies are now available (Abia et al. 2001, 2002), based on high-resolution spectra of a more extended sample of N stars. This has caused a strong revision in the quantitative *s*-element abundances. N stars were confirmed to be of near solar metallicity, but they show on average ⟨[ls/Fe]⟩ = +0.67 ± 0.10 and ⟨[hs/Fe]⟩ = +0.52 ± 0.29, which is significantly lower than estimated by Utsumi and is more similar to S star abundances (Smith & Lambert 1990; Busso et al. 2001). This revision allowed the extension to C(N) stars of the generally good agreement between observed *s*-process abundances and theoretical predictions of *s*-process nucleosynthesis in AGB stars (Gallino et al. 1998; Busso et al. 1999). Such comparisons confirm also for C(N) stars the existence of an intrinsic spread in

---

* In the following, "ls" refers to the light-mass *s*-elements Y and Zr, and, "hs" refers to the high-mass *s*-elements Ba, Nd, La, and Sm.

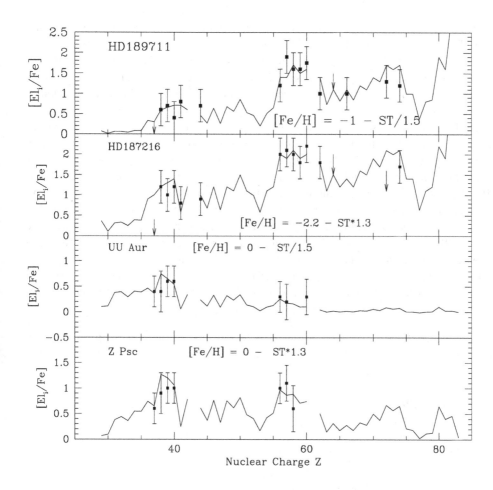

Fig. 7.6. Observed and theoretical distributions of *s*-elements at the surface of AGB stars, when the abundance ratio by number of C/O reaches unity. The models adopted are those discussed in §7.5.

the abundance of ^{13}C burnt, and allow us to place observed AGBs with different *s*-process and carbon enrichment along simple evolutionary sequences (Fig. 7.5). In these sequences only 1–2 pulses separate a typical S star from a C(N) star. During this short time interval the spectroscopic appearance changes drastically because C-based opacities start to dominate, but the *s*-element composition is only slightly affected. Later, as the star evolves along the AGB and TPs and TDU continue operating, more carbon and *s*-elements are dredged up into the envelope. Eventually, the C/O ratio may exceed unity by a consistent amount, before stellar winds terminate the evolution; but the simultaneous increase of mass loss rates forms a thick and dusty circumstellar envelope obscuring the star at optical wavelengths. This scenario might explain not only the large C abundances observed in some planetary nebulae (Parthasarathy 1999), but also the apparent *gap* in the level of *s*-element and carbon enhancement between N stars and post-AGB stars (Van Winckel & Reyniers 2000; Reyniers & Van Winckel 2004). We cannot "see" carbon stars with a C/O ratio largely exceeding unity. The fact that we now understand the evolutionary status and the composition of C(N) stars is confirmed by the possibility of obtaining satisfactory models of their detailed photospheric abundances, through *s*-process computations in AGB models that also account for dredge-up and mass loss. Examples of such models for C(N) stars of different metallicities are shown in Figure 7.6.

Another important result inferred from observations of C(N) stars and from the properties of the *s*-process path at reaction branchings is an estimate of the initial mass of C(N) stars. As noted previously, in low-mass stars the major neutron source is ^{13}C. If C(N) stars were of intermediate mass, ^{22}Ne would be favored as a neutron donor by the higher temperature in TPs. Because of the very different neutron density provided by the two neutron-producing reactions, different compositions are expected from them, especially for nuclei depending on the complex reaction branching at ^{85}Kr. The simplest example of this is displayed by the ratio Rb/Sr. As the isotope ^{87}Rb is strongly fed only at high neutron densities, this ratio increases for increasing n_n, hence for increasing contributions of the ^{22}Ne$(\alpha, n)^{25}$Mg reaction to neutrons. On this basis Abia et al. (2001) showed that, since C(N) stars observed so far always show low values of the Rb/Sr ratio, their neutron captures must be controlled by ^{13}C$(\alpha, n)^{16}$O and they must therefore be of low mass ($M \leq 3 M_\odot$).

Acknowledgements. We wish to thank the organizers for the opportunity to attend a very interesting conference. M.B. and R.G. acknowledge support from the MURST-FIRB program "The astrophysical origin of heavy elements beyond Fe."

References

Abia, C., et al. 2002, ApJ, 579, 817

Abia, C., Busso, M., Gallino, R., Domínguez, I., Straniero, O., & Isern, J. 2001, ApJ, 559, 1117

Abia, C., & Wallerstein, G. 1998, MNRAS, 293, 89

Allende Prieto, C., Lambert, D. L., & Asplund, M. 2002, ApJ, 573, 137

Anders, E., & Grevesse, N. 1989, Geochim. Cosmochim. Acta, 53, 197

Anders, E., & Zinner, E. 1993, Meteoritics, 28, 490

Arlandini, A., Käppeler, F., Wisshak, K., Gallino, R., Lugaro, M., Busso, M., & Straniero, O. 1999, ApJ, 525, 886

Becker, S. A., & Iben, I., Jr. 1979, ApJ, 232, 831

Bensby, T., Feltzing, S., & Lundström, L. 2004, in Carnegie Observatories Astrophysics Series, Vol. 4: Origin and Evolution of the Elements, ed. A. McWilliam & M. Rauch (Pasadena: Carnegie Observatories, http://www.ociw.edu/symposia/series/symposium4/proceedings.html)

Blanco, V. M., McCarthy, M. F., & Blanco, B. M. 1980, ApJ, 242, 938

Burbidge, E. M., Burbidge, G. R., Fowler, W. A., & Hoyle, F. 1957, Rev. Mod. Phys., 29, 547

Busso, M., Gallino, R., Lambert, D. L., Travaglio, C., & Smith, V. V. 2001, ApJ, 557, 802

Busso, M., Gallino, R., & Wasserburg, G. J. 1999, ARA&A, 37, 239

——. 2004, PASA, in press

Cameron, A. G. W. 1954, Phys. Rep., 93, 932

——. 1957, Report CRL-41, Chalk River, Canada

Cameron, A. G. W., & Fowler, W. A. 1971, ApJ, 164, 111

Castellani, V., Chieffi, A., & Straniero, O. 1990, ApJS, 74, 463

Castellani, V., Chieffi, A., Tornambè, A., & Pulone, L. 1985, ApJ, 296, 204

Castellani, V., Giannone, P., & Renzini, A. 1971, Ap&SS, 10, 340

Castellani, V., Marconi, M., & Straniero, O. 1998, A&A, 340, 160

Charbonnel, C. 2004, in Carnegie Observatories Astrophysics Series, Vol. 4: Origin and Evolution of the Elements, ed. A. McWilliam & M. Rauch (Cambridge: Cambridge Univ. Press), in press

Charbonnel, C., & Balachandran, S. C. 2000, A&A, 359, 563

Clayton, D. D. 1968, Principles of Stellar Evolution and Nucleosynthesis, (Chicago: Chicago Univ. Press)

Clayton, D. D., Fowler, W. A., Hull, T., & Zimmerman, B. A. 1961, Ann. Phys., 12, 331

Clayton, D. D., & Rassbach, M. E. 1967, ApJ, 148, 69

Clayton, D. D., & Ward, R. A. 1974, ApJ, 193, 397

Clegg, R. E. S., Lambert, D. L., & Bell, R. A. 1979, ApJ, 234, 188

Cristallo, S., Gallino, R., & Straniero, O. 2004, Mem. S. A. It., in press

Cristallo, S., Straniero, O., Gallino, R., Herwig, F., Chieffi, A., Limongi, M., & Busso, M. 2001, Nucl. Phys. A, 688, 217

Denissenkov, P. A., & Tout, C. A. 2003, MNRAS, 340, 722

Despain, K. H. 1980, ApJ, 236, 648

Domínguez, I., Tornambè, A., & Isern, J. 1993, ApJ, 419, 268

Forestini, M., & Charbonnel, C. 1997, A&AS, 123, 241

Frost, C., & Lattanzio, J. C. 1996, ApJ, 473, 383

Fusi-Pecci, F., & Renzini, A. 1975, A&A, 39, 413

Gallino, R., Arlandini, C., Busso, M., Lugaro, M., & Travaglio, C. 1998, ApJ, 497, 388

Gallino, R., Busso, M., & Lugaro, M. 1997, in Astrophysical Implications of the Laboratory Studies of Presolar Material, ed. T. Bernatowicz & E. Zinner (Woodbury, N.Y.: AIP), 115

Gallino, R., Busso, M., Picchio, G., Raiteri, C. M., & Renzini, A. 1988, ApJ, 334, L45

Goldberg, L., Muller, E. A., & Aller, L. H. 1960, ApJS, 5,1

Goriely, S., & Siess, L. 2001, A&A, 378, 25

Greenstein, J. L. 1954, in Modern Physics for Engineers, ed. I. Ridenour (New York: McGraw-Hill), 124

Herwig, F. 2000, A&A, 360, 952

Herwig, F., Blöcker, T., & Schönberner, D. 1997, A&A, 324, L81

Hollowell, D., & Iben, I., Jr. 1988, ApJ, 333, L25

Iben, I., Jr. 1975, ApJ, 196, 525

——. 1981, ApJ, 246, 278

——. 1991, in IAU Symp. 145, Evolution of Stars: The Photospheric Abundance Connection, ed. G. Michaud & A. Tutukov (Dordrecht: Kluwer), 257

Iben, I., Jr., & Renzini, A. 1982a, ApJ, 249, L79

——. 1982b, ApJ, 263, L23

——. 1983, ARA&A, 21, 271

Iben, I., Jr., & Truran, J. W. 1978, ApJ, 220, 980

Käppeler, F. 2001, in Nuclei far from Stability and Astrophysics, ed. D. N. Poenaru, H. Rebel, & J. Wentz (Nato Science Series II: Math. Phys. and Chemistry 17), 427

Käppeler, F., Beer, H., & Wisshak, K. 1989, Rep. Prog. Phys., 52, 945

Käppeler, F., Beer, H., Wisshak, K., Clayton, D. D., Macklin, R. L., & Ward. R. A. 1982, ApJ, 257, 821

Käppeler, F., Gallino, R., Busso, M., Picchio, G., & Raiteri, C. M. 1990, ApJ, 354, 630

Karakas, A. I., & Lattanzio, J. C. 2004, in Carnegie Observatories Astrophysics Series, Vol. 4: Origin and Evolution of the Elements, ed. A. McWilliam & M. Rauch (Pasadena: Carnegie Observatories, http://www.ociw.edu/symposia/series/symposium4/proceedings.html)

Kastner, J. H., Forveille, T., Zuckerman, B., & Omont, A. 1993, A&A, 275, 163

Keenan, P. C. 1993, PASP 105, 905

Langer, N., Heger, A., Wellstein, S., & Herwig, F. 1999, A&A, 346, L37

Lattanzio, J. C. 1998, in Stellar Evolution, Stellar Explosions and Galactic Chemical Evolution, ed. A. Mezzacappa (Bristol: Inst. of Physics), 299

Lattanzio, J., & Forestini, M. 1999, in IAU Symp. 191, Asymptotic Giant Branch Stars, ed. T. Le Bertre, A. Lèbre, & C. Waelkens (Dordrecht: Kluwer), 31

Lattanzio, J. C, Frost, C., Cannon, R, & Wood, P. R. 1996, Mem. S. A. It., 67, 729

Loup, C., Forveille, T., Omont, A., & Paul, J. F. 1993, A&AS, 99, 291

Luck. R. E., & Bond, H. E. 1991, ApJS, 77, 515

Lugaro, M., Herwig. F., Lattanzio, J. C., Gallino, R., & Straniero, O. 2003, ApJ, 586, 1305

Macklin, R. L., & Gibbons, J. H. 1965, Rev. Mod. Phys., 37, 166

Marengo, M., Busso, M., Silvestro, G., Persi, P., & Lagage, P. O. 1999, A&A, 348, 501

Marengo, M., Ivezić, Z., & Knapp, G. R. 2001, MNRAS, 324, 1117

Marigo, P. 2001, A&A, 370, 194

Mathews, G. J., & Ward, R. A. 1985, Rep. Progr. Phys., 48, 1371

Meyer, B. S. 1994, ARA&A, 32, 153

Miyaji, S., & Nomoto, K. 1987, ApJ, 318, 307

Mollá, M., Gavilán, M., & Buell, J. F. 2004, in Carnegie Observatories Astrophysics Series, Vol. 4: Origin and Evolution of the Elements, ed. A. McWilliam & M. Rauch (Pasadena: Carnegie Observatories, http://www.ociw.edu/symposia/series/symposium4/proceedings.html)

Nollett, K. M., Busso, M., & Wasserburg, G. 2003, ApJ, 582, 1036

Ott, U. 1993, Nature, 364, 25

Paczyński, B. 1970, Acta Astron., 20, 47

Parthasarathy, M. 1999, in IAU Symp. 191, Asymptotic Giant Branch Stars, ed. T. Le Bertre, A. Lèbre, & C. Waelkens (Dordrecht: Kluwer), 475

Raiteri, C. M., Gallino, R., Busso, M., Neuberger, D., & Käppeler, F. 1993, ApJ, 419, 207

Reimers, D. 1975, in Problems in Stellar Atmospheres and Envelopes, ed. B. Baschek, W. H. Kegal, & G. Traving (Berlin: Springer), 229

Reyniers, C., & Van Winckel, H. 2004, in Carnegie Observatories Astrophysics Series, Vol. 4: Origin and Evolution of the Elements, ed. A. McWilliam & M. Rauch (Pasadena: Carnegie Observatories, http://www.ociw.edu/symposia/series/symposium4/proceedings.html)

Ritossa, C., García-Berro, E., & Iben, I., Jr. 1996, ApJ, 460, 489

Sackmann, J. I., & Boothroyd, A. I. 1991, in IAU Symp. 145, Evolution of Stars: The Photospheric Abundance Connection, ed. G. Michaud & A. Tutukov (Dordrecht: Kluwer), 275

Schwarzschild, M., & Härm, R. 1965, ApJ, 142, 885

Seeger, P. A., Fowler, W. A., & Clayton, D. D. 1965, ApJS, 11, 121

Simmerer, J., Sneden, C., Woolf, V., & Lambert D. L. 2004, in Carnegie Observatories Astrophysics Series, Vol. 4: Origin and Evolution of the Elements, ed. A. McWilliam & M. Rauch (Pasadena: Carnegie Observatories, http://www.ociw.edu/symposia/series/symposium4/proceedings.html)

Smith, V. V., & Lambert, D. L. 1990, ApJS, 72, 387

Spite, M., et al. 2004, in Carnegie Observatories Astrophysics Series, Vol. 4: Origin and Evolution of the Elements, ed. A. McWilliam & M. Rauch (Pasadena: Carnegie Observatories, http://www.ociw.edu/symposia/series/symposium4/proceedings.html)

Straniero, O., Chieffi, A., Limongi, M., Busso, M., Gallino, R., & Alandini, C. 1997, ApJ, 478, 332

Straniero, O., Gallino, R., Busso, M., Chieffi, A., Raiteri, C. M., Limongi, M., & Salaris, M. 1995, ApJ, 440, L85

Straniero, O., Limongi, M., Chieffi, A., Domínguez, I., Busso, M., & Gallino, R. 2000, Mem. S. A. It., 71, 719

Suess, H. E., & Urey, H. C. 1956, Rev. Mod. Phys., 28, 53

Sweigart, A. 1999, in IAU Symp. 191, Asymptotic Giant Branch Stars, ed. T. Le Bertre, A. Lèbre, & C. Waelkens (Dordrecht: Kluwer), 533

Travaglio, C., Galli, D., Gallino, R., Busso, M., Ferrini, F. , & Straniero, O. 1999, ApJ, 521, 691

Travaglio C., Gallino, R., Busso, M., & Gratton, R. 2001, ApJ, 549, 346

Truran, J. W., & Iben, I., Jr. 1977, ApJ, 216, 797

Ulrich, R. K. 1973, in Explosive Nucleosynthesis, ed. D. N. Schramm & W. D. Arnett (Austin: Univ. of Texas Press), 139

Utsumi, K. 1970, PASJ, 22, 93

——. 1985, in Cool Stars with Excesses of Heavy Elements, ed. M. Jascheck & P. C. Keenan (Dordrecht: Reidel), 243

Van der Hoek, L. B. & Groenewegen, M. A. T. 1997, A&AS, 123, 395

Vanture, A. 1992, AJ, 104, 1986

Van Winckel, H., & Reyniers, C. 2000, A&A, 354, 135

Wallerstein, G., et al. 1997, Rev. Mod. Phys., 69, 995

Wallerstein, G., & Knapp, G. R. 1998, ARA&A, 36, 369

Ward, R. A. 1977, ApJ, 216, 540

Ward, R. A., & Newman, M. J. 1978, ApJ, 219, 195

Ward, R. A., Newman, M. J., & Clayton, D. D. 1976, ApJS, 31, 33

Wheeler, C. J., Sneden, C., & Truran, J. W. 1989, ARA&A, 27, 391

Zinner. E. 1997, in Astrophysical Implications of the Laboratory Studies of Presolar Material, ed. T. Bernatowicz & E. Zinner (Woodbury, N.Y.: AIP), 3

8

Models of chemical evolution

FRANCESCA MATTEUCCI
University of Trieste, Astronomy Department, Trieste, Italy

Abstract

The basic principles underlying galactic chemical evolution and the most important results of chemical evolution models are discussed. In particular, the chemical evolution of the Milky Way galaxy, for which we possess the majority of observational constraints, is described. Then, it is shown how different star formation histories influence the chemical evolution of galaxies of different morphological type. Finally, the role of abundances and abundance ratios as cosmic clocks is emphasized and a comparison between model predictions and abundance patterns in high-redshift objects is used to infer the nature and the age of these systems.

8.1   Introduction

In order to build a chemical evolution model, one needs to specify some basic parameters such as the boundary conditions, namely whether the system is closed or open and whether the gas is primordial or already chemically enriched. Then one needs the stellar birthrate function, which is generally expressed as the product of two independent functions, the star formation rate (SFR) and the initial mass function (IMF),

$$B(m,t) = \psi(t)\varphi(m), \tag{8.1}$$

where the SFR is assumed to be only a function of time and the IMF only a function of the stellar mass. The stellar evolution and nucleosynthesis are also necessary ingredients for modeling chemical evolution, and in particular we need to specify the stellar yields $[p_i(m)]$.

Then, one can include in the chemical evolution model some supplementary ingredients such as the the infall of extragalactic material, radial flows, and galactic winds, which can play a more or less important role depending on the galactic system under study. In particular, gas infall and radial flows can be important in describing the chemical evolution of spiral disks, whereas galactic outflows are probably important in elliptical galaxies where the SFR was very high in the past and there is no cold gas at the present time, as well as in dwarf galaxies, which possess a smaller potential well. In fact, galactic outflows, which eventually can transform into galactic winds, are determined by two main processes: the supernova (SN) feedback, namely how much energy is transferred from SNe into the interstellar medium (ISM), and the galactic potential well.

In this paper, I will recall the most popular approximations used to describe both the basic and the supplementary ingredients involved in galactic chemical evolution. Then I will describe detailed numerical models that can follow the evolution in space and time of

---

the abundances of the most abundant chemical species. I will show some applications of such models to the Milky Way galaxy and then to galaxies of different morphological type, focusing on the fact that abundances and abundance ratios are very useful tools to infer the history of star formation in galaxies, to impose constraints on stellar nucleosynthesis, and to derive information on the nature and age of high-redshift objects.

8.2   The Star Formation Rate

The SFR is one of the most important drivers of galactic chemical evolution: it describes the rate at which the gas is turned into stars in galaxies. Since the physics of the star formation process is still not well known, several parameterizations are used to describe the SFR. A common aspect to the different formulations of the SFR is that they include a dependence upon the gas density. Here I recall the most commonly used parameterizations for the SFR adopted so far in the literature.

An exponentially decreasing SFR provides an easy-to-handle formula:

$$SFR = \nu e^{-t/\tau_*},\tag{8.2}$$

with $\tau_* = 5 - 15$ Gyr in order to obtain a good fit to the properties of the solar neighborhood (Tosi 1988) and $\nu = 1 - 2\,\mathrm{Gyr}^{-1}$, being the so-called efficiency of star formation, which is expressed as the inverse of the time scale of star formation.

However, the most famous and most widely adopted formulation for the SFR is the Schmidt (1959) law:

$$SFR = \nu \sigma_{gas}^k,\tag{8.3}$$

which assumes that the SFR is proportional to some power of the volume or surface gas density. The exponent suggested by Schmidt was $k = 2$, but Kennicutt (1998) suggested that the best fit to the observational data on spiral disks and starburst galaxies is obtained with an exponent $k = 1.4 \pm 0.15$.

A more complex formulation, including a dependence also from the total surface mass density, which is induced by the SN feedback, was suggested by the observations of Dopita & Ryder (1994) who proposed the following formulation:

$$SFR = \nu \sigma_{tot}^{k_1} \sigma_{gas}^{k_2},\tag{8.4}$$

with $1.5 < k_1 + k_2 < 2.5$.

Kennicutt suggested also an alternative law to the Schmidt-like one discussed above, in particular a law containing the angular rotation speed of gas, Ω_{gas}:

$$SFR = 0.017 \Omega_{gas} \sigma_{gas} \propto R^{-1} \sigma_{gas}\tag{8.5}$$

A similar law for the SFR taking into account star formation induced by spiral density waves was proposed by Wyse & Silk (1989), and it can be expressed as (Prantzos 2003):

$$SFR = \nu V(R) R^{-1} \sigma_{gas}^{1.5},\tag{8.6}$$

where $V(R)$ is the rotational velocity in the disk and R is the galactocentric distance. It is worth noting that the SFRs expressed by Equations (8.4)–(8.6) contain a stronger dependence on the radial properties of the disk than the simple Schmidt law, and this characteristic is required to best fit the disk properties (see next sections).

8.3   The IMF

The most common parameterization for the IMF is that proposed by Salpeter (1955), which assumes a one-slope power law with $x = 1.35$, such that

$$\varphi(M) = cM^{-(1+x)} \tag{8.7}$$

is the number of stars with masses in the interval M to $M + dM$, and c is a normalization constant.

The IMF is generally normalized as

$$\int_0^\infty M\varphi(M)dM = 1. \tag{8.8}$$

More recently, multi-slope (x_1, x_2,...) expressions of the IMF have been adopted since they better describe the luminosity function of the main sequence stars in the solar vicinity (Scalo 1986, 1998; Kroupa, Tout, & Gilmore 1993). Generally, the IMF is assumed to be constant in space and time, with some exceptions such as the one suggested by Larson (1998), who adopts a variable slope:

$$x = 1.35(1 + m/m_1)^{-1}, \tag{8.9}$$

where m_1 is variable with time and associated with the Jeans mass. The effects of a variable IMF on the Galactic disk properties have been studied by Chiappini, Matteucci, & Padoan (2000), who concluded that only a very *ad hoc* variation of the IMF can reproduce the majority of observational constraints, thus favoring chemical evolution models with an IMF constant in space and time.

8.4   Infall and Outflow

Depending on the galactic system, the infall rate can be assumed to be constant in space and time, or, more realistically, the infall rate can be variable in space and time:

$$IR = A(R)e^{-t/\tau(R)}, \tag{8.10}$$

with $\tau(R)$ constant or varying along the disk. The parameter $A(R)$ is derived by fitting the present-day total surface mass density in the disk of the Galaxy, $\sigma_{tot}(t_G)$. Otherwise, for the formation of the Galaxy one can assume two independent episodes of infall during which the halo and perhaps part of the thick disk formed first, followed by the thin disk. As in the two-infall model of Chiappini, Matteucci, & Gratton (1997),

$$IR = A(R)e^{-t/\tau_H(R)} + B(R)e^{-(t-t_{max})/\tau_D(R)} \quad . \tag{8.11}$$

Here $\tau_H(R)$ is the time scale for the formation of the halo/thick disk and $\tau_D(R)$ is the time scale for the formation of the thin disk. The latter is assumed to increase linearly with galactocentric distance (Matteucci & François 1989; Chiappini et al. 1997; Boissier & Prantzos 1999). There are no specific prescriptions for the rate of gas outflow or galactic wind, but generally one simply assumes that the wind rate is proportional to the star formation rate through a suitable parameterization (Hartwick 1976; Matteucci & Chiosi 1983):

$$W = -\lambda SFR, \tag{8.12}$$

with λ being a free parameter.

8.5   Stellar Yields

The stellar yields are fundamental ingredients in galactic chemical evolution. In order to introduce the stellar yields, we recall some useful concepts and define the *yield per stellar generation.*

Under the assumption of "instantaneous recycling approximation," we define the yield per stellar generation of a given element i as (Tinsley 1980):

$$y_i = \frac{1}{1-R} \int_{m_1}^{\infty} m p_{im} \varphi(m) dm \tag{8.13}$$

where p_{im} is the stellar yield of the element i, namely the newly formed and ejected mass fraction of the element i by a star of initial mass m, and m_1 is the turn-off mass of the stellar generation. It is worth noting that the expression (8.13) is an oversimplification, and it does not have much meaning when considering chemical elements formed on long time scales. The quantity R is the returned fraction:

$$R = \int_{m_1}^{\infty} (m - M_{rem}) \varphi(m) dm, \tag{8.14}$$

where M_{rem} is the remnant mass of a star of mass m. In the past 10 years a large number of calculations of the stellar yields have become available for stars of all masses and metallicities. However, uncertainties in stellar yields are still present, especially in the yields of Fe-peak elements. This is due to uncertainties in the nuclear reaction rates, treatment of convection, mass cut, explosion energies, neutron fluxes and possible fall-back of matter on the proto-neutron star. Moreover, the ^{14}N nucleosynthesis and its primary and/or secondary nature are still under debate. The most recent calculations are summarized below.

- Low- and intermediate-mass stars ($0.8 \leq M/M_\odot \leq 8.0$) produce ^{4}He, C, N and some *s*-process ($A > 90$) elements (Marigo, Bressan, & Chiosi 1996; van den Hoeck & Groenewegen 1997; Meynet & Maeder 2002; Siess, Livio, & Lattanzio 2002; Ventura, D'Antona, & Mazzitelli 2002).
- Massive stars ($M \geq 10 M_\odot$) produce mainly α-elements (O, Ne, Mg, Si, S, Ca), some Fe-peak elements, *s*-process elements ($A < 90$), and *r*-process elements (Langer & Henkel 1995; Woosley & Weaver 1995; Thielemann, Nomoto, & Hashimoto 1996; Nomoto et al. 1997; Rauscher, Hoffman, & Woosley 2002; Limongi & Chieffi 2003).
- SNe Ia produce mainly Fe-peak elements (Nomoto et al. 1997; Iwamoto et al. 1999).
- Very massive objects ($M > 100 M_\odot$), if they exist, should produce mostly oxygen (Portinari, Chiosi, & Bressan 1998; Nakamura et al. 2001; Umeda & Nomoto 2002).

In Figures 8.1 and 8.2 we show a comparison between the yields, produced in massive stars, of two α-elements (O and Mg) and Fe, respectively, as predicted by different authors. As is evident from Figure 8.1, the O yields from different sources seem to be in good agreement with one another. The yields of Mg, especially the most recent ones, seem to be also in good agreement with each others, whereas for the yields of Fe (Fig. 8.2) there is not yet an agreement among different authors. Observational estimates of Fe produced in Type II supernovae (SNe II) (Elmhamdi, Chugai, & Danziger 2003) can help in constraining the Fe yields in massive stars.

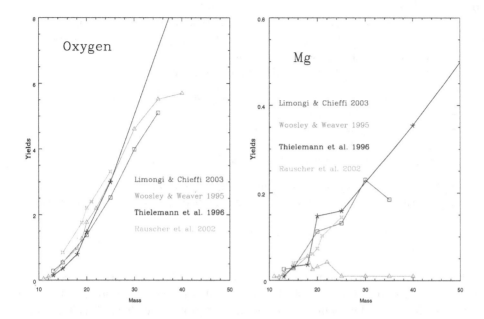

Fig. 8.1. Comparison between different yields of O and Mg from SNe II: open triangles, Woosley & Weaver (1995); open squares, Limongi & Chieffi (2003); stars, Thielemann et al. (1996); four-point stars, Rauscher et al. (2002).

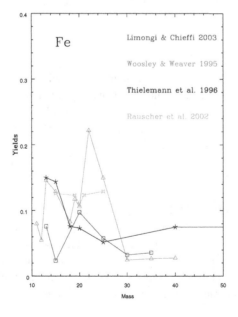

Fig. 8.2. Comparison between different yields of Fe from SN II: open triangles, Woosley & Weaver (1995); open squares, Limongi & Chieffi (2003); stars, Thielemann et al. (1996); four-point stars, Rauscher et al. (2002).

8.6   Analytical Models of Chemical Evolution

The simplest model of galactic chemical evolution is the so-called *Simple Model* for the evolution of the solar neighborhood. We define the solar neighborhood or solar vicinity as a cylinder centered on the Sun with 1 kpc radius.

The basic assumptions of the Simple Model can be summarized as follows:

- The system is one-zone and closed, no inflows or outflows
- The initial gas is primordial (no metals)
- The instantaneous recycling approximation holds
- $\varphi(m)$ is constant in time and space
- The gas is well mixed at any time (instantaneous mixing approximation)

Let X_i be the abundance of an element i and $\beta = \frac{M_{gas}}{M_{tot}}$ be the ratio between the mass of gas and the total mass of the system. If $X_i \ll 1$, which is generally true for metals, then we can write:

$$X_i = y_i \ln(\frac{1}{\beta}),\qquad(8.15)$$

which is the well-known solution for the Simple Model, where y_i is the yield per stellar generation as defined in Equation (8.13). In particular, the yield appearing in Equation (8.15) is usually referred to as the *effective yield*. If X_i is not much lower than 1, as is the case for X_{He} (Maeder 1992), a more precise expression for the solution of the Simple Model is given by

$$X_i = 1 - \beta^{y_i}\quad.\qquad(8.16)$$

It is worth noting that in the instantaneous recycling approximation we can generally assume

$$\frac{X_i}{X_j} = \frac{y_i}{y_j},\qquad(8.17)$$

namely that the abundance ratios are equivalent to the yield ratios, and this holds also in analytical models with infall and/or outflow. Clearly, Equation (8.17) can be used safely only for elements produced on short time scales, such as α-elements, but it fails if applied to elements produced on long time scales, such as Fe and N. These days the Simple Model is rarely used in describing the chemical evolution of the Milky Way since it does not reproduce the G-dwarf metallicity distribution (the G-dwarf problem) as well as the elements produced on long time scales such as Fe (Matteucci 2001).

8.7   Numerical Models of Chemical Evolution

In the last few years, a great number of models relaxing the instantaneous recycling approximation and the closed box assumptions, but retaining the constancy of the IMF and the instantaneous mixing approximation, have appeared in the literature. As an example of the basic equations adopted in such models, I show the formulation of Matteucci & Greggio (1986), which is based on the original formulations of Talbot & Arnett (1971) and Chiosi (1980).

Letting G_i be the mass fraction of gas in the form of an element i $[\sigma_i/\sigma_{tot}(t_G)]$, we can write

$$\dot{G}_i(t) = -\psi(t)X_i(t) + \int_{M_L}^{M_{Bm}} [\psi(t-\tau_m)Q_{m1i}X_i(t-\tau_{m2})d\mu]dm$$

$$+A\int_{M_{Bm}}^{M_{BM}} \phi(m) \cdot [\int_{\mu_{min}}^{0.5} f(\mu)\psi(t-\tau_{m2})Q_{mi}X_i(t-\tau_{m2})d\mu dm$$

$$+(1-A)\int_{M_{Bm}}^{M_{BM}} \psi(t-\tau_m)Q_{mi}X_i(t-\tau_m)\phi(m)dm$$

$$+\int_{M_{BM}}^{M_U} \psi(t-\tau_m)Q_{mi}X_i(t-\tau_m)\phi(m)dm + X_{A_i}IR(t) - X_iW(t), \tag{8.18}$$

where A is a constant parameter chosen in order to fit the present-day SN Ia rate and it lies in the range $0.05 < A < 0.09$. The SN Ia rate should be based on the existing theories on SN Ia progenitors, which I will summarize in the next section. In the equations above the Type Ia SNe (SNe Ia) are assumed to originate from C-O white dwarfs (WDs) in binary systems, and $f(\mu)$ represents the distribution of mass ratios in such binary systems ($\mu = \frac{m_2}{(m_1+m_2)}$). The quantities m_1 and m_2 are the primary and the secondary mass of the binary system, respectively. The primary star of the system is the exploding WD, which contributes to the nucleosynthesis products of the SN Ia explosion, whereas the secondary star is the clock of the system. In other words, the SN Ia occurs when the secondary evolves off the main sequence and transfers material on the companion. Therefore, τ_{m2} is the lifetime of the secondary star and the clock of the explosion. The integral that calculates the material restored by SNe Ia is made over the range of total masses of the binary systems that are supposed to give rise these SNe. In particular, $M_{Bm} = 3M_\odot$ is the minimum mass for such a binary system and $M_{BM} = 16M_\odot$ is the maximum (see Matteucci & Greggio 1986 and Matteucci & Recchi 2001 for details). The masses M_L and M_U represent the minimum mass contributing to galactic chemical enrichment at a given time and the maximum assumed mass for stars, respectively. The quantities Q_{mi} and Q_{m1i} are production matrices that take into account all the nucleosynthesis products ejected by a star of mass m and m_1, respectively (see Matteucci 2001 for details).

The star formation rate $\psi(t)$ and all the rates in this equation are expressed as functions of the fraction of gas [$G = \frac{\sigma_{gas}}{\sigma_{tot}(t_G)}$]. The quantities $IR(t)$ and $W(t)$ represent the rate of gas accretion and the rate of galactic outflow (wind), respectively, and X_{A_i} are the abundances of the accreting material, which are usually assumed to be primordial (no metals). Generally, in describing the solar neighborhood and the Galactic disk one assumes $W(t) = 0$, whereas the infall of mostly primordial material is most likely to be responsible for the formation of the disk.

8.7.1 SN Ia Progenitors

The "single degenerate" scenario is the classical scenario originally proposed by Whelan & Iben (1973), namely C-deflagration in a C-O WD reaching the Chandrasekhar mass limit, $M_{Ch} \approx 1.44M_\odot$, after accreting material from a red giant companion. The progenitors of C-O WDs lie in the range $0.8-8.0M_\odot$; therefore, the most massive binary system of two C-O WDs is the $8M_\odot+8M_\odot$ one. The clock of the system in this scenario is provided by the lifetime of the secondary star (i.e. the less massive one in the binary system).This implies that the minimum time scale for the appearance of the first SNe Ia is the lifetime of the most massive secondary star. In this case the time is $t_{SNIa_{min}} = 0.03$ Gyr (Greggio & Renzini 1983a; Matteucci & Greggio 1986; Matteucci & Recchi 2001).

The "double degenerate" scenario consists in the merging of two C-O WDs due to loss of angular momentum occurring as a consequence of gravitational wave radiation, which then explode by C-deflagration when M_{Ch} is reached (Iben & Tutukov 1984). In this case the minimum time scale for the appearance of SNe Ia is the lifetime of the most massive secondary star plus the gravitational time delay, which depends on the original separation of the two WDs and which is computed according to Landau & Lifshitz (1962), namely $t_{SNIa_{min}} = 0.03 + \Delta t_{grav} = 0.03 + 0.15$ Gyr (Tornambè & Matteucci 1986). We recall that the gravitational time delay can be as long as several Hubble times.

The model by Hachisu, Kato, & Nomoto (1999) is based on the classical scenario of Whelan & Iben (1973), but with a metallicity effect implying that no SNe Ia systems can form for [Fe/H] < −1.0 in the ISM. This implies a much longer time delay for the first SNe Ia to occur, also because the maximum masses of the secondary stars in the binary systems are assumed to be $\leq 2.6 M_\odot$. Therefore, in this case $t_{SNIa_{min}} = 0.33$ Gyr + metallicity delay due to the chemical evolution of the considered system. In this scenario, SNe Ia are not associated with young stellar populations, contrary to what is suggested by a recent search for SNe Ia in starburst galaxies (see Mannucci et al. 2003), which seem to favor the scenario where the progenitors of SNe Ia can be as high as 8 $M_\odot$.

Clearly, $t_{SNIa_{min}}$ is a very important parameter for computing galactic chemical evolution since the abundance patterns will depend strongly on this time scale, although the most important time scale is the one for which the SNe Ia have an impact on the abundances of the ISM. In successful models of galactic chemical evolution of the solar vicinity this time scale is 1.0–1.5 Gyr in the framework of the single degenerate model (Matteucci & Greggio 1986). In the other two scenarios this time is longer, especially in the one including the metallicity effect and it varies with different histories of star formation (see Matteucci & Recchi 2001 and § 8.13).

8.8   Different Approaches to the Formation and Evolution of the Galaxy

In the past years, several approaches to the calculation of the chemical evolution of the Galaxy were proposed.

(1) *Serial formation:* The halo, thick disk, and thin disk form in a sequence as a continuous process (e.g., Matteucci & François 1989).

(2) *Parallel formation:* The various Galactic components start forming at the same time and from the same gas but evolve at different rates. This approach predicts overlapping of stars belonging to the different components (e.g., Pardi, Ferrini, & Matteucci 1995), but it does not provide a simultaneously good fit to the G-dwarf metallicity distribution and the halo star metallicity distribution, as discussed in Matteucci (2001).

(3) *The two-infall approach:* The halo and disk form out of two separate infall episodes from extragalactic gas. Also in this case we predict an overlap in metallicity between the different Galactic components (e.g., Chiappini et al. 1997; Chang et al. 1999). A threshold density $(\sigma_{th} = 7 M_\odot$ pc$^{-2})$ in the star formation process is also included in the model of Chiappini et al. (1997).

(4) *The stochastic approach:* The assumption is made that in the early halo phases mixing was not efficient and pollution from single SNe would dominate the Galactic enrichment (Tsujimoto, Shigeyama, & Yoshii 1999; Argast et al. 2000; Oey 2000). In this case a

large spread in the abundances and abundance ratios at low metallicities is predicted. This predicted spread, however, is much larger than observed, especially for α-elements.

8.9   Observational Constraints

A good model of chemical evolution should be able to reproduce a maximum number of observational constraints, and the number of observational constraints should be larger than the number of free parameters, which are τ_H, τ_D, k_1, k_2, ν, the IMF slope(s), and the parameter describing the wind, if adopted.

The main observational constraints in the solar vicinity that a good model should reproduce (see Boissier & Prantzos 1999; Chiappini, Matteucci, & Romano 2001) are:

- The present-day surface gas density: $\Sigma_G = 13 \pm 3 M_\odot \, pc^{-2}$
- The present-day surface star density: $\Sigma_* = 43 \pm 5 M_\odot \, pc^{-2}$
- The present-day total surface mass density: $\Sigma_{tot} = 51 \pm 6 M_\odot \, pc^{-2}$
- The present-day SFR: $\psi_o = 2 - 5 M_\odot \, pc^{-2} \, Gyr^{-1}$
- The present-day infall rate: $0.3 - 1.5 M_\odot \, pc^{-2} \, Gyr^{-1}$
- The present-day mass function
- The solar abundances, namely the chemical abundances of the ISM at the time of birth of the solar system 4.5 Gyr ago and the present-day abundances
- The observed [X_i/Fe] vs. [Fe/H] relations
- The G-dwarf metallicity distribution
- The age-metallicity relation

And finally, a good model of chemical evolution of the Milky Way should reproduce the distributions of abundances, gas and star formation rate along the disk, as well as the average SN II and SN Ia rates along the disk [$1.2 \pm 0.8 \, (100 \, yr)^{-1}$ for SNe II and $0.3 \pm 0.2 \, (100 \, yr)^{-1}$ for SNe Ia].

8.10   Time-delay Model Interpretation

The difference in the time scales for the occurrence of SNe II and SNe Ia produces a time delay in the Fe production relative to the α-elements (Tinsley 1979; Greggio & Renzini 1983b; Matteucci 1986). On this basis we can interpret all the observed abundance ratios plotted as functions of metallicity. In particular, this interpretation is known as the time-delay model and can be easily illustrated by Figure 8.3, where we show the predictions of the two-infall model for the chemical evolution of the solar vicinity concerning the [O/Fe] vs. [Fe/H] relation. We show the standard case in which the contributions to Fe enrichment from both SNe II and SNe Ia are taken into account, as well as the cases where only one type of SNe at a time is assumed to contribute to Fe enrichment. Both data and models are normalized to the meteoritic solar abundances (Anders & Grevesse 1989). From Figure 8.3, it is evident that only SNe Ia as Fe producers would predict a continuous decrease of the [O/Fe] ratio from low to high metallicities (upper curve), whereas only SNe II would create a roughly constant [O/Fe] ratio. Therefore, only a model including both contributions in the percentages of 30% (SNe II) and 70% (SNe Ia) can reproduce the data.

The time-delay model, with the assumption of an IMF constant in time, can explain the different abundance patterns in the halo, disk, and bulge (see Matteucci 2001).

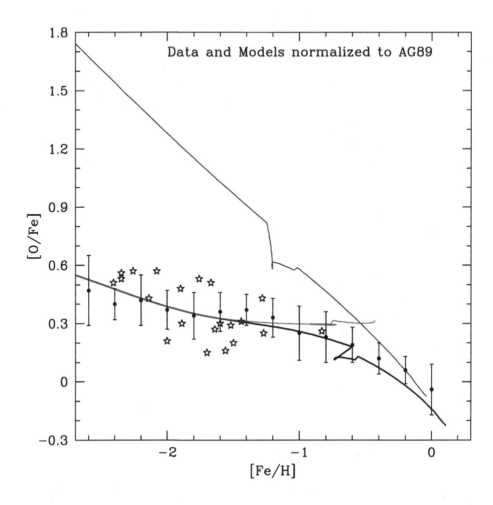

Fig. 8.3. Time-delay model. The models and the data are normalized to the solar meteoritic abundances of Anders & Grevesse (1989). The thick curve represents the predictions of the standard time-delay model where SNe Ia produce 70% of Fe and SNe II produce the remaining 30%. The figure is from Matteucci & Chiappini (2004). The data are all from Meléndez & Barbuy (2002).

8.11   Common Conclusions from Galaxy Models

Most of the more recent chemical evolution models agree on several important issues.

(1) The G-dwarf metallicity distribution can be reproduced only by assuming that the formation of the local disk occurred by infall of extragalactic gas on a long time scale, of the order of $\tau_D \sim 6-8$ Gyr (Chiappini et al. 1997; Boissier & Prantzos 1999; Chang et al. 1999; Alibès, Labay, & Canal 2001; Chiappini et al. 2001). Previous authors had already suggested a time scale of a few (1–3) Gyr for the formation of the Galactic disk (among others, Larson 1969; Lynden-Bell 1975; Yoshii & Saio 1979; Pagel 1989).

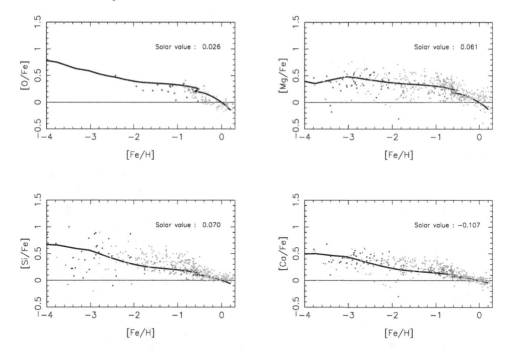

Fig. 8.4. Predicted and observed [α/Fe] vs. [Fe/H] in the solar neighborhood. The models and the data are from François et al. (2004). The models are normalized to the predicted solar abundances. The predicted abundance ratios at the time of the formation of the Sun are shown in each panel and indicate a good fit.

(2) The relative abundance ratios $[X_i/\text{Fe}]$ vs. [Fe/H], interpreted as being due to time delay between SNe Ia and SNe II, allow one to reproduce the observed relations (see Fig. 8.4) and to infer the time scale for the halo-thick disk formation (corresponding to [Fe/H] = −1.0), which should be of the order of $\tau_H \approx 1.5$ Gyr (Matteucci & Greggio 1986; Matteucci & François 1989; Chiappini et al. 1997). On the other hand, the external halo formed more slowly, perhaps on time scales of the order of 3–4 Gyr (see Matteucci & François 1992). In Figure 8.4 we show the predictions of the model developed by Chiappini et al. (1997), by adopting the yields of Woosley & Weaver (1995) for SNe II* and those of Nomoto et al. (1997) for SNe Ia (their case W7). The Mg yield in SNe Ia also had to be increased in order to fit the solar abundances (François et al. 2004). The problem of Mg underproduction in nucleosynthesis models is well known, as was pointed out by Thomas, Greggio, & Bender (1998). Figure 8.4 shows clearly that the agreement between the model predictions and data for α-elements is quite good and support the time-delay model.

(3) To fit abundance gradients and SFR and gas distributions along the disk, the disk should have formed inside-out (variable τ_D), and the SFR should be a strongly varying function of the galactocentric distance, as in Equations (8.4)–(8.6) (Matteucci & François 1989; Chiappini et al. 1997, 2001; Portinari & Chiosi 1999; Goswami & Prantzos 2000; Alibés et al. 2001).

---

* With the exception of the Mg yields, which have been artificially increased by a factor of 5 to obtain a good agreement with the solar abundance of Mg.

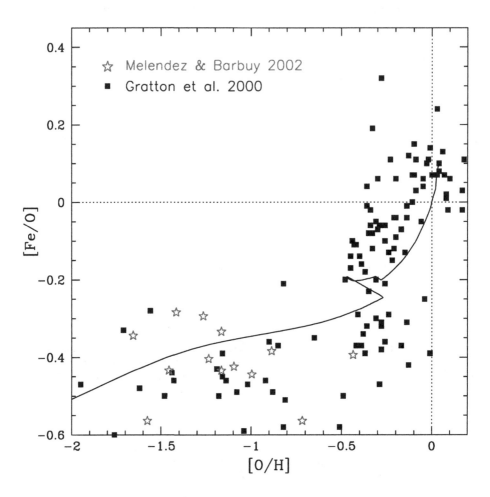

Fig. 8.5. Gap in the SFR. The model predictions for [Fe/O] vs. [Fe/H] are obtained by means of the two-infall model (see text). As one can see, a gap seems to be evident at [O/H] ≈ −0.3.

8.12   Specific Conclusions from Galaxy Models

The assumed threshold in the SFR (Chiappini et al. 1997) produces naturally a gap in the star formation process between the end of the halo-thick disk phase and the beginning of the thin disk phase. Such a gap lasts for ∼1 Gyr, and it seems to be indicated by the observations (Gratton et al. 2000). In particular, in Figure 8.5 we show the observed and predicted [Fe/O] vs. [O/H], where it appears that at around [O/H]= −0.3 there is a lack of stars and then the [Fe/O] ratio rises sharply. This is a clear indication, on the basis of the time-delay model, that there has been a period when only Fe was produced, in other words a halt in the SFR. This gap seems to be observed also in the [Fe/Mg] vs. [Mg/H] (Furhmann 1998). However, more data are necessary before drawing firm conclusions on this important point (see also Chiappini & Matteucci 2004).

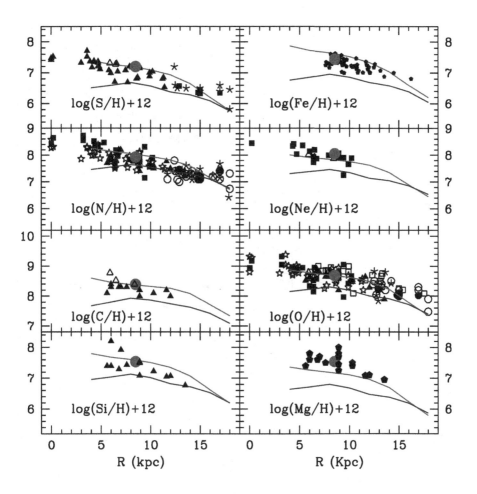

Fig. 8.6. Abundance gradients along the Galactic disk and their evolution in time, as predicted by the model of Chiappini et al. (2001), where all the references to the data can be found. The lower curve of each panel represents the model prediction after 2 Gyr from the beginning of star formation in the thin disk. The upper curve represent the predicted gradient at the present time.

No agreement on the behavior of gradients in time exists among different authors. In particular, some authors (e.g., Boissier & Prantzos 1999; Portinari & Chiosi 1999; Alibès et al. 2001) find a flattening of abundance gradients with time, whereas others (Matteucci & François 1989; Chiappini et al. 2001; see Fig. 8.6) predict a steepening of the abundance gradients, in agreement with results from chemo-dynamical models (Samland, Hensler, & Theis 1997). The difference between the two different approaches assumed by the various authors could perhaps reside in the different star formation and/or infall laws adopted for the Galactic disk (see Tosi 2000 for a discussion of this point). Data from planetary nebulae of different ages can help in solving this problem; recently, Maciel, Costa, & Uchida (2002)

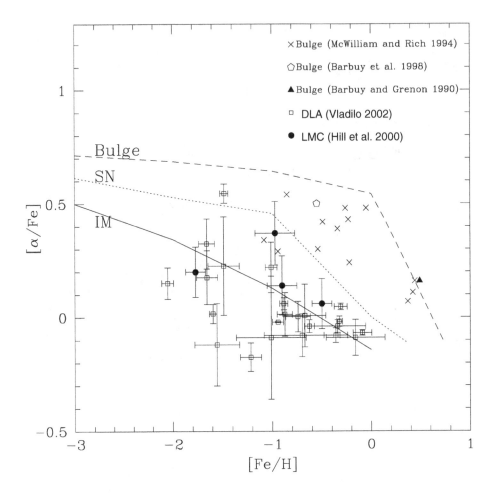

Fig. 8.7. The predicted [α/Fe] vs. [Fe/H] in different objects. Data for the Galactic bulge, the LMC, and damped Lyα systems are shown for comparison.

suggested a flattening of the gradients with time. In any case, all the models predict a very small evolution in the last 5 Gyr.

8.13   Star Formation Rate in Galaxies

It is very important to understand that different histories of star formation determine the abundance patterns in galaxies. In particular, the $[X_i/\text{Fe}]$ vs. [Fe/H] relations are strongly influenced by the SFR, which influences the temporal growth of the Fe abundance.

In other words, the time-delay model coupled with different star formation histories implies different time scales for the bulk of Fe production from SNe Ia, thus producing different [α/Fe] vs. [Fe/H] relations in different objects. The typical time scale for SN Ia enrichment can be defined as the time when the SN Ia rate reaches the maximum t_{SNIa} (Matteucci & Recchi 2001). This time scale depends upon the progenitor lifetimes, the IMF, and the SFR.

For an elliptical galaxy or a bulge of a spiral with high SFR, the maximum in the SN Ia rate is reached at $t_{SNIa} = 0.3-0.5$ Gyr, whereas for a spiral like the Milky Way a first maximum in the SN Ia rate is reached at $t_{SNIa} = 1.5$ Gyr and a second maximum, if one uses the two-infall model, at $t_{SNIa} = 4-5$ Gyr. Therefore, for the Milky Way 1–1.5 Gyr is the time at which SNe Ia are no longer negligible in the process of chemical enrichment, and it corresponds to the change in slope observed in the [el/Fe] vs. [Fe/H] relations (Fig. 8.4) and to the end of the halo phase. For an irregular galaxy, where the SFR is assumed to proceed more slowly than in the solar vicinity, the time scale for SN Ia enrichment is $t_{SNIa} = 7-8$ Gyr. Therefore, it is worth repeating that t_{SNIa} is different in different galaxies, since often in the literature it is adopted as a universal time scale of 1 Gyr! On the basis of that, we expect that objects where the SFR proceeds very fast, such as in the spheroids (bulges and ellipticals), the [α/Fe] ratio stays flat for a larger metallicity interval than in systems with slower star formation, such as the Milky Way, and that eventually the [α/Fe] ratios in irregulars, where the star formation rate has been less efficient than in spirals, decreases almost continuously, as shown in Figure 8.7. The models shown in Figure 8.7 contain the same nucleosynthesis prescriptions and differ in the star formation history.

8.14   High-redshift Objects

In Figure 8.7 we show also some observational data relevant to the Galactic bulge, the LMC, and some damped Lyα systems. As one can see, our predictions seem to reproduce the data for the Galactic bulge and for LMC. They also suggest that damped Lyα systems could be the progenitors of the present-day irregular galaxies, since most damped Lyα systems, once their abundances are corrected for the effect of dust, show low [α/Fe] ratios at low metallicities (Vladilo 2002; Calura, Matteucci, & Vladilo 2003). On the other hand, Lyman-break galaxies such as cB58 (Pettini et al. 2002) show an abundance pattern compatible with a young spheroid, either a galactic bulge or a small elliptical, experiencing a galactic outflow, as recently shown by Matteucci & Pipino (2002). Therefore, the [X_i/Fe] vs. [Fe/H] diagram represents a very useful tool to infer the nature of high-redshift objects, for which we know just the abundances and abundance ratios.

References

Alibés, A., Labay, J., & Canal, R. 2001, A&A, 370, 1103
Anders, E., & Grevesse, N. 1989, Geochim. Cosmochim. Acta, 53, 197
Argast, D., Samland, M., Gerhard, O. E., & Thielemann, F.-K. 2000, A&A, 356, 873
Boissier, S., & Prantzos, N. 1999, MNRAS, 307, 857
Calura, F., Matteucci, F., & Vladilo, G. 2003, MNRAS, 340, 59
Chang, R. X., Hou, J. L., Shu, C. G., & Fu, C. Q. 1999, A&A, 350, 38
Chiappini, C., & Matteucci, F. 2004, in Carnegie Observatories Astrophysics Series, Vol. 4: Origin and Evolution
    of the Elements, ed. A. McWilliam & M. Rausch (Pasadena: Carnegie Observatories,
    http://www.ociw.edu/symposia/series/symposium4/proceedings.html)
Chiappini, C., Matteucci, F., & Gratton, R. 1997, ApJ, 477, 765
Chiappini, C., Matteucci, F., & Padoan, P. 2000, ApJ, 528, 711
Chiappini, C., Matteucci, F., & Romano, D. 2001, ApJ, 554, 1044
Chiosi, C. 1980, A&A, 83, 206
Dopita, M. A., & Ryder, S. D. 1994, ApJ, 430, 163
Elmhamdi, A., Chugai, N. N., & Danziger, I. J. 2003, A&A, 404, 1077
François, P., Matteucci, F., Cayrel, R., Spite, M., Spite, F., & Chiappini, C. 2004, A&A, submitted
Fuhrmann, K. 1998, A&A, 338, 161
Goswami, A., & Prantzos, N. 2000, A&A, 359, 191
Gratton, R. G., Carretta, E., Matteucci, F., & Sneden, C. 2000, A&A, 358, 671

Greggio, L., & Renzini, A. 1983a, A&A, 118, 217

——. 1983b, Mem. Soc. Ast. It., 54, 311

Kennicutt, R. C., Jr. 1998, ApJ, 498, 541

Kroupa, P., Tout, C. A., & Gilmore, G. 1993, MNRAS, 262, 545

Hachisu, I., Kato, M., & Nomoto, K. 1999, ApJ, 522, 487

Hartwick, F. 1976, ApJ, 209, 418

Hill, V., François, P., Spite, M., Primas, F., & Spite, F. 2000, A&A, 364, L19

Iben, I., Jr., & Tutukov, A. V. 1984, ApJS, 54, 335

Iwamoto, K., Brachwitz, F., Nomoto, K., Kishimoto, N., Umeda, H., Hix, W. R., & Thielemann, F.-K. 1999, ApJS, 125, 439

Landau, L. D., & Lifshitz, E. M. 1962, Quantum Mechanics (London: Pergamon)

Langer, N., & Henkel, C. 1995, Space Sci. Rev., 74, 343

Larson, R. B. 1969, MNRAS, 145, 405

——. 1998, MNRAS, 301, 569

Limongi, M., & Chieffi, A., 2003, ApJ, 592, 404

Lynden-Bell, D. 1975, Vistas in Astronomy, 19, 299

Maciel, W., Costa, R. D. D., & Uchida, M. M. M. 2003, A&A397, 667

Maeder, A. 1992, A&A, 264, 105

Mannucci, F., et al. 2003, A&A, 401, 519

Marigo, P., Bressan, A., & Chiosi, C. 1996. A&A, 313, 545

Matteucci, F. 1986, ApJ, 305, L81

——. 2001, The Chemical Evolution of the Galaxy (Dordrecht: Kluwer)

Matteucci, F., & Chiappini, C. 2004, in preparation

Matteucci, F., & Chiosi, C. 1983, A&A, 123, 121

Matteucci, F., & François, P. 1992, A&A, 262, L1

——. 1989, MNRAS, 239, 885

Matteucci, F., & Greggio, L. 1986, A&A154, 279

Matteucci, F., & Pipino, A. 2002, ApJ, 569, L69

Matteucci, F., & Recchi, S., 2001, ApJ, 558, 351

McWilliam, A., & Rich, R. M. 1994, ApJS, 91, 7

Meléndez, J., & Barbuy, B. 2002, ApJ, 575, 474

Meynet, G., & Maeder, A. 2002, A&A, 390, 561

Nakamura, T., Umeda, H., Iwamoto, K., Nomoto, K., Hashimoto, M., Hix, W. R., & Thielemann, F.-K. 2001, ApJ, 555, 880

Nomoto, K., Iwamoto, K., Nakasato, N., Thielemann, F.-K., Brachwitz, F., Tsujimoto, T., Kubo, Y., & Kishimoto, N. 1997, Nuc. Phys. A, 621, 467

Oey, M. S. 2000, ApJ, 542, L25

Pagel, B. E. J. 1989, in Evolutionary Phenomena in Galaxies, ed. J. E. Beckman & B. E. J. Pagel (Cambridge: Cambridge Univ. Press), 201

Pardi, M. C., Ferrini, F., & Matteucci, F. 1995, ApJ, 444, 207

Pettini, M., Rix, S. A., Steidel, C. C., Adelberger, K. L., Hunt, M. P., & Shapley, A. E. 2002, ApJ, 569, 742

Portinari, L., & Chiosi, C. 1999, A&A, 350, 827

Portinari, L., Chiosi, C., & Bressan, A. 1998, A&A, 334, 505

Prantzos, N. 2003, in The Evolution of Galaxies III: From Simple Approaches to Self-Consistent Models, ed. G. Hensler et al. (Dordrecht: Kluwer), 381

Rauscher, T., Hoffman, R. D., & Woosley, S. E. 2002, ApJ, 576, 323

Salpeter, E. E. 1955, A&A, 121, 161

Samland, M., Hensler, G., & Theis, C. 1997, ApJ476, 544

Scalo, J. M. 1986, Fund. Cosmic Phys., 11, 1

——. 1998 in The Stellar Initial Mass Function, ed. G. Gilmore, I. Parry, & S. Ryan (San Francisco: ASP), 201

Schmidt, M. 1959, ApJ, 129, 243

Siess, L., Livio, M., & Lattanzio, J. 2002, ApJ, 570, 329

Talbot, R. J., & Arnett, D. W. 1971, ApJ, 170, 409

Thielemann, F.-K., Nomoto, K., & Hashimoto, M. 1996, ApJ, 460, 408

Tinsley, B. M. 1979, ApJ, 229, 1046

——. 1980, Fund. Cosmic Phys., 5, 287

Thomas, D., Greggio, L., & Bender, R. 1998, A&A, 296, 119

Tornambé, A., & Matteucci, F. 1986, MNRAS, 223, 69

Tosi, M. 1988 A&A197, 33

——. 2000, in The Chemical Evolution of the Milky Way: Stars versus Clusters, ed. F. Matteucci & F. Giovannelli (Dordrecht: Kluwer), 505

Tsujimoto, T., Shigeyama, T., & Yoshii, Y. 1999, ApJ, 519, L63

Umeda, H., & Nomoto, K. 2002, ApJ, 565, 385

van den Hoek, L. B., & Groenewegen, M. A. T. 1997 A&AS, 123, 305

Ventura, P., D'Antona, F., & Mazzitelli, I. 2002, A&A, 393, 21

Vladilo, G. 2002, A&A, 391, 407

Whelan, J., & Iben, I., Jr. 1973, ApJ, 186, 1007

Woosley, S. E., & Weaver, T. A. 1995, ApJS, 101, 181

Wyse, R. F. G., & Silk, J. 1989, ApJ, 339, 700

Yoshii, Y., & Saio, H. 1979, PASJ, 31, 339

9

Model atmospheres and stellar abundance analysis

BENGT GUSTAFSSON
Department of Astronomy and Space Physics, Uppsala University, Sweden

Abstract

Model atmospheres have now been used in the analysis of stellar abundances for more than 50 years. During this period, remarkable progress has been made in the understanding of the physics of stellar atmospheres and in their modeling. The advances made in the observation of stellar spectra are even more remarkable. The question addressed here is whether comparable progress also can be found in the accuracy of the resulting abundances. It seems that this is *not* the case to the extent one might have expected, and the reasons for this are discussed. A number of recent developments in model atmosphere construction and in basic atomic and molecular data may, however, suggest that we are now approaching a situation with significantly diminished systematic errors in stellar abundances.

9.1   Introduction: A Bird's-Eye Perspective with Bibliometrical Comments

Bengt Strömgren's (1940) classical paper "*On the Chemical Composition of the Solar Atmosphere*" introduced the model atmosphere technique in the analysis of stellar abundances (for some brief historical remarks, see, e.g., Aller 1960). However, it took some time before the method became common practice, in particular for late-type stars. Most probably, this reflected the complicated task of calculating model atmospheres and the uncertainties in observed equivalent widths and gf values, which then contributed significantly to the total errors in abundances. A simple search in the NASA ADS database shows that the number of analyses with models started increasing very considerably around 1970 (see Fig. 9.1). Obviously, the existence of "grids" of model atmospheres (Carbon & Gingerich 1969; Gustafsson et al. 1975; Kurucz 1979; etc.) considerably facilitated this development. Other equally important factors were new measurements and compilations of more accurate gf values (Blackwell et al. 1980; Wiese & Martin 1980, and references therein) and new efficient spectrometers with detector systems with linear response (Reticons, CCDs). Presently, on the order of 200 papers are published each year with abundance analyses based on model atmospheres and directly related topics.

The increase of papers found in the ADS with the words "model atmospheres" and "abundance" in the abstract of the paper is about 2% of all papers with "stars" in the abstract, and about 0.6% of the more than 700,000 papers in the database. These figures refer to the total number of papers published during the period 1948–2001, but we have found that the percentages per year have remained astonishingly constant since 1970, in spite of the huge increase of the total number of papers published annually. It is also interesting to see that the accumulated number of papers published since 1976 scales very well with the number

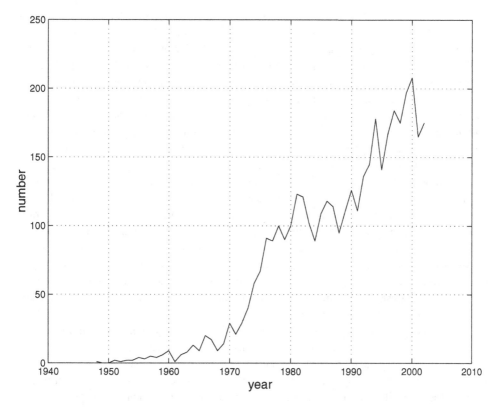

Fig. 9.1. Number of articles published per year with both "model atmosphere" and "abundance" in the text of the abstract, according to the NASA ADS database. The slower increase since 1980 may reflect the fact that the use of model atmospheres in abundance analysis had become standard by then, and hence why the word "model atmosphere" was not necessarily mentioned in the abstract.

of stars listed in the different editions of the Cayrel et al. catalog of field stars with [Fe/H] determinations (Cayrel de Strobel, Soubiran, & Ralite 2001, and references cited therein), which indicates that although many papers now each give abundances for many stars, this is compensated for by the fact that many papers contain analyses of stars for which some previous abundance determinations have already been made, maybe then only for iron or a few elements.

Another fact of some interest, at least when we have to deal with administrators and financing agencies, is the ratio of papers, again with "model atmosphere" and "abundance" in the abstract, relative to the total number of astronomers. If the latter is estimated by the number of individual members of the International Astronomical Union, we find a ratio that gradually increased from 0.8% in 1960 to 2.5% in 2000. In particular, there has been no tendency for the "productivity," measured in this way, to decrease during the latest decades. However, one should note that these crude measures neither give any indication of the very significant growth that has occurred in abundances published for new elements or for new types of stars, nor do they represent a great number of relevant papers without our key words in their abstract so that they were not caught in our crude search.

It is also interesting to see from which institutions the papers came. A simple comparison between affiliations of authors in 1970 and 2000 gave some clear indications. In 1970, more than 70% or the papers had authors from the USA. The others were mainly of French, German, British or Canadian origin. In 2000, the fraction of authors with US affiliation had diminished to 38%. For the rest, authors from the countries mentioned above still contributed significantly, but in addition to them astronomers from Spain, Sweden, Denmark, Japan and Australia, as well as a number of other countries, were active. A similar tendency is found within the USA, where astronomers from many institutions, not visible in the statistics for 1970, now contribute significantly, such as astronomers at Pennsylvania State University or University of Oklahoma. Also within the European countries this tendency is visible—e.g., in 1970, most of the papers were written by authors in Kiel and Hamburg. In 2000 authors from Berlin, Munich, Potsdam, Tübingen, and several other places also contributed. True enough, astronomy and astrophysics in general has been spread to an increasing number of places during recent decades. Still, I think it is a safe conclusion that the practice of abundance analyses has even spread more in the astronomical community during the last 30 years than many other specialties. It has become common practice; standard analyses do not require very specialized knowledge, due to generally available computer software for spectrum analysis, model atmosphere calculation and the derivation of abundances, and much easier access to suitable spectrometers.

This superficial overview, based on the ADS database, gives a rather healthy impression of a field in rapid expansion, which has now provided general and well tested tools for the study of stellar evolution, galactic history, and nucleosynthesis to the astronomical community. In what follows, we shall study whether this rosy picture is correct. We shall argue that further work is necessary if one wants to reach a level of accuracy in abundance analyses that in a reasonable way matches the development in observation technology and in the physical data on which the analysis is based.

9.2   Enormous Improvements—Better Abundances?

The techniques of abundance analysis of stellar spectra have developed immensely during the last 30 years. Thus, the largest telescopes provide more than a factor of 3 greater light-collecting area, the spectrometers with detectors are more than a factor of 50 more efficient, the volume and the quality of the required atomic and molecular data have improved by typically 1 or 2 orders of magnitude, the computers are more than a factor of 10^3 more rapid, the algorithms for some important types of calculations are more than a factor of 100 faster. One may also claim that the models are, as a result of the improvement in physical data and computing, more self-consistent.

The most obvious and most impressive of these developments is in spectrometer and detector technology, which has now made it possible to obtain high-resolution, high-S/N spectra for hundreds of thousands of stars, of a quality that was only obtained for the Sun and perhaps Arcturus and a few other stars 30 years ago. We shall here give some more detailed examples of further recent improvements, and then end the section with the question of whether these have contributed to significantly higher accuracy in abundances.

9.2.1   *Examples of Improved Physics Data*

Important new identifications and measures of wavelengths of spectral lines have been contributed in recent years. A good example is the laboratory work by Nave et al.

(1994) and Nave & Johansson (1995), where 28 new energy levels were identified and 818 were revised, leading to about 4000 new identifications of Fe I lines, many in the near-ultraviolet and infrared and suitable as diagnostics in stellar spectroscopy. Wavelengths and energy levels were measured with high accuracy.

Line lists with f-values for atomic lines from the Opacity Project (Seaton et al. 1994) in the TOP database (Cunto et al. 1993) with more than one million lines mark a very considerable step forward in accuracy and consistency. The lists by Kurucz (1993), though generally less accurate, also mark a very important contribution, containing 25 times as many lines, including lines from very complex atoms. Other useful lists of line data, such as VALD (Kupka et al. 1999), NIST (Reader et al. 2002), and CHIANTI (Dere et al. 2001) are compiled according to different principles and contribute data for a variety of purposes.

The TOP database also contains photoionization data from the Opacity Project. The XS-TAR database (Bautista & Kallman 2001) gives such data for iron from the Iron Project (Bautista 1997). These data are significant for the calculation of model atmospheres since bound-free opacities from elements like C, Mg, Al, Si, and Fe play significant roles in the ultraviolet spectral regions. Moreover, these cross-sections are important for calculation of the statistical equilibrium; photoionization affects the populations of different states in the atoms, and thus also the spectral lines.

For statistical equilibrium calculations, the cross-sections for atomic impact collisions with electrons and hydrogen atoms are still a major problem. Electron collision data may be found, for example, in the XSTAR database, but there is a general lack of reliable data for complex atoms and excited states. The hydrogen impact collision excitation and ionization rates for Li have recently been explored by Barklem & Belyaev (2004) and found to be significantly lower than the standard modification of Drawin's recipe.

Also concerning line broadening very significant steps forward have been taken in recent years. In particular, reliable damping data for collisions with hydrogen atoms have been calculated for many spectral lines by Anstee, Barklem, O'Mara, and Piskunov (see Barklem, Piskunov, & O'Mara 2000, and references cited therein). For Stark broadening, extensive work has been done by Dimitrijevic, Sahal-Brechot, and collaborators (see Dimitrijevic et al. 2003 for references). Improved calculations of Stark broadening data for hydrogen lines have been published by Stehlé & Hutcheon (1999).

Data for molecular lines are very significant for analyses of cool stars, both directly for the calculation based on a given model atmosphere of diagnostics in a spectral region, and indirectly for calculation of the structure of the model atmosphere. For the latter purpose global data are needed—the molecular opacity in general may be caused by millions of weak contributing lines. Also for the spectrum calculation the effects of molecular absorption, for example depressing the general spectrum around a particular diagnostics of interest, may well be due to a very great number of weak lines. Major contributions in calculating extensive molecular line lists have been made by U. G. Jørgensen and collaborators (1989, 1990, 1994, 1996, 2001) for CH, TiO, H_2O, C_3, HCN, and, more approximately, C_2H_2. Recently lists for CH_4 have been calculated (Borysow et al. 2004). Partridge & Schwenke (1997) have also contributed a line list for the vibration-rotation bands of H_2O with altogether 3×10^8 lines, and Schwenke (1998) and Plez (1998) produced almost as extensive lists for the numerous electronic bands of TiO. The situation for molecular data is thus much improved and has enabled detailed modeling of cool red giant atmospheres and spectra (e.g., Alvarez & Plez 1998; Allard, Hauschildt, & Schwenke 2000; Aringer, Kerschbaum, & Jørgensen

2002), but data for a number of important species, in particular polyatomic molecules, are still missing. Also, the existing data for several molecules, for example for water vapour in the near-infrared (see Allard et al. 2000; Aringer et al. 2002), are still not fully satisfactory.

9.2.2   Computers, Algorithms, Self-consistent Models

For his early calculations of line profiles with electronic computers, Heiser (1957) published some details on timing: the calculation of one single spectral line, represented by 15 frequency points and Voigt profiles in LTE, took 25 minutes on an IBM 650. With a modern work station we made the same calculation in 24 milliseconds, or about 60,000 times more rapidly. This agrees well with the famous Moore's law (Moore 1965), which suggests that the computer speed increases by a factor of 2 every 1–2 years.

However, in addition to this, the development of new algorithms for calculating radiative transfer in spectral lines and model atmospheres has been very important. A basic step forward was the development of the accelerated lambda iteration (ALI) methods for line transfer, by Scharmer (1981), Scharmer & Carlsson (1985), Werner & Husfeld (1985), Hamann (1986), and others, partly based on early work by Cannon and Rybicki. With these methods non-LTE problems may now be treated with thousands of atomic levels and transitions, as well as problems with velocity fields, partial redistribution, and in three space dimensions (cf. Auer, Fabiani Bendicho, & Trujillo Bueno 1994; Asplund, Carlsson, & Botnen 2003, and references therein).

The large number of contributing levels and spectral lines in realistic radiative transfer problems, not the least when model atmospheres are to be calculated, makes statistical representations necessary. Anderson (1989), Hubeny & Lanz (1995), and Hauschildt and collaborators (see Schweitzer et al. 2003, and references therein) have devised efficient schemes with "superlevels" and "superlines" for a consistent modeling of radiative transfer with non-LTE in heavy-blanketed atmospheres. For the calculation of the energy transfer in model atmospheres with complex spectra, a simple statistical method, called opacity sampling (OS), was developed by Peytremann (1974), Sneden, Johnson, & Krupp (1976), Ekberg, Eriksson, & Gustafsson (1986), and Helling & Jørgensen (1998). This is now used extensively for LTE model atmospheres; it is more flexible and easier to generalize to dynamical atmospheres and to non-LTE models than the previously used opacity distribution function (ODF) method. With the OS method and contemporary computers, LTE models with 10^5 frequency points, which is quite sufficient for a reliable representation of the radiative field through the model, may be calculated routinely.

A major step forward in the modeling of stellar atmospheres has been taken during the last two decades by Nordlund, Stein and collaborators (see Nordlund 1982; Stein & Nordlund 2000; Asplund et al. 2000a, and references therein), in calculating three-dimensional (3-D) models of convective atmospheres with consideration of radiation in some detail. Basically, the Navier-Stokes equations are solved in 3-D simultaneously with the equation of radiative transfer, assuming LTE (in most respects). The models tend to reproduce the observed stellar line shifts and line profiles with an astonishing precision (cf., e.g., Asplund et al. 2000b; Allende Prieto et al. 2002a), without relying on extra free fitting parameters like macro- and microturbulence, at least for solar-type stars. The models predict considerable temperature inhomogeneities in the stellar surface layers, which may severely affect the strengths of low-excitation atomic lines and molecular lines, and thus lead to strong revisions of abundances. Freytag is producing models of full stars—"a star in a box"—and concentrates on modeling red supergiants like Betelgeuse (Freytag 2003). His models are characterized by the appearance of very few huge "granulae," as proposed in a classical paper by Schwarzschild (1975). Gray (2001) has questioned the relevance of this result, on the argument that observed spectral line widths do not change very much with time. It remains to explore the observational implications of Freytag's models further, not the least for abundance analysis (see, however, Freytag, Steffen, & Dorch 2002).

Both for hot and cool stars with winds, very significant improvements have also been made. Spherically symmetric non-LTE models for hot stars with radiatively driven winds described by the hydrodynamic equations have been produced by the Munich group (Pauldrach, Hoffman, & Lennon 2001, and references therein). Models of hot stars, such as Wolf-Rayet stars and OB supergiants, with non-LTE radiation fields and with prescribed velocity fields have been calculated by several groups. These models often show good or satisfactory general agreement with observed spectra (see, e.g., the comparison between observed spectra of supergiants in the Magellanic Clouds and models published by Crowther et al. 2002), an agreement that suggests that fairly accurate abundances can be derived. For cool pulsating stars, the problems may be more severe, but a promising development is taking place. At least four groups are active in modeling atmospheres and winds of long-period variables [see, e.g., Wood 1979; Bessell, Scholz, & Wood 1996; Bowen 1988; Bowen & Willson 1991 (cf. also Willson 2000, and references therein); Fleischer, Gauger,& Sedlmayr 1992; Winters et al. 2000; Höfner & Dorfi 1997]. Höfner and collaborators (cf. Höfner et al. 1998; Höfner 1999; Andersen, Höfner, & Gautschy-Loidl 2003; Höfner et al. 2003; Sandin & Höfner 2003a, b) have produced self-consistent models with hydrodynamics, dust formation, and radiative transfer considered in detail. The spectra of these models clearly show much improved agreement with observed spectra, as compared with hydrostatic models or models with an oversimplified treatment of radiation. Also, the diagnostic power of emission lines of Mira stars are being explored in non-LTE calculations for dynamic models of the shocked outer atmospheres, with promising results (see Richter et al. 2003, and references therein).

9.2.3 Smaller Errors in Abundances?

Now, how are the improvements by orders of magnitude that have taken place in the underlying physics data and in observational data and computational resources reflected in the quality of the abundances derived? This may be studied by exploring how the abundance results and quoted errors have developed with time for certain stars. I have chosen two examples, the iron abundance of the archetypical extreme Population II star HD 140283, and the oxygen abundance of the Sun.

HD 140283 ($T_{eff} = 5700$ K, log $g = 3.7$, [Fe/H] $= -2.4$; Nissen et al. 2002) was one of the first stars analyzed—it is one of the subdwarfs studied by Chamberlain & Aller (1951) in their historic paper, laying the empirical ground for the view that nucleosynthesis is a stellar phenomenon. The first model atmosphere analysis of the star was made by Cohen & Strom (1968). The results of the different analyses of the star are presented in Figure 9.2, along with error bars when provided by the authors. We note that, in spite of the fact that the errors quoted have certainly diminished with time, the scatter between different determinations is still considerable and now significantly greater than the error bars given. It is questionable whether the scatter between different independent determinations has decreased during the last decades.

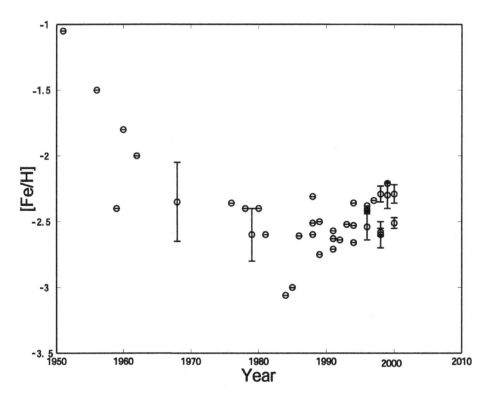

Fig. 9.2. The iron abundance determinations for HD 140283 as a function of time. Estimated errors are indicated when published.

There are several reasons for this remaining scatter. One is the uncertainty in the effective temperature scale—the temperatures given still range from 5500 K to 5800 K. Another is uncertainties in convection or convection parameters when the mixing-length theory is used. Effects of thermal inhomogeneities in the surface layers, caused by convection, and departures from LTE in the ionization equilibrium of iron are treated differently in the most recent studies and also contribute to the scatter.

Determinations of the solar oxygen abundance are displayed as a function of time in Figure 9.3. The first point here represents the determination in the classical study by Russell (1929). We see that an initially large spread has decreased, but during the last 30 years a considerable scatter has prevailed. There is a tendency now for error bars to shrink very considerably, and the estimated oxygen abundance itself has recently decreased significantly. The scatter reflects the use of different criteria, and of different model atmospheres. The most recent results are those of Allende Prieto, Lambert, & Asplund (2002b) and Asplund et al. (2003). In the latter paper different criteria—the forbidden [O I] $\lambda6300$ Å line, the [O I] infrared triplet lines, and the vibration-rotation as well as the rotation-rotation infrared lines of OH—are all used, and give concordant results for a 3-D convective model atmosphere. The only remaining deviating criterion is that of the ultraviolet OH (A-X) lines—possibly deviating due to the effects of unknown opacity sources.

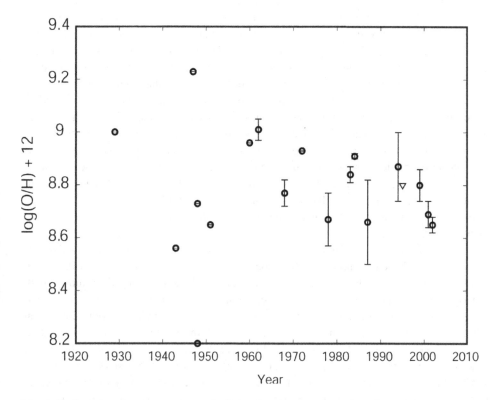

Fig. 9.3. Oxygen abundance determinations for the Sun, as a function of time. Estimated errors are indicated when published.

So, it seems that in spite of the general and considerable improvements in observations and modeling of stellar spectra, there is rather slow convergence in accuracy, as illustrated by these two examples. Against this conclusion one might advocate that the examples were chosen selectively—data for other stars might have shown more definitive convergence—or that the scatter admittedly has been great in the last two decades, but now, during the last two years, we are really improving (e.g., with 3-D models and non-LTE). In spite of the possibility that these explanations may be partly relevant, they certainly do not explain the full remaining scatter in the figures, nor the tendency for this scatter to stay roughly constant during recent decades.

Another possibility for the fairly constant scatter may be what I would call a "sheep-goat effect" (a term that I recently discovered is independently used, and with a different meaning, in parapsychology; cf. Lawrence 1993). The term "sheep effect" sometimes is used to explain the fact that different scientists tend to get rather similar results even if the systematic errors are severe, probably because they search for and eliminate systematic errors in their experiments and debug their computer programs only until they get the expected results. On the other hand, they may also welcome results that are not identical with those already published by others, in order to be able to contribute some new data, a behavior one may call a "goat effect" in its extreme form. The balance between these two, supposedly leading to a scatter in abundances on the order of 0.1–0.2 dex, is here called the "sheep-goat effect."

A less cynical and presumably more interesting explanation for the remaining and seemingly not very time-dependent scatter in the figures would be that severe systematic errors still remain in abundance determinations. Previous studies did not try to take these into consideration, so the earlier scatter reflects uncertainties in observed data, fundamental parameters, and standard model atmospheres, while some of the scatter that now shows up could be due to different attempts to handle convection, inhomogeneities, and non-LTE. In order to explore this possibility further, I studied error discussions in recent abundance analyses, and also approached a number of colleagues (many of them present at this Symposium) to find out of what systematic errors are considered most important, and what sizes these may have, in contemporary analyses. Here, the tentative result of this study will be sketched, without any detailed further references. It should be noted that the possible systematic errors due to non-LTE and inhomogeneities quantified below are still preliminary, and sometimes not much more than educated guesses.

In many papers, it is common to essentially limit the study of errors to those caused by uncertainties in fundamental parameters (T_{eff}, log g, [M/H], etc). Errors of typically 0.1 dex in abundances relative to hydrogen and relative to the solar abundances are often quoted, both for early- and late-type stars, as a result of such parameter uncertainties, although for hot supergiants and for cool M and C stars errors twice as large are sometimes given (Gustafsson 1989). Errors in measurement, in defining the continuum in crowded spectra, and in relating stellar abundances to that of the Sun (e.g., by using lines that are saturated in the star or in the Sun) may typically sum up to similar amounts, in particular for stars very different from the Sun. Errors due to non-LTE in an LTE treatment, or caused by the uncertainties in collision cross-sections, may typically be estimated to 0.1–0.2 dex, both for early-type main sequence stars and for solar-type stars. Finally, errors due to inhomogeneities and 3-D convection are typically 0.1–0.2 dex for solar-type stars, while they may be 2 times larger, or possibly even more for certain elements, for Population II stars of similar temperatures. All these estimates are rough; the true values depend on whether the line is molecular, atomic, or ionic, and what its excitation energy is. In general, strongly temperature-dependent lines imply larger errors. Certainly, the type of star also matters—the more different the star is from the well-explored solar-type stars, the more uncertain are the abundances.

Let us take one example, being somewhat more precise. For metal-poor solar-type stars, as well as Population II giants, one may typically estimate the effective temperature to an accuracy of about 100 K, and the logarithmic surface gravity to about 0.2 dex. These uncertainties introduce uncertainties in metal abundances of about 0.1 dex, where the effects of the temperature uncertainty dominate for lines of neutral atoms that are not in the dominating ionization stage, such as Fe I, Li I, Mg I, or Si I. For lines of dominant species, such as Fe II, Be II, Ba II, etc., the error in gravity leads to similar abundance errors. Errors in defining the continuum level are relatively small in these stars, although there may be a problem with the opacity in the ultraviolet—either for these stars or for the Sun (cf. Bell, Balachandran, & Bautista 2001, and references therein; however, see also Allende Prieto, Hubeny, & Lambert 2003)—which could affect the accuracy of abundances relative to the Sun and other metal-rich stars of Be, B and several other elements with useful spectral lines only in the ultraviolet, as well as nitrogen and oxygen abundances from the ultraviolet electronic systems of NH and OH, respectively. Another problem for the metal-poor stars is the very small number of lines of strength suitable for abundance analysis for many elements. Effects of departures from LTE have been studied for elements like Li, B, Na, and Fe I and found to lead to abundance corrections of typically 0.1–0.2 dex, and possibly more in some cases for metal-poor stars, primarily due to overionization by the strong ultraviolet radiation fields (e.g., Kiselman 1994; Gratton et al. 1999; Idiart & Thevenin 2000; Gehren et al. 2001; Takeda 2003; Christlieb et al. 2004). For abundance determinations from the oxygen triplet line there may be effects of similar size (Kiselman 2001). Finally, the effects of convective thermal inhomogeneities in the surface layers may strongly affect abundances determined from lines sensitive to the temperature, such as C, N and O abundances from the hydride lines, or abundances of Fe and other metals when determined from low-excitation lines of the neutral atom. Typical errors in such abundances, overestimated in standard analyses, may be on the order of 0.3 dex to as much as 1 dex (Asplund & García Pérez 2001; Asplund, private communication).

Obviously, we have about 3 to 4 different causes of systematic errors, each contributing uncertainties on the order of 0.1–0.2 dex, and for some abundance criteria and some stars even more. If these sources of error are not correlated or coupled—an assumption that may be doubtful (e.g., a positive feedback between convective thermal inhomogenities and photoionization is conceivable)—it thus seems realistic to ascribe uncertainties on the order of 0.2–0.3 dex to typical "absolute" abundances (i.e. relative to hydrogen and to the Sun).

9.2.4   *Fortunately, Differential Studies are Better!*

A very important fact, stressed for instance by Poul Erik Nissen at this meeting, is that abundance ratios such as [X/Fe] can be determined with much higher accuracy than absolute abundances, even with standard analysis using plane-parallel model atmospheres in LTE, provided that the abundance criteria are suitably chosen. For instance, one could base determinations of [Mg/Fe] or [Si/Fe] for solar-type stars on measurements of lines of Mg I, Si I, and Fe I. Similarly, [C/O] determinations could be based on highly excited lines from both atoms, with the expectation that systematic errors in fundamental parameters, model structure, and even in the ionization and excitation equilibria will cancel to a large extent. It is then important to try to use lines from both elements of about equal strengths and, if possible, excitation, so that the line contribution functions roughly sample the same layers in the stellar atmosphere. In general relatively weak (unsaturated) lines should be preferred since abundance values resulting from them are less dependent on the model structures and the atmospheric velocity fields (see Gustafsson 1983). Differential studies have indeed produced a number of significant and very interesting results on levels close to or below the systematic errors discussed above, such as the study of scatter in relative abundances by Edvardsson et al. (1993) and Nissen et al. (1994), as well as the interesting abundance differences between different populations at a given [Fe/H] found by Fuhrmann (1998, 1999), Nissen & Schuster (1997), and Feltzing, Bensby, & Lundström (2003; see also Bensby, Feltzing, & Lundström 2003 and Feltzing et al. 2004).

Another important fact to stress is that one may reduce the effects of systematic errors very considerably by intercomparing stars that are close in the T_{eff}-log g-[Fe/H] space. For example, Nissen & Schuster (1997) find errors in [Mg/Fe] of only 0.03 dex in their comparison of thick- and thin-disk stars, errors mainly determined by uncertainties in measured equivalent widths. The extent to which differential studies of these types are free of systematic errors still remains to be explored. Two factors are obviously of significance here. The first is the extent to which the element of interest—Mg in the example mentioned—is just a "trace element," i.e. of no significance for the atmospheric structure. For cooler stars Mg may, for instance, be an important electron donor, thus affecting the continuous H^- opacity. Such a possible effect may easily be handled in a classical analysis by using model atmospheres tailored to be self-consistent with the results of the abundance analysis. The only worry is then that the effect could couple to systematic errors that have not been considered, such as non-LTE overionization of Mg, which could in itself depend on the Mg abundance. The other factor to consider is *how* close in the T_{eff}-log g-[Fe/H] space stars have to be in order for the systematic errors to be reduced to unimportant levels. This question must be studied using more detailed model atmospheres, with the systematic effects (non-LTE, 3-D convection, etc.) modeled, at least schematically. An important question is then also whether more parameters than the classical trio, such as stellar rotation, mass, or age, are fundamental in determining the atmospheric structure and the spectrum.

One circumstance that is of major significance for accurate differential studies is the fact that spectrometers with high S/N and high resolution now provide a very wide spectral region in one exposure, so that optimal criteria can be chosen for accurate differential analyses. Also, different criteria can easily be used, and resulting abundances may be systematically compared.

9.3   Conclusions

We have found that there have been very significant improvements, by orders of magnitude when measurable, in most aspects of stellar abundance analysis. However, in absolute abundances there are still several types of systematic errors remaining, of at least 0.1 dex each and even considerably more, in many important cases. We have therefore advocated differential studies, as far as possible *strictly* differential between different but physically similar stars and different elements with similar types of abundance criteria, as being much more accurate.

The progress in 3-D convective models with radiative transfer, in non-LTE modeling of early-type stars with winds, in pulsating models of red giants with time-dependent radiative transfer and dust formation, in non-LTE models of cool stars where the statistical equilibrium equations are solved for many atoms and molecules and very many levels, has been brought to a stage at which detailed comparisons with spectra show very much improved agreements with observations as compared with standard, static LTE models. However, the fully *physically* consistent model atmospheres, with 3-D geometry, magnetohydrodynamics in full play, and non-LTE radiative transfer in satisfactory detail, still remain to be constructed.

Will it really be worth going to this degree of sophistication? There are different views expressed on this in the astronomical community. I would suggest that the answer is "yes," in view of the facts that the attempts and comparisons with non-classical models demonstrate systematic errors in absolute abundances that are as great as 0.1–0.2 dex or more, and that many new and interesting features in galactic abundance patterns have been disclosed in differential studies to exist at these levels or lower.

It seems that the theoretical development will within a decade lead us to a stage when standard analyses with static LTE model atmospheres will be looked upon as curve-of-growth analyses were regarded about 30 years ago—an old-fashioned tool that gives abundance indications but does does not match the quality of, and the effort behind, the observed spectra.

This progress should not be worrying for the stellar spectroscopist—the effort in applying the new theoretical developments will hopefully be no greater than it was to apply classical model atmospheres about 30 years ago. Also, there is a much more constructive attitude to meet the complexity that nature demonstrates to us in the significant details of stellar atmospheres—these beautiful plasma "landscapes" also show phenomena of structure formation that may be related to the study of the origins of structure and complexity in broad areas of natural sciences, which may bring us, and our field, closer to interesting developments in neighboring fields. For the observer, this also means that many features in spectra, not only abundance criteria like the equivalent widths of certain selected lines, should be brought into the analysis for checking and challenging the models. These features will include line profiles and strengths of numerous lines with diagnostic power for the analysis of velocity fields, temperature structure, thermal inhomogeneities, magnetic fields, etc. The development of modern instrumentation has made this possible, or even simple.

In his fundamental paper *"On the Composition of the Sun's Atmosphere"* from 1929 Henry Norris Russell wrote:

The hope that from the familiar qualitative spectrum analysis of the solar atmosphere a quantitative analysis might be developed is of long standing. Recent developments in spectroscopy and astrophysics have turned the hope into rational anticipation. The most precise method of investigation—the study of the detailed contours of individual lines, promises the most, but it will be some time before it can be applied to the multitudes of lines available ...

Russell ended his paper with the following remark:

In conclusion, it should be emphasized that the present work ... is of the nature of a reconnaissance of new territory. It is to be hoped that the determinations made here by approximate methods will be replaced within a few years by others of much greater precision, based on accurate measures of the contours and intensities of as many lines as possible. An extensive field of work is open, and it is hoped that much more may be done at this Observatory.

This statement is visionary enough to be repeated again, although I would replace the last three words with "in many places."

Finally, an ending side remark about the art of making and using model atmospheres. In the art museum in Pasadena, The Norton Simon Museum, a nice collection of art from many centuries is on display. Among the oil paintings there are several by Rembrandt Van Rijn. In the same room two particularly fascinating portraits face each other (Fig. 9.4). The first is from 1636–38, one of Rembrandt's many self portraits. Here the artist wears a characteristic beret, is about 30 years old but looks older and quite dignified—with a golden chain around his neck, a symbol of prestige. I fear that about 10 years ago we regarded model stellar atmospheres to be just as established.

The second portrait was painted about 10 years later, a portrait of a young boy, presumably Rembrandt's son Titus. It is a very charming picture of a vivid young person, a much more dynamic portrait, but unfinished. (In fact, it has been used in studies of Rembrandt's way of working.) It reminds me more of contemporary model atmospheres.

The user of model atmospheres, for example someone mainly interested in abundance analysis, may feel that such aesthetic aspects are less significant. Rightly so! But just as real children are dynamic, so are real stars. Careful scientific work often means preferring the

Fig. 9.4. *Left panel:* Self-portrait (c. 1636–40) by Rembrandt van Rijn at the Norton Simon Museum, Pasadena. *Right panel:* Portrait of a boy, presumed to be the artist's son Titus (c. 1645–50) by Rembrandt van Rijn, Norton Simon Museum, Pasadena.

unfinished attempt to catch reality over detailed static academic modeling, so far from the real world.

Acknowledgements. Many colleagues have kindly contributed to this paper by giving their views on the accuracies obtainable in abundance determinations, and the systematic errors remaining in contemporary analyses. Among these colleagues are Carlos Allende Prieto, Martin Asplund, Ann Boesgaard, Judith Cohen, Philip Dufton, Sofia Feltzing, Raffaele Gratton, Artemio Herrero, Andreas Korn, Poul Erik Nissen, John Norris, Sean Ryan, Monique Spite, George Wallerstein, and Gang Zhao. I thank them for their assistance; the responsibility for all numbers and simplifications behind those is, however, mine. Paul Barklem, Bengt Edvardsson, and Michelle Mizuno-Wiedner read the manuscript and suggested important improvements, Nils Ryde helped in preparing the diagrams.

References

Allard, F., Hauschildt, P. H., & Schwenke, D. 2000, ApJ, 540, 1005
Allende Prieto, C., Asplund, M., Lopez, R. J. G, & Lambert, D. L. 2002a, ApJ, 567, 544
Allende Prieto, C., Hubeny, I., & Lambert, D. L. 2003, ApJ, 591, 1192
Allende Prieto, C., Lambert, D. L., & Asplund, M. 2002b, ApJ, 573, L137
Aller, L. H. 1960, in Stellar Atmospheres, ed. J. L. Greenstein (Chicago and London: Univ. Chicago Press), 156
Alvarez, R., & Plez, B. 1998, A&A, 330, 1109
Andersen, A. C., Höfner, S., & Gautschy-Loidl, R. 2003, A&A, 400, 981
Anderson, L. S. 1989, ApJ, 339, 558
Aringer, B., Kerschbaum, F., & Jørgensen, U. G. 2002, A&A, 395, 915
Asplund, M., Allende Prieto, C., Lambert, D. L., & Sauval, A. J. 2004, in preparation
Asplund, M., Carlsson, M., & Botnen, A. V. 2003, A&A, 399, L31
Asplund, M., & García Pérez, A. E. 2001, A&A, 372, 601
Asplund, M., Ludwig, H.-G., Nordlund, Å., & Stein, R. F. 2000a, A&A, 359, 669

Asplund, M., Nordlund, Å., Trampedach, R., Allende Prieto, C., & Stein, R. F. 2000b, A&A, 359, 729
Auer, L., Fabiani Bendicho, P., & Trujillo Bueno, J. 1994, A&A, 292, 599
Barklem, P. S., & Belyaev, A. 2004, in preparation
Barklem, P. S., Piskunov, N., & O'Mara, B. J. 2000, A&AS, 142, 467
Bautista, M. A. 1997, A&AS, 122, 167
Bautista, M. A., & Kallman, T. R. 2001, ApJS, 134, 139
Bell, R. A., Balachandran, S. C., & Bautista, M. 2001, ApJ, 546, L65
Bensby, T., Feltzing, S., & Lundström, I. 2003, A&A, 397, L1
Bessell, M. S., Scholz, M., & Wood, P. R. 1996, A&A, 307, 481
Blackwell, D., Petford, A. D., Shallis, M. J., & Simmons, G. J. 1980, MNRAS, 191, 445
Borysow, A., Champion, J. P., Jørgensen, U. G., & Wenger, C. 2004, in Workshop in Stellar Atmosphere
   Modeling, ed. I. Hubeny, D. Mihalas, & K. Werner (San Francisco: ASP), in press
Bowen, G. H. 1988, ApJ329, 299
Bowen, G. H., & Willson, L. A. 1991, ApJ, 375, L53
Carbon. D. F., & Gingerich, O. 1969, Proc. 3rd Harvard-Smithsonian Conf. on Stellar Atmospheres, ed. O.
   Gingerich (Cambridge, Mass: MIT), 377
Cayrel de Strobel, G., Soubiran, C., & Ralite, N. 2001, A&A, 373, 159
Chamberlain, J. W., & Aller, L. H. 1951, ApJ, 114, 52
Christlieb, N., Gustafsson, B., Korn, A. J., Barklem, P. S., Beers, T. C., Bessell, M. S., Karlsson, T., &
   Mizuno-Wiedner, M. 2004, in preparation
Cohen, J., & Strom, S. E. 1968, ApJ, 151, 623
Crowther, P. A., Hillier, D. J., Evans, C. J., Fullerton, A. W., De Marco, O., & Willis, A. J. 2002, ApJ, 579, 774
Cunto, W., Mendoza, C., Ochsenbein, F., & Zeippen, C. J. 1993, A&A, 275, L5
Dere, K. P., Landi, E., Young, P. R., & DelZanna, G. 2001, ApJS, 134, 331
Dimitrijevic, M. S., Dacic, M., Cvetkovic, Z., & Sahal-Brechot, S. 2003, A&A, 400, 791
Edvardsson, B., Andersen, J., Gustafsson, B., Lambert, D. L., Nissen, P. E., & Tomkin, J. 1993, A&A, 275, 101
Ekberg, U., Eriksson, K., & Gustafsson, B. 1986, A&A, 167, 304
Feltzing, S., Bensby, T., Gesse, S., & Lundström, I. 2004, in Carnegie Observatories Astrophysics Series, Vol. 4:
   Origin and Evolution of the Elements, ed. A. McWilliam & M. Rauch (Pasadena: Carnegie Observatories,
   http://www.ociw.edu/symposia/series/symposium4/proceedings.html)
Feltzing, S., Bensby, T., & Lundström, I. 2003, A&A, 397, L1
Fleischer, A. J., Gauger, A., & Sedlmayr, E. 1992, A&A, 266, 321
Freytag, B. 2003, Proc. SPIE, 4838, 348
Freytag, B., Steffen, M., & Dorch, B. 2002, AN, 323, 213
Fuhrmann, K. 1998, A&A, 338, 161
——. 1999, Ap&SS, 265, 265
Gehren, T., Butler, K., Mashonkina, L., Reetz, J., & Shi, J. 2001, A&A, 366, 981
Gratton, R. G., Carretta, E., Eriksson, K., & Gustafsson, B. 1999, A&A, 350, 955
Gray, D. F. 2001, PASP, 113, 1378
Gustafsson, B. 1983, PASP, 95, 101
——. 1989, ARA&A, 27, 701
Gustafsson, B., Bell, R. A., Eriksson, K., & Nordlund, Å. 1975, A&A, 42, 407
Hamann, W.-R. 1986, A&A, 160, 347
Heiser, A. M. 1957, ApJ, 125, 470
Helling, C., & Jørgensen, U. G. 1998, A&A, 337, 477
Höfner, S. 1999, A&A, 346, L9
Höfner, S., & Dorfi, E. A. 1997, A&A, 319, 648
Höfner, S., Gautschy-Loidl, R., Aringer, B., & Jørgensen, U. G. 2003, A&A, 399, 589
Höfner, S., Jørgensen, U. G., Loidl, R., & Aringer, B. 1998, A&A, 340, 497
Hubeny, I., & Lanz, T. 1995, ApJ, 439, 875
Idiart, T., & Thevenin, F. 2000, ApJ, 541, 207
Jørgensen, U. G. 1990, A&A, 232, 420
——. 1994, A&A, 284, 179
Jørgensen, U. G., Almlöf, J., & Siegbahn, P. E. M. 1989, ApJ, 343, 554
Jørgensen, U. G., Jensen, P., Sørensen, G. O., & Aringer, B. 2001, A&A, 372, 249
Jørgensen, U. G., & Larsson, M. 1990, A&A, 238, 424
Jørgensen, U. G., Larsson, M., Iwamae, A., & Yu, B. 1996, A&A, 315, 204

Kiselman, D. 2001, NewAR, 45, 559

——. 1994, A&A, 286, 169

Kupka, F., Piskunov, N., Ryabchikova, T. A., Stempels, H. C., & Weiss, W. W. 1999, A&AS, 138, 119

Kurucz, R. L. 1979, ApJS, 40, 1

——. 1993, Kurucz CD-Rom No. 1. (Cambridge, Mass.: Smithsonian Astrophysical Observatory)

Lawrence, A. R. 1993, in Proceedings of the 36th Annual Convention of the Parapsychological Association, Parapsychological Association 1993, 75

Moore, G. E. 1965, Electronics, 38, No. 8

Nave, G., & Johansson, S. 1995, in Workshop on Laboratory and Astronomical High-resolution Spectra, ed. A. J. Sauval, R. Blomme, & N. Grevesse (San Francisco: ASP), 197

Nave, G., Johansson, S., Learner, R. C. M., Thorne, A. P., & Brault, J. W. 1994, ApJS, 94, 221

Nissen, P. E., Gustafsson, B., Edvardsson, B., & Gilmore, G. 1994, A&A, 285, 440

Nissen, P. E., Primas, F., Asplund, M., & Lambert, D. L. 2002, A&A, 390, 235

Nissen, P. E., & Shuster, W. J. 1997, A&A, 326, 751

Nordlund, Å. 1982, A&A, 107, 1

Partridge, H., & Schwenke, D. W. 1997, J. Chem. Phys., 106, 4618

Pauldrach, A. W. A., Hoffman, T. L., & Lennon, M. 2001, A&A, 375, 161

Peytremann, E. 1974, A&A, 33, 203

Plez, B. 1998, A&A, 337, 495

Reader, J., Wiese, W. L., Martin, W. C., Musgrove, A., & Fuhr, J. R. 2002, in NASA Laboratory Astrophysics Workshop, ed. F. Salama (NASA), 80

Richter, He., Wood, P. R., Woitke, P., Bolick, U., & Sedlmayr, E. 2003, A&A, 400, 319

Russell, H. N. 1929, ApJ, 70, 11

Sandin, C., & Höfner, S. 2003a, A&A, 398, 253

——. 2003b, A&A, 404, 789

Scharmer, G. 1981, ApJ, 249, 720

Scharmer, G., & Carlsson, M. 1985, J. Comp. Phys., 59, 56

Schwarzschild, M. 1975, ApJ, 195, 137

Schwenke, D. W. 1998, Faraday Discussion, 109, 321

Schweitzer, A., Hauschildt, P. H., Baron, E., & Allard, F. 2003, Workshop in Stellar Atmosphere Modeling, ed. I. Hubeny, D. Mihalas, & K. Werner (San Francisco: ASP), 339

Seaton, M. J., Yan, Y., Mihalas, D., & Pradhan, A. K. 1994, MNRAS, 266, 805

Sneden, C., Johnson, H. R., & Krupp, B. M. 1976, ApJ, 204, 281

Stehlé, C., & Hutcheon, R. 1999, A&AS, 140, 93

Stein, R. F., & Nordlund, Å. 2000, Solar Phys., 192, 91

Strömgren, B. 1940, Publ. Copenhagen Obs. Vol. 127

Takeda, Y. 2003, A&A, 402, 343

Werner, K., & Husfeld, D. 1985, A&A, 148, 417

Wiese, W. L., & Martin, G. A. 1980, NSRDS-NBS 68, 359

Willson, L. A. 2000, ARA&A, 38, 573

Winters, J. M., Le Berte, T., Jeong, K. S., Helling, C., & Sedlmayr, E. 2000, A&A, 361, 641

Wood, P. R. 1979, ApJ, 227, 220

10

The light elements: lithium, beryllium, and boron

ANN MERCHANT BOESGAARD
Institute for Astronomy, University of Hawaii

Abstract
A review of the rare light elements, Li, Be, and B, is presented. This includes a discussion of their origins, processes of destruction, and Galactic evolution. The basics of the determination of their abundances is also covered, which describes the observations, stellar parameter determinations, spectral synthesis, and associated errors. Highlights of some current observational results are presented.

10.1 Introduction

The rare light elements, lithium, beryllium, and boron, are extremely underabundant relative to their neighbors on the periodic table, light hydrogen and helium, and heavier carbon, nitrogen, and oxygen. Although Li, Be, and B are formed by nuclear fusion reactions inside stars at high temperatures, they are destroyed by other reactions at lower temperatures, i.e. closer to the stellar surface. Therefore, a method other than stellar nucleosynthesis is needed to form them. In the breath-taking paper on the origin of the elements, Burbidge, Burbidge, Fowler, & Hoyle (1957) covered the nucleosynthesis of elements by various processes, but called the process for forming ^{2}H, Li, Be, and B the x-process, where "x" was for "unknown." If these elements cannot be formed by fusing smaller nuclei together, then the alternative formation mechanism is by splitting heavier nuclei into smaller pieces. The idea of spallation was invented by Reeves, Fowler, & Hoyle (1970), and details were described by Meneguzzi, Audouze, & Reeves (1971). The basic process is that high-energy ($\sim$150 MeV) protons and neutrons bombard interstellar nuclei of C, N, and O, creating lighter isotopes. It has also been suggested that the "bullets" and the "targets" might be reversed near supernovae, where C, N, and O nuclei could be accelerated into the ambient interstellar gas including protons and neutrons (see, for example, Duncan et al. 1997, 1998; Lemoine, Vangioni-Flam, & Cassé 1998).

The easiest of the three light elements to observe is Li, and spectroscopic observations of Li have been made in a large array of stars of various ages and metallicities throughout the Hertzsprung-Russell diagram, as well as in the interstellar gas. In recent years there have been typically 100 papers a year with lithium in the title, a testament to the usefulness of the abundance of this one element. Due to the atomic and spectroscopic properties of Be and B, they have been less extensively observed. However, information on all three elements complement and supplement each other and can illuminate several issues of astrophysical interest.

In the study of light elements, the results are most often presented as plots of the light element abundance as a function of metallicity, [Fe/H], or as a function of stellar effective temperature, T_{eff}. This provides an important context through which we can derive the astrophysical interpretation. In some circumstances, the abundance of one light element can be plotted against another to elucidate the processes at work in stellar interiors.

This review presents an overview of the origin of the light elements and their destruction in stars, the basics of light-element observations and analyses, and some recent results.

10.2   Origins and Current Observational Status

10.2.1   *Lithium*

The element Li can be formed by several different processes. The original source is through the nuclear reactions that occur during the first 10^3 s after the Big Bang while the temperature is $> 5 \times 10^8$ K. Two different reactions produce the isotope ^{7}Li: ^{4}He(^{3}H,γ)^{7}Li and ^{4}He(^{3}He,γ)^{7}Be, where the ^{7}Be captures an electron to become ^{7}Li.

As discussed in the first section, another major source of Li results from spallation reactions between energetic cosmic rays and abundant atoms such as C, N, and O in the interstellar gas, as first proposed by Reeves et al. (1970). These reactions could take place in the ambient interstellar gas and in the vicinity of supernovae where energetic CNO nuclei as well as energetic protons would be found. Cosmic ray interactions can also result in $\alpha + \alpha$ production, leading to Li atoms (Steigman & Walker 1992; Prantzos, Cassé, & Vangioni-Flam 1993; Ramaty, Kozlovsky, & Lingenfelter 1996).

A potentially interesting source of ^{7}Li is the ν-process in Type II supernovae (Woosley et al. 1990; Timmes, Woosley, & Weaver 1995). Neutrino irradiation gives rise to both ^{7}Li and ^{7}Be primarily in the He shell. Most is destroyed subsequently, and an evaluation of the contribution to Li by this mechanism may come from the observations of ^{11}B, which is similarly produced (Woosley & Weaver 1995), along with those of ^{7}Li. Most models produce 2–5 times more B than Li. The models do not produce ^{6}Li, ^{9}Be, or ^{10}B.

Another possibility is that the light elements could be produced in hypernovae and in superbubbles (e.g., Parizot & Drury 1999; Parizot 2000). In order not to overproduce Li, Be, and B a limit of three hypernovae for every 100 supernovae has been set by Fields et al. (2002).

It was suggested by Cameron (1955) and discussed by Cameron & Fowler (1971) that freshly synthesized Li could be found on the stellar surface of red giant stars if it had been preserved as ^{7}Be long enough to rise to the surface before capturing an electron to become ^{7}Li. The ^{7}Be would be transported by convection to temperatures cooler than the Li destruction temperature. The reaction ^{3}He(α,γ)^{7}Be would occur at temperatures hotter than 10^7 K, and the reaction ^{7}Be($e^-\nu$)^{7}Li at temperatures cooler than 3×10^6 K. This mechanism was proposed to account for the high Li content in a few S and C stars. More observational evidence of high Li in asymptotic giant branch stars has been presented by Smith & Lambert (1990) and Smith et al. (1995) and attributed to the ^{7}Be transport method with hot-bottom convective envelope burning. However, this is not thought by Romano et al. (2001) to be an important contributor to Li in disk stars. For a review of Li production in asymptotic giant branch and red giant branch stars, see Sackmann & Boothroyd (2000). Another possible site for Li production is in novae (e.g., Hernanz et al. 1996; Romano, et al. 1999).

Abundances of Li have been plotted against metallicity ([Fe/H]) for both disk stars and halo stars. Figure 10.1 shows a composite diagram adapted from Ryan et al. (2001) and

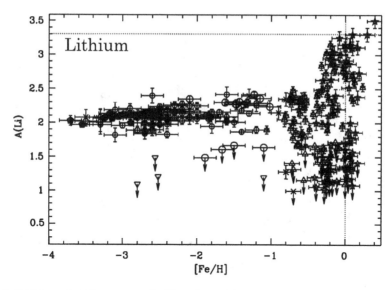

Fig. 10.1. Lithium abundances vs. [Fe/H], adapted from Ryan et al. (2001); see that paper for the original references and for the definition of the symbols. This sample includes stars hotter than 5600 K. The horizontal dotted line represents the meteoritic Li abundances, A(Li) = 3.31. The x- and y-axis scales for this and the next three figures are the same.

includes main sequence and turn-off stars with temperatures > 5600 K. (Unfortunately, it does not include the results of the 2001 work on 185 main sequence stars by Chen et al. 2001, which was published after the Ryan et al. 2001 paper.) The Li abundances are given as A(Li) = log N(Li/H) +12.00; this same notation is used for Be and B throughout. There are several features of note on this diagram. First is the apparent plateau in A(Li) at low metallicities, which is attributed to the production of Li during the Big Bang. (Controversy exists over the actual value of primordial Li deduced from this, whether there is a metallicity dependent slope, whether there is a dispersion in the plateau, whether there is depletion in the plateau stars, etc.) Secondly, there are several stars at low metallicities with low upper limits on the Li abundances. This is discussed in §10.5. Thirdly, stars near solar metallicity show an order of magnitude more Li than the plateau. And fourthly, there is a large range in A(Li). The Li-rich disk stars are preferentially younger and hotter main sequence stars (excluding Li-dip stars), while the disk stars with lower A(Li) are thought to have had their surface Li depleted via mixing of the photospheric material down to temperatures where Li is destroyed. For the coolest stars (only) this mixing can be done by the surface convection zone; other mixing mechanisms are needed for hotter stars. The rise in A(Li) that occurs between [Fe/H] = −0.4 and −0.2 is too steep to be explained by Galactic cosmic ray spallation production of Li. There is an apparent deficit of stars having high Li abundances with [Fe/H] between −1.0 and −0.4; this could be due to sampling bias or effective Li depletion from higher initial Li (also see Romano et al. 1999).

Another observational view of Li in halo stars comes from recent work on Li in globular cluster turn-off stars. Such stars are faint (V = 16 − 18 mag), and this work can only be done at high spectral resolution on the world's largest telescopes. Figure 10.2 shows the results for four globular clusters as a function of the Fe abundance for each cluster. These

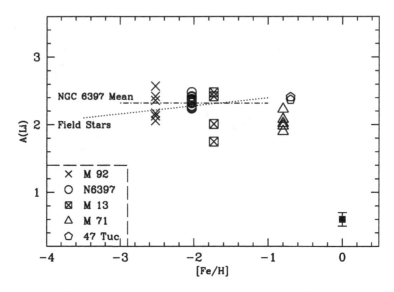

Fig. 10.2. Lithium abundances vs. [Fe/H] in globular cluster turn-off stars. The M92 results are from Boesgaard et al. (1998a), M13 and M71 are from Boesgaard et al. (2000), NGC 6397 from Bonifacio et al. (2002), and 47 Tuc from Pasquini & Molaro (1997). A typical error bar is shown in the lower right. The dot-dash line shows the mean found by Bonifacio et al. for NGC 6397, A(Li) = 2.32. The dotted line corresponds to the relation given for halo field stars by Ryan et al. (2001).

results are from Boesgaard et al. (1998a) for M92, Boesgaard et al. (2001) for M13 and M71 from Keck, from Bonifacio et al. (2002) for NGC 6397 from VLT, and Pasquini & Molaro (1997) for 47 Tuc. Bonifacio et al. (2002) conclude that their stars could all have the same Li abundance. However, for identical stars in M92 and M13 Boesgaard and colleagues find real star-to-star differences in the equivalent widths of the Li line, but not in the lines of any other elements. The red giants in globular clusters are very different from the turn-off stars in their Li content. Kraft et al. (1999) found one Li-rich red giant in the globular cluster M3, with A(Li) = 3.0; this star is thought to be ascending the red giant branch for the first time. Following this, Pilachowski et al. (2000) surveyed 261 giants in four globular clusters: M3, M13, M15, and M92. They found no other Li-rich giants, indicating that the percentage of Li-rich globular cluster giants is <1%.

10.2.2   *Beryllium*

The dominant, and perhaps only, method for the production of ^{9}Be is by Galactic cosmic ray spallation, albeit through a modified and expanded version of the original work of Reeves et al. (1970). Ramaty, Lingenfelter, & Kozlovsky (2000; see also Ramaty & Lingenfelter 1999) describe and discuss the three modern models for spallation. There is also the possibility of production of Be in hypernovae, mentioned above for Li.

Figure 10.3 is similar to Figure 10.1, but plots A(Be) against [Fe/H] for halo and disk stars. The data are from a number of different papers but the figure presents a good overview (see references in the figure caption). The increase in A(Be) with [Fe/H] is obvious, and the slope of the relationship from [Fe/H] ≈ −3.0 to ∼ −1.0 is close to 1. At the low-metallicity

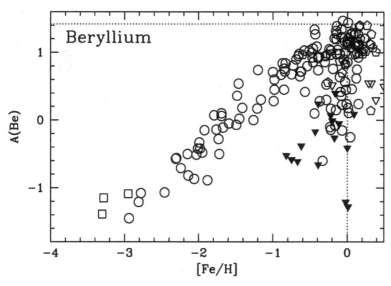

Fig. 10.3. Be abundances vs. [Fe/H]. Open circles and filled triangles (upper limits) are from Stephens et al. (1997), Boesgaard et al. (1999, 2001), and Boesgaard (2000). Open squares (low-metallicity stars) are from Primas et al. (2000a, b). The open pentagons and triangles (upper limits) are from Santos et al. (2002). The horizontal dotted line is the meteoritic Be abundance, A(Be) = 1.42.

end there may be a flattening toward a plateau in A(Be). The lowest-metallicity stars are preferentially the faintest, so those observations are the most difficult and time-consuming; observations are needed of Be in more stars with [Fe/H] <−3.0. It is useful to plot A(Be) vs. [O/H] (which has a steeper slope), as done by Boesgaard et al. (1999), to elucidate Galactic evolution of Be and to investigate details of spallation of Be. King (2001) has examined the Be and B evolution via comparisons with other α-elements, Mg and Ca, finding slopes between 1.1 and 1.3.

The error bars given in the original papers indicate that there may be a real spread in A(Be) at a given [Fe/H] in the halo stars. This could result from differing degrees of efficiency in the formation of Be by spallation in different parts of the Galaxy.

For some stars with [Fe/H] >−1.0 there are Be deficiencies, including some with measured Be abundances down by 2 orders of magnitude from the meteoritic A(Be). There is a large clustering of data points with A(Be) between 1.0 and 1.4 dex near solar metallicity. It is therefore difficult to disentangle the general increase of A(Be) with [Fe/H] in the disk stars from the spread due to some minor depletion. The upper envelope of data points seems to keep increasing; however, a flattening of the slope or a gentle curvature is possible.

10.2.3 Boron

The major source of B is Galactic cosmic ray production, in the various ways discussed by Ramaty et al. (2000), Ramaty & Lingenfelter (1999), and references therein. And, as with Li, the ν-process in supernovae can potentially produce [11]B. (For a review of this with emphasis on B, see Hartmann et al. 1999.) Perhaps B could be produced by hypernovae along with Li and Be.

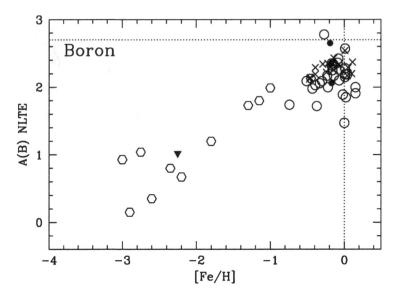

Fig. 10.4. Boron abundances from B I vs. [Fe/H]. The hexagons and the upper limit triangle are from Duncan et al. (1997, 1998), Primas et al. (1999), and García López et al. (1998); the crosses are from Boesgaard et al. (2004); the open circles are from Boesgaard et al. (1998b) and from new *HST* observations of Deliyannis et al. (in preparation); the solid circles are from Cunha et al. (2000). The horizontal dotted line is the meteoritic B abundance, A(B) = 2.70.

The trend of B with Fe (from B I observations only) is shown in Figure 10.4, which is on the same scale and similar to Figures 10.1 and 10.3 for Li and Be. These observations are all from *HST* so there are fewer B data points than Be points, and far fewer than Li. The local thermodynamic equilibrium (LTE) abundances have been corrected for non-LTE (NLTE) effects from the work of Kiselman & Carlsson (1996). An increase in A(B) with [Fe/H] can be seen in the halo stars, which is similar to the trend of Be with Fe. The slope of B vs. Fe in NLTE appears not to be as steep as the slope for Be vs. Fe, although the slope from the LTE results is very similar for both B and Be. There are two stars near [Fe/H] = −3 with A(B)$_{NLTE}$ near 1.0. The values of A(B)$_{LTE}$ for those two stars are near 0.0, which makes them consistent with the other stars in the LTE version of Figure 10.4. In LTE the slope is near 1.0, while in NLTE it is near 0.7 (Duncan et al. 1997). If the slopes for Be and Be are identical (as they are in LTE), this implies a common origin for the two elements. A shallower slope for B might imply that the ν-process is an important source for B.

As in Figure 10.3 for Be, the stars with [Fe/H] > 0.6 show a spread, possibly resulting from depletion of B in some of the stars. The points marked by crosses are stars that are undepleted in Be and thus expected to be undepleted in B. These stars lie along the upper envelope of the Be-Fe and the B-Fe relation (Boesgaard et al. 2004). Relative to those stars there are some half dozen stars that seem genuinely deficient in B by factors of 5 or more.

10.3   The Destruction of Li, Be, and B

All three light elements are susceptible to destruction in stellar interiors by nuclear reactions; some of these reactions are of the type (p,α), (p,γ), (α,n), and (α,γ). The dominant

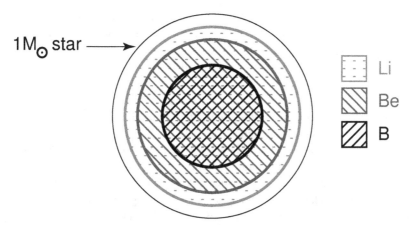

Fig. 10.5. One-solar mass model showing the region where all the Li has been destroyed, a smaller sphere where there is no Be, and a still smaller one that is devoid of B.

ones cause Li to be destroyed at temperatures of 2.5×10^6 K and higher, Be at 3.5×10^6 K and higher, and B at 5×10^6 K and higher. This leaves regions in the stellar interior that are deficient in Li, Be, and B. For a model one-solar mass star, only the outer shell of 2.5% by mass contains Li, only the outer 5% contains Be, while the outer $\sim$18% has B. This is illustrated in Figure 10.5 in a not-to-scale cross-section of the Sun.

The light elements can be circulated by various mixing mechanisms to the stellar interior— down to the temperatures where they are fused into other nuclear products. Among the most common proposed mechanisms are convection, microscopic diffusion, instabilities related to rotation, turbulence, gravity waves, and meridional circulation. In addition, sufficient surface mass loss can rid a star of the surface layer where the light element had been preserved. In that case the star would first lose all its Li, then all its Be, and finally all its B (if it lost $\sim$20% of its total mass).

For in-depth discussions of the destruction of light elements, see Deliyannis, Pinsonneault, & Charbonnel (2000), Pinsonneault, Charbonnel, & Deliyannis (2000), and Charbonnel, Deliyannis, & Pinsonneault (2000).

10.4   Abundance Determinations

10.4.1   Observations

Inasmuch as the abundances of the light elements are low, they are best observed in their respective resonance lines. For Li this is the resonance doublet of Li I λ6707, a ^2S–$^2P^0$ transition. The ionization potential of Li I is low at 5.39 eV, so Li is almost completely ionized in the photospheres of the stars we observe. However, the resonance line of Li II (a He-like atom) is at 199 Å, and the subordinate lines of Li II have extremely high excitation potentials. In some conditions a subordinate line of Li I λ6104 can be observed.

The most commonly observed Be features are the resonance lines of Be II, which has the same doublet structure as Li I, a ^2S–$^2P^0$ transition. These lines are in the ultraviolet just longward of the atmospheric cut-off: λ3130 and λ3131. For solar temperature stars and hotter, Be II is the dominant ionization state.

Three ionization states, B I, B II, and B III, can be used in the study of B abundances, depending on the stellar temperature. Neutral B is the dominant ion in solar-like stars, B II in A stars, and B III in hotter stars. All of the resonance lines are in the space ultraviolet, with B I λ2497, B II λ1362, and B III λ2066. The *Copernicus* satellite was used for early work on B II, some B results came from *IUE*, and *HST* is used for current observations of B I and B III.

10.4.2   *Stellar Parameters and Abundance Determinations*

The determination of the stellar parameters—effective temperature (T_{eff}), surface gravity (log g), metallicity ([Fe/H]), and microturbulent velocity (ξ)—is important in deriving the correct abundances. Temperatures usually come from one or multiple color indices or Balmer-line determinations (e.g., Fuhrmann, Axer, & Gehren 1994). Values of log g can come from ionization balance or from *Hipparcos* data, along with other fundamental stellar properties. Metallicities are typically from spectroscopic studies, but may come from photometric and spectroscopic indices. Some empirical calibrations can yield microturbulent velocities (e.g., Edvardsson et al. 1993).

Associated with the errors in each of these parameters are abundance errors. The commonly used spectral features for each of the three elements Li, Be, and B have different sensitivities to different parameters. For F and G dwarfs, for example, the Li abundance from Li I λ6707 is sensitive to T_{eff}, but insensitive to log g and [Fe/H]. A change of +100 K in temperature gives +0.09 dex in A(Li), while a change of 0.5 in log g is <0.01 dex, and a change in [Fe/H] of 0.10 is only 0.02 dex.

The Be II features are more sensitive to changes in log g than in temperature in the 5500–6500 K range where Be goes from dominant Be I to Be II. A change of +100 K results in a change of only +0.01 dex in A(Be), but a change in log g of 0.2 gives a change in A(Be) of 0.07 to 0.10, increasing with decreasing temperature. For Be an increase in [Fe/H] of +0.10 gives an increase in A(Be) of 0.04 dex near solar metallicities.

For the B I λ2497 resonance line, the abundance of B is sensitive to temperature, gravity and metallicity. For 100 K in T_{eff} the value of A(B) changes by 0.10 dex; for a change in log g of 0.2 there is a change in A(B) of 0.10 dex; a change in [Fe/H] of 0.10 results in a change in A(B) of 0.03 dex.

There can be small differences in the abundances due to the class of models used (e.g., Kurucz vs. OSMARCS). The potential effects of three-dimensional hydrodynamical vs. one-dimensional models are still being investigated (see, for example, Asplund et al. 1999; Asplund 2000). In addition, there can be differences in the abundances derived in LTE compared to NLTE. Calculations to make corrections for these effects have been done for Li by Carlsson et al. (1994) and for B by Kiselman (1994) and Kiselman & Carlsson (1996). Interestingly, the effects due to NLTE cancel out for Be (García López, Severino, & Gomez 1995; Kiselman & Carlsson 1995).

10.4.3   *Spectrum Synthesis to Derive Abundances*

Early work and some current work on Li and Be involved determining abundances from measured equivalent widths of Li I λ6707 and Be II λ3131. With more sophisticated routines such as MOOG (Sneden 1973) and SYNTHE (Kurucz 1993) spectrum synthesis methods can be used readily now. Examples of synthesized spectra for all three elements are given in Figures 10.6–10.8, adapted from Boesgaard et al. (2004). They show observed spectra and synthetic fits for three different abundances for two stars for each element. This

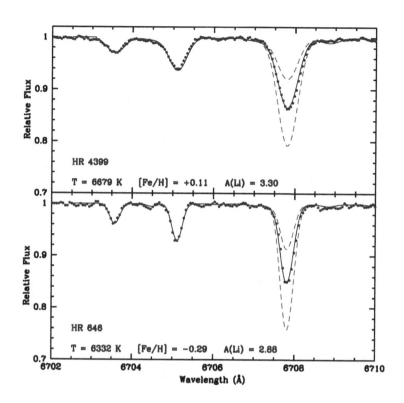

Fig. 10.6. Spectrum synthesis fits for the region of the Li I resonance doublet in two stars. The observed points are the dots from a Keck I HIRES spectrum of HR 4399 (*upper panel*) and from a McDonald spectrum of HR 646 from the 107-in telescope with the coudé cross-dispersed echelle spectrograph (*lower panel*). The solid line represents the best fit for Li abundance, while the dashed lines are a factor of 2 higher and lower in A(Li). The stellar parameters and Li abundances are given for each star. (Adapted from Boesgaard et al. 2004.)

technique is especially important for the badly blended features of Be II and B I. The Li doublet is blended and has hyperfine structure; it is therefore asymmetric, so the synthesis technique is preferable for the analysis of this element also.

10.5   Some Recent Results on Li

In the past five years there have been over 500 papers, reviews, and conference proceedings on the topic of lithium. A complete review is not possible, so only a few results are presented here. In IAU Symposium 198, "The Light Elements and their Evolution" (da Silva, de Medeiros, & Spite 2000) a section on lithium abundances is 131 pages long, and lithium appears throughout the 590-page volume.

10.5.1   A Second Li Feature

Most of the abundance work on Li has been done on the Li I $\lambda6707$ resonance doublet. The Li I $\lambda6104$ subordinate line was detected in red giant stars by Merchant (1967)

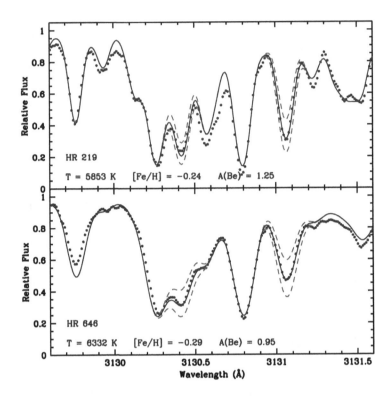

Fig. 10.7. Spectrum synthesis fits for the resonance doublet of Be II for two stars observed with CFHT and the gecko spectrograph. The dots are the observations, and the solid line represents the best fit for the Be abundance. The dashed lines are a factor of 2 higher and lower in Be abundance. The stellar parameters and Be abundances are given for each star. These two stars have similar metallicities, but differ in Be abundance by a factor of 2. (Adapted from Boesgaard et al. 2004.)

and Wallerstein & Sneden (1982). It was found by Duncan (1991) to give a more reliable Li abundance than Li I λ6707 in the T Tauri star BP Tau. The line is weak and appears at the long wavelength edge of a blend of Fe I, Ca I, and Fe I. This feature was detected in the metal-poor halo star, HD 140283, by Bonifacio & Molaro (1998). This star has [Fe/H] = −2.5, so the other lines in the blend are very weak. They derive a similar Li abundance from the resonance line and from the subordinate line, which is formed at deeper layers in the stellar atmosphere.

More recently, Ford et al. (2002a) studied the Li I λ6104 triplet feature in several halo stars. They have firm detections in seven of their 14 stars. The Li abundances determined from the λ6104 feature are slightly higher than those from the resonance line. However, the abundances found provide a good confirmation of the presence and abundances of Li from the λ6707 doublet. In a similar study Ford, Jeffries, & Smalley (2002b) used the spectrum synthesis technique to determine Li abundances from both λ6707 and λ6104 in several cool Pleiades stars. Of the 11 stars four had detectable Li features at 6104 Å, four had marginal

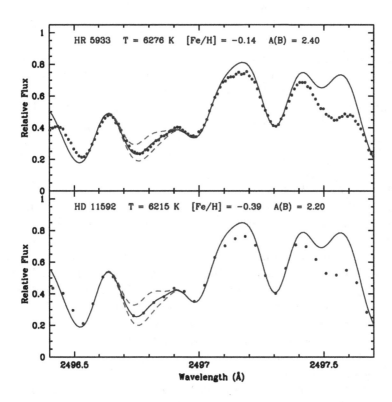

Fig. 10.8. Spectrum synthesis of B for two stars observed with *HST* and STIS. The star in the upper panel, HR 5933, was observed at high resolution and that in the lower panel, HD 11592, was observed at medium resolution. The dots are the observations. The solid line is the best fit, while the dashed lines are a factor of 2 higher and lower in B abundance. The stellar parameters and B abundances are given for each star. These stars have similar temperatures, but HD 11592 is lower in metallicity and in B. (Adapted from Boesgaard et al. 2004.)

detections, and in three $\lambda 6104$ was undetectable. The four definite detections gave A(Li) values higher on average by 0.42 dex compared to A(Li) from $\lambda 6707$. They discuss possible causes of this difference, including the use of one-dimensional model atmospheres and the presence of star spots in these young stars.

10.5.2 *Li-poor Halo Stars*

As can be seen in Figure 10.1 there are some stars with [Fe/H] < −1.0 that have upper limits only on A(Li); the upper limits indicate Li depletions, as they are clearly outliers in the plateau distribution, having abundances only half (or less) the plateau value. Ryan et al. (2002) propose that these stars are the halo field star counterpart to globular cluster blue stragglers. In their sample of 18 stars the Li-normal stars have sharp lines, while three of the four Li-poor stars have rotational broadening and are spectroscopic binaries. They suggest that there has been mass and angular momentum transfer and that the Li was destroyed before or during the transfer of mass. A similar phenomenon was found by Pritchett & Glaspey (1991) in their study of Li in blue stragglers in the old open cluster M67: the blue stragglers have no detectable Li. They suggest that blue stragglers are caused by binary mergers, mass transfer, or deep mixing.

An interesting suggestion for the Li-poor halo stars comes from Pinsonneault et al. (2002) who suggest that the strongly Li-depleted stars belong to a very small subset of very rapidly rotating stars. The Li depletion results from rotationally induced mixing, which destroys more Li in these rapid rotators. Similarly, the spread in Li in the globular clusters shown in Figure 10.2 could be due to a spread in initial rotation, as suggested by Boesgaard et al. (1998a).

10.5.3   New Li Cluster Work

The study of Li in open clusters dates to the major paper on Li by Herbig in 1965, where several cluster stars are included along with field stars. In 1965 the paper by Wallerstein, Herbig, & Conti focuses on Li in the main sequence Hyades stars. They showed clearly the drop-off in Li with cooler surface temperatures in the G dwarfs. For some F dwarfs they could determine only Li upper limits; in retrospect, this was the first prognostication of the now well-studied Li gap in cluster F stars, first revealed by Boesgaard & Tripicco (1986) in the Hyades some 20 years later. Several subsequent studies of the Hyades have resulted in the Li-temperature profile shown in Figure 10.9. Subsequently, there have been Li studies in several other open clusters; this literature is too plentiful to review here, and instead only two recent cluster Li papers will be mentioned.

Jeffries et al. (2002) have determined Li abundances in 29 F, G, and K dwarfs in NGC 6633, a slightly metal-poor open cluster ([Fe/H] = -0.10 ± 0.08) with an age similar to the Hyades. They make careful comparisons with four other young clusters and find that the mid-F stars have Li depletions similar to those at the same temperature in the Hyades (see Fig. 10.9), Praesepe, and Coma Ber. The cooler stars (late-G and K dwarfs) show less Li depletion than the Hyades stars of similar temperatures but similar depletions with respect to the Coma cluster stars of similar metallicity to NGC 6633. They suggest that the severity of the Li destruction in the coolest stars may be related to metallicity. One of the stars they observed in NGC 6633 is extraordinarily rich in Li. It is a *bona fide* member of the cluster with an effective temperature of 7086 ± 85 K. Deliyannis, Steinhauer, & Jeffries (2002) derive A(Li) = 4.29 ± 0.07 for J37 and suggest that this overabundance of Li results from radiatively driven upward diffusion, as predicted by Richer & Michaud (1993).

Randich, Sestito, & Pallavicini (2003) have found Li abundances for 11 G dwarfs in the oldest open cluster, NGC 188; three of the stars were also in the eight-star sample of Hobbs & Pilachowski (1988). They compare their results from NGC 188 with an age of 6–8 Gyr with those from M67 at 4–5 Gyr and the Hyades at 700 Myr. The Li depletions are only slightly larger than those in the Hyades. For the early-G stars they are similar to the upper envelope for M67, but do not decline as much toward cooler temperatures. They suggest some explanations for these results, including different initial Li abundance in different clusters, different Li depletions rates, and a cessation in Li depletion after an age of ~2 Gyr.

HYADES LITHIUM

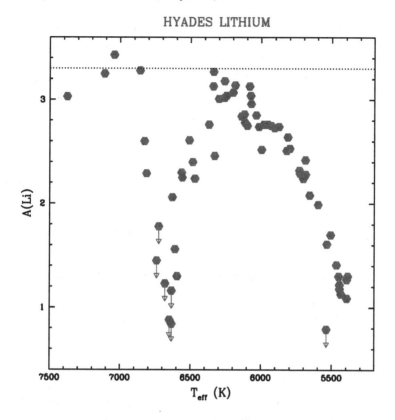

Fig. 10.9. Abundances of Li vs. temperature in the Hyades, showing the Li gap between $\sim 6450-6850$ K and the steep drop in A(Li) with decreasing temperature below 6000 K. The data are from Boesgaard & Budge (1988) and Thorburn et al. (1993).

10.5.4 Summary

Interesting new Li results continue to appear on a range of topics. The problem of low-metal stars with very depleted Li appears to be solved by the work of Ryan et al. (2002) who suggest that they are rotating binaries and the low-mass counterparts of blue stragglers.

New work has been done to determine Li abundances from the subordinate Li feature $\lambda6104$. In general, reasonable agreement is found with A(Li) from $\lambda6707$, but in some cases $\lambda6104$ gives values higher by 0.5 dex.

There are new Li results in open clusters. In a large study of the properties of NGC 6633, Jeffries et al. (2002) compare the results from this Hyades-age cluster with those from several other young clusters and discuss the differences and similarities. One early-F star in that cluster has A(Li) an order of magnitude larger than the meteoritic value, possibly caused by upward diffusion.

The Li results for the old open cluster NGC 188 show less depletion and less scatter than the slightly younger M67 in the G dwarfs. The amount of depletion is only slightly greater than the much younger Hyades. Randich et al. (2003) discuss the implications of those findings.

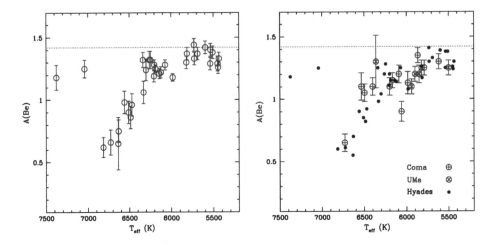

Fig. 10.10. *Left panel:* The Be dip in the Hyades, adapted from Boesgaard & King (2002). *Right panel:* The Be dip in Coma and UMa clusters shown with the Hyades data from Boesgaard et al. (2003b).

10.6   Some Recent Results on Be

10.6.1   Halo Be

As shown in Figure 10.3 there are only a few Be detections in stars with [Fe/H] values as low as −3.0. Three of these are new observations from the VLT with UVES (Primas et al. 2000a, b). The trend shown in Figure 10.4 of Boesgaard (2000) was an ever-decreasing A(Be) with decreasing [Fe/H]. The three new points change this perspective. There is the possibility that the slope is shallower than it seemed in the earlier data. Or possibly there is a plateau in A(Be) near −1.3. This would not necessarily imply the formation of Be by inhomogeneities in the Big Bang.

10.6.2   Be in Clusters

Boesgaard & King (2002) reported a major study of Be in 34 Hyades dwarfs well spaced in temperature from 5450 K to 7400 K. This covered the hot side of the Li dip, the Li dip itself, the Li plateau, and the cool star Li decline (see Fig. 10.9). They found a strong Be dip at the same temperature as the Li dip where Be is detected but depleted. In this dip Be is not as depleted as Li; this was not unexpected, inasmuch as Be is more robust with respect to destruction than Li. They found little or no Be depletion in the 13 stars cooler than 6000 K in spite of the large Li depletions (0.5 to >2.5 dex). This part of the work confirmed and expanded the results of García López, Rebolo, & Pérez de Taoro (1995) on four Hyades G stars. The left panel in Figure 10.10 shows the Hyades Be results of Boesgaard & King (2002).

Observations of Be were made in other young open clusters: the Coma Bernices cluster and UMa moving group by Boesgaard, Armengaud, & King (2003b) and Pleiades and α Per by Boesgaard, Armengaud, & King (2003a). Coma is a little younger than the Hyades (500 Myr vs. 700 Myr for the Hyades), but also shows both a Li and, now, a Be dip. There is no Be depletion in the G stars in any of the clusters even though they show large Li depletions

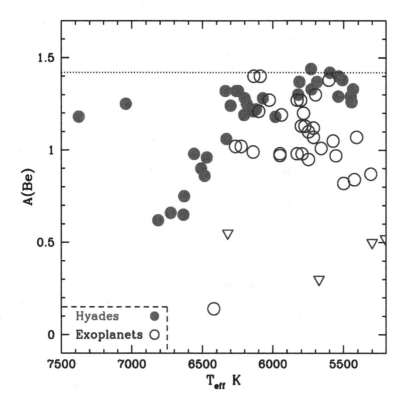

Fig. 10.11. Be abundances in stars with planets (open circles and triangles; Santos et al. 2002) compared with Be in the Hyades stars (filled circles; Boesgaard & King 2002). Inverted triangles represent upper limits.

with decreasing temperature. The Be results for Coma and UMa are shown in the right panel of Figure 10.10. The Pleiades and α Per are still younger at 50–70 Myr. There is only a mild Li dip in the Pleiades F dwarfs and only a weak decline in Li with decreasing temperature in the G dwarfs. The Be abundances in 14 Pleiades stars showed some possibly real scatter, but no trends with temperature. Like the Li depletion, the Be depletion does not reveal itself until an age of >300 Myr in the Li-Be dip in F dwarfs.

Another study of both Li and Be in clusters was done by Randich et al. (2002) with excellent data from the VLT + UVES on faint main sequence stars in M67. At 4–5 Gyr M67 is too old to have any F stars left on the main sequence in the Li-Be dip region. They observed five early-G dwarfs and found little or no Be depletion even though the stars were depleted in Li, concluding that the mixing process does not go to deep enough layers to destroy Be.

10.6.3   Be in Stars with Planets

A study of Be in 32 stars that harbor planets was made by Santos et al. (2002). Such stars have been found to be metal rich when compared to typical field stars, with an average [Fe/H] = +0.17 $\pm$ 0.20 (Gonzales et al. 2001). The enrichment could have enabled the stars to form protoplanetary disks and then planets, or the enrichment might be the result of accretion of metal-rich material. They find no clear differences in the distribution of Be (or Li) abundances compared to a set of non-planet stars. Figure 10.11 shows the comparison in A(Be) of the exoplanet stars with the Hyades, which also has an enhanced metallicity, near [Fe/H] = +0.13. Whereas Be in the Hyades is undepleted, several stars with planets have small to major Be depletions, perhaps due to their greater average age.

10.6.4   *Correlation of Li and Be*

Deliyannis et al. (1998) determined Li and Be abundances in 24 F disk dwarfs in the temperature range 6000–6700 K. Five stars were undepleted in both Li and Be, nine stars showed that the depletions were correlated, while the rest showed only upper limits of one or both elements. They examined several mixing mechanisms to try to explain the correlation and were able to rule out microscopic diffusion and surface mass loss. Rotationally induced mixing seemed best able to account for the observations; possibly gravity waves could contribute. This work was continued by Boesgaard et al. (2001), who made high-resolution observations of Be at CFHT of 46 stars, along with Li observations for 40 of them from Keck I and the UH 2.24-m telescope. In the temperature range from 5850–6680 K, which defines the Li-Be correlation, there are 27 stars with detectable and correlated Li and Be. There is a possibility of a slightly different relation between Li and Be for stars with T_{eff} = 5850–6300 K and those with T_{eff} = 6300–6680 K.

The recent work by Boesgaard & King (2002) and Boesgaard et al. (2003a, b) on Be in open clusters extends the field-star relation to cluster stars. The Hyades A(Li) vs. T_{eff} plot (see Fig. 10.9) gives a context for the Li-Be relation. The cool side of the Li dip is between about 6300 to 6650 K. Figure 10.12 shows the Li-Be correlation in field and cluster stars in this restricted temperature range. The slope of this relation is 0.44, so as Li declines by a factor of 10, Be is reduced by a factor of 2.75. The stars in the youngest clusters (e.g., Pleiades) are in the upper right with low depletions. The other clusters and the field stars are spread along the line. It is thought that the mixing to deeper regions in stellar interiors results from internal angular momentum transfer as the stars age.

10.6.5   *Summary*

The VLT/UVES results for three very metal-poor stars shows that Be is lower by a factor of 400 relative to the meteorites. According to the observations at low metallicities, it is possible that there is a Be plateau below [Fe/H] of −2.8 at a level of A(Be) = −1.3. This needs to be addressed through additional, but difficult, observations of Be in low-metallicity stars.

Observations of Be in open clusters have led to some interesting new results. The Hyades, Praesepe, and Coma clusters with ages of $(5-8) \times 10^8$ yr show a Be dip like the Li dip in the mid-F stars with temperatures about 6300–6800 K. The Be dip is not as deep as the Li dip. No Be dip is found for younger clusters like Pleiades and α Per at $(5-7) \times 10^7$ yr; clusters this young do not have a significant, if any, Li dip. For the G dwarfs, which show Li depletions of more than 100 times, there are no Be depletions. This is true in all the G stars observed including the old cluster M67.

A survey of Be in stars which have planets revealed a normal distribution in Be. These planet-harboring stars generally have higher than solar [Fe/H]. A primordial origin for the high metallicity may be implied by the Be results.

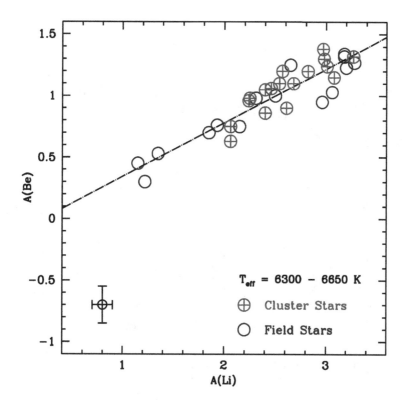

Fig. 10.12. The correlation of Li and Be on the cool side of the Li-Be dip for field and cluster stars. A typical error bar is shown in the lower left. The slope of the dashed line (least squares fit) is 0.44. The stars in the youngest clusters are in the upper right.

In the temperature range from 5900–6650 K there is a correlation between Li and Be. Stars on the cool side of the Li-Be dip (6300 K $< T_{\rm eff} <$ 6650 K) show a strong correlation in both field stars and cluster stars at all ages. This correlation is well matched by predictions of mixing induced by stellar rotation (e.g., Chaboyer, Demarque, & Pinsonneault 1995; Deliyannis & Pinsonneault 1997; Pinsonneault et al. 1999).

10.7   Some Recent Results on B

10.7.1   B Depletions
Although B is less prone to destruction than Li or Be, some stars do show the results of apparent nuclear destruction of B. The stars plotted in Figure 10.4 are those with B determinations from B I; some of those disk-metallicity stars show decreased B. Cunha, Smith, & Lambert (1999) have found a G star in the Orion Association to be B deficient and O rich. Cunha et al. (2000) show the trend, and anti-correlation, between B and O from three G stars and four B stars. Since Li in the G stars is independent of O, it seemed to them unlikely that the ν-process is a significant source of Li and B. They present a simple model that has spatial variations in the light element enrichments in the Orion cloud.

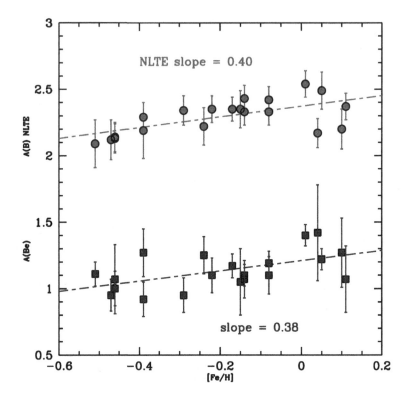

Fig. 10.13. Abundances of B and Be vs. [Fe/H] in Galactic disk stars that are undepleted in Be. Filled circles are the B results; filled squares are Be. These stars are from the upper envelope of disk stars in Figs. 10.3 and 10.4. The slopes are similar, and shallower than the slopes in the halo stars.

There are F and G field stars with apparent B deficiencies. Two examples are Procyon (Lemke, Lambert, & Edvardsson 1993) and ζ Her (Boesgaard et al. 1998b), both of which are very depleted in Li and Be. Procyon is in the Li-Be dip, while ζ Her is a fairly evolved subgiant. A newly identified Li-Be-B deficient star in the Li-Be dip region is HR 107. These stars have huge Li deficiencies—factors of 600 and more—and large Be deficiencies, up to 2 orders of magnitude.

Abundances of B in B-type stars have been determined by Venn et al. (2002) from the B III λ2066 resonance line. In their sample of seven main sequence stars they find two with undepleted B, one with B depleted but detectable, and four where they could only determine upper limits on the B abundance. Of the latter four, three are N rich. They include data on other hot stars from *HST*+GHRS (six stars) and *IUE* (20 stars, predominantly from Proffitt & Quigley 2001) with C, N, and (usually) O abundances in order to understand the B deficiencies and trends. The spread in B in this enlarged sample is 2 orders of magnitude, from A(B) = 2.9 to 0.9. The stars with the lowest B tend to have the highest N. They discuss the role of rotationally induced mixing and its influence on both B and N.

10.7.2 B in the Galactic Disk

Figure 10.4 shows NLTE B abundances in halo and disk F and G stars. In that plot the points represented by crosses are 20 Galactic disk stars from Boesgaard et al. (2004) that are undepleted in Be and, presumably therefore, in B. Although the trend with [Fe/H] is consistent for the halo and disk stars, examination of the disk stars alone indicates a shallower slope. Abundances of both B and Be for those 20 stars alone are shown in Figure 10.13. The slopes of the relationship with [Fe/H] are similar for both B and Be, which indicates a common origin in the disk, presumably spallation. Over a factor of 5 increase in [Fe/H], B and Be gradually increase by nearly a factor of 2. For these stars the B/Be ratio at [Fe/H] is about 14, whereas for the halo stars this ratio is $\sim 15-20$ (Duncan et al. 1998; García López et al. 1998).

10.7.3 Summary

Even though B is the least susceptible to destruction of the three light elements, there are some stars with apparent B deficiencies. Examples of this can be found among the N-rich B stars, F stars in the Li-Be dip, and in the Orion G dwarfs. There is a general increase in B with [Fe/H] with a slope near 0.7. Depending on the reliability of the corrections for NLTE effects in B, there may be some B-rich but metal-poor stars. The ratio of B/Be in the halo is about 15, close to the predictions of Galactic spallation. Disk stars that are undepleted in Be should be undepleted in B; a sample of 20 stars with [Fe/H] between −0.6 and +0.2 have similar relationships between A(B) vs. Fe and A(Be) vs. Fe, with slopes near 0.35 compared to halo stars with steeper slopes.

10.8 Concluding Remarks

We see that Li has an extraordinarily "complex" life with many possible "sources," from the Big Bang to several present-day production mechanisms. Both ^{6}Li and ^{7}Li are fragile elements, and there are many known "sinks" where they can be destroyed. Although there are departures from LTE for Li, the corrections are small, usually less than a few hundredths of a dex.

One the other hand, Be is the most "straightforward" of the light elements. (1) It has only one known source: Galactic cosmic ray spallation. (2) Only the isotope ^{9}Be is stable. (3) It is less susceptible to depletion than Li in stellar interiors. (4) There are minimal effects due to NLTE.

Boron can be characterized as "tough." (1) It is more robust against nuclear destruction. (2) It is difficult to observe because the resonance lines of B I, B II, and B III all occur in the satellite ultraviolet. (3) There seem to be at least two sources of production. (4) Substantial corrections for NLTE effects need to be made.

The abundances of these three elements, separately and in combination, have produced a great wealth of insights into a diverse array of astrophysical matters.

Acknowledgements. It was a pleasure to incorporate the thoughtful comments and suggestions of Dr. Constantine P. Deliyannis and Dr. Douglas K. Duncan. This work was supported by the National Science Foundation grant AST 0097945.

References

Asplund, M. 2000, in IAU Symp. 198, The Light Elements and their Evolution, ed. L. da Silva, R. de Medeiros, & M. Spite (San Francisco: ASP), 448

Asplund, M., Nordlund, A., Trampedach, R., & Stein, R. F. 1999, A&A, 346, L17

Boesgaard, A. M. 2000, in IAU Symp. 198, The Light Elements and their Evolution, ed. L. da Silva, R. de Medeiros, & M. Spite (San Francisco: ASP), 389

Boesgaard, A. M., Armengaud, E., & King, J. K. 2003a, ApJ, 582, 410

——. 2003b, ApJ, 583, 955

Boesgaard, A. M., & Budge, K. G. 1988, ApJ, 332, 410

Boesgaard, A. M., Deliyannis, C. P., King, J. R, Ryan, S. G., Vogt, S. S., & Beers, T. C. 1999, ApJ, 117, 1549

Boesgaard, A. M., Deliyannis, C. P., King, J. R., & Stephens, A. 2001, ApJ, 553, 754

Boesgaard, A. M., Deliyannis, C. P., Stephens, A., & King, J. R. 1998a, ApJ, 493, 206

Boesgaard, A. M., Deliyannis, C. P., Stephens, A., & Lambert, D. L. 1998b, ApJ, 492, 727

Boesgaard, A. M., & King, J. R. 2002, ApJ, 565, 587

Boesgaard, A. M., McGrath, E., Lambert, D. L., & Cunha, K. 2004, ApJ, in press

Boesgaard, A. M., Stephens, A., King, J. R., & Deliyannis, C. P. 2000, Proc. SPIE, 4005, 142

Boesgaard, A. M., & Tripicco, M. 1986, ApJ, 302, L49

Bonifacio, P., et al. 2002, A&A, 390, 91

Bonifacio, P., & Molaro, P. 1998 ApJ, 500, L175

Burbidge, E. M., Burbidge, G. R., Fowler, W. A., & Hoyle, F. 1957, Rev. Mod. Phys., 29, 547

Cameron, A. G. W. 1955, ApJ, 212, 144

Cameron, A. G. W., & Fowler, W. A. 1971, ApJ, 164, 111

Carlsson, M., Rutten, R. J., Bruls, J. H. M. J., & Shchukina, N. G. 1994, A&A, 288, 860

Chaboyer, B., Demarque, P., & Pinsonneault, M. H. 1995, ApJ, 441, 865

Charbonnel, C., Deliyannis, C. P., & Pinsonneault, M. H. 2000, in IAU Symp. 198, The Light Elements and their Evolution, ed. L. da Silva, R. de Medeiros, & M. Spite (San Francisco: ASP), 87

Chen, Y. Q., Nissen, P. E., Benoni, T., & Zhao, G. 2001, A&A, 371, 943

Cunha, K., Smith, V. V., & Lambert, D. L. 1999, ApJ, 519, 844

Cunha, K., Smith, V. V., Parizot, E., & Lambert, D. L. 2000, ApJ, 543, 850

da Silva, L., de Medeiros, R., & Spite, M. 2000, ed., IAU Symp. 198, The Light Elements and their Evolution (San Francisco: ASP)

Deliyannis, C. P., Boesgaard, A. M., Stephens, A., King, J. R., Vogt, S. S., & Keane, M. J. 1998, ApJ, 498, L147

Deliyannis, C. P., & Pinsonneault, M. H. 1997, ApJ, 488, 836

Deliyannis, C. P., Pinsonneault, M. H., & Charbonnel, C. 2000, in IAU Symp. 198, The Light Elements and their Evolution, ed. L. da Silva, R. de Medeiros, & M. Spite (San Francisco: ASP), 61

Deliyannis, C. P., Steinhauer, A., & Jeffries, R. D. 2002, ApJ, 577, L39

Duncan, D. K. 1991, ApJ, 373, 250

Duncan, D. K., Primas, F., Rebull, L. M., Boesgaard, A. M., Deliyannis, C. P., Hobbs, L. M., King, J. R., & Ryan, S. G. 1997, ApJ, 488, 338

Duncan, D. K., Rebull, L. M., Primas, F., Boesgaard, A. M., Deliyannis, C. P., Hobbs, L. M., King, J. R., & Ryan, S. G. 1998, A&A, 332, 1017

Edvardsson, B., Anderson, J., Gustafsson, B., Lambert, D. L., Nissen, P. E., & Tomkin, J. 1993, A&A, 275, 101

Fields, B. D., Daigne, F., Cassè, M., & Vangioni-Flam, E. 2002, ApJ, 581, 389

Ford, A., Jeffries, R. D., & Smalley, B. 2002b, A&A, 391, 253

Ford, A., Jeffries, R. D., Smalley, B., Ryan, S. G., Aoki, W., Kawanomoto, S., James, D. J., & Barnes, J. R. 2002a, A&A, 393, 617

Fuhrmann, K., Axer, M., & Gehren, T. 1994, A&A, 285, 585

García López, R. J., Lambert, D. L., Edvardsson, B., Gustafsson, B., Kiselman, D., & Rebolo, R. 1998, ApJ, 500, 241

García López, R. J., Rebolo, R., & Pérez de Taoro, M. R. 1995, A&A, 302, 184

García López, R. J., Severino, G., & Gomez, M. T. 1995, A&A, 297, 787

Gonzalez, G., Laws, C., Tyagi, S., & Reddy, B. E. 2001, AJ, 121, 432

Hartmann, D., Myers, J., Woosley, S. E., Hoffman, R., & Haxton, W. 1999, in LiBeB, Cosmic Rays, and Related X- and Gamma-Rays, ed. R. Ramaty et al. (San Francisco: ASP), 235

Herbig, G. H. 1965 ApJ, 141, 588

Hernanz, M., Josè, J., Coc, A., & Isern, J. 1996, ApJ, 465, L27

Hobbs, L. M., & Pilachowski, C. A. 1988, ApJ, 334, 734

Jeffries, R. D., Totten, E. J., Harmer, S., & Deliyannis, C. P. 2002, MNRAS, 336, 1109

King, J. R. 2001, PASP, 114, 25

Kiselman, D. 1994, A&A, 286, 169

Kiselman, D., & Carlsson, M. 1995, in The Light Element Abundances, ed. P. Crane (Berlin: Springer), 372
——. 1996, A&A, 311, 680
Kraft, R. P., Peterson, R. C., Guhathakurta, P., Sneden, C., Fulbright, J., & Langer, G. E. 1999, ApJ, 518, L53
Kurucz, R. L. 1993, in SYNTHE Spectrum Synthesis Programs and Line Data, CD-ROM No. 18. (Cambridge, Mass.: Smithsonian Astrophysical Observatory)
Lemke, M., Lambert, D. L., & Edvardsson, B. 1993, PASP, 105, 468
Lemoine, M., Vangioni-Flam, E., & Cassè, M. 1998, ApJ, 499, 735
Meneguzzi, M., Audouze, J., & Reeves, H. 1971, A&A, 15, 337
Merchant, A. E. 1967, ApJ, 147, 587
Parizot, E. 2000, A&A, 362, 786
Parizot, E., & Drury, L. 1999, A&A, 349, 673
Pasquini, L., & Molaro, P. 1997, A&A, 322, 109
Pilachowski, C. A., Sneden, C., Kraft, R. P., Harmer, D., & Willmarth, D. 2000, AJ, 119, 2895
Pinsonneault, M. H., Charbonnel, C., & Deliyannis, C. P. 2000, in IAU Symp. 198, The Light Elements and their Evolution, ed. L. da Silva, R. de Medeiros, & M. Spite (San Francisco: ASP), 74
Pinsonneault, M. H., Steigman, G., Walker, T. P., & Narayanan, V. K. 2002, ApJ, 574, 398
Pinsonneault, M. H., Walker, T. P., Steigman, G., & Narayanan, V. K. 1999, ApJ, 527, 180
Prantzos, N., Cassè, M., & Vangioni-Flam, E. 1993, ApJ, 403, 630
Primas, F., Asplund, M., Nissen, P. E., & Hill, V. 2000b A&A, 364, L42
Primas, F., Duncan, D. K., Peterson, R. C., & Thorburn, J. A. 1999, A&A, 343, 545
Primas, F., Molaro, P., Bonifacio, P., & Hill, V. 2000a A&A, 362, 666
Pritchett, C., & Glaspey, J. 1991, ApJ, 373, 105
Proffitt, C. R., & Quigley, M. F. 2001, ApJ, 548, 429
Ramaty, R., Kozlovsky, B., & Lingenfelter, R. E. 1996, ApJ, 456, 525
Ramaty, R., & Lingenfelter, R. E. 1999, in LiBeB, Cosmic Rays, and Related X- and Gamma-Rays, ed. R. Ramaty et al. (San Francisco: ASP), 104
Ramaty, R., Lingenfelter, R. E., & Kozlovsky, B. 2000, in IAU Symp. 198, The Light Elements and their Evolution, ed. L. da Silva, R. de Medeiros, & M. Spite (San Francisco: ASP), 51
Randich, S., Primas, F., Pasquini, L., & Pallavicini, R. 2002, A&A, 387, 222
Randich, S., Sestito, P., & Pallavicini, R. 2003, A&A, 399, 133
Reeves, H., Fowler, W. A., & Hoyle, F. 1970, Nature, 226, 727
Richer, J., & Michaud, G. 1993, ApJ, 416, 312
Romano, D., Matteucci, F., Molaro, P., & Bonifacio, P. 1999, A&A, 352, 117
Romano, D., Matteucci, F., Ventura, P., & D'Antona, F. 2001, A&A, 374, 646
Ryan, S. G., Gregory, S. G., Kolb, U., Beers, T. C., & Kajino, T. 2002, ApJ, 571, 501
Ryan, S. G., Kajino, T., Beers, T. C., Suzuki, T. K., Romano, D., Matteucci, F., & Rosolankova, K. 2001, ApJ, 549, 55
Sackmann, I.-J., & Boothroyd, A. I. 2000, in IAU Symp. 198, The Light Elements and their Evolution, ed. L. da Silva, R. de Medeiros, & M. Spite (San Francisco: ASP), 98
Santos, N. C., García López, R. J., Israelian, G., Mayor, M., Rebolo, R., Garcia-Gil, A., Pérez de Taoro, M. R., & Randich, S. 2002, A&A, 386, 1028
Smith, V. V., & Lambert, D. L. 1990, ApJ, 361, 69
Smith, V. V., Plez, B., Lambert, D. L., & Lubowich, D. A. 1995, ApJ, 441, 735
Sneden, C. 1973, ApJ, 184, 839
Steigman, G., & Walker, T. P. 1992, ApJ, 385, L13
Stephens, A., Boesgaard, A. M., King, J. R., & Deliyannis, C. P. 1997, ApJ, 491, 339
Thorburn, J. A., Hobbs, L. M., Deliyannis, C. P., & Pinsonneault, M. H. 1993, 415, 150
Timmes, F. X., Wooosley, S. E., & Weaver, T. A. 1995, ApJS, 98, 617
Venn, K. A., Brooks, A. M., Lambert, D. L., Lemke, M., Langer, N., Lennon, D. J., & Kennan, F. P. 2002, ApJ, 565, 571
Wallerstein, G., Herbig, G. H., & Conti, P. S. 1965 ApJ, 141, 610
Wallerstein, G., & Sneden, C. 1982, ApJ, 255, 577
Woosley, S. E., Hartmann, D., Hoffman, R., & Haxton, W. 1990, ApJ, 356, 272
Woosley, S. E., & Weaver, T. A. 1995, ApJS, 101, 181

11

Extremely metal-poor stars

JOHN E. NORRIS

Research School of Astronomy & Astrophysics, The Australian National University

Abstract

The discovery and analysis of extremely metal-poor stars have led to increasing insight into conditions when the Universe and Galaxy were young. We present the rationale for studying such objects, with a description of their systematic discovery, culminating in the recent analysis of an object having [Fe/H] = −5.3. We discuss the abundance patterns of several elements from Li through to the heavy neutron capture elements. Relatively few extremely metal-poor stars with [Fe/H] < −3.5 have been analyzed at high spectral resolution and high signal-to-noise ratio, but some 40% of that sample show astounding overabundances of some or all of the CNO group and the lighter α elements. At [Fe/H] $\approx$ −3.0, some stars show enormous enhancements of r-process elements, leading to age determinations. The diversity among the most metal-poor stars has yet to be fully understood.

11.1   Introduction

Extremely metal-poor stars provide fundamental clues to conditions at the earliest times. In the half century since the pioneering high-resolution chemical abundance analysis of spectra of metal-poor objects obtained using Carnegie's Mount Wilson 100-inch telescope by Chamberlain & Aller (1951), the field has continued to grow and is still attracting large allocations of observing time on the largest reflectors.

The rationale for the continued interest in these objects may be summarized as follows:

- They are the stars closest in time to the Big Bang.
- The abundance of lithium in metal-poor main sequence stars places constraints on Big Bang nucleosynthesis and the density of baryonic material in the Universe.
- Having formed at redshifts $z \gtrsim 4-5$, they probe conditions when the first heavy element-producing objects formed.
- They constrain our understanding of the explosions of the first supernovae/hypernovae.
- They provide clues on the manner in which the ejecta of the first supernovae were incorporated into later generations.
- They provide insight into the initial mass function at earliest times.
- They constrain our understanding of the manner in which the Galaxy formed.
- In some objects with large relative overabundances of the heavy neutron capture elements, the measurement of the abundances of thorium and uranium permits estimates of stellar ages, and hence that of the Galaxy.

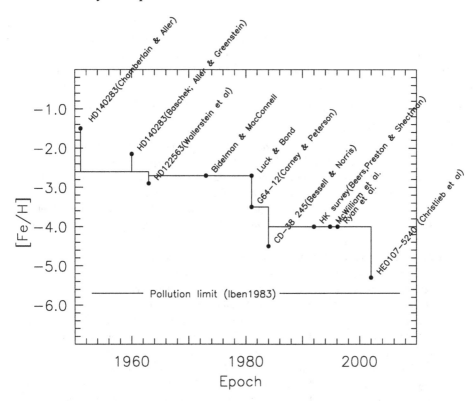

Fig. 11.1. [Fe/H] for the most metal-poor star then known as a function of epoch. The symbols refer to the values published by the authors, and are connected to currently accepted values (with the exception of HE 0107–5240).

11.2 Discovery

Chamberlain & Aller (1951) first explained the anomalous spectra of high-velocity stars in terms of lower chemical abundances. Sandage (1986) describes their work as "unbelievably revolutionary," providing as it did one of the fundamental parameters of stellar populations, at a time when the prevailing wisdom was that all stars had the same chemical properties. (He also nicely relates "a widely circulated underground story*'' in which "It is asserted that the authors had first obtained heavy element abundance deficiencies of ∼100 relative to the Sun (in fact, the modern value), but this seemed so outrageous that the authors themselves ... changed the atmospheric temperatures sufficiently to reduce the effect from a factor of 100 to a factor of ∼10–30, the value they ultimately published."

Figure 11.1 shows the steady downward march with time of the iron abundance of the most metal-poor object then known. The record is currently held by HE 0107–5240 (Christlieb et al. 2002) with [Fe/H] = –5.3. Also shown in the figure is the estimate by Iben (1983) of the abundance one would measure for a hypothetical red giant that formed at the earliest times from material containing no heavy elements, and which accreted material from the ambient interstellar medium as it orbited the Galaxy until the present epoch.

Extremely metal-poor stars are rare. In the solar neighborhood ∼one in 1000 stars belongs

---

* C. Sneden notes: "I personally heard this story from Chamberlain himself, so I think it is pretty accurate."

to the metal-poor halo population to which such objects belong. Further, simple models of chemical enrichment of a primordial cloud predict that at lowest abundance the number of metal-poor objects decreases by a factor of 10 for each factor of 10 decrease in metal abundance. Roughly speaking, then, one might expect some five in a million stars to have [Fe/H] < −3.5.

The discovery of extremely metal-poor objects has been a painstaking process, and has proceeded, essentially, in three ways:

(1) *Informed serendipity.* Two cases will illustrate the point. Attention was first drawn to the extremely metal-poor red giant CD–38°245 ([Fe/H] = −4.0) by Slettebak & Brundage (1971) in their search for A stars at the South Galactic Pole. The carbon-rich binary dwarf G77–61 ([Fe/H] = −5.5) was first extensively investigated astrometrically. It was only in subsequent studies that their extreme metal deficiency was appreciated (Bessell 1977; Gass, Wehrse, & Liebert 1988). (It would also seem reasonable to note that the extreme value claimed for G77–61 by Gass et al. requires confirmation.)

(2) *Systematic surveys of high-proper motion stars.* The low-resolution spectroscopic surveys of the high-proper motion catalogs of Giclas and Luyten, which selectively sample the halo population kinematically, provided the first estimates of the fraction of halo material having [Fe/H] < −3.0. See Ryan & Norris (1991) and Carney et al. (1994).

(3) *Systematic objective prism surveys for weak-lined objects with Schmidt telescopes.* Following the early work of Bidelman & MacConnell (1973), Schmidt telescopes have become the predominant source of extremely metal-poor candidates. Using follow-up intermediate-resolution spectroscopy of the Ca II K line (as proxy for the heavy elements), the HK surveys of Beers, Preston, & Shectman (1985, 1992) and Beers and co-workers (see Beers 1999), together with the HES survey of Christlieb and collaborators (see Christlieb at al. 1999) are now producing statistically significant numbers of such objects. At time of writing, these authors are discovering a few tens of objects (based on intermediate-resolution spectra ($R \approx 2000$) with [Fe/H] < −3.5 (Beers and Christlieb, private communications).

While the list of extremely metal-poor stars is growing rapidly, only a handful have published abundance analyses based on high-resolution ($R \approx 50,000$), high signal-to-noise ratio (S/N $\approx$ 100) spectra. (That said, see Spite et al. 2004 for an early report on data obtained with the VLT/UVES combination.) Table 11.1 presents the eight (and possibly nine) objects having [Fe/H] $\lesssim$ −3.5 for which high-quality data are available (from Gass et al. 1988; Norris, Beers, & Ryan 2000; Norris, Ryan, & Beers 2001; Aoki et al. 2002a; Christlieb et al. 2002). Note that six of the nine were discovered in the HK and HES surveys.

11.3   The Halo Metallicity Distribution Function

Figure 11.2 presents the metallicity distribution function for several halo samples. The rarity of the most metal-poor objects is clear in the figure: currently only 2–3 objects are known with [Fe/H] < −4.0. The data in the lower panels are still preliminary, but they rule out the simple model of Galactic chemical enrichment of a zero heavy element abundance system (Searle & Sargent 1972; Hartwick 1976; Ryan & Norris 1991)—the predicted number decrease by a factor of 10 per factor of 10 decrease in abundance is not seen. The data provide the starting point for more complicated models, such as those of Tsujimoto, Shigeyama, & Yoshii 1999); see also Oey (2004).

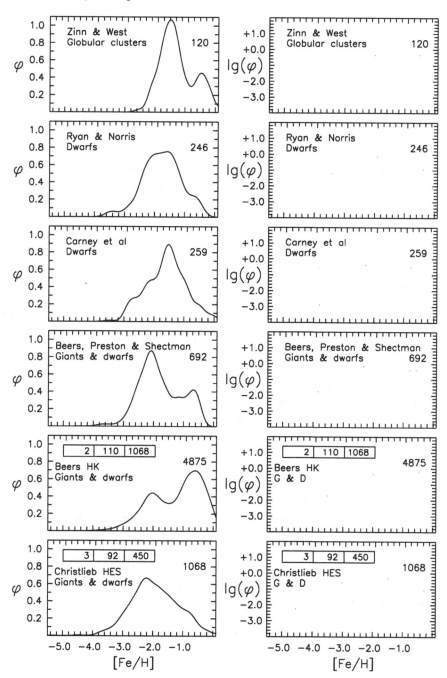

Fig. 11.2. The metallicity distribution function for several halo samples: globular clusters (Zinn & West 1984), high-proper motion objects (Ryan & Norris 1991; Carney et al. 1994), and Schmidt surveys (Beers et al. 1992; Beers 1999; Christlieb 2002, private communication). The data are presented as generalized histograms with Gaussian kernel having $\sigma = 0.15$. In each panel the total sample size is given on the right, while in the bottom two sets of panels the numbers in the boxes contain the sample size per abundance decade.

Table 11.1. *Stars with [Fe/H]* $\lesssim$ *−3.5 and Comprehensive Abundance Analysis*

Object	[Fe/H]	Technique	Type
CD–38°245	−4.0	Serendipity	Giant
CD–24°17504	−3.4	High proper motion	Dwarf
CS 22876–032	−3.7	Schmidt BPS	Dwarf
CS 22885–096	−3.7	Schmidt BPS	Giant
CS 22949–037	−3.8	Schmidt BPS	Giant
CS 22172–002	−3.6	Schmidt HK	Giant
CS 29498–043	−3.7	Schmidt HK	Giant
HE 0107–5240	−5.3	Schmidt HES	Giant
& possibly			
G 77–61	−5.5	Serendipity	Dwarf

11.4   Lithium

Spite & Spite (1982) first demonstrated that halo main sequence stars hotter than $T_{\mathrm{eff}} \approx 5500$ K appear to have the same lithium abundance—A(Li) = log N(Li)/N(H) + 12.0 ≈ 2.0. Subsequently, enormous effort has been expended to interpret this in terms of the primordial lithium abundance, and via Big Bang nucleosynthesis to determine the baryonic density of the Universe. To achieve the transition, one has to understand Li production and destruction mechanisms since the Big Bang. While this lies outside the scope of the present review, suffice it to say that one has to determine the amount of Li created by spallation in the interstellar medium and destroyed in and/or removed from the outer layers of main sequence halo stars by processes such as diffusion and rotational mixing.

There have been claims that A(Li) increases with increasing [Fe/H] at lowest abundance (Norris, Ryan, & Stringfellow 1994; Thorburn 1994; Ryan, Norris, & Beers 1999)—and counterclaims (Molaro, Primas, & Bonifacio 1995). Figure 11.3 shows data for an unbiased sample of field stars from Ryan et al. (1999), supporting the case that a positive correlation exists. The first reason for presenting this diagram is to emphasize the role of extremely metal-poor stars: the reader will need little convincing that further objects having [Fe/H] < −3.5 are needed to more strongly constrain the result.

A second point to take from Figure 11.3 is that there is little dispersion in A(Li). Ryan et al. (1999) claim that the dispersion is so small that it "limits permissible depletion by rotationally induced mixing models to less than 0.1 dex." For a counterargument, that the data permit a 0.2 dex median depletion, the reader should see Pinsonneault et al. (2002). (For recent, conflicting, claims on the Li abundance spread in near main sequence turn-off stars in globular clusters, see Bonifacio et al. 2002 and Boesgaard 2004.)

One should be alive to the fact that the Li abundances cited above are determined by using one-dimensional (1-D) model atmospheres and the assumption of local thermodynamic equilibrium (LTE). Asplund et al. (1999) and Aspund, Carlsson, & Botnen (2003) show that for models of near main sequence turn-off stars having [Fe/H] = −2.5, a change to 3-D decreases A(Li) by ∼0.3, while non-LTE modeling leads to an increase by a similar amount, and thus a net error of order only 0.05. It will be interesting to learn whether the cancellation applies for [Fe/H] = −4.0.

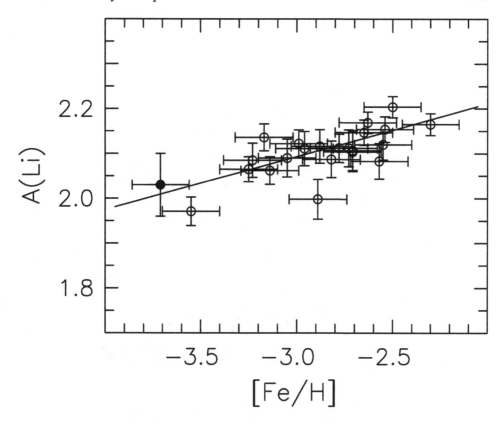

Fig. 11.3. A(Li) versus [Fe/H], from Ryan et al. (1999) and Norris et al. (2000).

Depending on one's view of the implications of abundance spread on the Spite Plateau for Li depletion, the dependence of A(Li) on [Fe/H], and model atmosphere assumptions, one comes to different values of the primordial lithium abundance. Ryan et al. (2000), for example, report $Li_p = 1.23^{+0.68}_{-0.32} \times 10^{-10}$, leading to $\Omega_b h^2 = 0.006–0.014$. This is lower than the value of $\Omega_b h^2 = 0.021 \pm 0.002$ reported by O'Meara et al. (2001) from primordial deuterium measurements, and the Spergel et al. (2003) value of $\Omega_b h^2 = 0.022 \pm 0.001$ from the *WMAP* experiment.

If one turns the argument around, and adopts the *WMAP* result and the Big Bang nucleosynthesis Li abundance ($Li_p \approx 4 \times 10^{-10}$) as given, one now has the option to use the Li observations to constrain production and destruction mechanisms, and the accuracy of the model atmosphere abundance analysis.

11.5 Relative Abundances versus [Fe/H]

Relative abundances, [M/Fe], contain important clues to the nature of the first heavy element-producing objects. In what follows, emphasis is given to the most metal-poor stars—those having [Fe/H] < –3.5—and what they have to tell us about these objects.

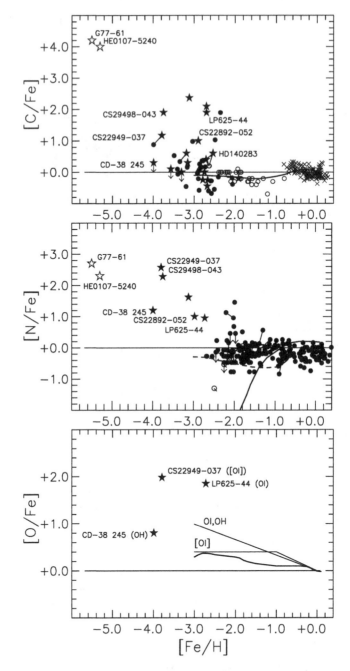

Fig. 11.4. Relative C, N, and O abundances. The sources of the data are given in Norris, Ryan, & Beers (1977), Norris (1999), and in the text. For oxygen, schematic representations are shown of recent results based on [O I] λ6300 Å, on the one hand, and the O I infrared triplet and ultraviolet OH lines, on the other. The full lines are from the theoretical models of Timmes, Woosley, & Weaver (1995), while the dashed line in the middle panel represents their result with the arbitrary inclusion of convective overshoot.

11.5.1 CNO

At the lowest values of [Fe/H], the results for the CNO group are astounding and totally unexpected, as demonstrated in Figure 11.4. Of the eight stars (excluding G77–61)* in Table 11.1, which have [Fe/H] < –3.5 (and which form an unbiased sample as far as CNO abundances are concerned), three have [C/Fe] > +1.0 and [N/Fe] > +2.0, while a fourth has [N/Fe] = +1.2. One of the three (CS 22949–037) has [O/Fe] = 2.0†. (For details, see Norris et al. 2001, 2002; Aoki et al. 2002a; Christlieb at al. 2002; Depagne et al. 2002).

We defer discussion of these results to § 11.6.2.

11.5.2 α Elements

Results for the α elements are shown in Figure 11.5, and are as presented by Norris et al. (2001), with additional material from Aoki et al. (2002a, CS 29498–043) and Christlieb et al. (2002, HE 0107–5240).

Of the eight extremely metal-poor stars in Table 11.1, CS 22949–037 and CS 29498–043 have [Mg/Fe] > 1.0. The overabundances for these objects are real and outside uncertainties in the analysis. Errors are large for [Si/Fe], but here, too, these objects appear to have [Si/Fe] > 1.0. They also have large enhancements of C, N, and O, which sets them apart from the remainder of the group. We shall return to these objects in § 11.6.2.

11.5.3 Iron-peak Elements

For data on the relative abundances for the iron-peak elements the reader is referred to McWilliam et al. (1995), Ryan et al. (1996), and Norris et al. (2001). Of particular importance is the result that for [Fe/H] ≲ –3.0, [Cr/Fe] and [Mn/Fe] decrease as [Fe/H] decreases, while [Co/Fe] increases. [Ni/Fe] maintains its solar value. These results were not predicted by supernova modeling, but Nakamura et al. (1999) and Umeda & Nomoto (2002) have investigated the effects in terms of the assumed mass cut for the ejection of material in supernova explosions, and the energy of explosion. The reader is referred to Umeda & Nomoto for their explanation of the observations in terms of the explosion of hypernovae with energies of 50×10^{51} erg. The discussion presents a nice example of the interplay between observation and theory.

11.5.4 Heavy Neutron Capture Elements

Results for the heavy neutron capture elements Sr and Ba are shown in Figure 11.6. The enormous spread in [Sr/Fe] of 2 dex at [Fe/H] ≈ –3.5 attests to the patchy nature of the enrichment of these elements. [Ba/Fe], on the other hand, does not show the same range, which suggests that more than one type of object is responsible for the enrichment of these elements. The reader is referred to McWilliam (1998), Blake et al. (2001), and Ryan et al. (2001) who discuss this in terms of the basic *r*-process necessary to explain the well-established patterns of abundance in stars with [Fe/H] < –2.5, and an ill-defined second one that favors production of elements of heavy neutron capture elements of lower atomic number.

Further evidence for patchy enrichment comes from [Eu/Fe] versus [Fe/H], as is summarized by Truran et al. (2002), down to [Fe/H] = –3.0. Fields, Truran, & Cowan (2002)

---

* While it is probably not unreasonable to say that most astronomers with an interest in the matter are reserving judgement on the reality of [Fe/H] = –5.5 for G77–61, given the similarity of its carbon and nitrogen abundances to those of some of the other stars with [Fe/H] < –3.5, confirmation of its iron abundance takes on a new light.

† It is worth noting that Israelian (2004) reports [O/Fe] = 3.2 in CS 29498–043.

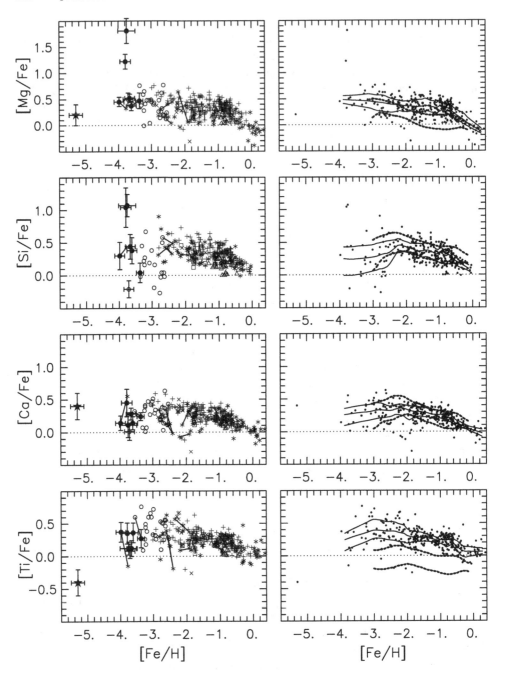

Fig. 11.5. Relative abundances for the α elements as a function of [Fe/H]. The data are as presented by Norris et al. (2001), with the addition of CS 29498–043 and HE 0107–5240 from Aoki et al. (2002a) and Christlieb et al. (2002), respectively. In the right panel, the thin lines (effectively) represent running means and quartiles, while the thicker lines are theoretical results from Timmes et al. (1995).

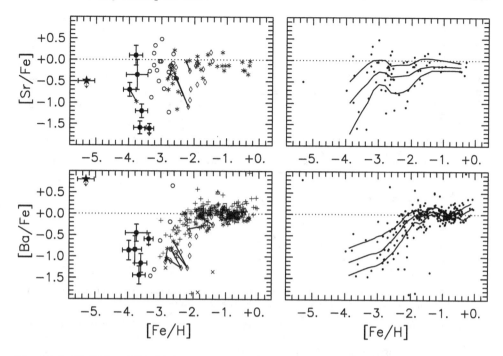

Fig. 11.6. [Sr/Fe] and [Ba/Fe] versus [Fe/H], from Norris et al. (2001), together with data from Aoki et al. (2002a, CS 29498–043) and Christlieb et al. (2002, HE 0107–5240). In the right panel the lines (effectively) represent running means and quartiles.

describe the results in terms of the different frequency of supernovae that produce *r*-process elements and iron. They note that data for [Fe/H] < –3.0 will be able to distinguish various possibilities. Given the intrinsic weakness of the europium lines available for measurement, however, this will not be a trivial task.

11.6 Abundance as a Function of Atomic Number

11.6.1 *Stars with Extreme Enhancements of r- and s-Process Elements*

One of the most exciting results to emerge from the study of extremely metal-poor stars has been the discovery of objects with [Fe/H] ≈ –3.0 with enormous overabundances of the heavy neutron capture elements in an *r*-process pattern—in particular CS 22892–052 (see Sneden et al. 1996, 2003) and CS 31082–001 (Cayrel et al. 2001). Figure 11.7 shows [M/H] versus atomic species for these stars, together with comparison data for the "normal" metal-poor red giant HD 122563. Also shown are results for LP 625–44, which exhibits huge *s*-process enhancement rather than the *r*-process pattern. (This can be seen from the overabundance of Ba relative to Eu in LP 625–44 in contrast to its relative underabundance in CS 22892–052 and CS 31082–001.)

Stars such as LP 625–44 are normally assumed to result from mass transfer of material from an asymptotic giant branch star across a binary onto the companion star that is now observed—giving insight into asymptotic giant branch nucleosynthesis at very low abundance (see Aoki et al. 2002b). It is not clear, however, that this provides a complete explanation of all metal-poor stars with *s*-process enhancement. Witness [O/Fe] = 1.8 in LP 625–44

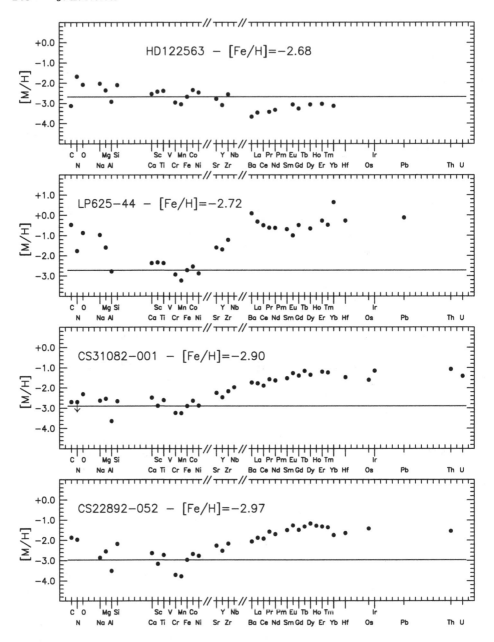

Fig. 11.7. Abundance relative to hydrogen for HD 122563 and the heavy neutron capture element-enhanced stars LP 625–44, CS 31082–001, and CS 22892–052. (Data from Lambert, Sneden, & Ries 1974; Sneden & Parthasarathy 1983; McWilliam et al. 1995; Ryan et al. 1996; Sneden et al. 1996; Norris et al. 1997, 2001; Aoki et al. 2002b; Hill et al. 2002.)

(Aoki et al. 2002b), which is difficult to explain in this manner, and the apparent radial velocity constancy of the strongly *s*-process-enhanced, metal-poor ([Fe/H] = –2.7) star LP 706–7 (Norris et al. 1997).

The reader is referred to Truran et al. (2002), Cayrel et al. (2004), and references therein, for details of age determinations based on Th and U abundances for the some half dozen *r*-process-enhanced objects currently known—which fall in the range 11–15 Gyr. The work of Wanajo et al. (2002) is also of considerable interest. They note that ages determined in this manner are very sensitive to the currently somewhat poorly understood modeling of the objects responsible for the *r*-process enhancements.

There is one final point from Figure 11.7 worth comment. In spite of the general similarity between the abundance patterns of CS 22892–052 and CS 31082–001, the behavior of C and N is quite different. In the latter, [C/Fe] has the solar value, while in the former [C/Fe] and [N/Fe] are overabundant by 1 dex. Canonical wisdom assumes that two processes have operated in CS 22892–052, and only one in CS 31082–001, but offers little insight into the matter. There is, however, suggestive but weak evidence that CS 22892–052 may be a binary (Preston & Sneden 2001).

11.6.2 Stars with Extreme CNO and Mg Enhancements

As noted in § 11.5.1, three of the most metal-poor stars (HE 0107–5240, CS 22949–037, and CS 29498–043) have enormous overabundances of some or all of C, N, and O, and the light α elements. Figure 11.8 compares their abundances, as a function of atomic species, with those of the "normal" extremely metal-poor star CD–38°245. The important point to note here is that the anomalously high abundances in the CNO-enhanced objects pertain only to elements less massive than Ca. The iron-peak and heavy neutron capture elements do not appear abnormal.

This behavior has led to suggestions that the peculiar abundance patterns seen in CS 22949–037 and CS 29498–043 are the result of ejecta from a supernova that has expelled only its outermost parts, with the iron-peak-enriched material falling back onto the exploding object (McWilliam et al. 1995; Norris et al. 2001, 2002; Depagne et al. 2002; Umeda & Nomoto 2003). The reader is referred to the paper by Nomoto (2004) in this volume. In the case of CS 22949–037, Umeda & Nomoto postulate a "mixing and fallback" model in which a 30 $M_\odot$ supernova, exploding with energy 20×10^{51} erg, completely mixes material within its 2.33–8.56 $M_\odot$ mass shell, with most of the shell falling back onto the star and only a small fraction ($f = 0.002$) being expelled. The model produces a good fit to the data except for N, where it fails completely, and also for Na.* In the case of N a further process is required, postulated to be subsequent CN cycle burning and mixing into the envelope of the star currently being observed. Nitrogen enhancements are common in globular cluster giants, but not in the field, and have been suggested to originate in this manner.

The nitrogen problem is an outstanding one. Inspection of the middle panel of Figure 11.4 suggests that canonical models fail to explain the apparently primary nature of nitrogen observed down to [Fe/H] = –2.5, and solutions such as rotation, convective overshoot, and the massive zero-heavy element hypernovae (massive pair instability supernovae) of Woosley & Weaver (1982) have been advocated (see Norris et al. 2002, and references therein).

One has to wonder if any of these agencies is playing a role in the enormous values of [N/Fe] found at [Fe/H] < –3.5. As noted by Carr, Bond, & Arnett (1984), the essential

---

* Umeda & Nomoto (2003) explain the abundance pattern of HE 0107–5240 in terms of a "mixing and fallback" model of a supernova of mass 25 $M_\odot$, explosion energy 0.3×10^{51} erg, mixing within a shell of 1.83–6.00 $M_\odot$, and an expelled mass fraction $f = 0.00002$. Note the large factor (60) between the explosion energies of the models for CS 22949–037 and HE 0107–5240.

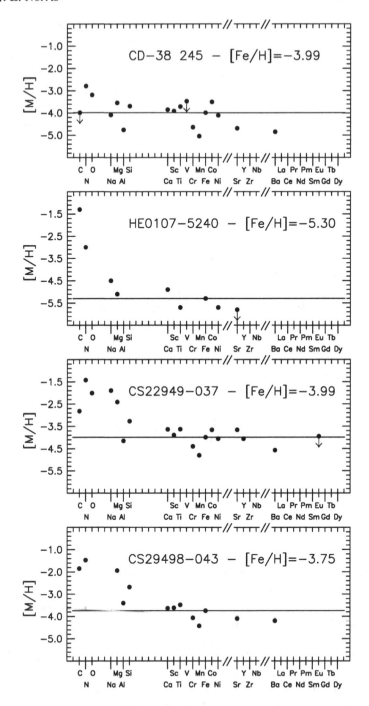

Fig. 11.8. Abundance relative to hydrogen for CD–38°245 and three extremely metal-poor, CNO-enhanced objects, HE 0107–5240, CS 22949–037, and CS 29498–043. (Data from McWilliam et al. 1995; Norris et al. 2001, 2002; Aoki et al. 2002a; Christlieb et al. 2002; Depagne et al. 2002; Bessell 2004.)

feature of the very massive ($M \gtrsim 200\ M_\odot$) objects is their potential to "pass carbon and oxygen from the helium-burning core through the hydrogen-burning shell, in such a way that it is CNO processed to nitrogen before entering the hydrogen envelope." Also of particular interest is the work of Fryer, Woosley, & Heger (2001), who consider the evolution of rotating zero-heavy element objects of mass 250 and 300 $M_\odot$. As noted by Norris et al. (2002) these models seem quite capable of explaining the observed C, N, and O abundances in CS 22949–037. That said, it should be noted it is not clear whether the observed α-element enhancements can be produced by such objects. Finally, the role of rotation in very metal-poor stars with $M \leq 60\ M_\odot$ in producing nitrogen enhancements has been considered by Meynet & Maeder (2002). These models seem incapable of explaining the large excesses being considered here.

11.7 Conclusion

In the half century since Chamberlain & Aller (1951) first reported the existence of metal-poor stars, objects having greater and greater deficiency have been discovered, with the most extreme case having [Fe/H] = –5.3—close to the observational limit that Iben (1983) suggests is set by accretion from the interstellar medium. The subject is still, however, in its infancy when it comes to understanding the details and implications of the abundance patterns of extremely metal-poor stars, defined by, say, [Fe/H] < –3.5. With the increasing number of extremely metal-poor candidates currently being discovered by the HK and HES surveys, together with the existing high-dispersion spectrographs on 8 and 10-meter telescopes, one should expect a rapid increase in our understanding of these objects in the coming decade, together with closer insight into the nature of the first heavy element-producing objects.

References

Aller, L. H., & Greenstein, J. L. 1960, ApJS, 5, 139
Aoki, W., et al. 2002b, PASJ, 54, 427
Aoki, W., Norris, J. E., Ryan, S. G., Beers, T. C., & Ando, H. 2002a, ApJ, 576, L141
Asplund, M., Carlsson, M., & Botnen, A. V. 2003, A&A, 399, L31
Asplund, M., Nordlund, A., Trampedach, R., & Stein, R. F. 1999, A&A, 346, L17
Baschek, B. 1959, ZsfAp, 48, 95
Beers, T. C. 1999, in The Third Stromlo Symposium: The Galactic Halo, ed. B. K. Gibson, T. S. Axelrod, & M. E. Putman (San Francisco: ASP), 202
Beers, T. C., Preston, G. W., & Shectman, S. A. 1985, AJ, 90, 2089
——. 1992, AJ, 103, 1987
Bessell, M. S. 1977, PASA, 3, 144
——. 2004, in CNO in the Universe, ed. C. Charbonnel, D. Schaerer, & G. Meynet (San Francisco: ASP), in press
Bessell, M. S., & Norris, J. 1984, ApJ, 285, 622
Bidelman, W. P., & MacConnell, D. J. 1973, AJ, 78, 687
Blake, L. A. J., Ryan, S. G., Norris, J. E., & Beers, T. C. 2001, Nucl. Phys. A, 688, 502
Boesgaard, A. M. 2004, in Carnegie Observatories Astrophysics Series, Vol. 4: Origin and Evolution of the Elements, ed. A. McWilliam & M. Rauch (Cambridge: Cambridge Univ. Press), in press
Bonifacio, P., et al. 2002, A&A, 390, 91
Carney, B. W., Latham, D. W., Laird, J. B., & Aguilar, L. A. 1994, AJ, 107, 2240
Carney, B. W., & Peterson, R. C. 1981, ApJ, 245, 238
Carr, B. J., Bond, J. R., & Arnett, W. D. 1984, ApJ, 277, 445
Cayrel, R., et al. 2001, Nature, 409, 691
——. 2004, in Carnegie Observatories Astrophysics Series, Vol. 4: Origin and Evolution of the Elements, ed. A. McWilliam & M. Rauch (Pasadena: Carnegie Observatories, http://www.ociw.edu/ociw/symposium/series/symposium4/proceedings.html)

Chamberlain, J. W., & Aller, L. H. 1951, ApJ, 114, 52

Christlieb, N., et al. 2002, Nature, 419, 904

Christlieb, N., Wisotzki, L., Reimers, D., Gehren, T., Reetz, J., & Beers, T. C. 1999, in The Third Stromlo
    Symposium: The Galactic Halo, ed. B. K. Gibson, T. S. Axelrod, & M. E. Putman (San Francisco: ASP), 259

Depagne, E., et al. 2002, A&A, 390, 187

Fields, B. D., Truran, J. W., & Cowan, J. J. 2002, ApJ, 575, 845

Fryer, C. L., Woosley, S. E., & Heger, A. 2001, ApJ, 550, 372

Gass, H., Wehrse, R., & Liebert, J. 1988, A&A, 189, 194

Hartwick, F. D. A. 1976, ApJ, 209, 418

Hill, V., et al. 2002, A&A, 387, 560

Iben, I., Jr. 1983, Mem.S.A.It., 54, 321

Israelian, G. 2004, in Carnegie Observatories Astrophysics Series, Vol. 4: Origin and Evolution of the Elements,
    ed. A. McWilliam & M. Rauch (Pasadena: Carnegie Observatories,
    http://www.ociw.edu/ociw/symposia/series/symposium4/proceedings.html)

Lambert, D. L., Sneden, C., & Ries, L. M. 1974, ApJ, 188, 97

Luck, R. E., & Bond, H. E. 1981, ApJ, 244, 919

McWilliam, A. 1998, AJ, 115, 1640

McWilliam, A., Preston, G. W., Sneden, C., & Searle, L. 1995, AJ, 109, 2757

Meynet, G., & Maeder, A. 2002, A&A, 390, 561

Molaro, P., Primas, F., & Bonifacio, P. 1995, A&A, 295, L47

Nakamura, T., Umeda, H., Nomoto, K., Thielemann, F.-K., & Burrows, A. 1999, ApJ, 517, 193

Nomoto, K. 2004, in Carnegie Observatories Astrophysics Series, Vol. 4: Origin and Evolution of the Elements,
    ed. A. McWilliam & M. Rauch (Pasadena: Carnegie Observatories,
    http://www.ociw.edu/ociw/symposia/series/symposium4/proceedings.html)

Norris, J. E. 1999, in The Third Stromlo Symposium: The Galactic Halo, ed. B. K. Gibson, T. S. Axelrod, & M. E.
    Putman (San Francisco: ASP), 213

Norris, J. E., Beers, T. C., & Ryan, S. G. 2000, ApJ, 540, 456

Norris, J. E., Ryan, S. G., & Beers, T. C. 1997, ApJ, 488, 350

——. 2001, ApJ, 561, 1034

Norris, J. E., Ryan, S. G., Beers, T. C., Aoki, W., & Ando, H. 2002, ApJ, 569, L107

Norris, J. E., Ryan, S. G., & Stringfellow, G. S. 1994, 423, 386

Oey, M. S. 2004, in Carnegie Observatories Astrophysics Series, Vol. 4: Origin and Evolution of the Elements, ed.
    A. McWilliam & M. Rauch (Pasadena: Carnegie Observatories,
    http://www.ociw.edu/ociw/symposia/series/symposium4/proceedings.html)

O'Meara, J. M., Tytler, D., Kirkman, D., Suzuki, N., Prochaska, J. X., Lubin, D., & Wolfe, A. M. 2001, ApJ, 552,
    718

Pinsonneault, M. H., Steigman, G., Walker, T. P., & Narayanan, V. K. 2002, ApJ, 574, 398

Preston, G., & Sneden, C. 2001, AJ, 122, 1545

Ryan, S. G., Aoki, W., Blake, L. A. J., Norris, J. E., Beers, T. C., Gallino, R., Busso, M., & Ando, H. 2001,
    Mem.S.A.It., 72, 337

Ryan, S. G., Beers, T. C., Olive, K. A., Fields, B. D., & Norris, J. E. 2000, ApJ, 530, L57

Ryan, S. G., & Norris, J. E. 1991, AJ, 101, 1865

Ryan, S. G., Norris, J. E., & Beers, T. C. 1996, ApJ, 471, 254

——. 1999, ApJ, 523, 654

Sandage, A. 1986, ARA&A, 24, 241

Searle, L., & Sargent, W. L. W. 1972, ApJ, 173, 25

Slettebak, A., & Brundage, R. K. 1971, AJ, 76, 338

Sneden, C., et al. 2003, ApJ, 591, 963

Sneden, C., & Parthasarathy, M. 1983, ApJ, 267, 757

Sneden, C., McWilliam, A., Preston, G. W., Cowan, J. J., Burris, D. L., & Armosky, B. J. 1996, ApJ, 467, 819

Spergel, D. N., et al. 2003, ApJS, 148, 175

Spite, F., & Spite, M. 1982, A&A, 115, 357

Spite, M., et al. 2004, in Carnegie Observatories Astrophysics Series, Vol. 4: Origin and Evolution of the
    Elements, ed. A. McWilliam & M. Rauch (Pasadena: Carnegie Observatories,
    http://www.ociw.edu/ociw/symposia/series/symposium4/proceedings.html)

Thorburn, J. A. 1994, ApJ, 421, 318

Timmes, F. X., Woosley, S. E., & Weaver, T. A. 1995, ApJS, 98, 617

Truran, J. W., Cowan, J. J., Pilachowski, C. A., & Sneden, C. 2002, PASP, 114, 1293

Tsujimoto, T., Shigeyama, T., & Yoshii, Y. 1999, ApJ, 519, L63

Umeda, H., & Nomoto, K. 2002, ApJ, 565, 385

——. 2003, Nature, 422, 871

Wallerstein, G., Greenstein, J. L., Parker, R., Helfer, H. L., & Aller, L. H. 1963, ApJ, 137, 280

Wanajo, S., Itoh, N., Ishimaru, Y., Nozawa, S., & Beers, T. C. 2002, ApJ, 577, 853

Woosley, S. E., & Weaver, T. A. 1982, in Supernovae: A Survey of Current Research, ed. M. J. Rees & R. J. Stoneham (Dordrecht: Reidel), 79

Zinn, R., & West, M. J. 1984, ApJS, 55, 45

12

Thin and thick Galactic disks

POUL E. NISSEN

Department of Physics and Astronomy, University of Aarhus, Denmark

Abstract

Studies of elemental abundances in stars belonging to the thin and the thick disk of our Galaxy are reviewed. Edvardsson et al. (1993) found strong evidence of [α/Fe] variations among F and G main sequence stars with the same [Fe/H] and interpreted these differences as due to radial gradients in the star formation rate in the Galactic disk. Several recent studies suggest, however, that the differences are mainly due to a separation in [α/Fe] between thin and thick disk stars, indicating that these populations are discrete Galactic components, as also found from several kinematical studies. Further evidence of a chemical separation between the thick and the thin disk is obtained from studies of [Mn/Fe] and the ratio between r- and s-process elements. The interpretation of these new data in terms of formation scenarios and time scales for the disk and halo components of our Galaxy is discussed.

12.1   Introduction

A long-standing problem in studies of Galactic structure and evolution has been the possible existence of a population of stars with kinematics, ages, and chemical abundances in between the characteristic values for the halo and the disk populations. Already at the Vatican Conference on Stellar Populations (O'Connell 1958), an *intermediate Population II* was introduced as stars with a velocity component perpendicular to the Galactic plane on the order of $W \approx 30$ km s^{-1}. Using the m_1 index of F-type stars, Strömgren (1966) later defined intermediate Population II as stars having metallicities in the range $-0.8 <$ [Fe/H] < -0.4, and from a discussion of the extensive $uvby$-β photometry of Olsen (1983), he concluded that the intermediate Population II consisted of old, 10–15 Gyr stars with velocity dispersions ($\sigma_U, \sigma_V, \sigma_W$) significantly higher than those of the younger, more metal-rich disk stars (Strömgren 1987, Table 2).

In a seminal paper, Gilmore & Reid (1983) showed that the distribution of stars in the direction of the Galactic South Pole could not be fitted by a single exponential, but required at least two disk components—a *thin disk* with a scale height of 300 pc and a *thick disk* with a scale height of about 1300 pc. They furthermore identified intermediate Population II with the sum of the metal-poor end of the old thin disk and the thick disk. Following this work, it has been intensively discussed if the thin and thick disks are discrete components of our Galaxy or if there is a more continuous sequence of stellar populations connecting the Galactic halo and the thin disk. For a comprehensive review and a discussion of possible formation scenarios, the reader is referred to Majewski (1993).

Quite a strong indication of the thin and thick Galactic disks as discrete populations with respect to kinematics and age came from the detailed abundance survey of Edvardsson et al. (1993). On the basis of the large *uvby-β* catalogs of Olsen (1983, 1988), main sequence stars in the temperature range $5600\,\mathrm{K} < T_{\mathrm{eff}} < 7000\,\mathrm{K}$ were selected and divided into 9 metallicity groups ranging from [Fe/H] ≈ -1.0 to $\sim +0.3$. In each metallicity group the ~ 20 brightest stars were observed. Hence, there is no kinematical bias in the selection of the stars. As shown by Edvardsson et al. (1993, Fig. 16b) and as first discussed by Freeman (1991), there is an abrupt increase in the W velocity dispersion of the stars when an age of 10 Gyr is passed. The same was found by Quillen & Garnett (2001), who reanalyzed the Edvardsson et al. sample using space velocities based on Hipparcos data (ESA 1997) and ages from Ng & Bertelli (1998). As seen from their Figure 2, the velocity dispersions are fairly constant for ages between 3 and 9 Gyr: $(\sigma_U, \sigma_V, \sigma_W) \simeq (35, 23, 18)\,\mathrm{km\,s^{-1}}$, corresponding to the thin disk, whereas for ages between 10 and 15 Gyr the dispersions are $(\sigma_U, \sigma_V, \sigma_W) \simeq (60, 50, 40)\,\mathrm{km\,s^{-1}}$, where the velocity dispersion $\sigma_W = 40\,\mathrm{km\,s^{-1}}$ corresponds quite well to the scale height of the Gilmore & Reid thick disk. About the same values were obtained by Nissen (1995) on the basis of the original Edvardsson et al. (1993) data. Furthermore, he derived rotational lags with respect to the local standard of rest (LSR), $V_{\mathrm{lag}} \simeq -10\,\mathrm{km\,s^{-1}}$ for the thin disk and $V_{\mathrm{lag}} \simeq -50\,\mathrm{km\,s^{-1}}$ for the thick disk.

Although the Edvardsson et al. data for the kinematics of stars in the solar neighborhood belonging to the thin and thick Galactic disks refer to 189 stars only, the values derived agree quite well with other recent investigations. For example, Soubiran, Bienaymé, & Siebert (2003) derive $(\sigma_U, \sigma_V, \sigma_W) = (63 \pm 6, 39 \pm 4, 39 \pm 4)\,\mathrm{km\,s^{-1}}$ and a rotational lag $V_{\mathrm{lag}} = -51 \pm 5\,\mathrm{km\,s^{-1}}$ for the thick disk based on Tycho-2 proper motions (Høg et al. 2000) and ELODIE (Baranne et al. 1996) spectra for a sample of 400 stars in directions toward the Galactic North Pole.

In the following, we review recent studies of the chemical composition of Galactic disk stars. As we shall see, there is increasing evidence that the thin and thick disks overlap in metallicity in the range $-0.8 <$ [Fe/H] < -0.4 but are separated in [α/Fe], where α refers to the α-capture elements. Furthermore, recent studies suggest that the two disk components are also separated in [Mn/Fe] and [Eu/Ba], i.e. the *r*- to *s*-process ratio. Hence, the chemical studies support the interpretation of the thin and thick disks as discrete components of our galaxy formed at separated epochs and having different evolution time scales.

12.2 The α-Capture Elements

It is well known that α-capture elements like O, Mg, Si, and Ca are overabundant by a factor of 2 to 3 relative to Fe in the large majority of metal-poor halo stars, i.e [α/Fe]= +0.3 to +0.5. In the disk [α/Fe] decreases with increasing [Fe/H] to zero at solar metallicity, an effect that is normally explained in terms of delayed production of iron by Type Ia supernovae (SNe) in the disk phase. As the release of Type Ia products occurs with a time delay of typically 1 Gyr, the metallicity at which [α/Fe] starts to decline depends critically on the star formation rate. Hence, [α/Fe] may be used as "a chemical clock" to date the star formation process in the Galaxy.

12.2.1 The Edvardsson et al. Survey

The Edvardsson et al. (1993) survey provided clear evidence for a scatter in [α/Fe] among disk stars with the same [Fe/H]. This is shown in Figure 12.1, where [α/Fe] is plotted as a function of [Fe/H]. [α/Fe] is defined as the average abundance of Mg, Si, Ca, and Ti

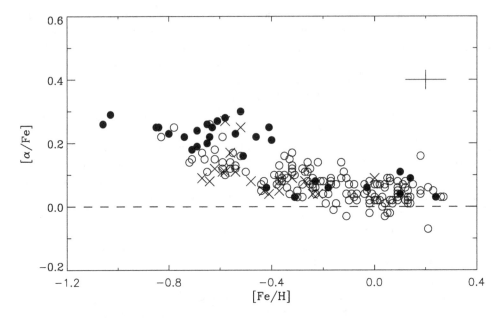

Fig. 12.1. [α/Fe] vs. [Fe/H] for the Edvardsson et al. (1993) stars. [α/Fe] is defined as $\frac{1}{4}$([Mg/Fe] + [Si/Fe] + [Ca/Fe] + [Ti/Fe]). Stars shown with filled circles have a mean galactocentric distance in their orbits $R_m < 7$ kpc. Open circles refer to stars with 7 kpc $< R_m < 9$ kpc, and crosses refer to stars with $R_m > 9$ kpc. Typical 1 σ error bars, referring to differential abundances at a given [Fe/H], are indicated.

with respect to Fe, and was measured with a differential precision of about 0.03 dex for stars having about the same metallicity. Such a high precision can be obtained when the selected stars belong to relatively narrow ranges in T_{eff} and gravity, like the Edvardsson et al. sample, and when the abundance ratios are derived from weak absorption lines having about the same dependence of T_{eff} and gravity, such as the Mg I, Si I, Ca I, Ti I and Fe I lines. In other cases, like [O/Fe], where the oxygen abundance is derived from the [O I] $\lambda6300$ line, the derived abundance ratio is more sensitive to errors in atmospheric parameters and the structure of the model atmospheres (e.g. 3D effects; Nissen et al. 2002). This is why oxygen abundances were not included when calculating the average α-element abundance. It should also be emphasized that the absolute abundances and the overall trend of [α/Fe] with [Fe/H] may be affected by non-LTE effects.

As seen from Figure 12.1, [α/Fe] for stars in the metallicity range –0.8 < [Fe/H] < –0.4 is correlated with the mean galactocentric distance R_m in the stellar orbit. Stars with $R_m > 9$ kpc tend to have lower [α/Fe] than stars with $R_m < 7$ kpc, and stars belonging to the solar circle lie in between. Assuming that R_m is a statistical measure of the distance from the Galactic center at which the star was born, Edvardsson et al. explained the [α/Fe] variations as due to a star formation rate that declines with galactocentric distance. In other words, Type Ia SNe start contributing with iron at a higher [Fe/H] in the inner parts of the Galaxy than in the outer parts. As we shall see in the following, the [α/Fe] variations may, however, also be interpreted in terms of systematic differences between thin and thick disk stars.

12.2.2 Recent Studies of α-Capture Elements

Gratton et al. (1996) were the first to point out that the variations in [α/Fe] could be interpreted in terms of systematic differences between the chemical composition of thin and thick disk stars. Later, Gratton et al. (2000) studied these differences in more detail; equivalent width data from Zhao & Magain (1990), Tomkin et al. (1992), Nissen & Edvardsson (1992), and Edvardsson et al. (1993) were reanalyzed in a homogeneous way and used to derive Fe/O and Fe/Mg ratios. When the stars are plotted in a [Fe/O] vs. [O/H] diagram, two groups of disk stars with [O/H]> -0.5 appear: thin disk stars with [Fe/O]> -0.25 and thick disk stars with [Fe/O]< -0.25. The two groups show a large degree of overlap in [O/H]. Gratton et al. interpreted this as evidence for a sudden decrease in star formation rate during the transition between the thick and thin disk phases, allowing Type Ia SNe to enrich the interstellar gas with Fe without any increase in O and Mg due to the absence of Type II SNe.

An even more clear chemical separation between thick and thin disk stars has been obtained by Fuhrmann (1998, 2000). For a sample of nearby stars with $5300 \, \mathrm{K} < T_{\mathrm{eff}} < 6600 \, \mathrm{K}$ and $3.7 < \log g < 4.6$, he derived Mg abundances from Mg I lines and Fe abundances from Fe I and Fe II lines. In a [Mg/Fe] vs. [Fe/H] diagram, stars with thick disk kinematics have [Mg/Fe] $\simeq +0.4$ and [Fe/H] between -1.0 and -0.3. The thin disk stars show a well-defined sequence from [Fe/H] $\simeq -0.6$ to $+0.4$ with [Mg/Fe] decreasing from $+0.2$ to 0.0. Hence, there is a clear [Mg/Fe] separation between thick and thin disk stars in the overlap region $-0.6 < $ [Fe/H] < -0.3 with only a few "transition" stars. This is even more striking in a diagram where [Fe/Mg] is plotted as a function of [Mg/H] (Fuhrmann 2000, Fig. 12). Fuhrmann's group of 16 thick disk stars have total space velocities with respect to the LSR in the range $85 \, \mathrm{km \, s^{-1}} < V_{\mathrm{tot}} < 180 \, \mathrm{km \, s^{-1}}$ and an average rotational lag of $V_{\mathrm{lag}} \simeq -80 \, \mathrm{km \, s^{-1}}$. It is unclear if this low value is a selection effect.

On the basis of stellar ages derived from evolutionary tracks in M_{bol}- $\log T_{\mathrm{eff}}$ diagrams, Bernkopf, Fiedler, & Fuhrmann (2001) claim that the maximum age of thin disk stars is about 9 Gyr, whereas the thick disk stars have ages between 12 and 14 Gyr. These data agree well with their suggestion that the systematic difference of [α/Fe] is due to a hiatus in star formation between the thick and thin disk phases. However, Bernkopf et al. (2001) derived ages for 7 stars only. Larger samples of thick and thin disk stars should be dated before firm conclusions regarding a hiatus in star formation can be drawn.

Further evidence of a higher [α/Fe] in thick disk stars than in thin disk stars has been presented by Prochaska et al. (2000), who made a thorough study of the chemical composition of 10 G-type stars having $-1.2 < $ [Fe/H] < -0.4 and maximum orbital distances from the Galactic plane greater than 600 pc. Interestingly, [O/Fe], [Si/Fe], and [Ca/Fe] show a decline with increasing [Fe/H], which may be interpreted as a signature of enrichment of Type Ia SNe in the thick disk. This would mean that the thick disk formed over a time scale ≥ 1 Gyr. Prochaska et al. argue that such a formation time scale would rule out most dissipational collapse scenarios for the formation of the thick disk.

Recently, Feltzing, Bensby, & Lundström (2003) have found strong evidence for the presence of Type Ia SNe in the thick disk (see also papers at this meeting by Feltzing et al. 2004 and Bensby, Feltzing, & Lundström 2004). From a sample of about 14 000 dwarf stars in the solar neighborhood with metallicities and ages derived by Feltzing, Holmberg, & Hurley (2001), they selected two samples with a high kinematical probability of belonging either to the thin or the thick disk. When plotted in a Toomre diagram* (Feltzing et al. 2003, Fig. 1)

---

* Sandage & Fouts (1987) appear to be the first to apply this type of diagram in a discussion of the escape velocity

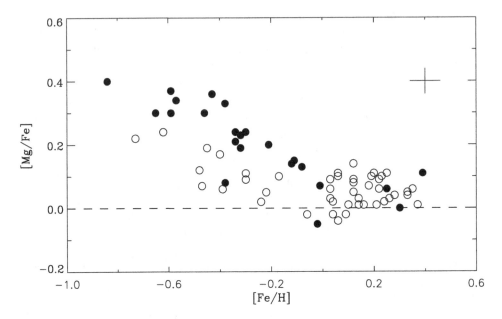

Fig. 12.2. [Mg/Fe] vs. [Fe/H] from Feltzing et al. (2003). Stars shown with filled circles have thick disk kinematics; open circles refer to thin disk stars.

it is seen that the thin disk stars have total space velocities $V_{tot} < +60 \, \text{km s}^{-1}$, whereas the thick disk stars are confined to the range $80 \, \text{km s}^{-1} < V_{tot} < 180 \, \text{km s}^{-1}$.

Interestingly, the thick disk stars of Feltzing et al. (2003) are distributed over the whole metallicity range from −1.0 to 0.0, and reach perhaps up to [Fe/H] $\simeq$ +0.4. Below [Fe/H] $\simeq$ −0.4, [α/Fe] in the thick disk stars is constant at a level of about 0.3 dex, and the thick disk is clearly separated from the thin disk in [α/Fe]. Above [Fe/H] = −0.4, [α/Fe] in the thick disk declines and the two disks merge together. This is seen for both Mg, Si, Ca, and Ti, but most clearly in [Mg/Fe], as shown in Figure 12.2. Hence, star formation in the thick disk went on long enough that Type Ia SNe started to enrich the gas out of which following generations of thick disk stars formed.

Attention is also drawn to a new work by Reddy et al. (2003). A sample of 181 F–G dwarfs were selected from the Olsen (1983, 1988) *uvby-β* catalogs, and the abundances of 27 elements were determined from high-resolution spectra. Parallaxes and proper motions were taken from the Hipparcos Catalogue (ESA 1997). Nearly all stars studied have thin disk kinematics. The α-elements, O, Mg, Si, Ca, and Ti, show [α/Fe] to increase slightly with decreasing [Fe/H] in the range −0.7 < [Fe/H] < 0.0. When compared with abundances for thick disk stars, mainly collected from Fulbright (2000), the thick disk stars have [α/Fe] about 0.15 dex higher than the thin disk stars in the overlap region around [Fe/H] $\approx$ −0.5. Hence, the new work of Reddy et al. supports the thick-thin α-element separation discussed in this section.

In view of these new results on the separation of [α/Fe] between thin and thick stars,

of the Galaxy and to name it the "Toomre energy diagram," recognizing that the representation was due to A. Toomre (1980, private communication).

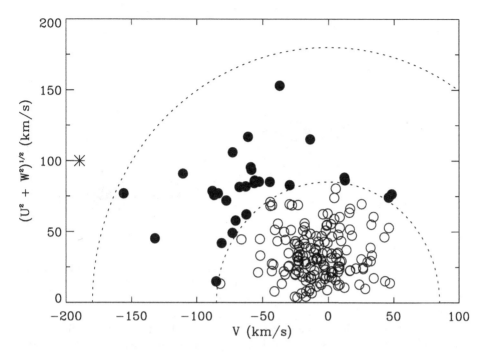

Fig. 12.3. Toomre diagram for the Edvardsson et al. stars. The two circles delineate constant total space velocities with respect to the LSR of $V_{tot} = 85$ and $180\,km\,s^{-1}$, respectively, as used by Fuhrmann (2000) to define a sample of thick disk stars. According to this definition, filled circles are thick disk stars, whereas open circles refer to thin disk stars. One star, HD 148816, shown by an asterisk, is classified as a halo star.

it is interesting to see if the scatter in $[\alpha/Fe]$ for the Edvardsson et al. (1993) sample can be interpreted in terms of thick-thin differences, instead of a correlation with galactocentric distance (Fig. 12.1). To investigate this, I have plotted the Edvardsson et al. stars in a Toomre diagram (Fig. 12.3) and divided them into thin and thick disk stars according to the kinematical definitions of Fuhrmann (2000). As seen from Figure 12.4, much of the scatter can indeed be explained in terms of thick-thin differences in $[\alpha/Fe]$. The separation is not quite as clear as seen in the diagrams of Fuhrmann (2000) and Feltzing et al. (2003), but this may be due to the fact that these authors have selected kinematically well-separated groups of stars, whereas the Edvardsson et al. sample is magnitude limited for a given metallicity bin and hence contains more stars with kinematics in the thick-thin transition region.

In a continuation of the work of Edvardsson et al., Chen et al. (2000) studied the chemical composition of 90 F and G dwarf stars. They do not find any clear $[\alpha/Fe]$ separation between thin and thick disk stars. As pointed out by Prochaska et al. (2000), this may, however, be due to the fact that they selected dwarf stars in the temperature range $5800\,K < T_{eff} < 6400$ K. Hence, the old, more metal-rich thick disk stars with $T_{eff} < 5700$ K are not included. The few thick disk stars in Chen et al. all have $[Fe/H] < -0.6$, i.e. they are lying in a metallicity region where $[\alpha/Fe]$ of the thin disk merges with $[\alpha/Fe]$ of the thick disk.

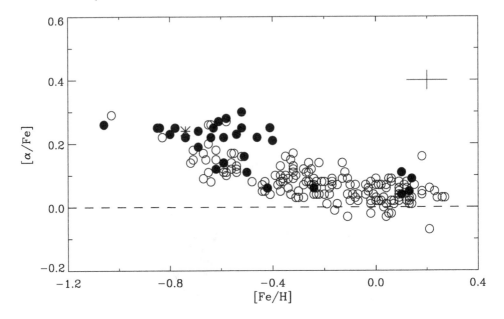

Fig. 12.4. [α/Fe] vs. [Fe/H] for the Edvardsson et al. (1993) stars. As classified in Fig. 12.3, filled circles are thick disk stars and open circles are thin disk stars. The asterisk shows HD 148816, the only halo star in the sample.

12.2.3 *A Comparison With* [α/Fe] *in Halo Stars*

Photometric and spectroscopic surveys of high-velocity, main sequence and sub-giant stars in the solar neighborhood by Nissen & Schuster (1991), Schuster, Parrao, & Contreras Martínez (1993), and Carney et al. (1996) have shown that the metallicity range $-1.5 < [Fe/H] < -0.5$ contains both halo stars having a small velocity component in the direction of Galactic rotation, $V_{rot} < 50$ km s^{-1} (where $V_{rot} = V + 225$ km s^{-1}), and thick disk stars with $V_{rot} \simeq 175$ km s^{-1}. Nissen & Schuster (1997) selected such two groups of stars with overlapping metallicities, and used high-resolution, high signal-to-noise ratio spectra to determine abundance ratios of O, Na, Mg, Si, Ca, Ti, Cr, Fe, Ni, Y, and Ba with a differential precision ranging from 0.02 to 0.07 dex for 13 halo stars and 16 thick disk stars. Figure 12.5 shows the results for [O/Fe] and [Mg/Fe] vs. [Fe/H]. The same pattern is seen for the other α-elements, Si, Ca, and Ti, although with a smaller amplitude for the abundance variations with respect to Fe. As seen, all thick disk stars have a near-constant [α/Fe] at a level of 0.3 dex, whereas the majority of the halo stars have lower values of [α/Fe].

As discussed by Nissen & Schuster (1997), there is a tendency for the α-poor halo stars to be on larger Galactic orbits than halo stars with the same abundance ratios as the thick disk stars. From this they suggest that the halo stars with "low-α" abundances have been formed in the outer part of the halo or have been accreted from dwarf galaxies, for which several models (Gilmore & Wyse 1991; Tsujimoto et al. 1995; Pagel & Tautvaišienė 1998) predict a solar α/Fe ratio at [Fe/H] $\simeq -1.0$ as a consequence of an early star formation burst followed by a long dormant period. Recently, Shetrone et al. (2003) and Venn et al. (2004) have found that stars in dwarf spheroidal and irregular galaxies with metallicities around

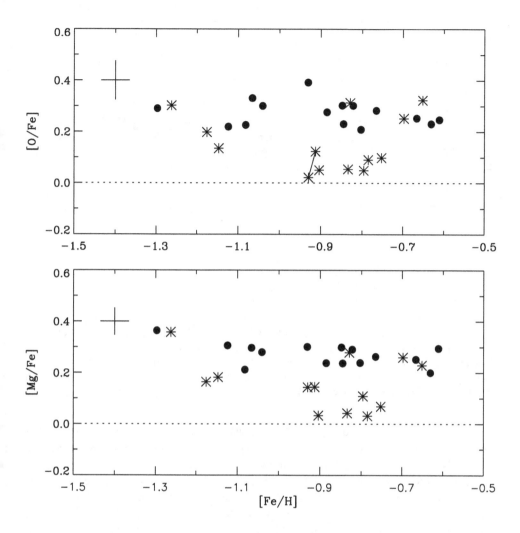

Fig. 12.5. [O/Fe] and [Mg/Fe] vs. [Fe/H] from Nissen & Schuster (1997). Filled circles refer to thick disk stars with a Galactic rotation velocity component $V_{rot} > 150\,km\,s^{-1}$, and asterisks refer to halo stars with $V_{rot} < 50\,km\,s^{-1}$. Two of the halo stars (connected with a line) are components in a spectroscopic binary.

[Fe/H] ≈ -1.0 indeed have a rather low [α/Fe] ratio, supporting the view that the low-[α/Fe] stars belong to an accreted halo component.

Jehin et al. (1999) found two additional halo stars at [Fe/H] $\simeq -1.1$ having low values of [α/Fe]. Among the more metal-poor halo stars with [Fe/H] < -1.4, α-poor stars are rare; only a couple of cases have been found (Carney et al. 1997; King 1997). A more systematic study by Stephens & Boesgaard (2002) of halo stars with unusual orbital properties, i.e. belonging to the "outer" or "high" halo, did not reveal any new α-poor stars, although a weak correlation between [α/Fe] and R_{apo} was detected.

Interestingly, the α-poor stars are also deficient in Na and Ni. Furthermore, there is a tight

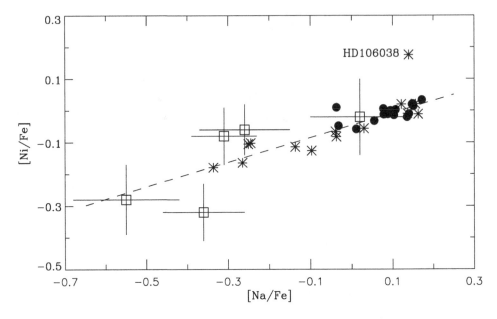

Fig. 12.6. [Ni/Fe] vs. [Na/Fe]. Filled circles refer to to thick disk stars and asterisks to halo stars with abundance ratios from Nissen & Schuster (1997). The dashed line is a fit to these data, excluding the peculiar star HD 106038. The squares with error bars are red giant stars in dwarf spheroidal galaxies with abundances determined by Shetrone et al. (2003) and with metallicities in the range −1.4 < [Fe/H] < −0.6.

correlation between [Ni/Fe] and [Na/Fe], as shown in Figure 12.6, except for one peculiar Ni-rich halo star, HD 106038, which is also very rich in Si and the *s*-process elements Y and Ba (Nissen & Schuster 1997). Furthermore, Figure 12.6 shows that dwarf spheroidal stars selected to have −1.4 < [Fe/H] < −0.6 tend to follow the relation delineated by the α-poor halo stars, hence supporting the idea that the α-poor stars are accreted from dwarf galaxies. The reason for the correlation between Na and Ni abundances is unclear, but it may be connected to the fact that the yields of both Na and the dominant Ni isotope (^{58}Ni) depend upon the neutron excess (Thielemann, Hashimoto, & Nomoto 1990).

12.3   Manganese and Zinc

Among the iron-peak elements, Cr and Ni follow Fe very closely (e.g., Chen et al. 2000), and there is no offset between thick and thin disk stars (Prochaska et al. 2000). Manganese, on the other hand, shows an interesting behavior. A detailed study of the trend of [Mn/Fe] in disk and metal-rich halo stars was published by Nissen et al. (2000) based on high-resolution observations of the Mn I λ6020 triplet. Nissen et al., however, applied outdated data for the hyperfine structure of the Mn I lines. Using modern hyperfine structure data, Prochaska & McWilliam (2000) found significant corrections to the [Mn/Fe] values of Nissen et al. (2000). In Figure 12.7 their revised data have been plotted with the same symbols as in previous figures for halo, thin disk, and thick disk stars. As seen, there is a steplike change in [Mn/Fe] at [Fe/H] ≃ −0.6. Thick disk stars with [Fe/H] below −0.6 have [Mn/Fe] ≃ −0.3, whereas thin disk stars with −0.8 < [Fe/H] < −0.2 have [Mn/Fe] ≃

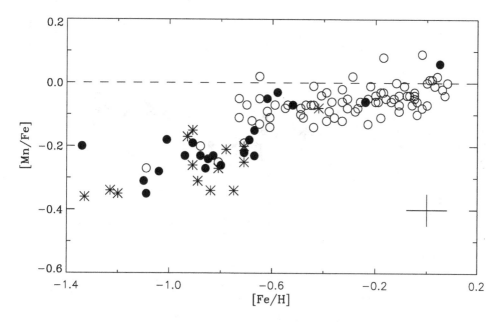

Fig. 12.7. [Mn/Fe] vs. [Fe/H] with data from Nissen et al. (2000), as corrected by Prochaska & McWilliam (2000). Open circles: thin disk; filled circles: thick disk; asterisks: halo stars.

−0.1. Due to the overlap in kinematics between the thick and the thin disk, the few stars with [Fe/H] > −0.6 classified as thick disk may, in fact, belong to the high-velocity tail of the thin disk. The three stars with [Fe/H] < −0.8 classified as thin disk have total space velocities with respect to the LSR close to the 85 km s^{-1} boundary that we have adopted as the separation velocity between the thick and the thin disk; i.e. they may belong to the thick disk. Altogether, the distribution of stars in Figure 12.7 may be interpreted as a separation in [Mn/Fe] between the thin and the thick disk.

The trend of [Mn/Fe] is close to mirror that of [α/Fe] with respect to the [X/Fe] = 0 line (compare Fig. 12.7 with Fig. 12.4). This suggests that Type Ia SNe is a main source for the production of Mn. On the other hand, the eight α-poor halo stars from Nissen & Schuster (1997), which are included in Figure 12.7, do not have higher [Mn/Fe] ratios than the thick disk stars, which one would have expected if Type Ia SNe were the main source of Mn production. Hence, it is not easy to understand the trend of [Mn/Fe]. Probably, the underabundance of Mn is partly caused by a metallicity-dependent yield due to a lower neutron excess in metal-poor stars (Timmes, Woosley, & Weaver 1995).

Zinc is an interesting element with a number of possible nucleosynthesis channels: neutron capture (*s*-processing) in low- and intermediate-mass stars, as well as explosive burning in Type II and Ia SNe (Matteucci et al. 1993). Furthermore, zinc is a key element in studies of elemental abundances of damped Lyα systems because Zn is practically undepleted unto dust (e.g., Pettini et al. 1999). In studies of damped Lyα systems, it is normally assumed that [Zn/Fe] ≃ 0.0 in Galactic stars, as found by Sneden, Gratton, & Crocker (1991) for the range −3.0 < [Fe/H] < 0.0, although with quite a high scatter. Prochaska et al. (2000) claim, however, that Zn is overabundant in thick disk stars, [Zn/Fe] ≃ +0.1. Recently, Mishenina et al. (2002) have published a survey of Zn abundances in 90 disk and halo stars based on

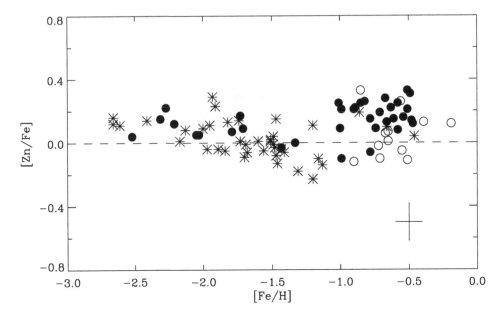

Fig. 12.8. [Zn/Fe] vs. [Fe/H] with data from Mishenina et al. (2002). Open circles: thin disk; filled circles: thick disk; asterisks: halo stars.

equivalent widths of the Zn I $\lambda\lambda$4722.2, 4810.5, 6362.4 lines in high-resolution spectra of dwarf and giant stars. Although the authors conclude that the data "confirms the well-known fact that the ratio [Zn/Fe] is almost solar at all metallicities," there is in fact a hint of interesting structure in their [Zn/Fe] trend. In Figure 12.8, I have plotted the data of Mishenina et al. (2002) using the total space velocity with respect to the LSR to separate the stars into the halo, thin, and thick populations, in the same way as in Figure 12.3. As seen, there is a tendency that thick disk stars in the metallicity range $-1.0 <$ [Fe/H] < -0.5 are overabundant in Zn by as much as [Zn/Fe] $\approx +0.2$. Furthermore, there may be a gradient in [Zn/Fe] as a function of [Fe/H] for the halo stars, with the highest [Zn/Fe] for the most metal-poor stars. Clearly, [Zn/Fe] in halo and disk stars should be further studied, if possible with smaller errors than those obtained by Mishenina et al. (2002).

An interesting detail from Figure 12.8 should be noted: stars classified as thick disk occur down to metallicities around [Fe/H] $\simeq -2.5$. Although it is difficult to distinguish between thick disk and halo stars due to their overlapping kinematics, it is interesting that studies of large samples of metal-poor stars selected without kinematical bias (Beers & Sommer-Larsen 1995; Chiba & Beers 2000) point to the existence of thick disk stars at a rate of $\sim$10% relative to the halo population in the range $-2.2 <$ [Fe/H] < -1.7 and $\sim$30% for $-1.7 <$ [Fe/H] < -1.0.

12.4   The *s*- and *r*-Process Elements

A very interesting set of papers on barium and europium abundances in cool dwarf stars have recently been published by Mashonkina & Gehren (2000, 2001) and Mashonkina et al. (2003). Their results are obtained from a non-LTE, differential model atmosphere analysis of high-resolution, high signal-to-noise ratio spectra of the Ba II $\lambda\lambda$5853, 6496

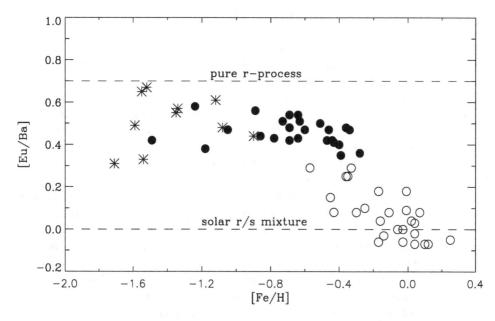

Fig. 12.9. [Eu/Ba] vs. [Fe/H] with data from Mashonkina & Gehren (2000, 2001) and Mashonkina et al. (2003). Open circles: thin disk; filled circles: thick disk; asterisks: halo stars.

lines and the Eu II λ4129 line, taking into account hyperfine structure effects for the Eu line. Their data for the [Eu/Ba] ratio are plotted in Figure 12.9. As seen, there is a rather clear separation between thick and thin disk stars. While thin disk stars have a solar r/s mixture at solar metallicity, thick disk stars and some halo stars approach a pure r-process ratio. The steplike change in Eu/Ba around [Fe/H] $\approx$ −0.5 from the thick to the thin disk suggests a hiatus in star formation before the thin disk developed, i.e. long enough to enable low-mass asymptotic giant branch (AGB) stars to produce Ba by the s-process.

In addition to the separation in [Eu/Ba] between thick and thin disk stars, Mashonkina et al. (2003) claim that the slight decline in [Eu/Ba] with increasing [Fe/H] for the thick disk stars is significant. If real, this suggests a rather long time scale (1.1 to 1.6 Gyr) for the formation of the thick disk according to the chemical evolution calculations of Travaglio et al. (1999). Finally, a significant dispersion in [Eu/Ba] is seen for the halo stars, which suggests a duration of the halo formation of about 1.5 Gyr. Interestingly, Mashonkina et al. (2003) also find a dispersion in [Mg/Fe] for the halo stars. At [Fe/H] $\approx$ −1.0, the average [Mg/Fe] in halo stars is lower than in thick disk stars, a result that agrees well with the findings of Nissen & Schuster (1997).

12.5 Conclusions

We have seen that disk stars in the metallicity range −0.8 < [Fe/H] < −0.4 have significant differences in the α-element/Fe abundance ratio, showing a variation of ∼ 0.2 dex in [Mg/Fe] and about 0.15 dex in [Si/Fe], [Ca/Fe], and [Ti/Fe]. These differences were originally detected by Edvardsson et al. (1993) and interpreted by them as due to a radial gradient in the star formation rate in the Galactic disk causing the enrichment of iron-peak elements by Type Ia SNe to start at a higher [Fe/H] in the inner disk than in the outer regions. More recent work by Gratton et al. (1996, 2000), Fuhrmann (1998, 2000), Prochaska et al. (2000), Feltzing et al. (2003), and Reddy et al. (2003) suggests, however, that the differences are due to a chemical separation between thin and thick disk stars. Thereby, these investigations indicate that the thin and thick disks are discrete components, as originally suggested by Gilmore & Reid (1983) from a study of the distribution of stars in the direction of the Galactic South Pole, and as also supported by kinematical studies of unbiased samples of stars (e.g., Soubiran et al. 2003). Further evidence of a chemical separation of thin and thick disk stars is seen in [Mn/Fe] (Nissen et al. 2000; Prochaska & McWilliam 2000), in [Eu/Ba] (Mashonkina & Gehren 2000, 2001; Mashonkina et al. 2003), and perhaps in [Zn/Fe] (Mishenina et al. 2002). Hence, the evidence for a two-component (thin and thick disk) interpretation of the [α/Fe], [Mn/Fe], and [Eu/Ba] differences is quite compelling, although one should note that some of the studies mentioned have selected stars with extreme kinematics to make it possible to classify stars as belonging to either the thin or the thick disk population. If volume-limited samples of stars are selected, one may see more stars with intermediate abundance ratios, as suggested by the work of Edvardsson et al. (see Fig. 12.4).

As mentioned in the introduction, the survey of Edvardsson et al. (1993) suggests that thick disk stars are older than thin disk stars. Bernkopf et al. (2001) determined isochrone ages for a few of the stars from Fuhrmann (1998, 2000) and obtained ages between 12 and 14 Gyr for the thick disk stars, whereas the oldest thin disk stars have ages around 9 Gyr. This points to a hiatus in star formation between the thick and thin disk phases, which nicely explains the abrupt decline in [α/Fe]; during the hiatus, Type Ia SNe started to enrich the interstellar gas with iron-peak elements, whereas the production of α-capture elements by Type II SNe stopped. Similarly, the decline in [Eu/Ba] is due to enrichment of the interstellar gas with Ba from low-mass AGB stars, while the high-mass stars had ceased to contribute Eu.

The work of Feltzing et al. (2003) provides evidence that [α/Fe] in thick disk stars starts to decline at a metallicity [Fe/H] $\geq$ −0.4, and Mashonkina et al. (2003) found a hint for a decline of [Eu/Ba] among thick disk stars at the same metallicity. Hence, we may see a signature of the occurrence of the products of Type Ia SNe and low-mass AGB stars in the thick disk at [Fe/H] $\simeq$ −0.4, suggesting that the thick disk phase lasted at least ∼1 Gyr.

As discussed in detail by Majewski (1993), there are two different classes of models for the formation of the thick disk: the pre-thin disk (top-down) models and the post-thin disk (bottom-up) models. Within the first class, the chemodynamical model of Burkert, Truran, & Hensler (1992) seems the most convincing. According to this model, the thick disk is a stage in the collapse of the Galaxy where a high star formation rate leads to a high energy input into the interstellar medium, halting the collapse and resulting in stars with a velocity dispersion $\sigma_W \approx 40\,\mathrm{km\,s^{-1}}$. As a result of metal enrichment, the cooling becomes more efficient, and the collapse continues forming the thin disk from inside-out. A difficulty with this model is that it predicts a thick disk phase with a duration of the order of 400 Myr only, i.e. shorter than estimated above from the signature of enrichment by Type Ia SNe and low-mass AGB stars in the thick disk. Also, it is not clear if the model would agree with a 1–2 Gyr hiatus in star formation between the thick and the thin disk.

Among the class of post-thin disk models, violent heating of the early thin disk due to merging of a major satellite galaxy (e.g., Quinn, Hernquist, & Fullagar 1993) is the most obvious possibility. The thick disk stars were originally formed in the ancient thin disk, with no tight limitations on the time scale for the chemical enrichment. After the merger, one can imagine that the star formation stopped for a while until the gas assembled again in a thin disk, causing the hiatus that is needed to explain the shift in [α/Fe] and [Eu/Ba] between the thick and the thin disk. Furthermore, if one assumes that the reestablished disk is formed from gas in the thick disk plus accreted metal-poor gas from the intergalactic medium, then one can explain why some thin disk stars have a lower metallicity than the maximum metallicity of the thick disk, i.e. the overlap of thin and thick disk stars in the metallicity range $-0.8 <$ [Fe/H] < -0.3.

As shown by Nissen & Schuster (1997), there is also an overlap in metallicity between halo and thick disk stars in the metallicity range $-1.4 <$ [Fe/H] < -0.6 with the majority of halo stars having lower [α/Fe], [Na/Fe] and [Ni/Fe] than thick disk stars. The low-[α/Fe] stars tend to be on larger Galactic orbits than halo stars having the same [α/Fe] as thick disk stars. The explanation may be that we have altogether two major components of our Galaxy: (1) a dissipative component consisting of the bulge, the *inner* halo, and the thick disk all formed in a collapse stage with a fast star formation rate enabling the metallicity to reach high values before Type Ia SNe and low-mass stars started to enrich the gas, and (2) an accreted *outer* halo plus the thin disk, where the star formation has proceeded on a longer time scale.

Much more work on stellar ages, kinematics, and abundances has to be carried out before we can be sure about the basic scenario for the formation and evolution of our Galaxy. Many of the results, quoted above, are based on studies of the chemical composition of kinematically selected samples of stars. Thereby, the conclusions may be affected by a kinematical bias, which, for example, exaggerates the chemical separation between thick and thin disk stars. To avoid this, it would be interesting to conduct an age-kinematics-abundance survey of, say, the 20 brightest stars in each of 25 metallicity groups spanning the range $-2.0 <$ [Fe/H] $< +0.5$ (i.e. a total of 500 stars), which should be selected to lie on the main sequence and in the temperature range $5000\,\text{K} < T_{\text{eff}} < 6500$ K. It would also be very interesting to make *in situ* studies of abundances and kinematics of stars in the inner and outer halo, and at various places in the thin and thick disk. Such studies may well give some surprises. Thus, Gilmore, Wyse, & Norris (2002) have recently conducted a low-resolution spectroscopic survey of ~ 2000 F–G stars situated 0.5–5 kpc from the Galactic plane, and have found evidence that the mean rotation velocity a few kpc away from the Galactic plane is $\sim 100\,\text{km s}^{-1}$ rather than the predicted $\sim 175\,\text{km s}^{-1}$ from the local thick disk population. Gilmore et al. propose that their outer sample is dominated by the debris stars from the disrupted satellite that formed the thick disk. Clearly, it would be very interesting to investigate the chemical composition of such stars in detail.

References

Baranne, A., et al. 1996, A&AS, 119, 373

Beers, T. C., & Sommer-Larsen, J. 1995, ApJS, 96, 175

Bensby, T., Feltzing, S., & Lundström, I. 2004, in Carnegie Observatories Astrophysics Series, Vol. 4: Origin and Evolution of the Elements, ed. A. McWilliam & M. Rauch (Pasadena: Carnegie Observatories, http://www.ociw.edu/symposia/series/symposium4/proceedings.html)

Bernkopf, J., Fiedler, A., & Fuhrmann, K. 2001, in Astrophysical Ages and Time Scales, ed. T. von Hippel, N. Manset, & C. Simpson (San Francisco: ASP), 207

Burkert, A., Truran, J. W., & Hensler, G. 1992, ApJ, 392, 651

Carney, B. W., Laird, J. B., Latham, D. W., & Aguilar, L. A. 1996, AJ, 112, 668

Carney, B. W., Wright, J. S., Sneden, C., Laird, J. B., Aguilar, L. A., & Latham, D. W. 1997, AJ, 114, 363

Chen, Y. Q., Nissen, P. E., Zhao, G., Zhang, H. W., & Benoni, T. 2000, A&AS, 141, 491

Chiba, M., & Beers, T. C. 2000, AJ, 119, 2843

Edvardsson, B., Andersen, J., Gustafsson, B., Lambert, D. L., Nissen, P. E., & Tomkin, J. 1993, A&A, 275, 101

ESA 1997, The Hipparcos and Tycho Catalogues, ESA SP-1200

Feltzing, S., Bensby, T., Gesse, S., & Lundström, I. 2004, in Carnegie Observatories Astrophysics Series, Vol. 4: Origin and Evolution of the Elements, ed. A. McWilliam & M. Rauch (Pasadena: Carnegie Observatories, http://www.ociw.edu/symposia/series/symposium4/proceedings.html)

Feltzing, S., Bensby, T., & Lundström, I. 2003, A&A, 397, L1

Feltzing, S., Holmberg, J., & Hurley, J. R. 2001, A&A, 377, 911

Freeman, K. C. 1991, in Dynamics of Disk Galaxies, ed. B. Sundelius (Göteborg: Göteborg Univ.), 15

Fuhrmann, K. 1998, A&A, 338, 161

Fuhrmann, K. 2000, http://www.xray.mpe.mpg.de/~fuhrmann

Fulbright, J. P. 2000, AJ, 120, 1841

Gilmore, G., & Reid, N. 1983, MNRAS, 202, 1025

Gilmore, G., & Wyse, R. F. G. 1991, ApJ, 367, L55

Gilmore, G., Wyse, R. F. G., & Norris, J. E. 2002, ApJ, 574, L39

Gratton, R., Caretta, E., Matteucci, F., & Sneden, C. 1996, in Formation of the Galactic Halo, ed. H. Morrison & A. Sarajedini (San Francisco: ASP), 307

——. 2000, A&A, 358, 671

Høg, E., et al. 2000, A&A, 355, L27

Jehin, E., Magain, P., Neuforge, C., Noels, A., Parmentier, G., & Thoul, A. A. 1999, A&A, 341, 241

King, J. R. 1997, AJ, 113, 2302

Majewski, S. R. 1993, ARA&A, 31, 575

Mashonkina, L., & Gehren, T. 2000, A&A, 364, 249

——. 2001, A&A, 376, 232

Mashonkina, L., Gehren, T., Travaglio, C., & Borkova, T. 2003, A&A, 397, 275

Matteucci, F., Raiteri, C. M., Busso, M., Gallino, R., & Gratton, R. 1993, A&A, 272, 421

Mishenina, T. V., Kovtyukh, V. V., Soubiran, C., Travaglio, C., & Busso, M. 2002, A&A, 396, 189

Ng, Y. K., & Bertelli, G. 1998, A&A, 329, 943

Nissen, P. E. 1995, in IAU Symp. 164, Stellar Populations, ed. P. C. van der Kruit & G. Gilmore (Dordrecht: Reidel), 109

Nissen, P. E., Chen, Y. Q., Schuster, W. J., & Zhao, G. 2000, A&A, 353, 722

Nissen, P. E., & Edvardsson, B. 1992, A&A, 261, 255

Nissen, P. E., Primas, F., Asplund, M., & Lambert, D. L. 2002, A&A, 390, 235

Nissen, P. E., & Schuster, W. J. 1991, A&A, 251, 457

——. 1997, A&A, 326, 751

O'Connell, D. J. K., ed. 1958, Stellar Populations, Vatican Observatory (Amsterdam: North Holland Publ. Comp.)

Olsen, E. H. 1983, A&AS, 54, 55

——. 1988, A&A, 189, 173

Pagel, B. E. J., & Tautvaišienė, G. 1998, MNRAS, 299, 535

Pettini, M., Ellison, S. L., Steidel, C. C., & Bowen, D. V. 1999, ApJ, 510, 576

Prochaska, J. X., & McWilliam, A. 2000, ApJ, 537, L57

Prochaska, J. X., Naumov, S. O., Carney, B. W., McWilliam, A., & Wolfe, A. M. 2000, AJ, 120, 2513

Quillen, A. C., & Garnett, D. R. 2001, in Galaxy Disks and Disk Galaxies, ed. J. G. Funes & E. M. Corsini (San Francisco: ASP), 87

Quinn, P. J., Hernquist, L., & Fullagar, D. P. 1993, ApJ, 403, 74

Reddy, B. E., Tomkin, J., Lambert, D. L., & Allende Prieto, C. 2003, MNRAS, 340, 304

Sandage, A., & Fouts, G. 1987, AJ, 93, 74

Schuster, W. J., Parrao, L., & Contreras Martínez, M. E. 1993, A&AS, 97, 951

Shetrone, M., Venn, K. A., Tolstoy, E., Primas, F., Hill, V., & Kaufer, A. 2003, AJ, 125, 684

Sneden, C., Gratton, R. G., & Crocker, D. A. 1991, A&A, 246, 354

Soubiran, C., Bienaymé, O., & Siebert, A. 2003, A&A, 398, 141

Stephens, A., & Boesgaard, A. M. 2002, AJ, 123, 1647

Strömgren, B. 1966, ARA&A, 4, 433

——. 1987, in The Galaxy, ed. G. Gilmore & B. Carswell (Dordrecht: Reidel), 229

Thielemann, F.-K., Hashimoto, M., & Nomoto, K. 1990, ApJ, 349, 222

Timmes, F. X., Woosley, S. E., & Weaver, T. A. 1995, ApJS, 98, 617

Tomkin, J., Lemke, M., Lambert, D. L., & Sneden, C. 1992, AJ, 104, 1568

Travaglio, C., Galli, D., Gallino, R., Busso, M., Ferrini, F., & Straniero, O. 1999, ApJ, 521, 691

Tsujimoto, T., Nomoto, K., Yoshii, Y., Hashimoto, M., Yanagida, S., & Thielemann, F.-K. 1995, MNRAS, 277, 945

Venn, K. A., Tolstoy, E., Kaufer, A., & Kudritzki, R. P. 2004, in Carnegie Observatories Astrophysics Series, Vol. 4: Origin and Evolution of the Elements, ed. A. McWilliam & M. Rauch (Pasadena: Carnegie Observatories, http://www.ociw.edu/symposia/series/symposium4/proceedings.html)

Zhao, G., & Magain, P. 1990, A&AS, 86, 65

13

Globular clusters and halo field stars

CHRISTOPHER SNEDEN[1], INESE I. IVANS[2], and JON P. FULBRIGHT[3]
(1) Department of Astronomy, University of Texas at Austin
(2) Department of Astronomy, California Institute of Technology
(3) The Observatories of the Carnegie Institution of Washington

Abstract
Abundance ratios of representative Fe-peak, α, neutron-capture, and proton-capture elements are gathered together for many globular clusters, several open clusters, and some representative samples of disk and halo stars. Comparisons of these abundances reveal generally very good agreement between clusters and the field, with some notable deviant clusters for each element. These deviations may be due to real abundance differences in a few cases, and in others may signal the need for new abundance analyses. Overall, with the exception of the light proton-capture elements, the element production history of globular clusters appears to be comparable to field stars of similar metallicities.

13.1   Introduction

Globular clusters are the most easily recognized constituents of the Milky Way Galaxy, and often are the only objects that can be detected in the halos of other galaxies. There are only 150 globular clusters listed in the summary catalog of Harris (1996)*, and the total mass contained in the cluster system is roughly only a percent of the mass sum of halo stars not included in clusters (which itself is a percent or less of the total mass of our Galaxy). But halo globular clusters are extremely old, and thus along with old field stars carry vital information on the formation and early chemical history of the Galaxy.

Many of the spectroscopic investigations of globular clusters over the past few decades have concentrated on determinations of reliable metallicity ($\equiv$ [Fe/H]) scales and on attempts to map out the abundance variations among light elements that are vulnerable to alterations from proton-capture fusion reactions in the interiors of low-to-intermediate-mass stars. A recent cluster metallicity scale study has been conducted by Kraft & Ivans (2003, 2004); see their papers for extended discussion and references to previous efforts (e.g., Zinn & West 1984; Carretta & Gratton 1997). Light element abundance variations in globular clusters have been discussed thoroughly several times in the past decade (e.g., Kraft 1994; Sneden 2000; Freeman & Bland-Hawthorn 2002), and will not be featured here.

Less attention has been given to abundance ratios of heavier elements, which usually can only be synthesized in massive stars. A fundamental question of early Galactic evolution has not been satisfactorily addressed to date: were the globular clusters and halo field stars formed from gas that experienced similar chemical enrichments? Are these two stellar groups truly members of the same population, or were they formed from fundamentally

---

* Available at http://physun.mcmaster.ca/~harris/WEHarris.html

different interstellar media? Here we will address this question from a purely observational perspective, by comparing relative abundance ratios [X/Fe] in globular clusters and halo field stars. No attempt will be made to examine all observed elements. Instead, representative examples from several element groups will be chosen for comment: (1) nickel and copper from the iron-peak elements, (2) calcium from the α-elements, (3) barium, lanthanum, and europium from the neutron-capture (n-capture) elements, and (4) briefly, oxygen, sodium, and aluminum from the proton-capture elements. These have been chosen because their observational uncertainties are relatively small, and they have been studied in a large number of stars in many globular clusters. We will examine trends in these [X/Fe] abundance ratios as functions of [Fe/H] metallicity, plotting the means and standard deviations for each globular cluster but displaying the values for individual stars of the halo field. Obviously anomalous points will be highlighted with the aim of triggering further observational studies to confirm or disprove the aberrations, and new theoretical studies to explain the agreements and disagreements between the chemical compositions of clusters and the halo field.

13.2   Cluster and Field Star Samples

We searched the literature for high-resolution abundance studies of globular clusters, concentrating on the results of the last 10–15 years that have been based on high signal-to-noise ratio spectra obtained with echelle spectrographs. Our selection of studies to be included here includes all of the large-sample ($\gtrsim$5 stars/cluster) investigations and most of the small-sample papers. In Table 13.1 we list the clusters included here, their overall metallicities on a uniform scale derived from Ca II infrared triplet measurements (Kraft & Ivans 2003, 2004)*, the number of stars analyzed per cluster, and the literature sources adopted. We do not pretend that the chosen set is complete, and for clusters that have been subjected to multiple analyses we employed our best judgment as to when to adopt the most recently published abundances and when to include both older and new abundance results. In the abundance plots of this paper, we will employ the cluster metallicities of Table 13.1, but will adopt without change the abundance ratios [X/Fe] reported in the individual papers listed in the table.

Our concentration on the most recent abundance studies of globular cluster stars unfortunately may suggest to the general reader that this subject was developed only in the late 1980s. Here we take proper note of the pioneering early efforts, made with analyses of photographic (thus relatively low signal-to-noise ratio) spectra obtained with typically 2–4 m telescopes beginning in the 1960s. The first detailed chemical composition study of globular cluster giants was that of Helfer, Wallerstein, & Greenstein (1959), who obtained spectra of one giant in each of M13 and M92. Many of their results, such as the metallicity difference between these two clusters, have not been challenged by subsequent investigations. In a series of papers, Cohen (1978, 1979, 1980, 1981) used the newly commissioned echelle spectrographs of the Kitt Peak and Cerro Tololo 4 m telescopes to perform the first systematic studies that allowed intercomparisons of the abundance patterns in several globular clusters. Peterson (1980) essentially introduced the now-familiar study of advanced, high-temperature proton-capture nucleosynthesis in globular clusters with her discovery of

---

* Kraft & Ivans (2003, 2004) derived their metallicity scale from clusters lying in the abundance range $-2.4 \leq$ [FeII/H] ≤ -0.7. So, for the more metal-rich clusters listed in Table 13.1 we adopt the Carretta et al. (2001) metallicity for NGC 6528 and the Cohen et al. (1999) metallicity for NGC 6553. Also, Pal 5 was not included in the Kraft & Ivans compilation, so we adopt here the value derived by Smith, Sneden, & Kraft (2002).

Table 13.1. *Cluster Metallicities and Abundance References*

NGC (Other)	[Fe/H][a]	No.	Reference
	GLOBULAR CLUSTERS		
104 (47 Tuc)	−0.70	4	Brown & Wallerstein (1992)
288	−1.41	13	Shetrone & Keane (2000)
362	−1.34	12	Shetrone & Keane (2000)
1904	−1.64	2	Gratton & Ortolani (1989)
2298	−1.97	3	McWilliam, Geisler, & Rich (1992)
3201	−1.56	18	Gonzalez & Wallerstein (1998),
		6	Covey et al. (2003)
4590 (M68)	−2.43	2	Gratton (1989)
4833	−2.06	2	Gratton (1989)
5272 (M3)	−1.50	23	Sneden et al. (2004)
5897	−2.09	2	Gratton (1987)
5904 (M5)	−1.26	36	Ivans et al. (2001)
		25	Ramírez & Cohen (2003)
6121 (M4)	−1.15	36	Ivans et al. (1999)
6205 (M13)	−1.60	17	Kraft et al. (1997), Sneden et al. (2004)
6254 (M10)	−1.51	14	Kraft et al. (1995)
6287	−2.20	3	Lee & Carney (2002)
6293	−2.00	2	Lee & Carney (2002)
6341 (M92)	−2.38	3	Shetrone (1996)
		29	Sneden et al. (2000a)
6352	−0.78	2	Gratton (1987)
6362	−1.15	2	Gratton (1987)
6397	−2.02	16	Castilho et al. (2000)
6528	+0.07	4	Carretta et al. (2001)
6541	−1.83	2	Lee & Carney (2002)
6553	−0.16	2	Cohen et al. (1999)
		5	Barbuy et al. (1999)
6656 (M22)	−1.71	7	Brown & Wallerstein (1992)
6715 (M54)	−1.47	5	Brown et al. (1999)
6752	−1.57	38	Yong et al. (2003)
6838 (M71)	−0.81	25	Ramírez & Cohen (2002)
7006	−1.48	6	Kraft et al. (1998)
7078 (M15)	−2.42	18	Sneden et al. (1997,2000a)
— (Pal 5)	−1.28	4	Smith, Sneden, & Kraft (2002)
— (Pal 12)	−0.95	2	Brown, Wallerstein, & Zucker (1997)
— (Rup 106)	−1.18	2	Brown, Wallerstein, & Zucker (1997)
	OPEN CLUSTERS		
1435 (M45)	+0.06	2	King et al. (2000)
2112	−0.09	2	Brown et al. (1996)
2243	−0.48	2	Gratton & Contarini (1994)
2264	−0.23	4	King et al. (2000)
6705 (M11)	+0.10	10	Gonzalez & Wallerstein (2000)
6819	+0.09	3	Bragaglia et al. (2001)
Mel 66	−0.38	2	Gratton & Contarini (1994)
Mel 71	−0.30	2	Brown et al. (1996)
γ Scl	+0.20	9	Edvardsson et al. (1995)

[a][Fe/H] values are from Kraft & Ivans (2003, 2004), except where noted in the text.

order-of-magnitude sodium abundance variations among red giant stars of M13. And Pila-chowski (1984, and references therein) analyzed more globular clusters, extended the same techniques to open clusters, and attempted to assemble the extant data of that era to gain some overall Galactic chemical evolution insights. The current state-of-the-art in globular cluster chemical composition studies owes much to these early works.

For comparison we gathered similar data for some open clusters, although these objects have been studied much less intensively than have globulars, and with usually smaller sample sizes. Table 13.1 lists the representative open clusters employed here; the metallicity values are those of the individual literature surveys.

The field star results emphasize large-sample surveys involving 50–200 stars. For high-metallicity ([Fe/H] > −1) disk stars we adopted the abundances of Reddy et al. (2003), supplemented by some Eu abundances of Woolf, Tomkin, & Lambert (1995) and La abundances of Simmerer et al. (2001, and in preparation). For a wider metallicity range, $0 \gtrsim$ [Fe/H] $\gtrsim$ −3, we employed the results of Fulbright (2000). And to expand the representation of very metal-poor ([Fe/H] < −2) stars in various plots, we added in data from the smaller samples of Gratton & Sneden (1991), McWilliam et al. (1995), Ryan, Norris, & Beers (1996), Burris et al. (2000), Johnson & Bolte (2001), Carretta et al. (2002), and Johnson (2002). We made no attempt to normalize the results of various studies to each other (for some elements there is little overlap among the studies). The reader is also cautioned that the field-star results often come from analyses of blue-region spectra, while those for globular cluster stars are generally from yellow-red region spectra. This raises concerns about comparing abundance results that have been obtained from different spectral features. Such scale considerations are beyond the scope of this general review but should be pursued in future inquiries.

13.3 Fe-peak Elements: Nickel and Copper

Elements in the atomic number range $21 \leq Z \leq 30$ comprise the Fe-peak domain, with the exception of Ti (Z = 22), which is also partially an α-element. Among these elements it is particularly easy to compare abundances of Ni and Fe, both of which have numerous neutral-species lines in most spectral ranges. Most comprehensive analyses of globular cluster stars include determinations of Ni abundances.

In Figure 13.1 we show mean abundances for 28 globular clusters and seven open clusters, with individual abundances for the field star samples described in the preceding sections. The metallicity range displayed in Figure 13.1 and most other figures in this paper is −3.2 < [Fe/H] < +0.3. Field stars certainly are known to exist at lower metallicities. For example, there are four stars in the Carretta et al. (2002) survey and eight stars in the McWilliam et al. (1995) survey with [Fe/H] $\leq$ −3.2. Recent surveys have discovered, or analyzed in detail, new extremely low-metallicity stars (e.g., Christlieb et al. 2002; Spite 2004). However, it is well known that no Galactic globular clusters have [Fe/H] < −2.5; the very metal-poor halo field star domain has no counterpart in the globular cluster system.

The data of Figure 13.1 show that [Ni/Fe] $\approx$ 0.0 in globular clusters, open clusters, and halo field stars. The cluster abundance σ values generally indicate the "quality" of the individual abundance studies: signal-to-noise ratio and resolution of the spectra, stellar sample sizes, and number of lines contributing to a given elemental abundance. No evidence to date can be found for true star-to-star [Ni/Fe] variations in any cluster.

Several clusters possess apparently subsolar Ni abundances: [Ni/Fe] $\approx$ −0.3. Two of the clusters, Pal 12 and Rup 106, are globulars that are unusual in several respects: (1) they are outer-halo clusters with galactocentric distances larger than most other clusters of Table 13.1; (2) their color-magnitude diagrams indicate ages younger by a few Gyr than typical globulars; and (3) they have abnormally low relative abundances of α-elements (see next section). The low [Ni/Fe] ratios of Pal 12 and Rup 106 may well be real, but there have only been abundance analyses of two (faint) stars in each of these two clusters (Brown

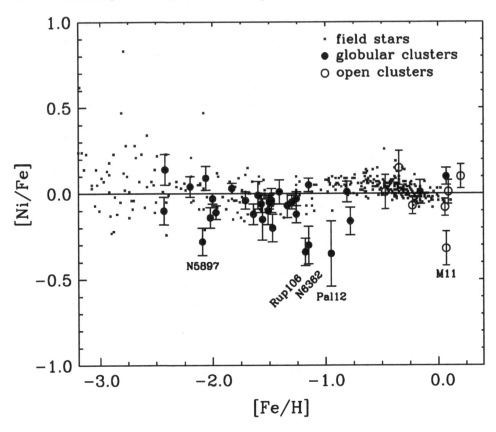

Fig. 13.1. Relative [Ni/Fe] abundance ratios as a function of stellar metallicity. The horizontal line at [Ni/Fe] = 0 indicates the solar [Ni/Fe] ratio. Labels of a few cluster names are placed to draw attention to clusters that deviate from the main trend. Very small filled squares represent field stars, filled circles are for globular clusters, and open circles are for open clusters. The error bars on the cluster points are sample standard deviations σ, and so represent the star-to-star abundance scatters within the clusters.

& Wallerstein (1997). Clearly this result warrants confirmation from future larger-sample studies. The [Ni/Fe] ratios of NGC 5897 and NGC 6362 are also based on just two stars per cluster, and come from the earliest survey chosen for inclusion here (Gratton 1987). The spectral resolving power of that study was only $R \equiv \lambda/\Delta\lambda \simeq 15,000$. These clusters also should be prime candidates for further abundance studies. Finally, the open cluster M11 appears to have a significant Ni abundance deficiency (Gonzalez & Wallerstein 2000). This result cannot be blamed on sample size, for their survey included 10 cluster giants. Nor should data quality be an issue, as they obtained high signal-to-noise ratio data at $R \approx 40,000$. Gonzalez & Wallerstein briefly discuss possible explanations for this anomalous abundance ratio, invoking different mixtures of Type I and II supernovae (SNe) or unusual mass cuts in Type II SNe. But there are no definitive scenarios for the low [Ni/Fe] ratios in M11, and a comparative study of this cluster with other open clusters would be welcome.

In Figure 13.2 we compare Cu abundances in globular cluster and field stars. Since Cu abundance studies are not plentiful, this figure (taken from Simmerer et al. 2003) considers

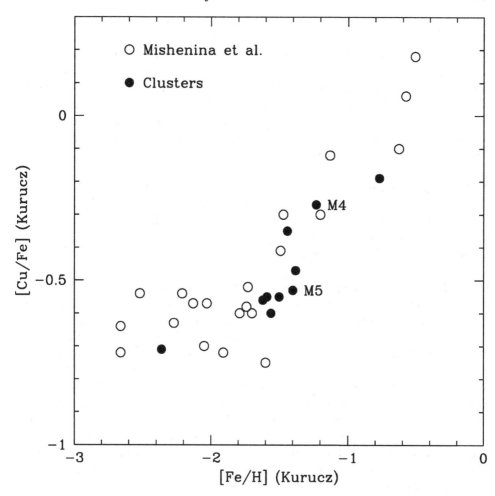

Fig. 13.2. Relative [Cu/Fe] abundance ratios as a function of stellar metallicity. The field star results (open circles) are those of Mishenina et al. (2002), while the globular cluster data are from Simmerer et al. (2003), who employed the spectroscopic data of earlier Lick-Texas group abundance studies in their derivation of Cu abundances.

only their internally homogeneous large-sample globular cluster analysis and the new field-star survey of Mishenina et al. (2002). It has been known for some time that [Cu/Fe] declines steeply with decreasing [Fe/H] in halo field stars. Older studies (e.g., Sneden et al. 1991) suggested that the decline is a nearly linear function of decreasing metallicity, reaching [Cu/Fe] ≈ −1.0 at [Fe/H] ≈ −3.0. Misheninia et al. show that the relationship of [Cu/Fe] with metallicity may be more complex, with a rapid decline until [Fe/H] ≈ −1.7, then possibly reaching a near-constant [Cu/Fe] ≈ −0.6 at lower metallicities. But whatever the exact details of the Cu abundance trend, it is clear from Figure 13.2 that no obvious differences exist between the field stars and globular clusters.

The nucleosynthesis of Cu is not simple (see discussions in Cunha et al. 2002 and Mishenina et al. 2002), and no single explanation for the variation of [Cu/Fe] with [Fe/H] has proved to be adequate. Cu can be generated by slow neutron bombardment (the *s*-process), but a clear positive correlation of Cu with heavier *n*-capture elements has yet to be shown. Metallicity-dependent Cu yields in Type II SNe can roughly mimic the observed trends. On the other hand, if Type Ia SNe produce Cu more efficiently than do Type II SNe, the increasing influence of Type Ia's at higher metallicities can also reproduce the [Cu/Fe] trends. These nucleosynthesis scenarios must now deal with a further observational constraint: clusters and field stars obey the same relationship. Either major Fe-peak element nucleosynthesis occurs prior to formation of both clusters and field stars, or the generation of Cu is a process that is very finely tuned to the Fe metallicity.

13.4   Alpha Elements: Calcium

The α-elements are light even-Z elements in the range $6 \leq Z \leq 22$, whose dominant isotopes can be imagined as collections of α particles. Neglecting C and O, whose abundances are also alterable in proton-capture reactions, among the spectroscopically observable α-elements Ca is chosen here to represent the whole group. The reasons are the same as those for Ni: spectroscopic accessibility leading to plentiful abundance data in clusters and field stars alike. In Figure 13.3 we show [Ca/Fe] ratios for cluster and field stars. For field stars this figure illustrates the well-known rise in relative Ca abundances among disk stars as metallicity decreases from solar values down to [Fe/H] $\approx$ −1.0, retaining its elevated value of [Ca/Fe] $\approx$ +0.3 to +0.4 throughout the entire halo-star low-metallicity regime.

For the most part globular clusters mimic this trend in α-element enhancement, but a few exceptions can be spotted easily. As mentioned previously, the outer-halo, younger clusters Rup 106 and Pal 12 are relatively deficient in α-elements (Brown et al. 1997), and the underabundances of Ca exhibited in Figure 13.3 are well beyond observational/analytical uncertainties. Ca also appears to be deficient in 47 Tuc, but that result comes from a single study of just four stars (Brown & Wallerstein 1992), and clearly that cluster needs reinvestigation. The relatively high Ca abundance in the very metal-poor cluster NGC 4833 comes from only two stars observed by Gratton (1987) and should not at this time be viewed as discrepant (the reader is reminded that the small error bar is only an internal scatter measure).

Some relatively metal-rich clusters deserve comment here. NGC 6528 is a solar-metallicity globular ([Fe/H] = +0.07), and yet it exhibits the high [Ca/Fe] value of the metal-poor halo clusters (Carretta et al. 2001). Apparently Si abundances are also elevated in NGC 6528, which, when coupled with the apparent Mn deficiencies, argues that this very metal-rich cluster material bears the same Type II nucleosynthesis signature as is seen in the very metal-poor clusters. The situation is similar in NGC 6553, another relatively metal-rich cluster, where Barbuy et al. (1999) and Cohen et al. (1999) derive overabundances of both Ca and Si. The α-element abundances of these clusters provide reminders that Fe metallicity is a risky surrogate clock in describing the chemical enrichment of our Galaxy.

The [Ca/Fe] ratios of M71 stars are 0.1 to 0.2 dex larger than field stars of this [Fe/H] $\approx$ − 0.8 metallicity regime (Ramírez & Cohen 2002). This result is based on a 25-star sample with representation from the cluster main sequence to the upper giant branch, and no variation in [Ca/Fe] is seen with evolutionary state. Moreover, a similar study of M5 by Ramírez & Cohen (2003) finds a mean value of [Ca/Fe] = +0.33, in excellent agreement with a mean [Ca/Fe] = +0.28 determined by Ivans et al. (2001) for 36 red giants in the same cluster. Small downward normalization shifts (<0.1 dex) could reduce the M71 [Ca/Fe] discrepancy with

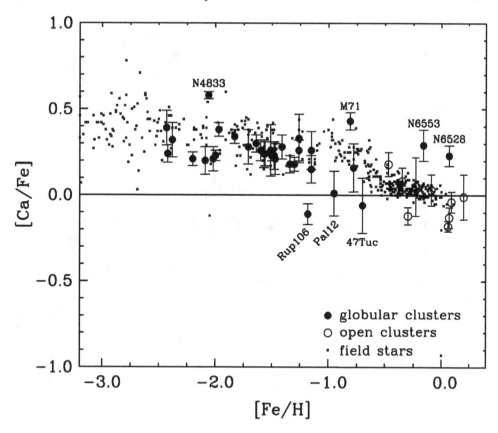

Fig. 13.3. Relative [Ca/Fe] abundance ratios as a function of stellar metallicity. The lines and symbols are as in Fig. 13.1.

field stars to the point where comment would be unwarranted, but for now we simply urge further [Ca/Fe] studies of metal-rich clusters to be undertaken.

Ratios of [Ca/Fe] may not give the entire story about the α-element abundances in globular clusters. Theoretical models of Type II SNe (e.g., Woosley & Weaver 1995) predict that the most massive Type II events heavily overproduce lighter α-elements like O and Mg, while the lower-mass events mainly produce the heavier α-elements Ca and Ti. Intracluster variations of O and Mg may mask any cluster-to-cluster variations in the ratios of these two elements with respect to Ca and Ti, but the [Si/Fe] and [Si/Ti] ratios do show some variations among both clusters and field stars of similar [Fe/H]. This is discussed in detail by Lee & Carney (2002) and Fulbright (2004). In brief, the variations in these ratios show matching correlations with respect to the kinematics (or at least the present galactocentric location) of the stars and clusters. Metal-poor stars and clusters closer to the Galactic Center show higher [Si/Fe] and [Si/Ti] ratios than similar-metallicity members of the outer halo.

13.5 Neutron-capture Elements: Europium, Barium, and Lanthanum

Europium (Z = 63) is the usual representative of the so-called r-process n-capture elements. These elements owe their production to rapid blasts of neutrons in some (as yet

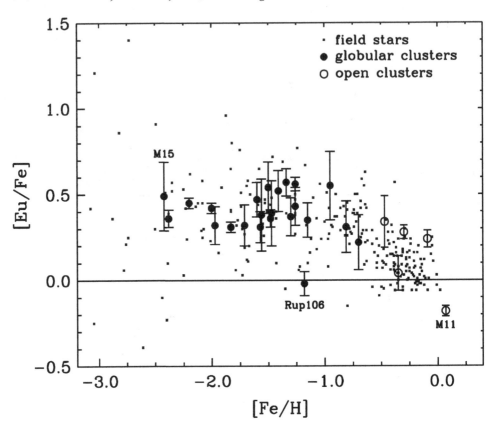

Fig. 13.4. Relative [Eu/Fe] abundance ratios as a function of stellar metallicity. The lines and symbols are as in Fig. 13.1.

ill defined) event(s) associated with the explosive deaths of massive stars. Most descriptions of the evolution of *r*-process elements in the Galaxy discuss Eu because useful transitions of Eu II occur in both blue and red spectral regions. No other *r*-process-dominated *n*-capture element (e.g., Gd, Dy, Ho) can be as easily observed in field and cluster stars.

In Figure 13.4 we correlate [Eu/Fe] ratios with [Fe/H] metallicities, and comparison of this figure with Figure 13.3 suggests that in field stars Eu abundances increase with decreasing metallicity in much the same fashion as do the α-elements, but with far more star-to-star scatter than is seen for Ca. The scatter itself increases with decreasing metallicity (see, e.g., Truran et al. 2002 and Cowan & Sneden 2004 for more extensive discussion and references to individual studies). But with the lone exception of Rup 106, mean [Eu/Fe] values in globular clusters exhibit little cluster-to-cluster variation, and the typical σ values of $\sim$0.15 dex show that Eu does not vary substantially within most clusters.

Rup 106 is an outlier in Figure 13.4 just as it was in Figures 13.1 and 13.3. However, Eu does not appear to be underabundant in Pal 12, the other outer-halo cluster studied by Brown et al. (1997). Although the Pal 12 [Eu/Fe] result is based just on one star, it is worth noting that Brown et al. derive higher abundances in Pal 12 than Rup 106 for all four *n*-capture elements (Zr, Ba, La, Eu) in common between the two clusters. The offset in Eu abundance

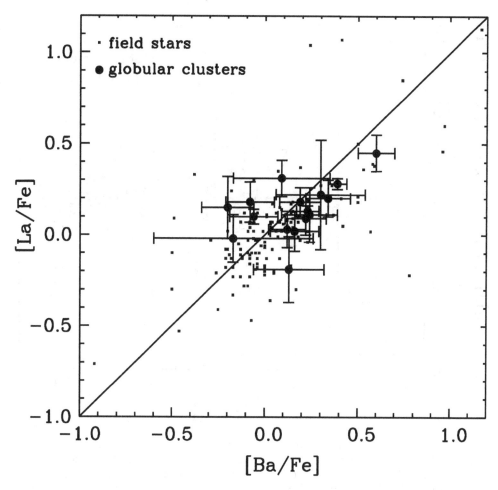

Fig. 13.5. [Ba/Fe] and [La/Fe] ratios in cluster and field stars. The symbols are as in Fig. 13.1; the diagonal line represents [Ba/Fe] = [La/Fe].

between Pal 12 and Rup 106 appears to be real. The most metal-poor cluster, M15, also deserves comment here. The mean [Eu/Fe] value shown in Figure 13.4 is from the relatively recent abundance study of 18 giant stars by Sneden et al. (1997). While the mean M15 [Eu/Fe] agrees well with other globular clusters, the star-to-star scatter ($\sigma = 0.20$) is larger than seen in most other clusters. The scatter is real, as can be seen in direct comparisons of spectra (Fig. 2 of Sneden et al.1997). M15 is the only globular cluster with confirmed general internal scatter in Eu abundances.

Ba and La in the solar system originated mostly in *s*-process *n*-capture synthesis; thus, their observed abundances should roughly rise and fall together. In Figure 13.5 we test this assertion by plotting these two abundances (when both are reported in the literature) for field and cluster stars. A correlation does exist, albeit with a discouraging amount of scatter. The analytical difficulties for Ba II and La II lines are different: Ba II lines are strong, often to the point of saturation, so that Ba abundances depend strongly on adopted microturbulent velocity values; La II lines are weak, often near the point of undetectability, so La abundances

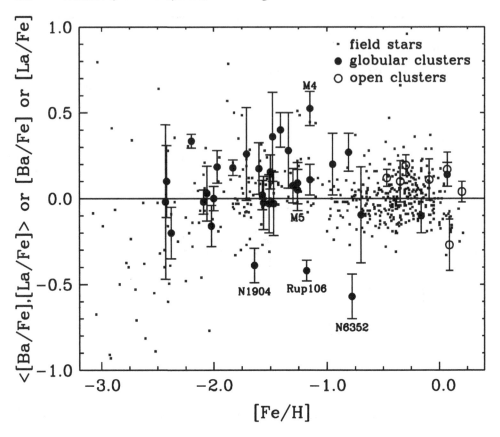

Fig. 13.6. Combined [Ba,La/Fe] ratios as a function of stellar metallicity. For stars with abundances reported for both elements, the ordinate is <[Ba/Fe],[La/Fe]>, and for stars with only Ba or La abundances, the ordinate is simply [Ba/Fe] or [La/Fe], as appropriate. The lines and symbols are as in Fig. 13.1.

depend on obtaining very good signal-to-noise ratio spectra. The observed star-to-star scatters in Ba and La do not appear to be clearly correlated in most globular clusters, suggesting that observational errors dominate the variations within individual clusters.

Therefore, we have chosen to average [Ba/Fe] and [La/Fe] abundance ratios wherever possible in order to mute the observational scatter, and in Figure 13.6 we present these averaged abundance ratios as a function of stellar metallicity. For stars with only [Ba/Fe] or [La/Fe] values available, those have been plotted instead. In field stars, Ba and La abundances have very large (and real) scatter at lowest metallicities, just as Eu does (Fig. 13.4). But unlike Eu, there is little discernible change in the mean [Ba,La/Fe] level with decreasing metallicity. Plots of [Eu/Ba] or [La/Ba], not shown here, confirm that on average metal-poor globular cluster stars have excess Eu abundances, and hence greater *r*-process contributions to the *n*-capture elements than does solar system material.

Relative *r*-process dominance has been confirmed in detail for M15. Sneden et al. (1997) found that Ba abundances varied from star to star, but mostly in lock-step with Eu, maintaining [Eu/Ba] ≈ +0.4 in all but one out of 18 stars of their sample. Most strong *n*-capture

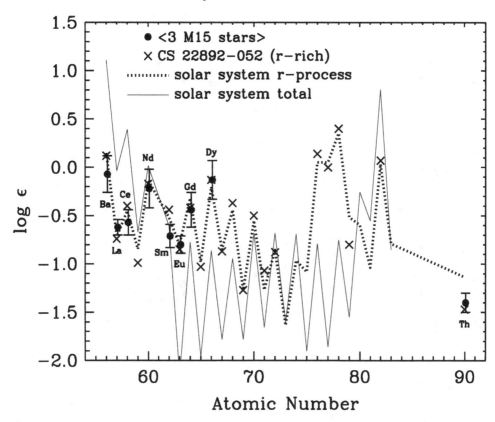

Fig. 13.7. Observed absolute abundances, $\log \epsilon(X) \equiv \log_{10}(N_X/N_H) + 12.0$ of heavy *n*-capture elements in the *r*-process-rich field giant CS 22892–052 (Sneden et al. 2003) and in M15 (the mean of three stars; Sneden et al. 2000b). Also shown are two scaled solar system abundance curves (Burris et al. 2000), one for the total elemental abundances and one for the amount of each element estimated to be due to *r*-process-only synthesis.

transitions in cool stars occur at $\lambda < 5000$ Å, and so Sneden et al. (2000b) reobserved three of these M15 giants in the blue spectral region. In Figure 13.7 we show a comparison of M15 *n*-capture abundances with those of the *r*-process-rich field giant CS 22892–052 and with two solar system abundance curves. These data clearly show that an *r*-process-dominated abundance pattern exists, confirming and strengthening the role of high-mass stellar explosive nucleosynthesis in the generation of M15's overall heavy element abundance pattern.

13.6 Proton-capture Elements: Oxygen, Sodium, and Aluminum

Nearly all globular clusters appear to possess star-to-star abundance variations in at least some of the light elements C, N, O, Na, Mg, and Al. These variations do not occur at random, as there are distinct correlations and anticorrelations among these element abundances. These relationships lead to an assertion that at some time(s), in some place(s) in globular cluster environments large amounts of stellar mass have been subjected to various proton-capture nuclear fusion cycles: C-N, O-N, Ne-Na, and Mg-Al.

Space available here does not permit a comprehensive discussion of the large observa-

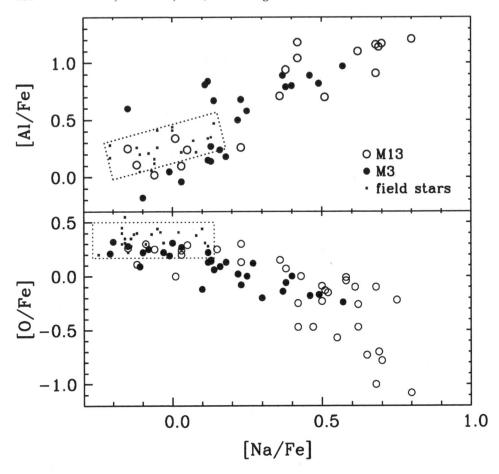

Fig. 13.8. Correlations of relative O and Al abundances with Na abundances in M3 and M13 (Sneden et al. 2004), and in field stars (Fulbright 2000 for the Na and Al data of the top panel; Gratton et al. 2000 for the O and Na data of the bottom panel). The data for O are only those derived from analyses of [O I] lines, which were also employed for the cluster abundances. Dashed-line boxes have been drawn in each panel to emphasize the locations of the field-star points.

tional literature on proton-capture element abundances in globular clusters; the reader is referred again to the reviews cited in §13.1. Here we simply note three observational facts that have led to ongoing debate on the origin of these cluster abundance variations. First, the existence of O-Na and Mg-Al anticorrelations (e.g., Fig. 9 of Kraft et al. 1997) leads to fusion-region temperature estimates of $\gtrsim 40 \times 10^6$ K. Such temperatures are not predicted to exist in the interiors of the low-mass red giants we now observe in globular clusters. Second, some proton-capture element abundance variations have been observed in stars at or near globular cluster main sequences (e.g., Briley & Cohen 2001). Third, some evidence exists for changes in some of these elemental abundances with luminosity in a few clusters (Kraft et al. 1997). These facts do not lead to simple conclusions that the light element abundance variations either were wholly created in the stars we see today, or are the products of a previous generation of intermediate-mass stars in the clusters.

Our discussion is limited to a reemphasis on the fundamental differences in proton-capture abundances between cluster and field stars. In Figure 13.8 we show the ranges of O, Na, and Al abundances in the well-studied cluster pair M3 and M13, and the corresponding much smaller ranges of these elements in the halo field. This figure is only one of many similar ones that could have been displayed, as many clusters have proton-capture abundance variations. To date, no confirmed case of a halo field star possessing "highly processed" light element abundances (e.g., large depletions of C, O, and possibly Mg, accompanied by large enhancements of N, Na, and Al) has been reported in the literature. Whatever the final outcome of the debate on the origin of these abundance effects in globular clusters, it will undoubtedly involve the higher stellar densities in the clusters that do not now exist in the general halo field (and, apparently, never did exist).

13.7 Summary and Conclusions

A review of representative heavy elements in globular clusters and in individual halo field stars has been conducted. We conclude that the nucleosynthesis events that created the Fe-peak elements and α-elements in these two populations were essentially the same. Some differences exist in detail among the *n*-capture elements, but *on average* little differences are found between field and cluster stars. The clearest distinction between these populations remains within the proton-capture element group.

Acknowledgements. We thank all of our colleagues in the Lick-Texas collaboration on globular cluster spectroscopy for their contributions to the work described here. The helpful suggestions of the referee are appreciated. This research has been supported in part by NSF grants AST-9987162 and AST-0307495 to CS. Research for III is currently supported by NASA through Hubble Fellowship grant HST-HF-01151.01-A from the Space Telescope Science Institute, which is operated by the Association of Universities for Research in Astronomy, Incorporated, under NASA contract NAS5-26555.

References

Barbuy, B., Renzini, A., Ortolani, S., Bica, E., & Guarnieri, M. D. 1999, A&A, 341, 539
Bragaglia, A., et al. 2001, AJ, 121, 327
Briley, M. M., & Cohen, J. G. 2001, AJ, 122, 242
Brown, J. A., & Wallerstein, G. 1992, AJ, 104, 1818
Brown, J. A., Wallerstein, G., Geisler, D., & Oke, J. B. 1996, AJ, 112, 1551
Brown, J. A., Wallerstein, G., & Gonzalez, G., 1999, AJ, 118, 1245
Brown, J. A., Wallerstein, G., & Zucker, D. 1997, AJ, 114, 180
Burris, D. L., Pilachowski, C. A., Armandroff, T. E., Sneden, C., Cowan, J. J., & Roe, H. 2000, ApJ, 544, 302
Carretta, E., Cohen, J. G., Gratton, R. G., & Behr, B. B. 2001, AJ, 122, 1469
Carretta, E., & Gratton, R. G. 1997, A&AS, 121, 95
Carretta, E., Gratton, R., Cohen, J. G., Beers, T. C., & Christlieb, N. 2002, AJ, 124, 481
Castilho, B. V., Pasquini, L., Allen, D. M., Barbuy, B., & Molaro, P. 2000, A&A, 361, 92
Christlieb, N., et al. 2002, Nature, 419, 904
Cohen, J. G. 1978, ApJ, 223, 487
——. 1979, ApJ, 231, 751
——. 1980, ApJ, 241, 981
——. 1981, ApJ, 247, 869
Cohen, J. G., Gratton, R. G., Behr, B. B., & Carretta, E. 1999, ApJ, 523, 739
Covey, K. R., Wallerstein, G., Gonzalez, G., & Vanture, A. D. 2003, PASP, 115, 819

Cowan, J. J., & Sneden, C. 2004, in Carnegie Observatories Astrophysics Series, Vol. 4: Origin and Evolution of the Elements, ed. A. McWilliam & M. Rauch (Cambridge: Cambridge Univ. Press), in press

Cunha, K., Smith, V. V., Suntzeff, N. B., Norris, J. E., Da Costa, G. S., & Plez, B. 2002, AJ, 124, 379

Edvardsson, B., Pettersson, B., Kharrazi, M., & Westerlund, B. 1995, A&A, 293, 75

Fulbright, J. P. 2000, AJ, 120, 1841

———. 2004, in Carnegie Observatories Astrophysics Series, Vol. 4: Origin and Evolution of the Elements, ed. A. McWilliam & M. Rauch (Pasadena: Carnegie Observatories, http://www.ociw.edu/ociw/symposia/series/symposium4/proceedings.html)

Gonzalez, G., & Wallerstein, G. 1998, AJ, 116, 765

———. 2000, PASP, 112, 1081

Gratton, R. G. 1987, A&A, 179, 181

Gratton, R. G., & Contarini, G. 1994, A&A, 283, 911

Gratton, R. G., & Ortolani, S. 1989, A&A, 211, 41

Gratton, R. G., & Sneden, C. 1991, A&A, 241, 501

Harris, W. E. 1996, AJ, 112, 1487 (http://physun.mcmaster.ca/ harris/mwgc.dat)

Helfer, H. L., Wallerstein, G., & Greenstein, J. L. 1959, ApJ, 129, 700

Ivans, I. I., Kraft, R. P., Sneden, C., Smith, G. H., Rich, R. M., & Shetrone, M. 2001, AJ, 122, 1438

Ivans, I. I., Sneden, C., Kraft, R. P., Suntzeff, N. B., Smith, V. V., Langer, G. E., & Fulbright, J. P. 1999, AJ, 118, 1273

Johnson, J. A. 2002, ApJS, 139, 219

Johnson, J. A., & Bolte, M. 2001, ApJ, 544, 888

King, J. R., Soderblom, D. R., Fischer, D., & Jones, B. F. 2000, ApJ, 533, 944

Kraft, R. P. 1994, PASP, 106, 553

Kraft, R. P., & Ivans, I. I. 2003, PASP, 115, 143

———. 2004, Carnegie Observatories Astrophysics Series, Vol. 4: Origin and Evolution of the Elements, ed. A. McWilliam & M. Rauch (Pasadena: Carnegie Observatories, http://www.ociw.edu/ociw/symposia/series/symposium4/proceedings.html)

Kraft, R. P. Sneden, C., Langer, G. E., Shetrone, M. D., & Bolte, M. 1995, AJ, 109, 2586

Kraft, R. P., Sneden, C., Smith, G. H., Shetrone, M. D., & Fulbright, J. P. 1998, AJ, 115, 1500

Kraft, R. P., Sneden, C., Smith, G. H., Shetrone, M. D., Langer, G. E., & Pilachowski, C. A. 1997, AJ, 113, 279

Lee, J.-W., & Carney, B. W. 2002, 124, 1511

McWilliam, A., Geisler, D., & Rich, R. M. 1992, PASP, 104, 1193

McWilliam, A., Preston, G. W., Sneden, C., & Searle, L. 1995, AJ, 109, 2757

Mishenina, T. V., Kovtyukh, V. V., Soubiran, C., Travaglio, C., & Busso, M. 2002, A&A, 396, 189

Peterson, R. C. 1980, ApJ, 237, L87

Pilachowski, C. A. 1984, ApJ, 281, 614

Ramírez, S. V., & Cohen, J. G. 2002, AJ, 123, 3277

———. 2003, AJ, 125, 224

Reddy, B. E., Tomkin, J., Lambert, D. L., & Allende Prieto, C. 2003 MNRAS, 340, 304

Ryan, S. G., Norris, J. E., & Beers, T. C. 1996, ApJ, 471, 254

Shetrone, M. D. 1996, AJ, 112, 1517

Shetrone, M. D., & Keane, M. J. 2000, AJ, 119, 840

Simmerer, J., Sneden, C., Ivans, I. I., Kraft, R. P., Shetrone, M. D., & Smith, V. V. 2003, AJ, 125, 2018

Simmerer, J. A., Sneden, C., Woolf, V. M., & Lambert, D. L 2001, BAAS, 199, 9110

Smith, G. H., Sneden, C., & Kraft, R. P. 2002, AJ, 123, 1502

Sneden, C. 2000, in 35th Liège Int. Ap. Coll.: The Galactic Halo, from Globular Clusters to Field Stars, ed. A. Noels et al. (Liège Belgium: Institut d'Astrophysique et de Géophysique), 159

Sneden, C., et al. 2003, ApJ, 591, 936

Sneden, C., Gratton, R. G., & Crocker, D. A. 1991, A&A, 246, 354

Sneden, C., Johnson, J., Kraft, R. P., Smith, G. H., Cowan, J. J., & Bolte, M. S. 2000b, ApJ, 536, 85

Sneden, C., Kraft, R. P., Guhathakurta, R., Peterson, R. C., & Fulbright, J. P., 2004, AJ, submitted

Sneden, C., Kraft, R. P., Shetrone, M. D., Smith, G. H., Langer, G. E., & Prosser, C. F. 1997, AJ, 114, 1964

Sneden, C., Pilachowski, C. A., & Kraft, R. P. 2000a, AJ, 120, 1351

Spite, M. 2004, Carnegie Observatories Astrophysics Series, Vol. 4: Origin and Evolution of the Elements, ed. A. McWilliam & M. Rauch (Pasadena: Carnegie Observatories, http://www.ociw.edu/ociw/symposia/series/symposium4/proceedings.html)

Truran, J. W., Cowan, J. J., Pilachowski, C. A., & Sneden, C. 2002, PASP, 114, 1293

Woolf, V. M., Tomkin, J., & Lambert, D. L. 1995, ApJ, 453, 660
Woosley, S. E., & Weaver, T. A. 1995, ApJ, 448, 315
Yong, D., et al. 2003, private communication
Zinn, R., & West, M. J. 1984, ApJS, 55, 45

14

Chemical evolution in ω Centauri

VERNE V. SMITH

Department of Physics, University of Texas at El Paso

Abstract

The globular cluster ω Centauri displays evidence of a complex star formation history and peculiar internal chemical evolution, setting it apart from essentially all other globular clusters of the Milky Way. In this review we discuss the nature of the chemical evolution that has occurred within ω Cen and attempt to construct a simple scenario to explain its chemistry. We conclude that its chemical evolutionary history can be understood as that which can occur in a small galaxy or stellar system ($M \approx 10^7 M_\odot$), undergoing discrete star formation episodes occurring over several Gyr, with substantial amounts of stellar ejecta lost from the system.

14.1   Introduction

Due to its distinctive chemical evolutionary history, the single globular cluster ω Cen certainly deserves mention in a meeting on the origin of the elements. Being the most massive known Galactic globular cluster, with a mass of $M \approx 3 \times 10^6 M_\odot$ (Merritt, Meylan, & Mayor 1997), it is almost as massive as a small dwarf spheroidal galaxy, such as Sculptor ($M \approx 6.5 \times 10^6 M_\odot$; Mateo 1998). Indeed, it has been argued that ω Cen is the surviving remnant of a larger system, such as a small galaxy, which was captured into a retrograde Galactic orbit in the distant past (Majewski et al. 2000; Gnedin et al. 2002). It should be mentioned that ω Cen was the topic of a recent meeting (in 2001), where all aspects of its nature were discussed; a large number of topics were covered, not just its chemical evolution, and this broader view of the cluster is well presented in the proceedings of that meeting (van Leeuwen, Hughes, & Piotto 2002).

ω Cen displays a number of fascinating traits that need to be noted in order to place it within the larger context of chemical evolution in various types of stellar systems or populations. First, ω Cen is the only known globular cluster to exhibit a large degree of chemical self-enrichment in all elements studied. Its iron abundance, for example, ranges from [Fe/H] ≈ -2.0, at the low end, up to ~ -0.40, where the bracket notation is defined as [A/B] $= \log(N_A/N_B)_{\text{ProgramObject}} - \log(N_A/N_B)_{\text{Sun}}$. The signature of this abundance spread was first detected as a large width in $(B-V)$ color of the giant branch in ω Cen from the photographic color-magnitude diagram of Woolley (1966). This giant branch "width" was confirmed photoelectrically by Cannon & Stobie (1973), who suggested that the color spread was due to varying metallicity. The inferred range in the heavy-element abundances (specifically the calcium abundance) was verified spectroscopically by Freeman & Rodgers (1975) using low-resolution spectra.

A second important trait of ω Cen is that the abundance distribution of the elements evolved in a most peculiar way as the metallicity (e.g., the Fe or Ca abundance) increased. As noted by Lloyd Evans (1977), the Ba II $\lambda 4554$ Å line in ω Cen giants is considerably stronger than in 47 Tuc giants of similar luminosity and metallicity. In a following paper, Lloyd Evans (1983) identified a cool population of giants in ω Cen with strong ZrO bands, a characteristic of the s-process heavy-element-rich MS and S stars. Both Ba and Zr are produced in slow neutron capture (s-process) nucleosynthesis that occurs during shell He-burning thermal pulses in asymptotic giant branch (AGB) stars: recent extensive reviews covering the s-process and AGB evolution can be found in Wallerstein et al. (1997) and Busso, Gallino, & Wasserburg (1999). The ω Cen MS and S stars identified by Lloyd Evans are of lower luminosity than typical Galactic or Magellanic Cloud MS and S stars, which are AGB stars undergoing thermal pulses and the dredge-up of ^{12}C and s-process neutron capture elements. Since the giants found to be s-process rich in ω Cen are not luminous enough to be thermally pulsing AGB stars (and, thus, could not have self-enriched their own atmospheres with ^{12}C and s-process elements), Lloyd Evans (1983) argued that these chemically peculiar stars, which tend to be the more metal-rich stars, formed from gas that had been heavily enriched in s-process elements. Such an enrichment would presumably have been driven by extensive pollution from a previous population of AGB stars.

Subsequent high-resolution spectroscopic abundance studies of ω Cen giants by Francois, Spite, & Spite (1988), Paltoglou & Norris (1989), and Vanture, Wallerstein, & Brown (1994) found that the s-process elemental abundances (such as Y, Zr, Ba, or Nd) increase as the overall metallicity (i.e. [Fe/H]) increases; the s-process increase is enormous relative to other elements (such as Ca, Fe, or Ni). These detailed abundance studies confirm the Lloyd Evans hypothesis that there is a large s-process component involved in the overall chemical evolution within ω Cen.

A third trait that defines the character of ω Cen is that it is not only chemically peculiar, but also dynamically complex. Based on Ca abundances derived from low-resolution spectra, Norris, Freeman, & Mighell (1996) identified two distinct populations: a "metal-poor" component, with [Ca/H] ≈ -1.4 and containing about 80% of the stars, and a "metal-rich" one comprising the other 20%, with [Ca/H] ≈ -0.9. Later work by Norris et al. (1997), using radial velocities of the cluster members, revealed that the metal-poor and metal-rich populations were kinematically distinct. The metal-poor component is rotating (with $V_{rot} \approx$ 5 km s^{-1}), while the metal-rich component is not. The metal-rich population is also more centrally condensed and has a lower velocity dispersion (or is kinematically cooler). Norris et al. (1997) interpreted these observations as being due to some type of merger in the proto-ω Cen environment.

More complexity was added to the two-population abundances and kinematics from Norris et al. (1996, 1997) by the discovery of increasingly metal-rich components (a third and a fourth) by Pancino et al. (2000), based on red giant branch (RGB) morphology. These additional metal-rich populations comprise about 5% of all ω Cen members. These newly discovered metal-rich stars became even more interesting when Ferraro, Bellazzini, & Pancino (2002) found them to display coherent, bulk motion with respect to the rest of the cluster stars. A subpopulation with its own distinct space motion would suggest a merger, and probably a recent one, as playing a role in the evolution of ω Cen to its present state. This potentially fascinating result has been questioned by Platais et al. (2003), who point out that a color term has introduced systematic effects in the proper motions that could be responsible for the apparent bulk motion of the most metal-rich ω Cen members. This question remains open and no doubt will lead to new observations.

Recent accurate color-magnitude diagrams have also revealed details of the star formation history of ω Cen than can shed light on the nature of its chemical evolution. Hughes & Wallerstein (2000) used Strömgren photometry to derive an age-metallicity relation. They found that the metal-poor stars were typically 3 Gyr older than the metal-rich population, and then argue that this indicates that ω Cen enriched itself over this time scale. Hilker & Richtler (2000) also studied a large sample of ω Cen stars using Strömgren photometry and came to essentially the same conclusions as Hughes & Wallerstein (2000), although Hilker & Richtler suggest a slightly longer time scale of $\sim$6 Gyr for chemical enrichment. A very large color-magnitude diagram study (with 130,000 stars) by Lee et al. (2002) points to discrete, multiple stellar populations in ω Cen—not necessarily a continuous distribution of ages. Their modeling of the color-magnitude diagram suggests four distinct populations in ω Cen. This result may then agree quite nicely with the abundance distributions from Suntzeff & Kraft (1996), Norris et al. (1996), or Pancino et al. (2000). Although Norris et al. (1996) only identified two populations, while Suntzeff & Kraft (1996) pointed to a metal-rich tail, the two most metal-rich components in ω Cen only comprise a tiny fraction ($\sim$5%) of the total number of members.

We will now focus the remainder of this review on trying to understand the detailed nature of chemical evolution in ω Cen, within the framework of the metallicities, kinematics, ages, and star formation history as discussed above. A simple picture of its chemical evolutionary history will be sketched, with the predictions from this simple model compared to the observed abundances.

14.2   The Abundance Distribution and Chemical Enrichment

Both Suntzeff & Kraft (1996) and Norris et al. (1996) present relatively large samples of metallicities in ω Cen members derived from low-resolution spectra of the Ca II infrared (IR) triplet lines. Both studies endeavored to obtain unbiased samples of stars, with no selection criteria based on color or abundance. Norris et al. (1996) focused on giants brighter than $M_V = -1$ and presented a sample containing 521 stars. Suntzeff & Kraft (1996) studied two separate samples, one of subgiants with V magnitudes similar to those of the horizontal branch (199 members), as well as a sample of bright giants (144 members). In both studies, the Ca II equivalent widths were transformed onto metallicity scales by using calibrating globular clusters (47 Tuc, M4, NGC 6752 and NGC 6397 for Norris et al. and 47 Tuc, M71, M4, and NGC 6397 for Suntzeff & Kraft), with Suntzeff & Kraft (1996) tying their scale to the cluster [Fe/H] values, while Norris et al. (1996) opted to tie their scale to their own calibration in the clusters of [Ca/H].

The resulting metallicity distributions from Suntzeff & Kraft (1996) and Norris et al. (1996) are shown in Figure 14.1. Both studies find essentially the same distribution, but calibrated on different scales ([Fe/H] and [Ca/H]). There is a sharp cutoff in the number of stars toward low metallicities, with the sharpness set by the observational uncertainty. There is also a well-defined high-metallicity "tail" containing about 20% of the members. Suntzeff & Kraft (1996) tried fitting the metallicity distribution with a simple one-zone model of chemical evolution with instantaneous recycling. No satisfactory fit to the ω Cen metallicity distribution could be found from such a model; large effective yields from primary nucleosynthesis products are needed to fit the metal-rich tail, but such yields result in too broad

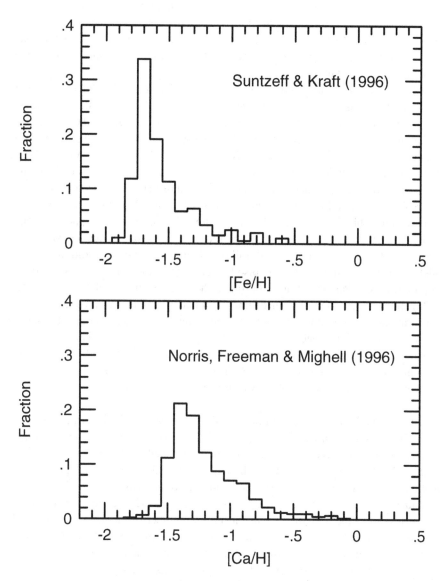

Fig. 14.1. Abundance distributions for samples of ω Cen members from Suntzeff & Kraft (1996) and Norris et al. (1996). The two "metallicity" distributions are tied to different scales, with [Fe/H] for Suntzeff & Kraft and [Ca/H] for Norris et al., but their shapes are very similar. Norris et al. (1996) fit two distinct populations to their distribution, which basically agrees with the metal-rich tail noted by Suntzeff & Kraft (1996).

of a peak in the metallicity distribution. Lower effective yields, which can fit the width of the main metallicity peak, fail to fit the metal-rich tail. Suntzeff & Kraft do point out that these two problems can be overcome if two generations of star formation are considered. Norris et al. (1996) consider slightly different types of models, where the cluster enrichment has occurred within a cloud of gas having some initial heavy-element enrichment, followed by further enrichment from stellar ejecta, with the total metallicity distribution being fit by two such components. Such a model can fit the [Ca/H] distribution, indicating two rather distinct populations in ω Cen. This population picture fits nicely into the later kinematic work from Norris et al. (1997), which finds two kinematic components that correlate with the metallicity, as discussed here in §14.1.

One possible consistent picture for the chemical evolution within ω Cen that fits the metallicity distributions discussed above would contain an initial population of stars with a very narrow range of metallicity: Suntzeff & Kraft (1996) estimate σ[Fe/H]$\leq$ 0.07 dex, based upon the sharpness of the low-metallicity cutoff. This narrow metallicity range is similar to what is found in other globular clusters, where typically σ[Fe/H]$\leq$ 0.05 dex (Suntzeff 1993). This initial generation of stars then pollutes the immediate interstellar medium (ISM) within the proto-ω Cen, and a second generation of stars is formed from this enriched material, with the new stars forming from different amounts of enriched material, giving rise to the broad high-metallicity tail. This simple picture does not address the cause of the second star formation episode, or whether such a scenario might be related to some sort of merger event, but it can, in very broad terms, account for the overall metallicity distribution. Additional support for a model using discrete star formation episodes comes from the more recent work on the RGB morphology from Pancino et al. (2000) and the color-magnitude diagram from Lee et al. (2002): both studies identify four distinct populations. As the most metal-rich third and fourth components account for only $\sim$5% of the members, just considering a two-epoch star formation model, as suggested by Suntzeff & Kraft (1996) and Norris et al. (1996), is probably adequate as a start. Such a model will be considered by us in §14.4; however, before applying this model, a more detailed discussion of the nature of the various types of elements involved in the chemical evolution within ω Cen is in order.

14.3   Abundance Ratios and the Nature of Chemical Evolution in ω Cen

Within a given stellar population, chemical evolution is driven by nucleosynthesis averaged over the stellar mass range and subsequent dispersal of this processed material back into the ISM. This heavy-element enrichment over time depends on such processes as star formation history, internal stellar evolution and nucleosynthesis as a function of mass, how stars return their processed ejecta back into the ISM, and whether some of the stellar ejecta can be lost from the system. By increasing the number of elements considered in an abundance analysis, we can obtain a more detailed picture of chemical evolution within ω Cen. In particular, elements that arise from different types of nucleosynthetic processes occurring in stars of differing masses provide more constraints on the chemical evolution.

In its simplest form, one might consider three basic stellar groups as contributing to most of a population's chemical evolution:

- High-mass stars ($M \geq 8 - 11 M_\odot$) that explode as supernovae of Type II (SNe II) and that contribute much of the heavy-element enrichment, such as O, Mg, Si, Ca, Ti, and some Fe, as well as the heavy neutron-rich, rapid neutron capture elements, personified, for example, by the element europium. Such stars can contribute their processed ejecta to a population's chemical enrichment on fairly short time scales ($\sim 10^7 - 10^8$ yr).
- Low- and intermediate-mass stars ($M \approx 1.0 - 8.0 M_\odot$) that evolve onto the RGB and AGB and lose much of their mass via low-velocity stellar winds as red giants. Such winds contain material processed through the stellar interior that has been mixed to the surface via various red giant dredge-up episodes. The red giant winds from AGB stars might

contribute significant yields of ^{12}C or ^{14}N, and substantial amounts of the heavy elements produced by the slow capture of neutrons, the *s*-process, as typified by such elements as Y, Zr, Ba, or La. These types of stars will contribute to chemical evolution over fairly long time scales of $\geq 10^8$–10^9 yr.

- Supernovae of Type Ia (SNe Ia), which almost certainly result from mass transfer in a binary driving a white dwarf over the Chandrasekhar mass limit. Such supernovae are expected to provide very large mass yields of Fe, and these systems can dominate Fe production in a stellar population over long time scales of $\sim$1 Gyr.

This is an admittedly thin sketch of stellar nucleosynthesis, but this is a limited review and the above points allow us to now investigate a few of the interesting points concerning chemical evolution within ω Cen in light of basic stellar nucleosynthesis.

The ability to detect and analyze a number of different elements, some of quite low abundance or represented by weak spectral lines, requires high-resolution spectra, with $R \equiv \lambda/\Delta\lambda \geq 18,000$, and preferably even higher, with $R = 35,000 - 60,000$. The results from high-resolution spectroscopic studies that are discussed here come from a number of studies that combine both high spectral resolution and high signal-to-noise ratio (S/N $\approx$ 50–100 or better). The largest such study to date is that of Norris & Da Costa (1995), who studied some 40 red giant members, but sacrificed a bit in S/N. Smaller samples, but with better S/N, include those of Francois et al. (1988), Smith, Cunha, & Lambert (1995), Smith et al. (2000), or Vanture, Wallerstein, & Suntzeff (2002). All of the studies mentioned above include a wide range of elements, while additional high-resolution analyses by Brown & Wallerstein (1993) or Pancino et al. (2002) contain elements produced by SNe II and SNe Ia, but do not include analyses of the *s*-process elements. Cunha et al. (2002) probe some 40 ω Cen giants, but focus on their copper abundances. All of these high-resolution results will be combined and discussed in the following two subsections, with the goal being to define the nature of the nucleosynthesis that is needed to explain the chemical enrichment observed in ω Cen.

14.3.1 The s-Process and AGB Stars

As first pointed out by Lloyd Evans (1983), the metal-enriched stars in ω Cen exhibit an overabundance of *s*-process elements. This increase is illustrated quantitatively in Figure 14.2, where we plot [Y/Fe] (top panel), [La/Fe] (middle panel), and [Eu/Fe] (bottom panel) versus A(Fe), where A(x) = log [N(x)/N(H)] + 12, from a number of different studies. We note that we have examined each abundance study to ascertain the sources of their respective gf-values and have put all of these different results on a common scale, when possible (we also adopt a solar Fe abundance of A(Fe) = 7.50). Yttrium and lanthanum are picked to represent the *s*-process, with Y (Z = 39) being a "light" *s*-process element and La (Z = 57) being a "heavy" *s*-process element. The abundance ratio of light to a heavy *s*-process elements is a diagnostic of the exposure to neutrons experienced by the material. Both elements are also spectroscopically well-observed species. Europium is included as a monitor of *r*-process nucleosynthesis. It is thought that most of the *s*-process abundances result from nucleosynthesis in AGB stars, while the *r*-process elements are synthesized in SNe II (for a review of these neutron capture processes and their relation to stellar evolution, see Wallerstein et al. 1997). Sample error bars are shown in the middle panel of Figure 14.2 to illustrate approximate internal abundance uncertainties in each study (adding error bars to each point would overwhelm the plotted points). All of the high-resolution studies

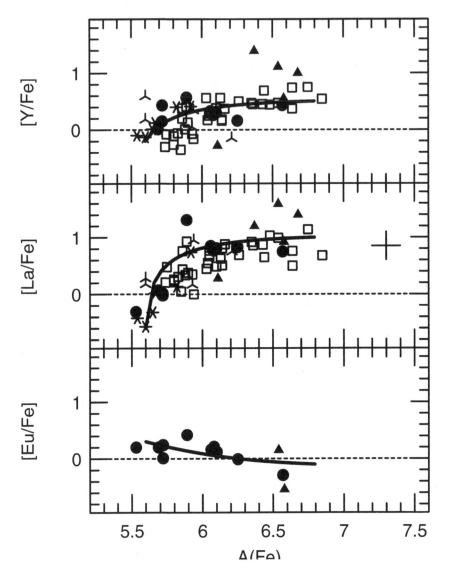

Fig. 14.2. Heavy-element abundance ratios in ω Cen showing light and heavy s-process elements Y and La, respectively, in the top and middle panels, along with the r-process element Eu in the bottom panel. The 6-pointed asterisks are from Francois et al. (1988), open squares from Norris & Da Costa (1995), 3-pointed symbols from Smith et al. (1995), filled circles from Smith et al. (2000), and filled triangles from Vanture et al. (2002). The set of error bars shown in the middle panel illustrates a typical internal uncertainty. As Fe increases by a factor of 10 in abundance, there is an enormous increase in [La/Fe] ($\sim$100 times), a large increase in [Y/Fe] ($\sim$10 times), and no increase (and perhaps a slight decrease) in [Eu/Fe]. This is a strong signature of s-process enrichment from low-metallicity, low-mass AGB stars. The solid curves in all three panels are what is expected from a chemical evolutionary scenario discussed in §14.4.

highlighted in these discussions use similar quality spectra and similar analysis techniques. As the Fe abundance increases in ω Cen, the *s*-process component increases dramatically (Y and La), with no such increase in the *r*-process (Eu). The enormous [*s*-process/Fe] abundance ratios found in the general population of the more metal-rich ω Cen stars have not been found in general stellar populations associated with the Milky Way halo or disk. Recent abundances derived in red giants of the Sagitarrius dwarf galaxy by Smecker-Hane & McWilliam (2004), however, do show large *s*-process to iron ratios as observed in ω Cen. The solid curves in Figure 14.2 are predictions from a simple two-component star formation plus mixing scheme of chemical evolution that is discussed in §14.4.

A key point of Figure 14.2 is that the increase in [La/Fe] is larger than the increase in [Y/Fe], meaning that the overall ratio of La/Y in the *s*-process material that has enriched ω Cen has a value of [La/Y] ≈ +0.4 to +0.5. This excess of La over Y is a signature of low-mass, low-metallicity AGB stars driving the *s*-process; Busso et al. (1999) provide a detailed review of the various types of *s*-process abundance distributions expected from AGB stars of various masses and metallicities.

Smith et al. (2000) compared the heavy-element abundances in ω Cen with predictions from models of AGB nucleosynthesis at the appropriate metallicities and found that 1.5 $M_\odot$ models yield the best overall fits to the *s*-process abundance distributions in the *s*-process-enriched members. The heavy-element abundances in the ω Cen stars thus point to low-mass AGB stars as dominating the *s*-process enrichment in this cluster. The lifetimes of these lower-mass stars are of order 3 Gyr, pointing to a protracted period of star formation, evolution, and chemical enrichment in ω Cen. This time scale agrees with the results from Hughes & Wallerstein (2000) and Hilker & Richtler (2000), who find age spreads of 3–6 Gyr based upon Strömgren photometry and main sequence isochrones (for Hughes & Wallerstein), and red giant isochrones (for Hilker & Richtler).

14.3.2 The α-Elements and Contributions from Supernovae

As reviewed by McWilliam (1997), it has been shown observationally by many investigations over many years that even-Z elements, such as O, Mg, Si, S, Ca, and Ti, are overabundant relative to Fe in almost all Galacic metal-poor stars. This mix of elements is often referred to collectively as the α-elements, although this collectivization should not be taken to imply that these elements are all produced in uniform ratios in a single type of object. All of the α-elements are produced in SNe II (e.g., Woosley & Weaver 1995), but their respective yields depend on such variables as stellar mass or metallicity. The general trend found in the majority of metal-poor Galactic disk and halo stars is that, for iron abundances of [Fe/H]≤ −1.0, the values of [α/Fe] ≈ +0.2 to +0.6 and are roughly constant with metallicity, but having different values of [α/Fe] for the different elements. At iron abundances [Fe/H]≥−1.0, there is a quasi-linear decrease in [α/Fe] toward 0.0 as [Fe/H] approaches 0.0. The behavior of [α/Fe] versus [Fe/H] is usually interpreted to result from the time delay between SN II contributions to chemical evolution relative to the contributions from SNe Ia. The beginning of the decrease in the values of [α/Fe] is then due to the onset of SNe Ia, which begin to add large amounts of Fe into a population's pool of heavy elements.

Abundances from three elements of the "α family" (Si, Ca, and Ti) in ω Cen are shown in Figure 14.3, plotted as [x/Fe] versus A(Fe). The top panel shows [Si/Fe], the middle [Ca/Fe], and the bottom [Ti/Fe] taken from a number of studies noted in the figure caption. Missing from this figure are [O/Fe] and [Mg/Fe] because, as discussed in both Norris & Da Costa

(1995) and Smith et al. (2000), ω Cen red giants show large ranges in, and anti-correlated behavior between, [O/Fe] and [Na/Fe], which is often seen in other globular clusters but not in the field red giants; for a review of this effect, see Kraft (1994). This abundance pattern reveals the effects of H-burning by both the ON part of the CNO cycles, as well as the Ne-Na cycle. It is still not clear to what extent these O and Na (as well as Al and Mg) abundance variations observed in the globular cluster-like populations may be due to red giant mixing within the giant itself, or patterns that were imprinted on the currently observed stars by a previous stellar generation. Whatever the fundamental cause, oxygen, sodium, aluminum, and magnesium abundances in ω Cen may have been altered by nuclear processes that have nothing to do with SNe II; thus, O and Mg are omitted from Figure 14.3.

In the middle panel are shown horizontal bars that schematically represent the approximate metallicities [in A(Fe)] of ω Cen subpopulations as identified by Pancino et al. (2000). The most metal-poor component is labeled RGB-MP, with RGB-Int being the intermediate-metallicity group and the extremely metal-rich population (identified by Pancino et al.) is RGB-a. Recall that RGB-MP and RGB-Int correspond to the metal-poor and metal-rich ω Cen stars identified by Norris et al. (1996) and Suntzeff & Kraft (1996), with these two subpopulations accounting for about 95% of the ω Cen members. In the metallicity range corresponding to RGB-MP and RGB-Int, the α-element to iron abundance ratios look indistinguishable from those of Galactic halo stars. Over the range of iron abundances of A(Fe) = 5.5 to 6.5 there are no strong trends of [Si/Fe], [Ca/Fe], or [Ti/Fe], and the respective values of [X/Fe] for these elements in ω Cen are essentially the same as found for the majority of Galactic metal-poor field halo stars between the same metallicity limits. There may be slight increases in the values of each of these [α/Fe] ratios as A(Fe) increases (with slopes of $\sim$+0.1 to +0.2 dex per dex), although the significance of these possible trends needs to be assessed with a larger, homogeneous analysis. The positive values of [α/Fe] in the ω Cen subpopulations RGB-MP and RGB-Int indicate chemical evolution in an environment in which SNe II control the nucleosynthesis (with little, or no significant, contribution from SNe Ia).

The situation for the RGB-a subpopulation may be different, as there are indications from the results of Pancino et al. (2002), as shown in Figure 14.3, that [Si/Fe] and [Ca/Fe] may decrease relative to the values for RGB-MP and RGB-Int. This may indicate that the RGB-a members are showing measurable amounts of Fe produced from SNe Ia. If so, this is probably not due to a simple time scale difference between chemical evolution in RGB-MP and RGB-Int compared to RGB-a. Recall from §14.3.1 that the substantial build-up in the s-process abundances from RGB-MP to RGB-Int requires time scales of $\sim$1–3 Gyr. This time scale estimate based on low-mass stellar evolution to the AGB is also in agreement with the color-magnitude main sequence turn-off results from Hughes & Wallerstein (2000) and Hilker & Richtler (2000). Such times are longer than expected for SNe Ia to affect chemical evolution. In their study of copper abundance in 40 ω Cen members, Cunha et al. (2002) discuss this time scale discrepancy between AGB and SN Ia chemical enrichment and speculate on two possible reasons. One suggestion is that the stellar density in ω Cen was so large that binary systems that would lead eventually to SNe Ia were disrupted. Another possibility is that SN Ia ejecta have larger kinetic energies than SN II ejecta, leading to lower rates of effective enrichment from SNe Ia compared to SNe II. In order to solve this puzzle, a larger and more homogeneous analysis should be conducted to compare, in detail, the

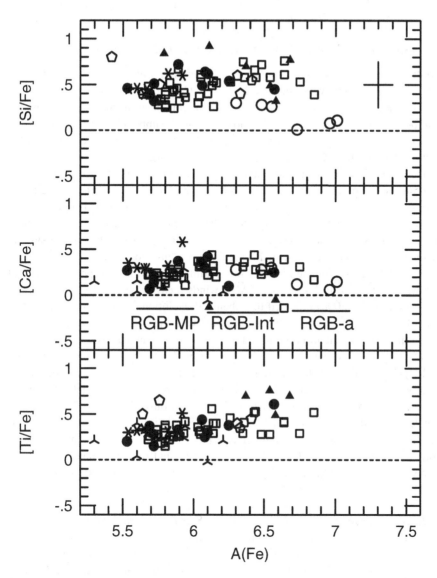

Fig. 14.3. Sample elements that are expected to be produced in SNe II (Si, Ca, and Ti) compared to Fe. The 6-pointed stars are from Francois et al. (1988), open squares Norris & Da Costa (1995), open pentagons from Brown & Wallerstein (1993), 3-pointed symbols from Smith et al. (1995), filled circles from Smith et al. (2000), open circles from Pancino et al. (2002), and filled triangles from Vanture et al. (2002). A typical internal uncertainty is shown by the error bars in the top panel. In the middle panel, the horizontal lines represent the approximate Fe abundances of the subpopulations identified by Pancino et al. (2000): RGB-MP (for metal-poor red giants), RGB-Int (for the intermediate-metallicity giants), and RGB-a (for the more metal-rich subpopulation). In the RGB-MP and RGB-Int red giants, all three of the α-elements shown are overabundant relative to Fe, as found in most Galactic halo stars. The RGB-a members may show hints of decreasing values in [Si/Fe] and [Ca/Fe], suggestive of nucleosynthesis from SNe Ia.

abundance distributions that characterize the various subpopulations in ω Cen—especially a study concentrating on the RGB-a members.

14.4   A Simple Model for the Chemical Evolution of ω Cen

14.4.1   α-Element and s-Process Abundances and Selective Retention of Stellar Ejecta

The increase in the overall Fe abundance in ω Cen, coupled to the enhanced ratios of [Si/Fe], [Ca/Fe], or [Ti/Fe]m points to chemical enrichment from SNe II. The possible decrease in both [Si/Fe] and [Ca/Fe] in the most metal-rich subpopulations of ω Cen suggests the eventual appearance of nucleosynthesis products from SNe Ia. In the context of SN enrichment, one puzzle in ω Cen is how to understand the large s-process (or AGB) component in its chemical enrichment. An initial mass function (IMF) deficient in high-mass stars is certainly one possibility. Another hypothesis, perhaps less extreme than altering the IMF, was explored by Smith et al. (2000) and Cunha et al. (2002). This hypothesis is that in the lowest-mass systems that undergo self-enrichment or internal chemical evolution, stellar ejecta or winds are retained preferentially within the system, or lost to the system, depending on their ejection velocities. In this picture, the high-velocity enriched ejecta from either SNe II or SNe Ia are retained inefficiently within the gravitational potential well of the system when compared to low-velocity AGB winds (heavily enriched in the s-process elements). Smith et al. (2000) modeled this chemical evolution numerically for a system with an initial mass of $10^7 M_\odot$ in which stars formed with the standard Salpeter IMF. The constraint on mass-ejecta retention required to fit the ω Cen abundances was that 10% of SN II ejecta, along with less than 10% of SN Ia ejecta (with SNe Ia becoming active after a time of 1.2 Gyr), were retained for future incorporation into new stars. The s-process-rich AGB ejecta were retained completely, however. In this simple model, the yields for the α-elements were taken from Woosley & Weaver (1995) and convolved with a Salpeter mass function to determine the chemical compositions of the SN II ejecta. Yields for the s-process were taken from the AGB models discussed in Smith et al. (2000). This straightforward exercise reproduces the general trends of SNe (e.g., Si or Ti) and AGB abundances in ω Cen, with the only novel assumption being that SN ejecta are much less efficiently retained than AGB ejecta.

14.4.2   Metallicity-dependent SN II Yields and the Star Formation History

Additional constraints on the nature of chemical evolution in ω Cen are provided by elements that show, or are predicted to have, metallicity-dependent SN II yields, such that [element/Fe] values will depend upon the metallicity of the parent SNe. Cunha et al. (2002) studied copper in 40 ω Cen giants and compared their results to those for field stars in the Galactic disk and halo from Sneden, Gratton, & Crocker (1991) and Mishenina et al. (2002). In the Galactic field stars, [Cu/Fe] shows a steady increase with increasing [Fe/H] over the range of −2.8 to −1.2 in [Fe/H], with a slope of ∼+0.12 dex per dex in [Cu/Fe] versus [Fe/H]. Sneden et al. (1991) attribute this general increase in [Cu/Fe] at low [Fe/H] as due to metallicity-dependent yields for Cu from SNe II. Within the interval of −1.2 to −1.0 in [Fe/H], [Cu/Fe] increases rapidly from −0.5 to +0.0, and then stays at this value up to a solar iron abundance. This steep increase in [Cu/Fe] is probably due to Cu production from SNe Ia. In contrast to the Galactic field, Cunha et al. (2002) find that [Cu/Fe] is constant in ω Cen, at ∼−0.55, from [Fe/H] = −2.0 up to −0.80. At the higher iron abundances, where [Fe/H] = −1.0 to −0.8, the ω Cen stars are falling significantly below the Galactic field trend (by ∼–0.5 dex in [Cu/Fe] at [Fe/H] = –0.8). Pancino et al. (2002) also derived [Cu/Fe] in six ω Cen stars over the range of –1.2 to –0.4 in [Fe/H]. Their copper abundance values overlap those of Cunha et al. (2002) nicely and confirm the low values of [Cu/Fe] in ω Cen. The most metal-rich giant in the Pancino et al. sample is trending upwards in [Cu/Fe] and may signal the measurable addition of SN Ia ejecta, as suggested for some of their results for [Si/Fe] and [Ca/Fe]. Recently, Simmerer et al. (2003) have sampled [Cu/Fe] in 10 "mono-metallicity" globular clusters and find that its behavior with [Fe/H] in these more typical globular clusters is indistinguishable from the Galactic field. Simmerer et al. (2003) conclude that the slow increase in [Cu/Fe] with [Fe/H] at low metallicities results from metallicity-dependent yields from SNe II (as indicated by earlier studies; e.g, Sneden et al. 1991). In addition, they suggest that the rapid increase from [Cu/Fe] ≈ –0.5 to +0.0 between [Fe/H] = –1.2 to –1.0 is probably due to the substantial input of Cu from SNe Ia (as argued by Matteucci et al. 1993).

Adding to the interesting differences in [Cu/Fe] between ω Cen and the Galactic field stars are the first abundances obtained for fluorine by Cunha et al. (2003). These initial results provide tantalizing hints that fluorine behaves similarly to copper in a comparison of ω Cen stars with field-star samples from the Galaxy and the Large Magellanic Cloud (LMC). In the case of fluorine, Cunha et al. (2003) found that a sample of five ω Cen giants displayed values of [F/O] that were significantly lower than the [F/O] ratios found in in Galactic and LMC field red giants having the same oxygen abundances. They also found that the trend in [F/O] versus A(O) defined by the Galactic and LMC stars agreed reasonably well with the predictions of chemical evolutionary models from Timmes, Woosley, & Weaver (1995) and Alibes, Labay, & Canal (2001), in which fluorine production is driven primarily by neutrino spallation off of ^{20}Ne during SN core collapse, as described by Woosley et al. (1990). These models predict a metallicity-dependent decline in [F/O] versus A(O) of about –0.30 dex per dex.

Cunha et al. (2003) suggested a scenario to explain the behavior of [F/O] in ω Cen (as well as [Cu/Fe]), in comparison to the field-star behaviors in the Galaxy and the LMC (for fluorine), by considering the possible star formation history of ω Cen. Their working assumption is that ω Cen underwent a few (2–4) well-separated star formation episodes many Gyr ago: this assumption is motivated by the morphology of the color-magnitude diagram (e.g., Lee et al. 2002). In the simplest case of two star formation episodes (that seem to account for ∼95% of the members, as found in the [Ca/H] distribution from Norris et al. 1996), some fraction of ejecta from the first stellar generation is retained with the ω Cen proto-system (with only a small fraction of SNe II retained, even less from SNe Ia, but 100% of AGB ejecta). This retained ejecta is then mixed with a reservoir of very metal-poor gas; this reservoir may represent the infall, or merging, of a primordial (or less chemically evolved) cloud that might itself induce the second episode of star formation. If the newly added metal-poor gas and the ejecta gas are mixed inhomogeneously, stars of the second generation will have a spread of heavy-element abundances, as observed in ω Cen. Abundance ratios, on the other hand, will be nearly constant in all stars as long as the infalling (or merging) gas cloud is severely underabundant in those elements comprising a given ratio. If an abundance ratio from SNe II is metallicity dependent, such as [Cu/Fe] or [F/O], then this ratio will appear anomalous (and will be ∼ constant with respect to [Fe/H] or [O/H]) when compared to stellar populations that form from gas that has undergone many star formation episodes in which stellar ejecta become well mixed with the ISM. In a sense, the abundance ratios carry the "memory" of the metallicities of the SNe II that formed the bulk of the ejecta out of which they formed. A comparison using abundance ratios insensitive to SN II metallicity (such as [Si/Fe] or [Ca/Fe]) will appear normal, as is observed in ω Cen.

14.4.3   *Putting Together the Pieces of the Abundance Puzzle*

The basic picture of an initial population of stars in the proto-ω Cen environment evolving and depositing ejecta, followed by a second stellar generation forming from a mixture of this ejecta and added metal-poor gas, can be tested for consistency by examining abundance ratios. In this scenario, the more metal-rich members are stars that formed from larger fractions of ejecta from the first population. The limiting case would be a star that formed from pure first generation ejecta. Thus, an approximate abundance distribution for the ejecta (or lower-limit elemental abundances to this gas) can be taken to be the abundances defined by the most metal-rich members of the second stellar generation. The additional metal-poor gas added to this mixture is taken to have much lower heavy-element abundances than the ejecta. The exact composition of this added gas is unknown, but limits to its metallicity can be deduced from the sharp low-metallicity cut-offs in the distributions derived by both Suntzeff & Kraft (1996) and Norris et al. (1996); see Figure 14.1. Both studies found that the sharpness (in either [Fe/H] or [Ca/H]) of this cut-off is set by their respective observational uncertainties: in principle, the low-metallicity cut-off could be a step function. This sharp boundary in the metallicity distribution at low values indicates that the gas in the proto-ω Cen system was "pre-enriched" and well mixed, as discussed in the chemical models employed by Norris et al. (1996). It is thus reasonable to use, as the composition of the added metal-poor gas that is incorporated into the second generation, this pre-enriched material that was already part of the proto-ω Cen system. In this case, stars that formed from differing amounts of these two reservoirs of material should follow a "mixing line" going from the most metal-poor to the most metal-rich stars. Referring back to Figure 14.2, where the heavy-element ratios of [Y/Fe], [La/Fe], and [Eu/Fe] are shown, the solid curves show such mixing lines, with the initial abundances taken as those defined by the metal-poor population. As discussed in Smith et al. (2000), the heavy-element abundance distribution in the most metal-poor ω Cen stars is characterized by an r-process distribution, but these abundances are then transformed into an s-process distribution as [Fe/H] increases. The s-process distribution, with [La/Y] $\approx$ +0.5 and [La/Eu] $\approx$ 1.2, is indicative of nucleosynthesis from low-mass, low-metallicity AGB stars, as reviewed by Busso et al. (1999). The mixing lines shown in Figure 14.2 are fair representations of the observed abundances of Y, La, and Eu.

Further consistency checks can be conducted using other abundance ratios. The transition from an r-process-dominated heavy-element distribution in the metal-poor, first stellar generation to a second generation with a strong s-process component should appear as a changing [La/Y] ratio (the [La/Eu] or [Y/Eu] ratios would be better, but most of the (few) Eu abundances derived in ω Cen stars are from the relatively small sample from Smith et al. 2000). The top panel of Figure 14.4 illustrates [La/Y] versus A(Fe) from a number of studies of ω Cen: note the very rapid rise in [La/Y] as a function of the iron abundance. The solid curve is the mixing line as set by the same curves used in Figure 14.2. Again, the simple two-component mixing model for chemical evolution in ω Cen provides an adequate description of the observed abundances.

The bottom panel of Figure 14.4 shows the results for [Cu/Fe] in ω Cen from Cunha et al.

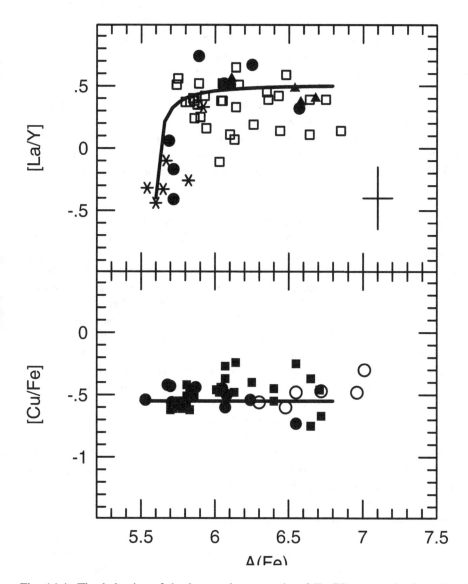

Fig. 14.4. The behavior of the heavy-element ratio of [La/Y] versus the iron abundance (top panel) and that of [Cu/Fe] (bottom panel). In the top panel, the 6-pointed stars are from Francois et al. (1988), open squares from Norris & Da Costa (1995), filled circles from Smith et al. (2000), and filled triangles from Vanture et al. (2002). In the bottom panel, the filled squares are from Cunha et al. (2002) and open circles from Pancino et al. (2002). A typical set of error bars are shown in the top panel. The increase in [La/Y] as A(Fe) increases is very steep (in the top panel), illustrating a transition from a heavy-element distribution dominated by the *r*-process to one dominated by the *s*-process. The solid curve is defined by the chemical evolutionary picture discussed in the text, and this curve tracks the behavior of the La/Y ratio well. The [Cu/Fe] values in ω Cen (bottom panel) show no measurable change as A(Fe) increases, unlike Galactic halo stars, which show increasing values of [Cu/Fe] over this metallicity range. If copper in these metal-poor populations is dominated by metallicity-dependent SN II yields, then the horizontal line is what is expected from the simple model of chemical evolution invoked here to understand ω Cen. In terms of the copper to iron abundance ratios, this model explains the observations.

(2002) and Pancino et al. (2002). In Galactic field halo stars (Sneden et al. 1991; Mishenina et al. 2002) and other globular cluster stars (Simmerer et al. 2003), values of [Cu/Fe] increase steadily as A(Fe) increases from ∼4.7 to 6.5; this increase is interpreted as being due to metallicity-dependent SN II yields (e.g., Simmerer et al. 2003). No such increase is observed in ω Cen, and in the picture of a single-metallicity initial stellar population producing the SN II products from which a second generation of stars are born, a constant Cu/Fe ratio would be predicted, with the value of this ratio set by the metallicity of the first generation of SNe II. Such a constant copper to iron ratio is illustrated by the straight line in Figure 14.4, and this horizontal line is clearly a very good fit to the behavior of copper with iron. In addition, the value of [Cu/Fe] ≈ –0.55 is the value found in field halo stars with A(Fe) ≈ 5.5 (see Mishenina et al. 2002), which corresponds to the lowest metallicity ω Cen stars.

14.4.4 *An Overview of Chemical Evolution and Star Formation in ω Cen*

In summary, we advocate a picture of chemical evolution in ω Cen that is driven by a small number of distinct, time-separated star formation events; this picture is motivated by the color-magnitude morphology from, for example, Hughes & Wallerstein (2000), Hilker & Richtler (2000), or Lee et al. (2002). The large s-process abundance component, relative to elements produced in either SNe II or SNe Ia, in ω Cen's chemical enrichment points to AGB stars as playing a major role in chemical evolution (relative to the Galaxy). This could result from a different IMF, although we prefer to suggest that selective retention of stellar ejecta is the cause, with SN ejecta being retained less efficiently than AGB winds. Continuing star formation in ω Cen then results from the mixing of this stellar ejecta with additional gas arriving from outside of the spatial regions defined by the stars. We have taken this gas to be metal-poor, relative to the stellar ejecta, and find that the observed abundance trends can be fit by such a picture (the solid curves in Figures 14.2 and 14.4). We have focused on describing the first two generations in ω Cen for two reasons: (1) these first two stellar generations seem to account for 95% of the ω Cen members (Suntzeff & Kraft 1996; Norris et al. 1996; Pancino et al. 2000), and (2) the most metal-rich stars have yet to be analyzed in detail in terms of their abundance distributions.

The chemical evolution discussed above is similar to a picture suggested by Freeman (2002). In this scenario, ω Cen is the surviving nucleus of a small galaxy, with its abundance spread being established by the infall of enriched stellar ejecta from the surrounding galaxy into the nucleus. This is a variation from the picture presented above in the sense that enriched material is not retained within the spatial extent of the stars, but arrives from outside. The Freeman (2002) scenario may explain more easily the observed kinematic and angular momentum properties of ω Cen's subpopulations. Certainly larger samples of stars and more detailed abundance and kinematic studies will shed light on the details of the history of ω Cen.

14.5   Conclusions

ω Cen represents a fascinating and valuable object to those interested in studying chemical evolution across a variety of environments and populations. Although classified as a globular cluster, its complex star formation and chemical enrichment histories, plus retrograde Galactic orbit, place it in a different category than the more typical "mono-metallicity" globular clusters. It is almost certainly the remnant of a captured small galaxy that experienced, before its capture, a chemical enrichment history unlike that of any other known Galactic disk or halo population. Because ω Cen is relatively nearby, compared to other small galaxies of the Local Group, it can be used as a comparison template and laboratory in which to probe different aspects of chemical evolution that may occur in some small galaxies.

Acknowledgements. VVS acknowledges support for chemical abundance work from the National Science Foundation (AST99–87374) and NASA (NAG5–9213).

References

Alibes, A., Labay, J., & Canal, R. 2001, A&A, 370, 1103

Brown, J. A., & Wallerstein, G. 1993, AJ, 106, 133

Busso, M., Gallino, R., & Wasserburg, G. J. 1999, ARA&A, 37, 239

Cannon, R. D., & Stobie, R. S. 1973, MNRAS, 162, 207

Cunha, K., Smith, V. V., Lambert, D. L., & Hinkle, K. H. 2003, AJ, 126, 1305

Cunha, K., Smith, V. V., Suntzeff, N. B., Norris, J. E., Da Costa, G. S., & Plez, B. 2002, AJ, 124, 379

Ferraro, F. R., Bellazzini, M., & Pancino, E. 2002, ApJ, 573, L95

Francois, P., Spite, M., & Spite, F. 1988, A&A, 191, 267

Freeman, K. C. 2002, in ω Centauri: A Unique Window into Astrophysics, ed. F. van Leeuwen, J. D. Hughes, & G. Piotto (San Francisco: ASP), 423

Freeman, K. C., & Rodgers, A. W. 1975, ApJ, 201, L71

Gnedin, O. Y., Zhao, G., Pringle, J. E., Fall, S. M., Livio, M., & Meylan, G. 2002, ApJ, 568, L23

Hilker, M., & Richtler, T. 2000, A&A, 362, 895

Hughes, J. D., & Wallerstein, G. 2000, AJ, 119, 1225

Kraft, R. P. 1994, PASP, 106, 553

Lee, Y.-W., Rey, S.-C., Ree, C. H., & Joo, J.-M. 2002, in ω Centauri: A Unique Window into Astrophysics, ed. F. van Leeuwen, J. D. Hughes, & G. Piotto (San Francisco: ASP), 305

Lloyd Evans, T. 1977, MNRAS, 181, 591

——. 1983, MNRAS, 204, 975

Majewski, S. R., Patterson, R. J., Dinescu, D. I., Johnson, W. Y., Ostheimer, J. C., Kunkel, W. E., & Palma, C. 2000, in The Galactic Halo: From Globular Cluster to Field Stars, ed. A. Noels et al. (Liége: Institut d'Astrophysique et de Geophysique), 619

Mateo, M. L. 1998, ARA&A, 36, 435

Matteucci, F., Raiteri, C. M., Busso, M., Gallino, R., & Gratton, R. 1993, A&A, 272, 421

McWilliam, A. 1997, ARA&A, 35, 503

Merritt, D., Meylan, G., & Mayor, M. 1997, AJ, 114, 1074

Mishenina, T. V., Kovtyukh, V. V., Soubiran, C., Travaglio, C., & Busso, M. 2002, A&A, 396, 189

Norris, J. E., & Da Costa, G. S. 1995, ApJ, 447, 680

Norris, J. E., Freeman, K. C., Mayor, M., & Seitzer, P. 1997, ApJ, 487, L187

Norris, J. E., Freeman, K. C., & Mighell, K. J. 1996, ApJ, 462, 241

Paltoglou, G., & Norris, J. E. 1989, ApJ, 336, 185

Pancino, E., Ferraro, F. R., Bellazzini, M., Piotto, G., & Zoccali, M. 2000, ApJ, 534, L83

Pancino, E., Pasquini, L., Hill, V., Ferraro, F. R., & Bellazzini, M. 2002, ApJ, 568, L101

Platais, I., Wyse, R. F. G., Hebb, L., Lee, Y.-W., & Rey, S.-C. 2003, ApJ, 591, L127

Simmerer, J., Sneden, C., Ivans, I. I., Kraft, R. P., Shetrone, M. D., & Smith, V. V. 2003, AJ, 125, 2018

Smecker-Hane, T. A., & McWilliam, A. 2004, ApJ, submitted

Smith, V. V., Cunha, K., & Lambert, D. L. 1995, AJ, 110, 2827

Smith, V. V., Suntzeff, N. B., Cunha, K., Gallino, R., Busso, M., Lambert, D. L., & Straniero, O. 2000, AJ, 119, 1239

Sneden, C., Gratton, R., & Crocker, D. A. 1991, A&A, 246, 354

Suntzeff, N. B. 1993, in The Globular Cluster-Galaxy Connection, ed. G. H. Smith & J. P. Brodie (San Francisco: ASP), 167

Suntzeff, N. B., & Kraft, R. P. 1996, AJ, 111, 1913

Timmes, F. X., Woosley, S. E., & Weaver, T. A. 1995, ApJS, 98, 617

van Leeuwen, F., Hughes, J. D., & Piotto, G., ed. 2002, ω Centauri: A Unique Window into Astrophysics (San Francisco: ASP)

Vanture, A. D., Wallerstein, G., & Brown, J. A. 1994, PASP, 106, 835

Vanture, A. D., Wallerstein, G., & Suntzeff, N. B. 2002, ApJ, 564, 395

Wallerstein, G., et al. 1997, Rev. Mod. Phys., 69, 995

Woolley, R. R. 1966, Royal Observatory Annals No. 2

Woosley, S. E., Hartmann, D. H., Hoffman, R. D., & Haxton, W. C. 1990, ApJ, 356, 272

Woosley, S. E., & Weaver, T. A. 1995, ApJS, 101, 181

15

Chemical composition of the Magellanic Clouds, from young to old stars

VANESSA HILL

GEPI, Observatoire de Paris-Meudon

Abstract

I review the current state of our knowledge of the detailed chemical composition of the Magellanic Clouds, concentrating on the best probes of detailed elemental abundances, namely individual stars observed by means of high-resolution spectroscopy, probing stellar population of all ages from the oldest (>10 Gyr) stellar generations, intermediate-age populations (1–10 Gyr), and young massive stars, complemented by H II region abundances.

15.1 Introduction

As two of our closest neighbor galaxies, the Large and the Small Magellanic Clouds (LMC and SMC) provide a close-by but distinct environment to be compared to our Galactic disk, bulge, or halo. Their young and metal-poor population has no counterpart in the solar neighborhood, making them a unique test-bench for (massive) star evolution, while their distinctive star formation histories, in comparison to that of our Galactic disk, bulge, or halo, provide a perfect opportunity to understand the mechanisms of chemical enrichment in irregular galaxies. In this review I will concentrate on this second aspect, briefly going over the constraints that may be derived from *detailed chemical composition* of Magellanic objects, from the young populations that give a picture of the end-point chemical evolution of each of these galaxies (H II regions and massive stars), to the intermediate and old population that help understand how the chemical enrichment, and particularly, the relative importance of the different nucleosynthetic sources such as Type II supernovae (SNe II), Type Ia supernovae (SNe I), and asymptotic giant branch (AGB) stars have proceeded with time.

Open questions that one would like to answer include: At which epochs did the main episodes of star formation happen? How important have been the bursts (if any) and their relation to the Milky Way-LMC-SMC interactions? What is the relative contribution to the chemical enrichment of SNe Ia and SNe II? Have winds played any significant role in removing metals from these relatively small galaxies? Can the effects of the bursts be seen in the chemical evolution? Are the most ancient stellar populations in the Magellanic Clouds similar to our Milky Way halo stars (could they be drawn from the same population, as would be expected if the halo is made of disrupted satellites of Magellanic-like systems)?

15.2 Present-day Composition

The *present-day* chemical enrichment state reached by the Magellanic Clouds can be probed using two main indicators: H II regions offer insight on the gas still present (and

Table 15.1. *Abundances in H II regions of the Magellanic Clouds (after Garnett 1999)
compared to Orion (Esteban et al. 1998).*

	C	N	O	Ne	S	Ar
$\log(X/H)_{Orion}$[1]	8.49	7.78	8.72	7.89	7.17	6.8
$\log(X/H)_{LMC}-\log(X/H)_{Orion}$	−0.59	−0.88	−0.32	−0.29	−0.47	−0.60
$\log(X/H)_{SMC}-\log(X/H)_{Orion}$	−1.09	−1.28	−0.72	−0.69	−0.87	−0.90

[1] log (X/H) on a scale where log N(H) = +12.

forming stars) in these galaxies, while young massive stars, which are very numerous in both systems, provide accurate measurement of abundances of stars formed in the last tens of million years. Both indicators are rather bright, and have been widely used for abundance measurement since the 1970s and the 1980s, respectively. I do not intend to review here extensively all the numerous and remarkable observational and analytical efforts that were conducted by many different teams over the world, but rather to summarize the most striking findings of these studies.

15.2.1   Nebular Abundances

Although the abundances of O, N, Ne, S, and Ar derived from optical spectra of Magellanic H II regions are virtually unchanged since the review by Dufour (1990), it is worth noting that the abundances of carbon have taken a new turn with the advent of UV spectroscopy from space—*IUE* as a first step (Dufour, Shields, & Talbot 1982), and then *HST* (Garnett et al. 1995; Kurt et al. 1999; Garnett, Galarza, & Chu 2000). From UV spectra, transitions of C^{+2}, N^{+2}, and O^{+2} and C^+, N^+, and O^+ can be observed, allowing one to reduce systematic uncertainties on N/O and C/O (by comparing lines of similar excitation energies and ionization stages). Table 15.1 summarizes the abundances in H II regions in both Clouds, following the review of Garnett (1999), and using the Orion nebula (Esteban et al. 1998) as the Galactic reference.

Inspection of the tabulated O, Ne, S, and Ar abundances reveals, at first glance, the well-known metal deficiency of the LMC by a factor of ∼3 (−0.4 dex) relative to the solar neighborhood, and of the SMC by a factor of 6 (−0.8 dex) is obvious for O, Ne, S, and Ar. On the other hand, carbon in the SMC, and nitrogen in both Clouds, seem to be overdeficient by significant factors.

In fact, these low N/O ratios [log (N/O) = −1.5], taken at face value, place the LMC and the SMC among the galaxies with the lowest N/O ratios known (e.g., Pilyugin, Thuan, & Vílchez 2003; their Fig. 4), with a level of nitrogen close to the *plateau*-like behavior defined by the lowest metallicity irregular galaxies, which is commonly interpreted as a primary nitrogen signature. It is quite unexpected that a galaxy that has been actively forming stars over an extended period of time (the LMC contains significant stellar populations ranging from pre-main sequence stars to ages of 15 Gyr, with a possible enhanced star formation period 2–5 Gyr ago) has still managed to enrich itself so little in nitrogen (relative to oxygen); even adopting intermediate-mass stars as the production site for nitrogen (as proposed by Henry, Edmunds, & Köppen 2000), nitrogen is released into the gas after a moderate time delay (∼250 Myr), so that the intermediate-mass stars formed a few billion years ago in the

LMC and the SMC should have contributed fully to the nitrogen enrichment of their host galaxy. Only if the oxygen seen in the Magellanic Clouds had been massively produced in the last 250 Myr (by a burst of star formation) could this low N/O be observed. But in that case, the O/Fe ratio should also reflect this sudden increase of oxygen production, contrary to what is observed (see §15.4.1).

A word of caution should, however, be said concerning nebular abundance uncertainties: different indicators do not always provide the same answer. As an illustration, recent work by Tsamis et al. (2003) on Galactic and Magellanic H II regions have shown that the collisionally excited recombination lines traditionally used to deduce oxygen abundances are lower than those deduced from the supposedly more robust optical recombination lines by factors of ~2 to 3. In the same work, the authors also show that oxygen abundances deduced from infrared fine structure lines, which are temperature insensitive, are compatible with those deduced from the collisionally excited recombination lines, thus ruling out the suspected temperature fluctuations as a source of the discrepancies between collisionally excited and optical recombination lines. Thus, it is safe to keep in mind that systematic effects can play an important role in nebular abundance determinations (especially CNO). In the particular case of the LMC N/O ratio, using the infrared collisionally excited recombination lines (from the dominating ions N^{2+} and O^{2+}), would bring the galaxy up and away from the primary nitrogen plateau value by ~0.2 dex (Tsamis et al. 2003).

15.2.2   Abundances from Young Stars

The brightest stars of both Magellanic Clouds (supergiants of all spectral types from B to K) have been studied in detail by means of high-resolution spectroscopy since the mid-1980s and mainly in the 1990s, allowing to derive abundances for many more elements than are available from H II regions. A summary of the most recent of these supergiants studies is given in §15.2.2.2. More recently still, with the advent of high-throughput spectrographs and 8–10 m-class telescopes, B-type stars still on the main sequence ($V \approx 15$ mag in the LMC) could also be studied, leading to very interesting new insights, in particular on the abundance of C and N, which are affected by internal mixing in evolved stars and therefore very difficult to interpret in giants and supergiants. The following section gives an account of these new studies.

15.2.2.1   Carbon, Nitrogen, and Oxygen from Hot Stars

It is of particular interest to compare the C, N, and O abundances as probed by the nebular and stellar abundances. Only in unevolved stars can this comparison be meaningful, assuring that none of the CN-cycled material from the central part of the star has yet been dredged-up to the stellar atmosphere. In recent years, B giants (B IV) (Rolleston et al. 1996; Korn et al. 2000; Rolleston, Trundle, & Dufton 2002) and B main sequence stars (LMC: Korn et al. 2002; SMC: Rolleston et al. 2003) have been observed in both Clouds. References to earlier works on B supergiants in both Clouds can be found in these papers. These two groups have taken two types of approaches to deal with the systematic uncertainties (including significant NLTE effects) that can hamper abundance determinations in these hot stars: Rolleston et al. have used a strictly differential approach between Magellanic and Galactic stars of similar temperatures, while Korn et al. chose to take a direct NLTE approach.

The resulting oxygen abundances derived both for giants and dwarfs are in very good agreement with the nebular abundances, yielding [O/H] = −0.4 and −0.7 dex, respectively, for the LMC and SMC. The carbon abundances, however, even in the dwarfs, seem to be *higher* than the nebular abundances by ∼0.4 dex in the SMC (the carbon abundance of the SMC dwarfs, relative to its Galactic counterpart, is −0.65, to be compared to −1.09 for the SMC H II regions, relative to Orion), and by a hardly detectable amount in the LMC. However, uncertainties still present in both stellar and nebular carbon abundance determination could explain a large part of the effect (Korn et al. 2002). The nitrogen abundance, on the other hand, is in very good agreement between the dwarfs and H II regions, consistently pointing toward a very low N/O ratio for both Magellanic Clouds, as discussed in §15.2.1. In giants (and even more in supergiants), the nitrogen abundance increases very rapidly from this initial value, as expected from mixing processes in the star, bringing freshly CN-cycled material toward the surface.

15.2.2.2   Other Elements

A wealth of different elements (over 15 different elements) can be probed in supergiant stars of spectral types A to K, and there are many papers on the abundances measured from such stars in the Magellanic Clouds: samples ranging from three to 10 F supergiants and Cepheid variables were observed in each field of each Cloud by several groups (Russell & Bessell 1989; Spite, Spite, & Barbuy 1989; McWilliam & Williams 1991; Luck & Lambert 1992; Hill, Andrievsky, & Spite 1995; Luck et al. 1998; Andrievsky et al. 2001), six K supergiants in the SMC (Hill, Barbuy, & Spite 1997), and a sample of 10 A-type supergiants in the SMC (Venn 1999). F–G and K supergiants in young populous clusters (rather dense and rich clusters of a few 10^7 yr) in the LMC (Richtler, Spite, & Spite 1989; Hill & Spite 1999) and in the SMC (Spite et al. 1986; Spite, Spite, & Richtler 1991; Hill 1999) have also been studied. No significant differences are found between the abundance ratios measured in the clusters and the field, although some clusters (NGC 330 in the SMC and NGC 1818 in the LMC) seem to be slightly more metal poor than their surrounding field counterparts.

The abundance ratios of representative samples of A to K supergiants in the field of the LMC and SMC are plotted in Figure 15.1 as a function of atomic number. The abundances are plotted *relative* to their solar neighborhood counterparts, to reduce systematic errors linked to the analysis of these luminous objects. The most striking results of Figure 15.1 are, on the one hand, the very good agreement of abundance ratios in the LMC and SMC with the solar abundance ratios (marked as the y = 0 line in the plots, together with two dashed lines showing typical uncertainties) for most elements from oxygen to zinc (Z ≤ 30), *including the α-elements*, and, on the other hand, the overabundance of the heavy neutron capture elements from Ba to Eu (Z ≥ 56), by factors of ∼2 and 3, respectively, in the LMC and SMC. This overabundance of the heaviest elements was noted since the very first studies of SMC and LMC cool supergiants (e.g., Spite et al. 1986), and came as a surprise: among the elements showing this enhancement is the *r*-process-dominated element Eu, which suggests that the *r*-process has been particularly active in the Magellanic Clouds (Luck & Lambert 1992; Russell & Dopita 1992; Hill et al. 1997). On the other hand, the *r*-process is thought to be produced by massive stars (SNe, binary neutron star coalescence, etc.), and therefore expected to be accompanied by α-element enhancements, which are not observed in the Magellanic Clouds. In §15.4.2, a different explanation for the heavy neutron capture elements enhancements is put forward.

To summarize briefly this section, one could say that, despite the likely very different

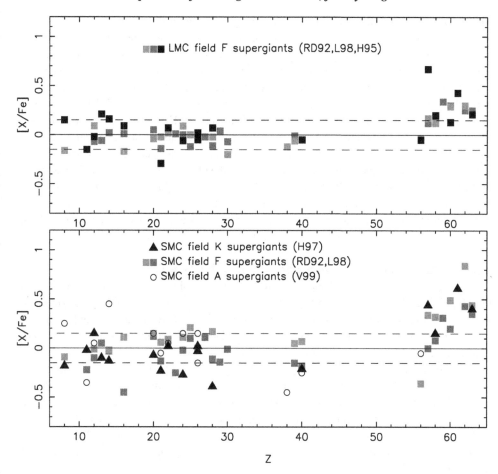

Fig. 15.1. Abundance ratios of young warm and cool supergiants in the LMC (*upper panel*) and SMC (*lower panel*), *normalized to their solar neighborhood counterparts*. Symbols are explained on the figure, with references as follows: RD92, Russell & Dopita (1992); L98, Luck et al. (1998); H95, Hill et al. (1995); H97, Hill et al. (1997); V99, Venn (1999).

evolutionary paths taken by each of the three systems (SMC, LMC, and solar neighborhood) to reach their present-day chemical composition, and despite the different overall metallicities reached (1/5, 1/3, and 1 times solar metallicity), the three systems have very similar abundance ratios in a wide range of elements. There are two notable exceptions. N and, to a lesser extent, C are underabundant (the N/O ratio is down by a factor of 4), while the heavy neutron capture elements are overabundant by factors of 2 to 3, respectively, in the LMC and the SMC.

15.2.3   *Chemical Evolution Models*

At this point, it is worth mentioning the attempts to compute chemical evolution models of the Magellanic Clouds, to reproduce the *present-day* abundance ratios and predict abundance ratios to be expected as a function of time (or metallicity) in each of the two Clouds. Two families of models can be considered, depending on which present-day oxygen over iron ratio they aimed to reproduce: (1) the *underabundance* of O/Fe found in Magellanic supergiants when directly compared to what the solar O/Fe ratio was thought to be at the time*, or (2) the *solar* O/Fe ratio, which is found when comparing the same supergiants to their solar neighborhood counterparts. Category (1) includes: Gilmore & Wyse (1991), who showed that a bursty star formation could either enhance or lower O/Fe ratios depending on which epoch from the burst the galaxy is observed (see §15.4.1); Tsujimoto et al. (1995), who used a steeper initial mass function (IMF) slope to lower the massive SN II contribution and therefore lower the O/Fe ratio (they also designed a bursty model, using the same steep IMF); Pilyugin (1996) and de Freitas Pacheco (1998), who advocate α-enriched winds to achieve the low O/Fe ratio. On the other hand, the semi-analytical models of Pagel & Tautvaišienė (1998) fall into category (2), reproducing solar O/Fe ratio at the evolution end-point, and predicting abundance ratios for two distinct cases of star formation histories, continuous or including two separated bursts of star formation. Examples of these predictions are displayed in Figures 15.2 and 15.3.

15.3   Star Formation History and Age-metallicity Relations

It is clear that, to understand correctly through which chemical enrichment path the Magellanic Clouds proceeded to reach the end-point evolution probed by the youngest stellar generations and the interstellar medium, one needs to understand how and when star formation and its associated nucleosynthesis took place. Star formation history reconstruction from deep and detailed color-magnitude diagrams (CMDs) has proven to be extremely powerful in nearby galaxies where stars can be resolved down to the main sequence, as it is the case for the LMC and the SMC. Unfortunately, in systems like the Magellanic Clouds where the star formation has been quite extended in time, the age-metallicity degeneracy in CMDs is not easy to overcome unless some information on the metallicity distribution at each epoch is injected independently (the so-called age-metallicity relation). Here is not the right place to review all the attempts that have been made in this direction, but a couple of key points can be reminded:

(1) Star formation histories deduced from deep *HST* images in different regions of the LMC all detected a significant increase of star formation (burst) at relatively recent times (few Gyr ago), with various detailed characteristics (exact epoch and intensity). A nice overview of these results can be gathered from Holtzman et al. (1997), Geha et al. (1998), and Smecker-Hane et al. (2002). For the SMC, no such detailed studies have been performed to date.

(2) Age-metallicity relations in the LMC and SMC were for a long time probed only from clusters that provide relatively easy age (from deep CMDs in the best cases, and integrated colors otherwise) and metallicity determinations, using either narrow-band (Strömgren or Washington systems) photometry or low-resolution calcium triplet spectroscopy of individual stars (for the LMC, see Geisler et al. 1997; Olzsewski et al. 1991, and references therein; for the SMC, see Da Costa & Hatzidimitriou 1998, and references therein). The two cluster age-metallicity relations are quite distinct: the SMC shows a smooth distribution of cluster ages with slowly rising metallicities, whereas the LMC shows a prominent gap in age between the old (>10 Gyr) globular clusters and the young ones (<4–5 Gyr), in which no (or very few) clusters are to be found (see Sarajedini 1998). Lately (in the LMC only), age-metallicity

---

* The solar log (O/H) ≈ 8.9 advocated until recently was higher than the O abundance derived from B stars and supergiants in the solar neighborhood. Instead, the current log (O/H) ≈ 8.69 derived by Allende Prieto, Lambert, & Asplund (2001) is in agreement with the local B stars and K–F supergiants.

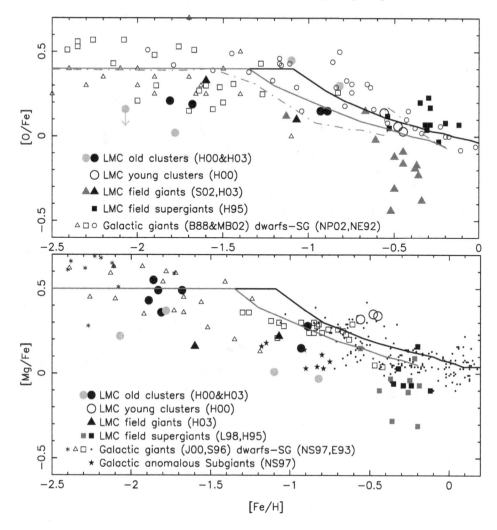

Fig. 15.2. Abundance of the α-elements O and Mg and for LMC stars, compared with Milky Way halo and disk stars. The symbols are explained on each plot, and use the following notation for LMC objects (big or filled symbols): H00, Hill et al. (2000); H03, Hill et al. (2004); S02, Smith et al. (2002); L98, Luck et al. (1998); H95, Hill et al. (1995). Overplotted as black and gray lines are chemical evolution models by Pagel & Tautvaišienė (1995, 1998) for the Galaxy and the LMC, respectively (see text for explanation). The references for Galactic comparison stars are: NE92, Nissen & Edvardsson (1992); B88, Barbuy et al. (1988); MB02, Melendez & Barbuy (2002); NP02, Nissen et al. (2002); E93, Edvardsson et al. (1993); NS97, Nissen & Schuster (1997); J00, Johnson (2002); S96, Shetrone (1996); G91, Gratton & Senden (1991); G94, Gratton & Senden (1994).

relations for field stars have appeared, from planetary nebulae (Dopita et al. 1997) or from red giant branch (RGB) stars (Bica et al. 1998; Cole, Smecker-Hane, & Gallagher 2000; Dirsch et al. 2000), showing that (not surprisingly) no such age gap exists in the LMC field.

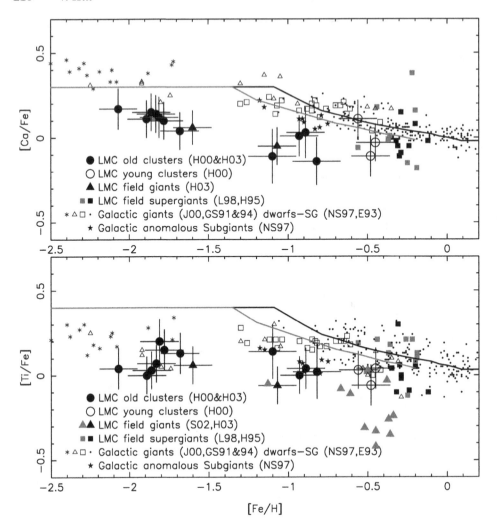

Fig. 15.3. Abundance of the α-elements Ca and Ti and for LMC stars, compared with Milky Way halo and disk stars. Symbols and references as in Fig. 15.2

15.4   Detailed Composition of the Older Populations

A new path can be taken to unravel the history of the Magellanic Clouds, by measuring abundance ratios in stars of various ages, thus probing the chemical content of the galaxies *at different epochs along their evolution*. This can readily be done using high-resolution spectrographs on the 8–10 m-class telescopes to observe red giants (which span a large age and metallicity range) in both Magellanic Clouds. So far, the only published high-resolution analysis of red giants in the Magellanic Clouds concerns the LMC. The only work dealing with field red giants is that of Smith et al. (2002), who analyzed from near-infrared spectra (Phoenix spectrograph on Gemini South), a sample of 12 LMC giants in the metallicity range $-0.3 < \text{[Fe/H]} < -1.1$, for which they derive abundances of C, N, O, Na,

Table 15.2. *Summary of characteristics of the clusters observed.*

	Age (1)	[Fe/H]$_{CaT}$ (2)	[Fe/H]$_{HRspec}$ (3)	R (4)	
NGC 1841	14	−2.11	−2.07 ± 0.10	14°	S
NGC 2257	14.3	−1.80	−1.86 ± 0.07	8°	NE
NGC 2210	14	−1.97	−1.75 ± 0.07	4.5°	E
ESO 121-SCO3	9.2	−0.93	−0.91 ± 0.11	10°	N
NGC 1978	2	−0.42	−0.91 ± 0.15	3.5°	N
NGC 1866	0.2	...	−0.51 ± 0.08	4°	N

(1): Age in Gyr. (2): Metallicity estimate from the calcium triplet infrared lines at low resolution, from Olszewski et al. (1991). (3): Iron abundance derived from high-resolution spectra (Hill et al. 2000, 2004). (4): Distance to the LMC Bar.

Sc, Ti, and Fe*. Because it is difficult to select old and metal-poor field stars from CMDs (other than statistically), a very appealing way to select an old and metal-poor component of the Magellanic Clouds is by using globular clusters. This is the path that has been followed so far by our group, using the UVES high-resolution echelle spectrograph mounted on VLT, to measure the abundances of up to 20 different elements in individual RBG stars in LMC clusters. We have selected globular clusters in a wide age-range, from the young NGC 1866 to the oldest and most metal-poor clusters (see Table 15.2) for which we have analyzed 2–4 RGB stars (Hill et al. 2000, 2004). Two LMC field RGB stars ([Fe/H] = −1.1 and −1.6) have also been observed for comparison purposes. Johnson (2004) also observed RBG stars in four old clusters, three in the LMC (NGC 1898 and NGC 2019, close to the LMC Bar, and Hodge 11) and NGC 121 in the SMC, but her results are not reported here.

It is not obvious that globular clusters in a galaxy should reflect faithfully the behavior of the whole galaxy. Counting them to trace star formation along an age axis has proved to be quite biased (see §15.3). The star formation episodes that gave rise to the bulk of the intermediate-age stellar population of the LMC does not seem to have left any globular cluster relics behind: most probably, no clusters massive enough to survive longer than a few billion years where formed during that period. On the other hand, abundance ratios in Galactic globular cluster stars trace very well those of the halo field stars (except for a few of the lighter elements; see §15.4.1), and we will therefore use the Magellanic Cloud clusters as representatives of the chemical composition of the LMC gas at the time when they were formed.

From the combined data set (clusters and field stars), it now becomes possible to examine the trends of abundance ratios with time, or metallicity, in the LMC (hopefully, such data will soon also become available in the SMC). The following sections attempt to do so, concentrating on two particularly appealing groups of elements: the α-elements and the neutron capture elements.

15.4.1 *Oxygen and other α-Elements*

The so-called α-elements are produced in massive stars during the hydrostatic burning phase for the lighter elements (O, S, Mg, and Si), but also during the explosive phase of

* And now also fluorine: Cunha et al. (2003).

the SNe II (Si, Ca, and Ti). They are obvious chemical evolution tracers as they are released very promptly after a star formation episode, thanks to the short lifetime of their progenitors. The α-elements are also very useful to compare to iron, since as soon as a few SNe Ia start contributing to the chemical enrichment ($\sim$1 Gyr after the star formation episode), most of the iron is produced by SNe Ia. Hence, the ratio [α/Fe] probes star formation time scales and SN II/SN Ia rates.

Figures 15.2 and 15.3 display the LMC abundance ratios for old and intermediate stars (Hill et al. 2000, 2004; Smith et al. 2002), together with comparison samples in the Galactic disk and halo (see figure caption for references) and a selection of young LMC supergiants, normalized to their solar neighborhood counterparts (B stars were excluded from the plot on the basis of their rather uncertain iron abundance measurements). Also plotted on the figures are chemical evolution models for the Galaxy and the LMC assuming a continuous star formation rate over the last 14 Gyr (Pagel & Tautvaišienė 1995, 1998). To illustrate the expected response of α/Fe ratios to a burst of star formation, we also plotted (Fig. 15.2, upper panel) their LMC chemical evolution model assuming two bursts of star formation separated by 9 Gyr of very low activity (dot-dashed gray line): the sudden bump around [Fe/H] = -0.5 is due to the prompt α-elements output from SNe II of the second burst of star formation, some 2 Gyr ago.

Before commenting further on oxygen and magnesium, it is necessary to recall that in Galactic globular clusters abundance anomalies of O, Na, Mg, and Al are often observed on the RGB (and also now at the turn-off; e.g., Gratton et al. 2001), in the form of chemical inhomogeneity among the stars of a given cluster, which define O-Na and Mg-Al anticorrelations. In the LMC oldest globular cluster, this seems to be also the case (Hill et al. 2000), and the two metal-poor stars that appear as faint gray filled circles in Figure 15.2 are stars that are probably oxygen (and Mg?) poor due to these anomalies, and should likely not be considered representative of the LMC chemical evolution. (The other two more metal-rich cluster stars that appear as faint gray circles are members of NGC 1978 that displayed strong TiO bands, introducing extra uncertainties in the analysis.)

Setting aside these more uncertain data points, several comments can be made from Figures 15.2 and 15.3:

(1) The oldest, most metal-poor population in the LMC (represented by the clusters, as black circles), defines O/Fe and Mg/Fe ratios that are compatible with those of the Galactic halo, although slightly on the lower side in the case of oxygen. Calcium and titanium, on the other hand, are clearly below the halo counterparts. If these low values for some α-elements are confirmed with larger samples, it would set interesting constraints on the formation mechanisms of halos: in canonical views, lower α/Fe ratios in objects as old as $\sim$14 Gyr cannot be due to contributions from SNe Ia but have to be linked to the relative numbers of very massive SNe II to less massive SNe II (producing relatively fewer α-elements than the more massive ones), which contributed to the early chemical enrichment. This relative number (an *effective IMF*) could be altered either by changing the IMF itself, or by selective winds blowing preferentially α-enriched material out of the galaxy, or by *statistically truncating* the high-mass end of the IMF (see Shetrone 2004). Most of these explanations, however, do not seem very plausible for the LMC: the IMF measured in Magellanic Clouds clusters and associations is compatible with a Salpeter IMF (Massey, Johnson, & DeGoia-Eastwood 1995; Hunter et al. 1997), winds would have to be very strong to escape the potential of the LMC, and the statistical truncation of the IMF, while a valid idea when the mass involved in

star formation episodes is very small, does not seem very appealing in a system of the size and mass of the LMC, where present star-forming regions are huge.

(2) For metallicities above [Fe/H] = −1, the O and Ti behavior becomes very confused: while the cluster stars (and two field stars) from our own work seem to define very constant [O/Fe] and [Ti/Fe] ratios (around +0.15 dex and 0.0 dex, respectively) merging into the solar O/Fe and Ti/Fe ratios defined by the young supergiants* of Hill et al. (1995) and Luck et al. (1998), the field RGB stars of Smith et al. (2002, gray triangles) define a very strong decrease of the O/Fe and Ti/Fe ratio with increasing metallicity. This difference could be due to several effects, among them the use of two distinct oxygen indicators (Smith et al. 2002 use infrared OH lines, whereas we have used the [O I] forbidden line), but the most probable cause for the disagreement lies in the temperature calibration of the stars. For their sample, Smith et al. (2002) estimate that the uncertainty that they have on temperature is of the order of 100 K (order-of-magnitude estimate of the possible systematic offset between various $(J-K)$ color-temperature relations), which, in turn, translates into an uncertainty of +0.32 dex on their derived [O/Fe] ratios and +0.26 dex on their [Ti/Fe] ratios, while [Fe/H] only decreases by −0.10 dex. Assuming a temperature scale hotter by 100 K would therefore bring their whole sample back into agreement with the rest of the data, both for oxygen and titanium (the published data do not contain Mg or Ca abundances for this sample).

(3) On this hypothesis, one would conclude that the [O/Fe], [Ca/Fe], and [Ti/Fe] ratios are rather flat at all metallicities, defining a slope shallower than predicted by the Pagel & Tautvaišienė (1998) models. It is only in the Mg diagram that a significant evolution of the [Mg/Fe] ratio with metallicity is detected.

(4) An alternative view is developed in Smith et al. (2002; their Fig. 10), who take their [O/Fe] at face value and conclude that the run of [O/Fe] with increasing metallicity defines a sequence *always below* the one defined by the Galactic disk, which they successfully reproduce with a simple closed-box analytical model by lowering the SN II rate by a factor of 3, and the SN Ia rate by a factor of 2, compared to the rates needed to reproduce the Galactic disk O/Fe (in effect, increasing the SN Ia/SN II ratio).

Finally, no detection of the effects of bursts in the chemical evolution diagrams has occurred so far, but no conclusion can be drawn at this stage, owing to small statistics.

15.4.2   *Neutron Capture Heavy Elements*

Elements heavier than the iron peak are synthesized by neutron capture, which, depending on the neutron capture rate (and therefore the neutron density), is either called slow (*s*-process) or rapid (*r*-process). The *s*-process channel can happen in or in between thermal pulses of AGB stars, whereas the *r*-process requires much more extreme neutron exposures and are thought to take place either in SN II explosions (classical or in ν-driven winds) or in binary neutron stars, all hypotheses requiring massive progenitor stars (see Cowan & Sneden 2004 for a review on the *r*-process). Therefore, the time scale on which the *r*-process starts contributing to the chemical enrichment of a system is expected to be much shorter than for the *s*-process, as it is indeed observed in Galactic halo stars that are still dominated by the *r*-process up to metallicities of at least −2, or even −1 (Burris et al. 2000; Johnson 2002).

We have seen in §15.2.2.2 that the *present-day* neutron capture element abundances in the LMC (and even more strikingly in the SMC) are enhanced compared to what is observed

---

* Normalized to their solar neighborhood counterparts.

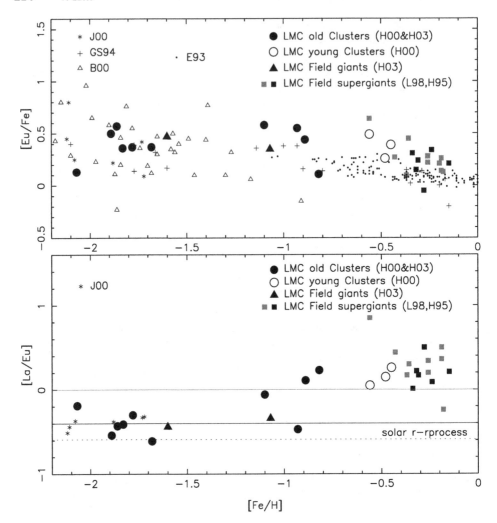

Fig. 15.4. Abundance of the *r*-process element [Eu/Fe] and the *r*-/*s*-process ratio indicator [La/Eu] for LMC stars, compared with Milky Way halo and disk stars. Symbols and references as in Fig. 15.2, except for the addition of B00 (Burris et al. 2000).

in our Galaxy, and the cause of this enhancement is not well understood. The epoch (or metallicity) to which this enhancement can be traced back yields precious information to understand the process responsible for it.

The upper panel of Figure 15.4 shows the abundance ratio of europium (an element produced almost entirely by the *r*-process) to iron, as a function of [Fe/H] for the LMC sample (mainly globular clusters this time, since Smith et al. 2002 did not have access to any heavy element transitions in their infrared spectra), while the lower panel displays the abundance ratio of lanthanum to europium, a diagnostic of the *s*-/*r*-process ratio (as soon as the *s*-process starts contributing, La is dominated by *s*-process). A few remarks arise from the inspection of these plots:

(1) At the lowest metallicities (oldest globular clusters), both the overall [Eu/Fe] and the *s*-/*r*-process ratio are indistinguishable from that of the Galactic halo. This shows that, at early times, the *r*-process is the process dominating the neutron capture element synthesis (the pure *r*-process [La/Eu] ratio is displayed in the figure as black full and dashed lines, following the solar *r*-process decompositions of Burris et al. 2000 and Arlandini et al. 1999), and that the efficiency of the *r*-process in the LMC and Galactic halos were the same (giving rise to a same [Eu/Fe] ratio).

(2) The clear positive slope of the [La/Eu] ratio versus metallicity is the signature of an increasing *s*-process contribution with increasing metallicity, reaching a solar *s*-/*r*-process ratio (gray line) at metallicities around and above −1. It is interesting to note that this increase is already visible in the cluster ESO 121-SC03 (two points at [Fe/H] ≈ −0.8), which is a 9 Gyr cluster.

(3) The [La/Eu] ratio at the highest metallicities and youngest ages, reaches values that are above the solar ratio, suggesting that the *s*-process becomes even more efficient in the LMC than in our own Galaxy. In this case, the *s*-process would be responsible for the strong neutron capture enhancement in the present-day matter of the LMC, originating from AGBs rather than the massive stars needed to produce the *r*-process.

15.5   Conclusions

We have reviewed the current state of our knowledge of the chemical composition of the Magellanic Clouds, concentrating on the best probes of detailed elemental abundances.

We have seen that despite the likely very different evolutionary paths taken by the SMC and the LMC (compared to our own Galaxy) to reach their present-day chemical composition, and despite the different overall metallicities reached (1/5 and 1/3 solar, respectively), both Magellanic Clouds have reached very similar abundance ratios in a wide range of elements, apart from two notable exceptions: nitrogen and, to a lesser extent, carbon are underabundant (the N/O ratio is down by a factor of 4), while the heavy neutron capture elements are overabundant by factors of 3 and 2, respectively, in the SMC and the LMC.

At the other end of evolution, the oldest population in the LMC also seems to be very like the Milky Way halo stars, except for the notable exception of calcium and titanium (and maybe oxygen to a lesser extent). The situation in between these two extremes so far suffers from a lack of statistics, with only one study of 12 field RGB stars (Smith et al. 2002), and two intermediate-age clusters (Hill et al. 2000) in the LMC. Two alternative views are emerging from the two samples, favoring an evolution more strongly dominated by SNe Ia (compared to the Galactic disk) that leads to lower α/Fe ratios in all intermediate and young populations in one case, and in the other case, a remarkably flat variation of these α/Fe ratios along the whole evolution.

References

Allende Prieto, C., Lambert, D. L., & Asplund, M. 2001, ApJ, 556, L63
Andrievsky, S. M., Kovtyukh, V. V., Korotin, S. A., Spite, M., & Spite, F. 2001, A&A, 367, 605
Arlandini, C., Käppeler, F., Wisshak, K., Gallino, R., Lugaro, M., Busso, M., & Straniero, O. 1999, ApJ, 525, 886
Barbuy, B. 1988, A&A, 191, 121
Bica, E., Geisler, D., Dottori, H., Clariá, J. J., Piatti, A. E., & Santos., J. F. C. 1998, AJ, 116, 723
Burris, D. L., Pilachowski, C. A., Armandroff, T. E., Sneden, C., Cowan, J. J., & Roe, H. 2000, ApJ, 544, 302
Cole, A. A., Smecker-Hane, T. A., & Gallagher, J. S. 2000, AJ, 120, 1808

Cowan, J. J., & Sneden, C. 2004, in Carnegie Observatories Astrophysics Series, Vol. 4: Origin and Evolution of the Elements, ed. A. McWilliam & M. Rauch (Cambridge: Cambridge Univ. Press), in press

Da Costa, G. S., & Hatzidimitriou, D. 1998, AJ, 115, 1934

de Freitas Pacheco, J. A. 1998, AJ, 116, 1701

Dirsch, B., Richtler, T., Gieren, W. P., & Hilker, M. 2000, A&A, 360, 133

Dopita, M. A., et al. 1997, ApJ, 474, 188

Dufour, R. J. 1990, in Evolution in Astrophysics: IUE Astronomy in the Era of New Space Missions, 117

Dufour, R. J., Shields, G. A., & Talbot, R. J. 1982, ApJ, 252, 461

Edvardsson, B., Andersen, J., Gustafsson, B., Lambert, D. L., Nissen, P. E., & Tomkin, J. 1993, A&A, 275, 101

Esteban, C., Peimbert, M., Torres-Peimbert, S., & Escalante, V. 1998, MNRAS, 295, 401

Garnett, D. R. 1999, in IAU Symp. 190, New Views of the Magellanic Clouds, ed. Y.-H. Chu et al. (Dordrecht: Kluwer), 266

Garnett, D. R., Galarza, V. C., & Chu, Y. 2000, ApJ, 545, 251

Garnett, D. R., Skillman, E. D., Dufour, R. J., Peimbert, M., Torres-Peimbert, S., Terlevich, R., Terlevich, E., & Shields, G. A. 1995, ApJ, 443, 64

Geha, M. C., et al. 1998, AJ, 115, 1045

Geisler, D., Bica, E., Dottori, H., Claria, J. J., Piatti, A. E., & Santos, J. F. C. 1997, AJ, 114, 1920

Gilmore, G., & Wyse, R. F. G. 1991, ApJ, 367, L55

Gratton, R. G., et al. 2001, A&A, 369, 87

Gratton, R. G., & Sneden, C. 1991, A&A, 241, 501

——. 1994, A&A, 287, 927

Henry, R. B. C., Edmunds, M. G., & Köppen, J. 2000, ApJ, 541, 660

Hill, V. 1999, A&A, 345, 430

Hill, V., et al. 2004, in preparation

Hill, V., Andrievsky, S., & Spite, M. 1995, A&A, 293, 347

Hill, V., Barbuy, B., & Spite, M. 1997, A&A, 323, 461

Hill, V., François, P., Spite, M., Primas, F., & Spite, F. 2000, A&A, 364, L19

Hill, V., & Spite, M. 1999, Ap&SS, 265, 469

Holtzman, J. A., et al. 1997, AJ, 113, 656

Hunter, D. A., Light, R. M., Holtzman, J. A., Lynds, R., O'Neil, E. J., & Grillmair, C. J. 1997, ApJ, 478, 124

Johnson, J. A. 2002, ApJS, 139, 219

——. 2004, in Carnegie Observatories Astrophysics Series, Vol. 4: Origin and Evolution of the Elements, ed. A. McWilliam & M. Rauch (Pasadena: Carnegie Observatories, http://www.ociw.edu/ociw/symposia/series/symposium1/proceedings.html)

Korn, A. J., Becker, S. R., Gummersbach, C. A., & Wolf, B. 2000, A&A, 353, 655

Korn, A. J., Keller, S. C., Kaufer, A., Langer, N., Przybilla, N., Stahl, O., & Wolf, B. 2002, A&A, 385, 143

Kurt, C. M., Dufour, R. J., Garnett, D. R., Skillman, E. D., Mathis, J. S., Peimbert, M., Torres-Peimbert, S., & Ruiz, M.-T. 1999, ApJ, 518, 246

Luck, R. E., & Lambert, D. L. 1992, ApJS, 79, 303

Luck, R. E., Moffett, T. J., Barnes, T. G., & Gieren, W. P. 1998, AJ, 115, 605

Massey, P., Johnson, K. E., & DeGoia-Eastwood, K. 1995, ApJ, 454, 151

McWilliam, A., & Williams, R. E. 1991, in IAU Symp. 148, The Magellanic Clouds, ed. R. Haynes & D. Milne (Dordrecht: Kluwer), 391

Meléndez, J., & Barbuy, B. 2002, ApJ, 575, 474

Nissen, P. E., & Edvardsson, B. 1992, A&A, 261, 255

Nissen, P. E., Primas, F., Asplund, M., & Lambert, D. L. 2002, A&A, 390, 235

Nissen, P. E., & Schuster, W. J. 1997, A&A, 326, 751

Olszewski, E. W., Schommer, R. A., Suntzeff, N. B., & Harris, H. C. 1991, AJ, 101, 515

Pagel, B. E. J., & Tautvaišiené, G. 1995, MNRAS, 276, 505

——. 1998, MNRAS, 299, 535

Pilyugin, L. S. 1996, A&A, 310, 751

Pilyugin, L. S., Thuan, T. X., & Vílchez, J. M. 2003, A&A, 397, 487

Richtler, T., Spite, M., & Spite, F. 1989, A&A, 225, 351

Rolleston, W. R. J., Brown, P. J. F., Dufton, P. L., & Howarth, I. D. 1996, A&A, 315, 95

Rolleston, W. R. J., Trundle, C., & Dufton, P. L. 2002, A&A, 396, 53

Rolleston, W. R. J., Venn, K., Tolstoy, E., & Dufton, P. L. 2003, A&A, 400, 21

Russell, S. C., & Bessell, M. S. 1989, ApJS, 70, 865

Russell, S. C., & Dopita, M. A. 1992, ApJ, 384, 508

Sarajedini, A. 1998, AJ, 116, 738

Shetrone, M. D. 1996, AJ, y112, 1517

——. 2004, in Carnegie Observatories Astrophysics Series, Vol. 4: Origin and Evolution of the Elements, ed. A. McWilliam & M. Rauch (Cambridge: Cambridge Univ. Press), in press

Smecker-Hane, T. A., Cole, A. A., Gallagher, J. S., & Stetson, P. B. 2002, ApJ, 566, 239

Smith, V. V., et al. 2002, AJ, 124, 3241

Spite, F., Spite, M., & Richtler, T. 1991, A&A, 252, 557

Spite, M., François, P., Spite, F., Cayrel, R., & Richtler, T. 1986, A&A, 168, 197

Spite, M., Spite, F., & Barbuy, B. 1989, A&A, 222, 35

Tsamis, Y. G., Barlow, M. J., Liu, X.-W., Danziger, I. J., & Storey, P. J. 2003, MNRAS, 338, 687

Tsujimoto, T., Nomoto, K., Yoshii, Y., Hashimoto, M., Yanagida, S., & Thielemann, F.-K. 1995, MNRAS, 277, 945

Venn, K. A. 1999, ApJ, 518, 405

16

Detailed composition of stars in dwarf spheroidal galaxies

MATTHEW D. SHETRONE
University of Texas, McDonald Observatory

16.1   Introduction

Simulations of galaxy formation in a hierarchical scenario suggest that the halo of the Milky Way (MW) may have formed from many smaller protogalaxies and that the protogalaxy building blocks farthest from the center of the MW gravitational potential may have turned into small galaxies such as the Local Group dwarf spheroidal (dSph) system. The age and abundance estimates of the dSph systems suggest that these dSph systems have metallicities and ages similar to that of the MW halo. High-resolution abundance analysis of dSph giants is a fairly new field of study that grew out of the advent of 8–10 m class telescopes. The dSph systems are generally old, and thus the brightest stars available for high-resolution study are the giants near the tip of the giant branch. This type of study is similar to the early studies of globular clusters. The similarity with globular clusters cannot be overemphasized. Many of the advances in chemical evolution come from using the globular clusters as templates of how chemical evolution occurs. However, nearly all globular clusters are mono-metallic and thus only exhibit the abundance patterns one might expect for a specific metallicity (and a specific star formation rate). From early studies of the most nearby dSph galaxies, it has been known that they are not mono-metallic and thus could serve as even more important test cases for our knowledge of chemical and stellar evolution.

In this review the research done to date on high-resolution dSph abundances is summarized, including several not-yet-refereed studies, and some connection to our knowledge of chemical evolution and hierarchical galaxy formation is drawn. The current tally of surveyed galaxies is given in Table 16.1. As can be seen from Table 16.1 different researchers take different tactics in their approach to investigating dSph abundances. The Shetrone et al. teams' spectra have just a few spectra in each dSph galaxy with moderate signal-to-noise ratio (S/N) spectra, while the Smecker-Hane and Bonifacio teams have concentrated on a large sample in the closest dSph with high S/N in each spectrum. Both approaches have merit and yield different insights into the chemical evolution of the dSph system. For this review the dSph galaxies have been divided into two categories—simple and complex—based upon their implied star formation histories (e.g., Dolphin 2002). Draco, Ursa Minor, Sextans, and Sculptor we classify as simple because the implied star formation history has a single burst followed by a decline in star formation. Leo I, Fornax, Carina, and Sagittarius (Sgr) we classify as complex because their implied star formation histories contain more than a single burst of star formation. Other than the overall metallicity typified by Fe, we will concentrate on five classes of elements: the light α elements (represented by O and Mg), the heavy α elements

Table 16.1. *The High-resolution dSph Sample*

dSph	# Stars	Observatory	Reference
Sculptor	5	VLT UVES	Shetrone et al. (2003)
	4	VLT UVES	Geisler et al. (2004)
Ursa Minor	3	Keck HIRES	Shetrone, Bolte, & Stetson (1998)
	3	Keck HIRES	Shetrone, Côté, & Sargent (2001)
Draco	4	Keck HIRES	Shetrone, Bolte, & Stetson (1998)
	2	Keck HIRES	Shetrone, Côté, & Sargent (2001)
	1	Keck HIRES	Fulbright, Rich, & Castro (2004)
Sextans	5	Keck HIRES	Shetrone, Côté, & Sargent (2001)
Sagittarius	2	VLT UVES	Bonifacio et al. (2000)
	14	Keck HIRES	Smecker-Hane & McWilliam (2004)
	10	VLT UVES	Bonifacio & Caffau (2003)
			Bonifacio et al. (2003)
Carina	5	VLT UVES	Shetrone et al. (2003)
Fornax	3	VLT UVES	Shetrone et al. (2003)
Leo I	2	VLT UVES	Shetrone et al. (2003)

(represented by Ca and Ti), the light odd-Z elements (represented by Na), a few iron-peak elements (represented by Mn and Cu), and the neutron capture elements (represented by Y, La, Ba, and Eu).

16.2 Dangers and Caveats

One of the main dangers in constructing a coherent data set from different analyses is the zeropoints created by the observing and analysis techniques of different researchers. Because the dSph stars are observed with 8 m-class telescopes and telescope time is quite precious, there is almost no overlap between samples. However, there is a single exception: in the sample of Shetrone et al. (1998) there was a Draco star observed with very low metallicity, D119. This star was later reobserved by Fulbright et al. (2004). The Fulbright spectrum of D119 had much higher S/N ($\sim$100) than that of Shetrone et al. ($\sim$40). The techniques employed to get effective temperature, metallicity and surface gravity are different for these two authors. Despite all of these differences the abundance ratios are in reasonable agreement: within 1 σ for Fe, Ca, Ti, and Cr, and 2 σ for Na, Mg, and Ni. What is important about the results from the higher-S/N data in Fulbright et al. are the smaller error bars and interesting upper limits that can be used in chemical evolution analysis.

One of the easiest, although tedious, corrections that has to be made when combining the analysis of different authors is the choice of oscillator strengths. This is easy when the same lines are used by different authors but nearly impossible when different line sets are used. In this review some attempt will be made to bring the line lists into a common system. Examples of where this will have a strong impact include the Mg I λ5528 Å abundances (Shetrone et al. 2001 corrected upward by 0.14 dex and Geisler et al. 2004 and Bonifacio et al. 2000 corrected downward by 0.1 dex), the Mg I λ5711 Å abundances (Geisler et al. 2004 corrected downward by 0.11 dex), the Ca I λ6156 Å abundances (Bonifacio et al. 2000 corrected upward by 0.29 dex), the Ca I λ6161 Å abundances (Bonifacio et al. 2000 corrected upward by 0.25 dex), the Ca I λ6166 Å abundances (Bonifacio et al. 2000 corrected upward by 0.24 dex), and the Ca I λ6508 Å abundances (Bonifacio et al. 2000 corrected upward by 0.39 dex). Where there is little overlap in the choice of lines, the person combining the data sets must be wary and look for abundance jumps or inconsistencies in the abundance trends.

Hyperfine splitting (HFS) is also a correction that not all analyses include. For example, Shetrone et al. (1998), Bonifacio et al. (2000), and Shetrone et al. (2001) did not include HFS for Ba, Eu, Cu or Mn, while Smecker-Hane & McWilliam (2004) and Shetrone et al. (2003) did. HFS affects different elements differently and even different lines within the same species are affected differently; for example, Mn HFS is quite large and should be included for all lines, while the HFS for the red Ba lines are not very large, and the blue lines of Ba have larger HFS corrections. Generally speaking strong lines are affected more than weak lines, and thus metal-poor stars have smaller HFS corrections than metal-rich stars. For this review, corrections for HFS will be made when access to the data, published or unpublished, is available.

Differences in analysis techniques and model atmospheres used by the different researchers can also lead to zeropoint differences in the different data sets. In the work of Shetrone et al. (2003) some attempt was made to quantify the abundance differences using the AT-LAS/WIDTH vs. MARCS/MOOG; they found that the abundances changed by typically 0.06 dex, with no change being larger than 0.12 dex. Shetrone et al. (1998, 2001, 2003) have employed MARCS/MOOG, Smecker-Hane & McWilliam (2004) have employed AT-LAS/MOOG, and Bonifacio et al. (2000, 2004) and Bonifacio & Caffau (2003) have employed ATLAS/WIDTH. No attempt will be made in this review to compensate for differences in analysis techniques.

Non-local thermodynamic equilibrium (NLTE) modeling is not a standard part of the abundance analysis of dSph stars at the present time. This is in part due to the comparative nature of the analysis. The researchers compare the abundances of the dSph stars to similar analysis of MW stars, and since these MW stars were not analyzed with NLTE, neither are the dSph stars. No attempt to make NLTE corrections has been made in this review.

16.3   Simple Systems

The best studied among the simple systems is Sculptor (Scl), with five stars from Shetrone et al. (2003) and four stars from Geisler et al. (2004), and as such we begin our review by looking in detail at this galaxy as a template for the other simple systems. Figure 16.1 exhibits the heavy α elements (Mg and O) abundance ratios plotted against metallicity. The solid line represents a MW toy model where Fe yields of Type II supernovae (SNe II) and Type Ia supernovae (SNe Ia) are included and a constant star formation rate is assumed. The yields in this toy model are set by matching the MW abundances and do not rely on the theoretical SN yields. When the SNe Ia begin to contribute, the slope changes but the α abundance continues to rise because the star formation rate has not changed and SNe II continue to contribute α elements. The most metal-poor Scl dSph stars fall on the MW toy model, while the more metal-rich stars fall below that model. The dashed line represents a Scl toy model (we do not intend to imply that this model is correct and only use it for discussion purposes). The Scl toy model is similar to the MW toy model except that at an arbitrary point ([Fe/H] = −1.8, which is approximately the peak metallicity for many of the simple dSph systems) the star formation rate is dropped by a factor of 10 (another arbitrary

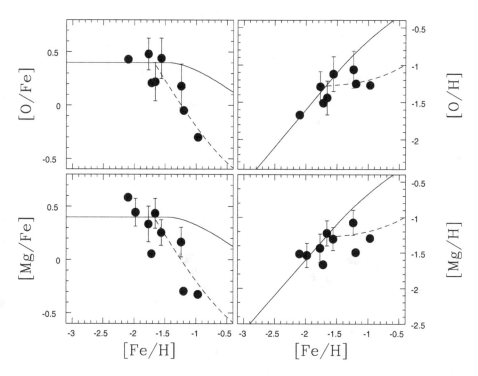

Fig. 16.1. The filled circles with error bars are from Shetrone et al. (2003), and the ones without error bars are from Geisler et al. (2004). The solid line represents the MW toy model and MW mean. The dashed line represents a Scl toy model.

value set to fit Fig. 16.1 only). In all other respects the Scl toy model will use the same parameters as the MW toy model (i.e. yields set to fit the MW abundance pattern). The rapid decline in [O/Fe] and [Mg/Fe] after [Fe/H] = −1.8 is due to the onset of SNe Ia, and since SNe Ia create very little light α elements and large amounts of Fe, these ratios decline. This can also be seen in the right panels of Figure 16.1, where the α abundances are plotted against metallicity. The total amount of O and Mg in the most metal-rich stars is nearly the same as the stars that have nearly 6 times lower [Fe/H]. Figure 16.2 shows a similar plot for the heavy α elements. In this figure the Ca and Ti abundances are not constant over the [Fe/H] range −1.8 to −1.0: there is an increase in the total amount of Ca and Ti. This is possible if SNe Ia produce a small amount of the heavy α elements (e.g., Iwamoto et al. 1999).

In Figure 16.3 the Eu abundances and s-process to r-process ratio ([Ba/Eu]) are plotted against the metallicity. Again the solid line represents the MW toy model, and the dashed line represents the Scl toy model. In both models the s-process time scale is taken to be the same as the SN Ia time scale (the Eu yields in this toy model do not include any metallicity terms), and no contribution from the s-process is included in Eu (Eu is nearly a pure r-process element even at [Ba/Eu] = 0). In the MW toy model the Eu abundance continues to increase as the SNe Ia begin to contribute Fe because the star formation rate is held constant and the SNe II continue to produce Eu. In the middle panel the ratio of s-process to r-process

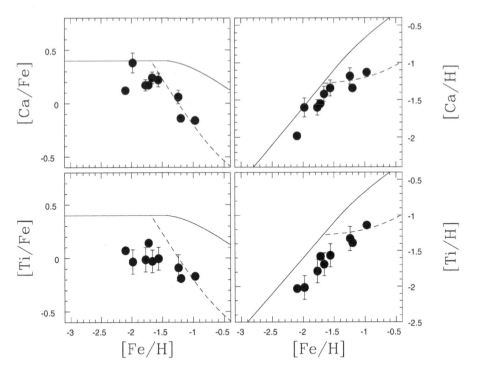

Fig. 16.2. The symbols are the same as in Fig. 16.1.

(Ba to Eu) are shown, and the models show a clear transition from a nearly pure *r*-process ratio at low metallicity to a mixture of *r*-process and *s*-process at higher metallicity. Recall that at [Fe/H] = −1.8 the star formation rate in the Scl toy model drops by a factor of 10, and thus the Eu abundance (largely produced by SNe II) should not increase. The most metal-rich star has a very large Eu abundance and is potentially self-contaminated or contaminated from a binary companion. In the Scl toy model, [Ba/Eu] rapidly rises, as the asymptotic giant branch (AGB) stars begin to contribute *s*-process material, and fits the Scl data reasonably well. The lowest panel of Figure 16.3 suggests that the Eu and light α elements are produced on the same time scales, as is also seen in the MW.

The understanding of the origin of Y is poorly understood in the MW. In very metal-poor stars observations suggest that there may be an additional source of Y (and Sr) in addition to the *r*-process that produced the heavier *r*-process elements (e.g., Ryan, Norris, & Beers 1996; Burris et al. 2000). This has led to models that split the production of light *r*-process elements and heavy *r*-process elements (e.g., Qian & Wasserburg 2001). Because of the complex nature of the Y abundances we have not tried to create a simple model and instead plot the MW abundances as a reference in Figure 16.4. For the more metal-poor stars the Y abundances behave in a way similar to the MW halo, although the [Ba/Y] abundances may be high with respect to the halo. For the more metal-rich Scl stars, where the *s*-process should begin to contribute, the results are a bit mixed. Two of the Y abundances are systematically lower than the MW halo, and [Ba/Y] is correspondingly larger, while one is similar to the MW. With only three stars it is difficult to determine what is the actual trend. In Figure 16.4

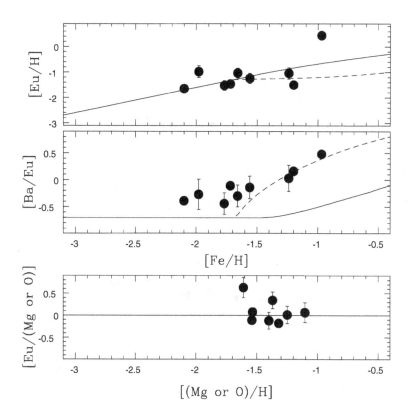

Fig. 16.3. The symbols are the same as in Fig. 16.1.

all of the simple dSph data are shown, and the trend of higher [Ba/Y] at higher metallicities becomes clear. The average [Ba/Y] for the simple dSph stars with [Fe/H] < -1.8 is 0.46 dex, while the average [Ba/Y] for the simple dSph stars with [Fe/H] > -1.8 is marginally larger at 0.62 dex. As was pointed out by Smecker-Hane & McWilliam (2004), these high values for heavy s-process to light s-process are typical for metal-poor AGB stars (e.g., Busso et al. 2001). What remains a mystery is why the pure r-process [Ba/Y] in the dSph is 0.46 dex, while the MW halo stars have an average [Ba/Y] of -0.08 dex, based on the Fulbright (2002) data.

In our Scl toy model we predict that there is significant contribution of SNe Ia in the most metal-rich stars. This leads to the prediction that the Mn and Cu abundances for these stars should be significantly enhanced *if* SNe Ia are responsible for the upturn seen at [Fe/H] in the MW. Figure 16.5 exhibits [Mn/Fe] and [Cu/Fe] plotted against the metallicity. The solid line represents the MW toy model, the dotted line represents a pure SN II contribution based on the metallicity-dependent yields of Woosley & Weaver (1995), and the dashed line represents the Scl toy model. The Scl toy model uses the same prescription as the MW toy model, whereby Cu and Mn are produced in SNe Ia. Clearly the Scl toy model is a poor fit to the data, while the metallicity-dependent SN II model is somewhat better. A similar argument can be made for any of the other simple dSph Mn and Cu abundances,

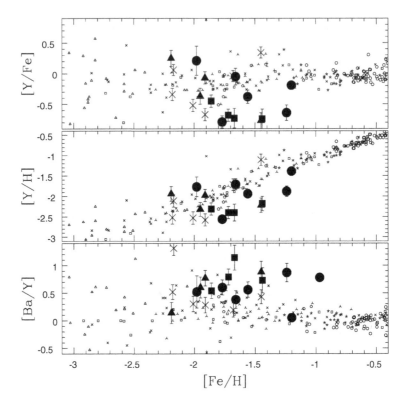

Fig. 16.4. The filled circles are the same as in Fig. 16.1. The large symbols with error bars are from Shetrone et al. (2001): the filled squares are Draco stars, filled triangles are Sextans stars, and the crosses are Ursa Minor stars. The small symbols without error bars are from the literature MW field halo sample: Gratton & Sneden (1988, 1991, 1994), Edvardsson et al. (1993), McWilliam et al. (1995), Nissen & Schuster (1997), Stephens (1999), Burris et al. (2000), Prochaska et al. (2000), and Stephens & Boesgaard (2002).

although care should be given to the Shetrone et al. (2001) data since no correction for HFS was made that would bring the Mn and Cu abundances *down* preferentially for the strongest lines (i.e. the most metal-rich stars). These data eliminate the possibilities of metal-poor SNe Ia as a significant contributor to the rise in [Mn/Fe] seen in the MW. Further constraints on the nucleosynthesis of Mn are introduced by examining the Mn abundances in the more metal-rich Sgr dSph and MW bulge stars (McWilliam, Rich, & Smecker-Hane 2003). Using these different environments, they conclude that Mn yields from both SNe Ia and SNe II are metallicity dependent. Detailed modeling will be required to eliminate one of these two possibilities.

The Scl Na abundance ratios generally fall at the low end of the MW abundance distribution. Smecker-Hane & McWilliam (2004) have attributed the low Na abundances in the Sgr dSph galaxies to metallicity-dependent SN II yields and an excess of Fe from SNe Ia that should not produce any Na (e.g., Thielemann, Nomoto, & Yokoi 1986), the latter being dominant in Sgr. This explanation works well for the more metal-rich dSph stars, but leads

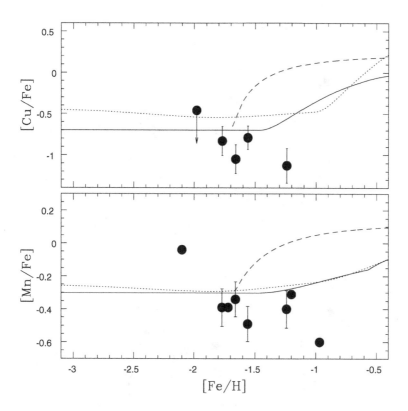

Fig. 16.5. The symbols are the same as in Fig. 16.1. The dotted line is a pure SN II toy model with yields from Woosley & Weaver (1995).

to the question of why the more metal-poor stars in the simple dSphs do not have halo-like Na abundances. Taking the entire simple dSph abundance set between [Fe/H] = −3 to −1.9, we find an average [Na/Fe] = −0.30 (σ = 0.24), while the Fulbright (2002) halo stars in the same metallicity range have an average [Na/Fe] = −0.05 (σ = 0.27). The dSph stars fall near the bottom of the MW [Na/Fe] distribution, although a slight trend with metallicity cannot be ruled out due to the low number of stars near [Fe/H] = −3.

The toy model fails to explain one important feature of the abundances in Scl: the heavy α elements among the most metal-poor stars in Scl fall systematically below the prediction. This can be seen in the left panes of Figure 16.2, where the [Ca/Fe] and [Ti/Fe] abundance ratios are plotted against the metallicity. The SN II model predictions are based on MW halo abundances. Only a single star in the Scl sample has a Ca abundance that falls in the mid range of what is found in the MW halo, and that star has the largest error bar in the sample. All of the other Scl abundances fall at the bottom of the MW Ca distribution at all metallicities. This sub-halo abundance pattern can be seen in the other simple dSph systems as well (see Fig. 16.6). In previous investigations the α elements were grouped together to form an "α" abundance, and the difference between the light α elements and the heavy α elements was not stated explicitly. Fulbright et al. (2004) suggested the split between the

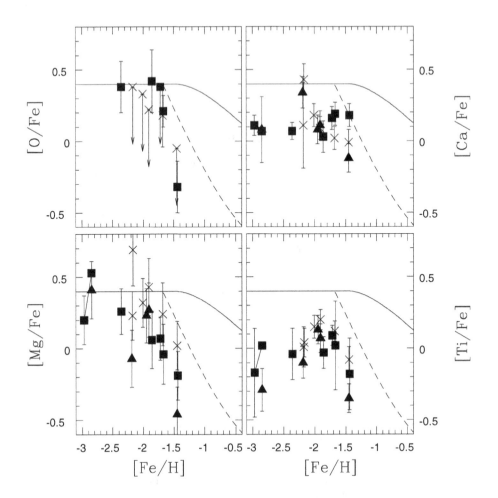

Fig. 16.6. The large symbols with error bars are the same as in Fig. 16.4. The large filled square without an error bar connected to the large filled square with an error bar is D119 from Fulbright et al. (2004). The solid and dashed lines are same as in Fig. 16.1.

light and heavy α elements is mimicked by the most massive SN II yields, as seen in the Woosley & Weaver (1995) yields. Thus, preferentially keeping high-mass SN II yields by changing the initial mass function, invoking selective mass loss from the galaxy, or creating a different prompt inventory for dSph galaxies could create the observed split between the light and heavy α elements.

A comparison of the Scl abundance patterns with the literature dSph abundances for the other simple dSph galaxies reveals remarkable uniformity. One exception is the neutron capture elements. Unfortunately, the Eu abundances from Shetrone et al. (2001) do not include HFS and thus should be corrected before using them. For example, the most metal-rich star in Ursa Minor should have a HFS correction of 0.26 dex downward, which would bring [Ba/Eu] to −0.46 dex. Once corrected, two of the remaining three simple dSph systems

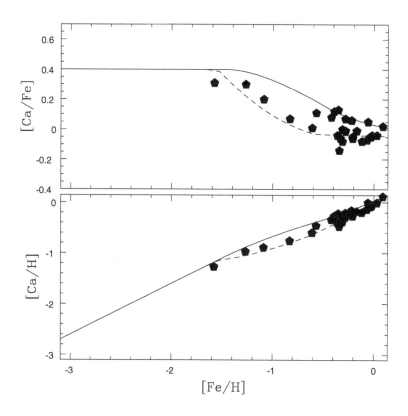

Fig. 16.7. The filled pentagons are from Bonifacio et al. (2000, 2004), Bonifacio & Caffau (2003), and Smecker-Hane & McWilliam (2004). The Ca abundances from Bonifacio have been corrected upward. The solid lines represent the MW toy model and WM mean. The dashed line represents a Sgr toy model.

can be fit with a similar toy model with a rapid decline in the star formation rate, but not necessarily at the same metallicity or of the same magnitude. The third simple system, Ursa Minor, exhibits a fairly flat and low [Ba/Eu] abundance trend. While this implies the neutron capture elements are dominated by the *r*-process, we should also note that the most metal-rich Ursa Minor star mentioned above has a very large *r*-process enhancement in comparison to MW halo stars of similar metallicity. We prefer not to draw a conclusion about the Ursa Minor galaxy neutron capture history because of the small sample size and the odd features of this star.

16.4   Complex Systems

The best studied among the complex systems is Sagittarius (Sgr), with two stars from Bonifacio et al. (2000), 14 stars from Smecker-Hane & McWilliam (2004), and 10 stars from Bonifacio & Caffau (2003) and Bonifacio et al. (2004). We begin by looking in detail at this galaxy as a template for the other complex systems. In Figure 16.7 the Ca abundances and abundance ratios are plotted against the metallicity. The solid line represents the MW toy model described previously. On average the Ca abundances fall below the MW toy

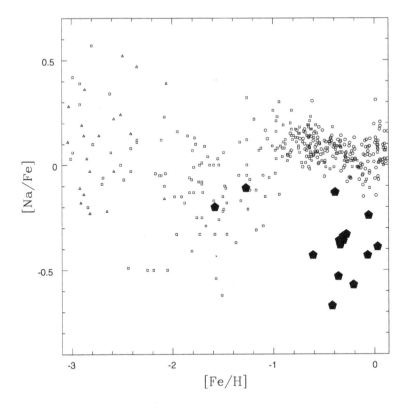

Fig. 16.8. The large filled pentagons are from Bonifacio et al. (2000) and Smecker-Hane & McWilliam (2004). The small symbols are the MW sample: Gratton & Sneden (1988), Edvardsson et al. (1993), McWilliam et al. (1995), Prochaska et al. (2000), and Fulbright (2002).

model at every metallicity. The dash line represents a Sgr toy model constructed in a way similar to the Scl toy model described in the previous section. The Sgr toy model has a initial burst followed by a 25% decline in the star formation rate, and then a second burst of equal strength to the initial burst. These, values were chosen to fit the Ca abundances, and this model is for purely illustrative purposes. The Sgr toy model predictions for the Ca abundance ratios exhibit a decline at [Fe/H] = −1.5, as the SNe Ia begin to contribute extra Fe. The Ca abundance ratios then level out as the second burst adds extra SN II-minted Ca. In contrast to the split seen between light α elements and heavy α elements in the most metal-poor stars among the simple dSph galaxies, the most metal-poor star in Sgr has high Ca and Ti abundances (+0.31 and +0.33, respectively). This seems to imply that Sgr has initial chemical evolution conditions more similar to the MW than to the simple dSph galaxies.

One metal-poor Sgr star observed by Smecker-Hane & McWilliam (2004) exhibits the deep mixing abundance pattern, significantly enhanced Al and Na (see Kraft 1994 for a review). While this is only a single star, it should be noted that out of the hundreds of MW halo stars and tens of simple dSph galaxy stars surveyed none exhibit the deep mixing

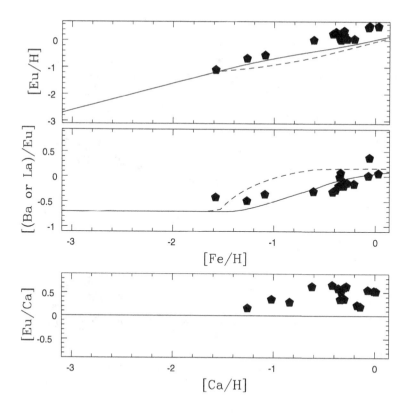

Fig. 16.9. The symbols are the same as in Fig. 16.7.

abundance pattern. If a significant fraction of the metal-poor stars exhibit the deep mixing pattern then this could place interesting constraints on the formation of the MW halo. Further discussion will have to wait for more extensive surveys of the metal-poor Sgr field and globular cluster stars.

Figure 16.8 shows the Na abundance ratios for Sgr dSph stars, excluding the star that exhibits the deep mixing pattern mentioned previously. The two remaining metal-poor Sgr stars have abundances similar to those found in the MW halo. This is in stark contrast to simple dSph stars that have systematically lower Na abundances. The metal-rich Sgr stars have Na abundance ratios lower than those found in the MW disk. These low Na abundance ratios are similar to those found in the simple dSph stars despite the lower metallicity of the simple dSph stars. As mentioned earlier, Smecker-Hane & McWilliam (2004) have attributed the low Na abundances in the Sgr dSph galaxy to an excess of Fe from SNe Ia, which should not produce any Na (e.g., Thielemann et al. 1986). The excess of Fe produced from SNe Ia is consistent with the low α abundances in these metal-rich dSph stars.

The predictions for the neutron capture elements are shown in Figure 16.9. The Sgr toy model was created using the same prescription for chemical evolution as used in both the Scl and MW toy models. The Sgr toy model underpredicts the amount of Eu actually found in the metal-rich Sgr stars. At metallicity [Fe/H] = −0.5 the toy model predicts too

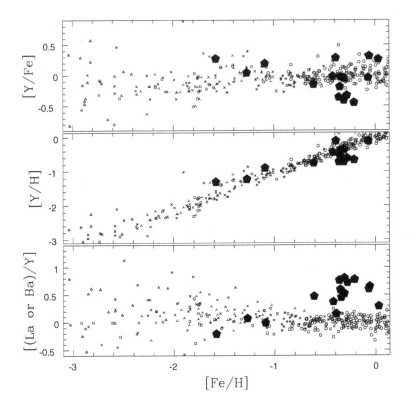

Fig. 16.10. The large pentagons are the same as in Fig. 16.7. The small symbols are the MW sample and the same as in Fig. 16.4.

large a contribution from AGB star *s*-processed material than is actually detected in the stars. This may only indicate that the metallicity at which the AGB stars begin to contribute La is higher than modeled—that the early chemical evolution was faster than modeled and more similar to the MW. Other interpretations of this poor fit of the model could be due to poor modeling of the *s*-process (e.g., a lack of metal-dependent yields), or the anomalously large Eu abundances. At the solar ratio of *s*-process to *r*-process material ([La/Eu] = 0), Eu has a tiny $\sim 5\%$ contribution from the *s*-process. This amounts to a meager 0.02 dex increase to a pure *r*-process Eu abundance. Thus, the overabundance of Eu is not likely to be caused by an *s*-process contribution for the metal-rich Sgr stars, since their [La/Eu] ratios are approximately 0. To investigate this further, we plot the Eu to Ca abundance ratio against Ca abundance in the bottom pane of Figure 16.9. As mentioned previously this ratio is roughly zero at all metallicities in the MW, while in the Sgr dSph it is super solar at all metallicities with a small trend toward larger ratios at higher metallicities. In our toy model the Eu abundances are made in the same proportion with the α elements. A divergence between the α elements and the *r*-process elements seen in this and perhaps other dSph may be a clue to the origins of the *r*-process site.

Figure 16.10 exhibits the Sgr Y abundances, abundance ratios, and the heavy *s*-process

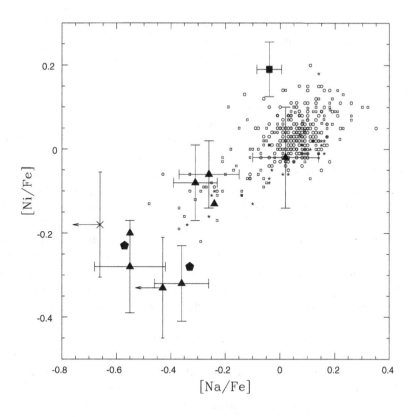

Fig. 16.11. The large symbols the dSph sample with the metallicities in the range −1.5 < [Fe/H] < 0.0 from: Bonifacio et al. (2000), Shetrone et al. (2001, 2003), and Geisler et al. (2004). The small symbols are from the MW sample in the same metallicity range: Gratton & Sneden (1988), Edvardsson et al. (1993), Nissen & Schuster (1997), Stephens (1999), Prochaska et al. (2000), Fulbright (2002), and Stephens & Boesgaard (2002).

to light s-process ratio ([La/Y]). The metal-poor Sgr stars have [La/Y] ratios that are similar to those of the halo; however, at this metallicity the La is mostly of r-process origin. The [La/Y] ratio among the more metal-rich Sgr stars is significantly enhanced with respect to the MW stars of similar metallicity. Smecker-Hane & McWilliam (2004) argue that the underabundance of Y is due to the metallicity-dependent origin of Y and La *and* a delay in the incorporation of the AGB s-process yields. They argue that the metal-rich generation of AGB stars did not contribute to the chemical evolution of that generation, but rather, the high [La/Y] ratio suggests the contribution from low-metallicity AGB stars. If true, this breaks the instant-recycling rule used in many chemical evolution models. Among the metal-rich Sgr stars [Y/Fe] is only slightly lower than the MW mean at the same metallicity. However, the lower panel of Figure 16.9 suggests that there is a high r-process contribution at a given metallicity and La has both an r-process and s-process contribution, so some of the enhanced [La/Y] contribution could be due to an enhanced r-process contribution and unrelated to the s-process.

The sample sizes for the other complex dSph stars are very small, but comparing the trends seen in the dSph with those found in Leo I, Carina, and Fornax reveals some significant differences: (1) The metal-poor stars exhibit Na abundance ratios similar to the simple dSph stars and below those found in the Sgr dSph; (2) the single metal-rich star in Fornax has enhanced α abundance ratios with respect to the halo and the Sgr dSph.

16.5   Connection to the Milky Way

Several lines of evidence suggest that the Galactic halo is, at least partially, composed of accreted dwarf galaxies. These include the current assimilation of the Sgr dwarf (Ibata et al. 1997; Dohm-Palmer et al. 2000; Newberg et al. 2002), and possibly ω Cen (Majewski et al. 2000, assuming that it is a stripped dwarf galaxy). Since the halo's metallicity distribution peaks near [Fe/H] $= -1.8$ and those stars show a higher heavy α to iron ratio than the dSph stars, clearly a large percentage of the halo cannot have been produced from dSphs similar to the simple systems reviewed in this work. In Fulbright (2002), fewer than 10% of the local metal-poor ([Fe/H] < -1.2) stars in that sample have heavy α to iron abundance ratios similar to those found in the dSph sample. However, by subdividing that sample by total space velocity, Fulbright found that the highest-space velocity stars have systematically lower α to iron abundance ratios. Nissen & Schuster (1997) conducted a detailed abundance analysis of a nearby sample of disk and halo stars with similar metallicities to study the disk-halo transition. Their sample was chosen to get an equal number of disk and halo stars as defined by the stars' rotation. Of their 13 chosen halo stars, eight show an unusual abundance pattern: low α element to iron ratio, low [Ni/Fe] abundances, and low [Na/Fe] abundances. These odd halo stars also exhibited larger R_{max} and z_{max} orbital parameters than the other halo stars sampled. Nissen & Schuster (1997) suggest that these anomalous stars may have their origins in disrupted dSph. In Figure 16.11 the dSph stars with metallicities between -1.5 and -0.5 are shown along with literature MW field stars in the same metallicity range. The dSph stars systematically fall among the stars that Nissen & Schuster (1997) predicted may have an origin in dSph galaxies. This seems to lend support to the idea put forward by Nissen & Schuster (1997) that a large fraction ($> 50\%$) of the metal-rich halo stars may have their origin in disrupted dSph like those studied in this work. This still leaves the question of the origin of the metal-poor halo stars and the fraction of the metal-poor halo stars that formed through monolithic collapse versus accretion of dSph galaxies.

ω Cen is a system once categorized as a globular cluster, but recent studies suggest that it may actually be a captured galaxy (e.g., Majewski et al. 2000). ω Cen exhibits a large metallicity spread, with the more metal-rich stars exhibiting very large s-process element enhancements, implying an age spread and self-pollution. Unlike the simple dSph systems, a large fraction of the metal-poor ω Cen giants exhibit a deep mixing abundance pattern. Could the Sgr dSph be similar to ω Cen in this regard?

16.6   A History of Conclusions

(1) Most dSph abundance patterns do not look like that of the average MW halo; thus, it would be difficult to make the majority of the MW halo from building blocks that look like today's dSph galaxies (Shetrone et al. 1998, 2001).

(2) Abundances derived from color-magnitude diagrams can be dramatically fooled by the age-metallicity relationship, as seen in the Sgr dSph. The Sgr dSph may be much younger and more metal rich that previously thought (Bonifacio et al. 2000).

(3) Studies of the outer halo suggest that it may have abundances similar to those found in the simple dSph galaxies (Nissen & Shuster 1997; Fulbright 2002; Stephens & Boesgaard 2002).

(4) Both complex and simple dSph systems show significant evolution from a SN II abundance pattern to a SN Ia abundance pattern (Shetrone et al. 2001, 2003; Smecker-Hane & McWilliam 2004).

(5) Evidence from metal-poor dSph stars with a clear indication of SN Ia contamination indicates that Mn and Cu are not produced in large quantities in metal-poor SNe Ia (McWilliam et al. 2003; Shetrone et al. 2003).

(6) There appears to be a split between the light and heavy α elements in the most metal-poor simple dSph stars (Fulbright et al. 2004; this work).

(7) The Sgr dSph metal-poor stars do not exhibit the split between the light and heavy α elements, suggesting abundances more like those found in the MW globular clusters, including a star exhibiting a deep mixing abundance pattern (Smecker-Hane & McWilliam 2004).

References

Bonifacio, P., et al. 2004, in preparation

Bonifacio, P., & Caffau, E. 2003, A&A, 399, 1183

Bonifacio, P., Hill, V., Molaro, R., Pasquini, L., Di Marcantonio, R., & Santin, P. 2000, A&A, 359, 663

Burris, D. L., Pilachowski, C. A., Armandroff, T. E., Sneden, C., Cowan, J. J., & Roe, H. 2000, ApJ, 544, 302

Busso, M., Gallino, R., Lambert, D. L., Travaglio, C., & Smith, V. V 2001, ApJ, 557, 802

Dohm-Palmer, R. C., Mateo, M., Olszewski, E., Morrison, H., Harding, P., Freeman, K. C., & Norris, J. 2000, AJ, 120, 2496

Dolphin, A. 2002, MNRAS, 332, 91

Edvardsson, B., Andersen, J., Gustafsson, B., Lambert, D. L., Nissen, P. E., & Tomkin, J. 1993, A&A, 275, 101

Fulbright, J. P. 2002, AJ, 123, 404

Fulbright, J. P., Rich, R. M., & Castro, S. 2004, in preparation

Geisler, D., Smith, V. V., Wallerstein, G., Gonzalez, G., & Charbonnel, C. 2004, in preparation

Gratton, R. G., & Sneden, C. 1988, A&A, 204, 193

——. 1991, A&A, 241, 501

——. 1994, A&A, 287, 927

Ibata, R. A., Wyse, R. F. G., Gilmore, G., Irwin, M. J., & Suntzeff, N. B. 1997, AJ, 113, 634

Iwamoto, K., Brachwitz, F., Nomoto, K., Kishmoto, N., Umeda, H., Hix, W. R., & Thielemann, F. 1999, ApJS, 125, 439

Kraft, R. P. 1994, PASP, 106, 553

Majewski, S. R., Patterson, R. J., Dinescu, D. I., Johnson, W. Y., Ostheimer, J. C., Kunkel, W. E., & Palma, C. 2000, in The Galactic Halo: From Globular Cluster to Field Stars, ed. A. Noels et al. (Liége: Institut d'Astrophysique et de Geophysique), 619

McWilliam, A., Preston, G. W., Sneden, C., & Searle, L. 1995, AJ, 109, 2757

McWilliam, A., Rich, R. M., & Smecker-Hane, T. 2003, ApJ, 592, L21

Newberg, H. J., et al. 2002, ApJ, 569, 245

Nissen, P. E., & Schuster, W. J. 1997, A&A, 326, 751

Prochaska, J. X., Naumov, S. O., Carney, B. W., McWilliam, A., & Wolfe, A. M. 2000, AJ, 120, 2513

Qian, Y. Z., & Wasserburg, G. J. 2001, ApJ, 559, 925

Ryan, S. G., Norris, J. E., & Beers, T. C. 1996, ApJ, 471, 254

Shetrone, M. D., Bolte, M., & Stetson, P. B 1998, AJ, 115, 1888

Shetrone, M. D., Côté, P., & Sargent, W. L. W. 2001, ApJ, 548, 592

Shetrone, M. D., Venn, K. A., Tolstoy, E., Primas, F., Hill, V., & Kaufer, A. 2003, AJ, 125, 684

Smecker-Hane, T. A., & McMilliam, A. 2004, ApJ, submitted

Stephens, A. 1999, AJ, 117, 1771

Stephens, A., & Boesgaard, A. M. 2002, AJ, 123, 1647

Thielemann, F. K., Nomoto, K., & Yokoi, K. 1986, A&A, 158 17

Woosley, S. E., & Weaver, T. A. 1995, ApJS, 101, 181

17

The evolutionary history
of Local Group irregular galaxies

EVA K. GREBEL

Max-Planck Institute for Astronomy, Heidelberg, Germany
Astronomical Institute of the University of Basel, Switzerland

Abstract

Irregular (Irr) and dwarf irregular (dIrr) galaxies are gas-rich galaxies with recent or on-going star formation. In the absence of spiral density waves, star formation occurs largely stochastically. The scattered star-forming regions tend to be long-lived and migrate slowly. Older populations have a spatially more extended and regular distribution. In fast-rotating Irrs high star formation rates with stronger concentration toward the galaxies' center are observed, and cluster formation is facilitated. In slowly or nonrotating dIrrs star formation regions are more widely distributed, star formation occurs more quiescently, and the formation of OB associations is common. On average, Irrs and dIrrs are experiencing continuous star formation with amplitude variations and can continue to form stars for another Hubble time.

Irrs and dIrrs exhibit lower effective yields than spirals, and [α/Fe] ratios below the solar value. This may be indicative of fewer Type II supernovae and lower astration rates in their past (supported by their low present-day star formation rates). Alternatively, many metals may be lost from the shallow potential wells of these galaxies due to selective winds. The differences in the metallicity-luminosity relation between dIrrs and dwarf spheroidals (which, despite their lower masses, tend to have too high a metallicity for their luminosity as compared to dIrrs) lends further support to the idea of slow astration and slow enrichment in dIrrs. The current data on age-metallicity relations are still too sparse to distinguish between infall, leaky-box, and closed-box models. The preferred location of dIrrs in the outer parts of galaxy groups and clusters and in the field as well as the positive correlation between gas content and distance from massive galaxies indicate that most of the dIrrs observed today probably have not yet experienced significant interactions or galaxy harassment.

17.1 Introduction

In this contribution, I will focus on the evolutionary histories of irregular (Irr) and dwarf irregular (dIrr) galaxies, including their chemical evolution. The name "irregular" refers to the irregular, amorphous appearance of these galaxies at optical wavelengths, where the light contribution tends to be dominated by scattered bright H II regions and their young, massive stars. Irrs are typically gas-rich galaxies that lack spiral density waves as well as a discernible bulge or nucleus. Many Irrs are disk galaxies and appear to be an extension of late-type spirals. The most massive disky Irrs with residual spiral structure are also called Magellanic spirals; e.g., the Large Magellanic Cloud (LMC) (Kim et al. 1998) is a barred Magellanic spiral. Looser and more amorphous Irrs like the Small Magellanic Cloud (SMC) are sometimes also referred to as Magellanic irregulars (or barred Magellanic irregulars if a bar is present); see de Vaucouleurs (1957). A different system of subdivisions was suggested by van den Bergh in his DDO luminosity classification system (van den Bergh 1960, 1966). DIrrs are simply less massive, less luminous Irrs; the distinction between the two is a matter of definition rather than physics. Typical characteristics of dIrrs are a central surface brightness $\mu_V \lesssim 23$ mag arcsec^{-2}, a total mass of $M_{tot} \lesssim 10^{10} M_\odot$, and an H I mass of $M_{HI} \lesssim 10^9 M_\odot$. Solid body rotation is common among the Irrs and more massive dIrrs, while low-mass dIrrs do not show measurable rotation; here random motions dominate. A typical characteristic of Irrs and dIrrs alike is ongoing or recent star formation. The star formation intensity may range from burst-like, strongly enhanced activity, to slow quiescent episodes. Irrs and dIrrs can continue to form stars over a Hubble time (Hunter 1997).

Substantial progress has been made in the photometric exploration of the star formation histories of Irrs and dIrrs over the past decade, largely thanks to the superior resolution of the *Hubble Space Telescope*. There is still very little known about the progress of the chemical evolution in these galaxies as a function of time, but the advent of 6-m to 10-m class telescopes and their powerful spectrographs is beginning to change this situation. The most detailed information is available for nearby Irr and dIrr galaxies in the Local Group, most notably the LMC, which is the Irr galaxy closest to the Milky Way.

17.2 Distribution and Census of Irregulars in the Local Group

The Local Group, our immediate cosmic neighborhood, resembles other nearby galaxy groups in many ways, including in its galaxy content, structure, mass, and other properties (e.g., Karachentsev et al. 2002a, b). It is our best local laboratory to study galaxy evolution at the highest possible resolution and in the greatest possible detail. The Local Group contains two dominant spiral galaxies surrounded by a large number of smaller galaxies. Thirty-six galaxies are currently believed to be members of the Local Group if a zero-velocity surface of 1.2 Mpc is adopted (Courteau & van den Bergh 1999; Grebel, Gallagher, & Harbeck 2003)*. The smaller galaxies in the Local Group include a spiral galaxy (M33), 11 gas-rich Irr and dIrr galaxies (including low-mass, so-called transition-type galaxies that comprise properties of both dIrrs and dwarf spheroidals), four elliptical and dwarf elliptical galaxies, and 17 gas-deficient dwarf spheroidal (dSph) galaxies. For a listing of the basic properties of these galaxies, see Grebel et al. (2003). Their three-dimensional distribution is illustrated in Grebel (1999; Fig. 3). Recent reviews of Local Group galaxies include Grebel (1997, 1999, 2000), Mateo (1998), and van den Bergh (1999, 2000).

DIrrs are the second most numerous galaxy type in the Local Group. While new dwarf members of the Local Group are still being discovered (e.g., Whiting, Hau, & Irwin 1999), these tend to be gas-deficient, low-mass dSph galaxies, which have intrinsically low optical surface brightnesses and cannot be found from their H I 21 cm emission lines. The Irr and dIrr census of the Local Group appears to be complete.

Irrs and dIrrs are found in galaxy groups and clusters as well as in the field and exhibit little concentration toward massive galaxies in contrast to early-type dwarfs. This morphological segregation is clearly seen in the Local Group and in nearby groups (Fig. 17.1). It becomes even more pronounced in galaxy clusters, where the distribution of Irrs shows the least concentration of all galaxy types toward the cluster core (e.g., Conselice, Gallagher, &

---

* Note that recent kinematic estimates suggest an even smaller radius of (0.94 ± 0.10) Mpc for the zero-velocity surface (Karachentsev et al. 2002c), which reduces the above number of Local Group dwarf galaxies by two.

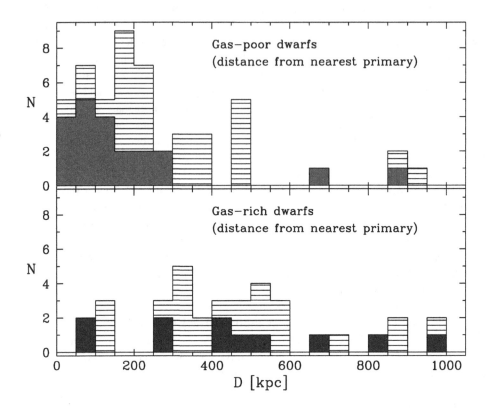

Fig. 17.1. Morphological segregation in the Local Group (filled histograms; see Grebel 2000) and in the M81 and Cen A groups (dashed histograms; input data from Karachentsev et al. 2002a, b). Note the pronounced concentration of gas-poor, early-type dwarfs around the nearest massive primary galaxy, while the gas-rich, late-type dwarfs show less concentration and are more widely distributed. This may be a signature of the impact of environmental effects, such as gas stripping.

Wyse 2001, and references therein), which has been attributed to continuing infall of Irrs and subsequent harassment. Conversely, in very loose groups or "clouds" (such as the Canes Venatici I Cloud) that are still far from approaching dynamical equilibrium, an overabundance of Irrs and dIrrs is observed as compared to early-type dwarfs (Karachentsev et al. 2003a), indicative of a lack of interactions.

17.3   The Interstellar Medium of Local Group Irregulars

17.3.1   *The Magellanic Clouds*

The Magellanic Clouds are the two most massive Irrs in the Local Group, and the only two Irrs in immediate proximity to a massive spiral galaxy. Their distances from the Milky Way are 50 kpc (LMC) and 60 kpc (SMC), respectively. They are the only two Local Group Irrs that are closely interacting with each other (and with the Milky Way). According to the earlier definition, the SMC qualifies as a dIrr.

The global distribution of neutral hydrogen within the Magellanic Clouds and other comparatively massive Irrs tends to show a regular, symmetric appearance, in contrast to their visual morphology. On smaller scales, the H I is flocculent and exhibits a complicated fractal pattern full of shells and clumps (e.g., Kim et al. 1998; Stanimirovic et al. 1999). The lack of correlation between the H I shells and the optically dominant H II shells suggests that H I shells live longer than the OB stars that caused them initially (Kim et al. 1999). The H I associated with H II regions is usually more extended than the ionized regions. The fractal structure of the neutral gas is self-similar on scales from tens to hundreds of pc (Elmegreen, Kim, & Staveley-Smith 2001) and appears to result from the turbulent energy input caused by winds of recently formed massive stars and supernova explosions.

With $0.5 \times 10^9 \, M_\odot$ (Kim et al. 1998) the LMC's gaseous component contributes about 9% to its total mass, while it is $\sim 21\%$ in the SMC (H I mass of $4.2 \times 10^8 \, M_\odot$; Stanimirovic et al. 1999). In comparison to the Milky Way, the gas-to-dust ratio is roughly 4 times lower in the LMC (Koornneef 1982) and about 30 times lower in the SMC (Stanimirovic et al. 2000), implying a smaller grain surface area per hydrogen atom, fewer coolants, and thus a reduced H_2 formation efficiency (Dickey et al. 2000; Stanimirovic et al. 2000; Tumlinson et al. 2002). Indeed, the total diffuse H_2 mass is only $8 \times 10^6 \, M_\odot$ in the LMC and $2 \times 10^6 \, M_\odot$ in the SMC, which corresponds to 2% and 0.5% of their H I masses, respectively (Tumlinson et al. 2002). Also the reduced CO emission from both Clouds (3–5 times lower than expected for Galactic giant molecular clouds) is indicative of the high UV radiation field in low-metallicity environments and hence high CO photodissociation rates (Israel et al. 1986; Rubio, Lequeux, & Boulanger 1993). While high dust content is correlated with high H_2 concentrations, H_2 does not necessarily trace CO or dust (Tumlinson et al. 2002).

Photoionization through massive stars is the main contributor to the optical appearance of the interstellar medium (ISM) at $\sim 10^4$ K in the Clouds and other gas-rich, star-forming galaxies. The LMC has a total Hα luminosity of 2.7×10^{40} erg s^{-1}; 30% to 40% is contributed by diffuse, extended gas (Kennicutt et al. 1995). In the LMC nine H II supershells with diameters > 600 pc are known (Meaburn 1980). Their rims are marked by strings of H II regions and young clusters/OB associations. The standard picture for supershells suggests that these are expanding shells driven by propagating star formation (e.g., McCray & Kafatos 1987). However, an age *gradient* consistent with this scenario was not detected in the largest of these supershells, LMC4 (Dolphin & Hunter 1998). Nor are other LMC supershells expanding as a whole, but instead appear to consist of hot gas confined between H I sheets and show localized expansion. Supershells in several other galaxies neither show evidence for expansion (e.g., Points et al. 1999), nor the expected young massive stellar populations (Rhode et al. 1999). In contrast, the three H I supershells and 495 giant shells in the SMC appear to be expanding (Staveley-Smith et al. 1997; Stanimirovic et al. 1999).

The hot, highly ionized corona of the LMC with collisionally ionized gas (temperatures $\gtrsim 10^5$ K) (Wakker et al. 1998) is spatially uncorrelated with star-forming regions. A hot halo is also observed around the SMC, but here clear correlations with star-forming regions are seen. This corona may be caused in part by gas falling back from a galactic (i.e., SMC) fountain (Hoopes et al. 2002). The O VI column density exceeds the corresponding Galactic value by 1.4 (Hoopes et al. 2002), consistent with the longer cooling times expected at lower metallicities (Edgar & Chevalier 1986).

The Magellanic Clouds, which only have a deprojected distance of 20 kpc from each other, interact with each other and with the Milky Way. Apart from an impact on the structure and star formation histories of these three galaxies (e.g., Hatzidimitriou, Cannon, & Hawkins 1993; Kunkel, Demers, & Irwin 2000; Weinberg 2000; van der Marel et al. 2002), this has given rise to extended gaseous features surrounding the Magellanic Clouds (Putman et al. 2003, and references therein). Part of these are likely caused by tidal interactions, but ram pressure appears to have played an important role as well (Putman et al. 1998; Mastropietro et al. 2004). Metallicity determinations for gas in the Magellanic Stream, which is trailing behind the Magellanic Clouds and subtends at least $10° \times 100°$ on the sky, confirm that the gas is not primordial (Lu et al. 1998; Gibson et al. 2000). The H_2 detected in the leading arm of the Stream may originally have formed in the SMC (Sembach et al. 2001). No stars are known to be connected with the Magellanic Stream (Putman et al. 2003).

Another prominent H I feature is the "Magellanic Bridge" or InterCloud region ($10^8 M_\odot$; Putman et al. 1998), which connects the LMC and SMC. Cold (20 to 40 K) H I gas has been detected in the Bridge (Kobulnicky & Dickey 1999), and recent star formation occurred there over the past 10 to 25 Myr (Demers & Battinelli 1998). Intermediate-age stars are also present in parts of the Bridge (carbon stars: Kunkel et al. 2000, and references therein). Higher ionized species with temperatures up to $\sim 10^5$ K show an abundance pattern suggesting depletion into dust (Lehner et al. 2000). Interestingly, the metallicities of young stars in the Bridge were found to be [Fe/H] ≈ -1.1 dex (Rolleston et al. 1999), 0.4 dex below the mean abundance of the young SMC population, which is inconsistent with the proposed tidal origin 200 Myr ago (Murai & Fujimoto 1980; Gardiner & Noguchi 1996).

17.3.2   *More Distant Dwarf Irregular Galaxies*

The other Local Group dIrrs are more distant from the dominant spirals, and fairly isolated. Interactions may still occur, but if this happens the interaction partners tend to be gas clouds rather than galaxies. Generally, star formation activity and gas content decrease with galaxy mass, but the detailed star formation histories and ISM properties of the dIrrs present a less homogeneous picture.

NGC 6822, a dIrr at a distance of ~ 500 kpc, is embedded in an elongated H I cloud with numerous shells and holes. Its total H I mass is $1.1 \times 10^8 M_\odot$, $\sim 7\%$ of its total mass. The masses of individual CO clouds reach up to $(1-2) \times 10^5 M_\odot$ (Petitpas & Wilson 1998), while the estimated H_2 content is 15% of the H I mass (Israel 1997), and the dust-to-gas mass ratio is $\sim 1.4 \times 10^{-4}$ (Israel, Bontekoe, & Kester 1996). NGC 6822 contains many H II regions. Its huge supershell (2.0×1.4 kpc) was likely caused by the passage of and interaction with a nearby $10^7 M_\odot$ H I cloud and does not show signs of expansion (de Blok & Walter 2000). The older stars in IC 10 describe an elliptical, extended halo (Letarte et al. 2002) distinct from the elongated H I distribution. The latter, however, is traced closely by a population of young blue stars (~ 180 Myr) that appear to have formed following the interaction with the passing H I cloud (de Blok & Walter 2003; Komiyama et al. 2003) some 300 Myr ago. In NGC 6822, the H I distribution is thus only slightly more extended than the stellar loci.

The H I of IC 10 (distance 660 kpc) is 7.2 times more extended than its Holmberg radius (Tomita, Ohta, & Saitō 1993). While the inner part of the neutral hydrogen of IC 10 is a regularly rotating disk full of shells and holes, the outer H I gas is counter-rotating (Wilcots & Miller 1998). IC 10 is currently experiencing a massive starburst, which is possibly triggered and fueled by an infalling H I cloud (Saitō et al. 1992; Wilcots & Miller 1998). IC 10 contains a nonthermal superbubble that may be the result of several supernova explosions

(Yang & Skillman 1993). The masses of the CO clouds in IC 10 appear to be as high as up to $5 \times 10^6 \, M_\odot$ (Petitpas & Wilson 1998), which would indicate that more than 20% of this galaxy's gas mass is molecular. Owing to the high radiation field and the destruction of small dust grains, the ratio of far-infrared [C II] to CO 1–0 emission is a factor 4 larger than in the Milky Way (Bolatto et al. 2000), resulting in small CO cores surrounded by large [C II]-emitting envelopes (Madden et al. 1997). Two H_2O masers were detected in dense clouds in IC 10, marking sites of massive star formation (Becker et al. 1993). The internal dust content of IC 10 is high, and its properties prompted Richer et al. (2001) to suggest that this galaxy should actually be classified as a blue compact dwarf.

Less detailed information is available for the ISM in the other Local Group dIrrs, which do not appear to be involved in ongoing interactions and which are evolving fairly quiescently. The H I in these dIrrs may be up to 3 times more extended than the optical galaxy and is clumpy on scales of 100 to 300 pc. The most massive clumps reach $\sim 10^6 \, M_\odot$. H I concentrations tend to be close to H II regions. Some dIrrs contain cold H I clouds associated with molecular gas. The total H I masses are usually $< 10^9 \, M_\odot$, and less than $10^7 \, M_\odot$ for transition-type dwarfs. The center of the H I distribution coincides roughly with the optical center of the dIrrs, although the H I may show a central depression surrounded by an H I ring or arc (e.g., SagDIG, Leo A), possibly a consequence of star formation, or the H I may be off-centered (e.g., Phoenix; St-Germain et al. 1999). In low-mass dIrrs there are no signatures of rotation, but these may be obscured by expanding shells and bubbles. Further details are given in Lo, Sargent, & Young (1993), Young & Lo (1996, 1997), Elmegreen & Hunter (2000), and Young et al. (2003).

Lower gravitational pull and the lack of shear in the absence of differential rotation imply that H I shells may become larger and are long-lived (Hunter 1997). Diameters, ages, and expansion velocities of the H I shells increase with later Hubble type (Walter & Brinks 1999) and scale approximately with the square root of the galaxy luminosity (Elmegreen et al. 1996). Shell-like structures, H I holes, or off-centered gas may be driven by supernovae and winds from massive stars following recent star formation episodes or tidal interactions.

For a review on nebular abundances in Irrs, see the contribution by Garnett (2004). Here it should only be mentioned that the effective yields in Irrs computed from gas-phase abundances are lower than those in the main stellar disks of spirals. Lower effective yields are also correlated with lower rotational velocities (Garnett 2002). This is interpreted as preferential metal loss through winds in the more shallow potential wells of Irrs and dIrrs, but may also be due to lower astration levels (e.g., Pilyugin & Ferrini 2000). For a review of the general ISM properties in Local Group dwarf galaxies, see Grebel (2002a).

17.4 Large-scale Star Formation and Spatial Variations

The dwarf galaxies in the Local Group vary widely in their star formation and enrichment histories, times and duration of their major star formation episodes, and fractional distribution of ages and subpopulations. Indeed, when studied in detail, no two dwarf galaxies turn out to be alike, not even if they are of the same morphological type or have similar luminosities (Grebel 1997). On the other hand, in spite of their individual differences, they do follow certain common global correlations such as increasing mean metallicity with luminosity (§17.6.2).

17.4.1 Large-scale Star Formation

The ISM properties of Irrs and dIrrs outlined in the previous section already show that there are spatial variations in star formation history and other characteristics within these galaxies. In general, dwarf galaxies of all types show a tendency for the younger populations to be more centrally concentrated (and possibly more chemically enriched), whereas older populations are more extended (Grebel 1999, 2000; Harbeck et al. 2001). In Irr and dIrr galaxies, H II regions tend to be located within the part of the galaxy that shows solid body rotation and are usually even more centrally concentrated (Roye & Hunter 2000). Star-forming regions may, however, be found out to six optical scale lengths, indicating that star formation is truncated at lower gas density thresholds than in spirals (Parodi & Binggeli 2003). In dIrrs dominated by chaotic motions, the degree of central concentration of recent star formation is lower, whereas fast-rotating Irrs tend to exhibit the highest central concentrations. The same trend also holds for the star formation activity: low-mass dIrrs with no measurable rotation also have lower star formation rates (Roye & Hunter 2000; Parodi & Binggeli 2003).

How does star formation progress in irregular galaxies? Irrs and the more massive dIrrs usually contain multiple distinct regions of concurrent star formation. These regions often remain active for several 100 Myr, are found throughout the main body of these galaxies (see above), and can migrate. This is illustrated in Figure 17.2 for the LMC, where the large-scale star formation history of the last ~ 250 Myr (approximately one rotation period) is shown (see Grebel & Brandner 1998 for full details). Note how some of the active regions have continued to form stars over extended periods and propagated slowly, whereas others only became active during the past 30 Myr. The star formation complexes resemble superassociations and may span areas of a few hundred pc (Grebel & Brandner 1998). In supershells, typical time scales for continuing star formation on length scales of 0.5 kpc range from 15 to 30 Myr, usually without showing clear signs of spatially directed propagation with time (see also Grebel & Chu 2000 and §17.3.1). CO shows a strong correlation with H II regions and young (< 10 Myr) clusters, but only little with older clusters and supernova remnants (Fukui et al. 1999; cf. Banas et al. 1997). Massive CO clouds have typical lifetimes of ~ 6 Myr and are dissipated within ~ 3 Myr after the formation of young clusters (Fukui et al. 1999; Yamaguchi et al. 2001). Spatially resolved star formation histories have also been derived for two dIrr galaxies just beyond the Local Group covering the past 500–700 Myr. They reveal similar long-lived, gradually migrating zones of star formation (Sextans A: Van Dyk, Puche, & Wong 1998; Dohm-Palmer et al. 2002; GR 8: Dohm-Palmer et al. 1998), as seen in the more massive Magellanic Clouds.

In low-mass dIrrs one usually observes only one single low-intensity star-forming region. DIrrs and transition-type dIrr/dSph galaxies tend to be fairly quiescent, often having experienced the bulk of their star formation at earlier times. (In fact, transition-type dwarfs resemble dSphs in their gradually declining star formation rates; see Grebel et al. 2003 for details.) Evidence for migrating star formation is found in low-mass dIrrs as well, albeit on smaller scales owing to the smaller sizes of these galaxies (e.g., Phoenix: Martínez-Delgado, Gallart, & Aparicio 1999).

17.4.2 Intermediate-age and Old Stellar Populations

Irrs and massive dIrrs tend to show extended halos of intermediate-age stars (ages ~ 1 to ~ 10 Gyr), which can be conveniently traced by carbon stars (e.g., Letarte et al. 2002).

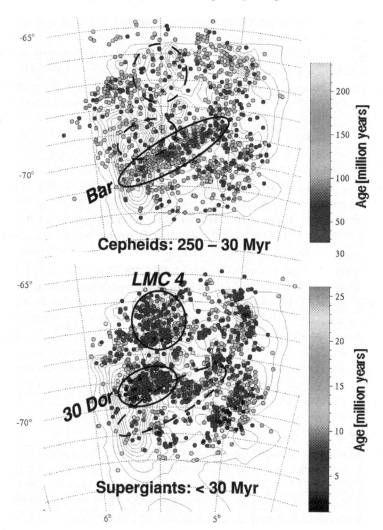

Fig. 17.2. Large-scale star formation patterns in the Large Magellanic Cloud spanning the past ∼ 250 Myr. The individual dots correspond to age-dated Cepheids (*upper panel*) and supergiants (*lower panel*). A few prominent features like the LMC bar, supershell LMC 4, and 30 Doradus are marked by solid and dashed lines. Note how star formation migrated along the LMC's bar and finally vanished in its southernmost past, and how other regions such as 30 Doradus and LMC 4 only became strongly active over the past ∼ 30 Myr. Within the time scales depicted here, which incidentally correspond to roughly one rotation period, stars are not expected to have migrated far from their birthplaces. (Adapted from Grebel & Brandner 1998.)

In the Magellanic Clouds, the density distributions of different populations ages become increasingly more regular and extended with increasing age (e.g., Cioni et al. 2000; Zaritsky et al. 2000), whereas the young populations are responsible for the irregular appearance of these two galaxies. The centroids of the different populations do not always coincide. Features resembling stellar bars are found in many dIrrs, which do not necessarily coincide with the peak H I distribution or its centroid. In low-mass dIrrs there are not enough intermediate-age tracers such as C stars to say much about the distribution of these populations (see, e.g., Battinelli & Demers 2000); the number of C stars decreases with absolute galaxy luminosity and also with galaxy metallicity (see Groenewegen 2002 for a recent review and census of C stars in the Local Group).

All Irr and dIrr galaxies examined in detail so far show clear evidence for the presence of old (> 10 Gyr) populations, a property that they appear to share with all galaxies whose stellar population have been resolved. For instance, deep ground-based imaging of the "halos" of dIrrs led to the detection of old red giant branches (e.g., Minniti & Zijlstra 1996; Minniti, Zijlstra, & Alonso 1999). In closer dIrrs, horizontal branch stars have been detected in field populations (e.g., IC 1613: Cole et al. 1999; Phoenix: Holtzman, Smith, & Grillmair 2000; WLM: Rejkuba et al. 2000; Leo A: Dolphin et al. 2002) and in globular clusters (e.g., WLM: Hodge et al. 1999). Horizontal branch stars are unambiguous tracers of ancient populations. In the closest dIrrs and Irrs even the old main sequence turn-offs have been resolved, allowing differential age dating. Interestingly (with the possible exception of the SMC), the oldest datable populations in all nearby dwarf galaxies turn out to be indistinguishable in age from each other and from the Milky Way, indicating a common epoch of early star formation (e.g., Olsen et al. 1998; Johnson et al. 1999; see Grebel 2000 for a full list of references). Apart from the recent interaction in NGC 6822 (§17.3.2), the old populations also are usually the most extended ones. However, their fractions vary: in some cases they only constitute a tiny portion of the stellar content of their parent galaxy.

17.4.3   Modes of Star Formation

The ISM in dIrrs is highly inhomogeneous and porous, full of small and large shells and holes. The global gas density tends to be significantly below the Toomre criterion for star formation (van Zee et al. 1997). Stochastic, star formation may be driven by homogeneous turbulence, which creates local densities above the star formation threshold (e.g., Stanimirovic et al. 1999). Self-propagating stochastic star formation (Gerola & Seiden 1978; Gerola, Seiden, & Schulman 1980; Feitzinger et al. 1981) can lead to structures of sizes of up to 1 kpc, in which star formation processes remain active for 30–50 Myr, or to the formation of long-lived spiral features if an off-centered bar is present (Gardiner, Turfus, & Putman 1998). In the absence of shear, star formation continues along regions of high H I column density, fueled by the winds of recently formed stars and supernovae explosions.

Dense gas concentrations may, however, also remain inactive for hundreds of Myr, and there are not usually obvious triggers for the onset of star formation (see Dohm-Palmer et al. 2002). This may be different in nonquiescently evolving, starbursting dIrrs like IC 10: gas accretion or other interactions may be triggering the starburst (see §17.3.2). The existence of isolated dIrrs with continuous star formation outside of groups shows that external triggers are not needed. Quiescently evolving dIrrs exhibit widely distributed star formation and have very small color gradients, whereas starbursting dIrrs show much more concentrated star formation and strong color gradients (van Zee 2001). The analysis of 72 dIrr galaxies in nearby groups and in the field revealed that the radial distribution of star-forming regions follows on average an annulus-integrated exponential distribution, and that secondary star-

Table 17.1. *Star Clusters in the Local Group Irr and dIrr Galaxies*

Galaxy	Type	D_{Sp} [kpc]	M_V [mag]	N_{GC}	S_N	[Fe/H] [dex]	N_{OC}
LMC	Ir III-IV	50	−18.5	~13	0.5	−2.3, −1.2	$\gtrsim$4000
SMC	Ir IV/IV-V	63	−17.1	1	0.1	−1.4	$\gtrsim$2000
NGC 6822	Ir IV-V	500	−16.0	1	0.4	−2.0	~20
WLM	Ir IV-V	840	−14.4	1	1.7	−1.5	$\geq$1
IC 10	Ir IV:	250:	−16.3	0	0	—	$\gtrsim$13
IC 1613	Ir V	500	−15.3	0	0	—	$\gtrsim$5
Phe	dIrr/dSph	405	−12.3	[4:]	[48:]	—	?
PegDIG	Ir V	410	−11.5	0	0	—	$\lesssim$3
LGS 3	dIrr/dSph	280	−10.5	0	0	—	$\lesssim$13

Notes: Only galaxies known to contain star clusters are listed. D_{Sp} denotes the distance to the nearest spiral galaxy (M31 or Milky Way, Col. 3). N_{GC} and N_{OC} (Cols. 5 & 8) list the number of globular clusters and open clusters, respectively. Note that the globular cluster suspects in Phoenix are highly uncertain. S_N (Col. 6) is the specific globular cluster frequency. When two values are listed in Col. 7 (metallicity), these indicate the most metal-rich and most metal-poor globular clusters. For more details, a full list of galaxies with star clusters in the Local Group, and references, see Grebel (2002b).

forming peaks at larger distances are consistent with internal triggering via stochastic, self-propagating star formation (Parodi & Binggeli 2003).

Quiescent dIrrs tend to form OB associations, while massive starbursts can lead to the formation of more compact star clusters. The number of massive clusters tends to correlate with galaxy mass (i.e., roughly with luminosity; see, e.g., Parodi & Binggeli 2003). For instance, in the dIrr NGC 6822 on average one cluster is formed per 6×10^6 years (a much smaller number than in the more massive LMC), while an OB association forms every 7×10^5 years, similar to the LMC (Hodge 1980). The distinctive, well-separated peaks in the formation rate of populous clusters in the LMC, however, seem to be caused by close encounters with the Milky Way and the SMC (Gardiner, Sawa, & Fujimoto 1994; Girardi et al. 1995; Lin, Jones, & Klemola 1995); it is surprising that no corresponding enhancement in the SMC's fairly continuous cluster formation rate is seen. Generally, old globular clusters are rare in dIrrs. For the cluster census in Local Group Irr and dIrr galaxies, see Table 17.1.

On a global, long-term scale, star formation in dIrrs has essentially occurred continuously at a constant rate with amplitude variations of 2–3 (Tosi et al. 1991; Greggio et al. 1993), is largely governed by internal, local processes, and will likely continue for another Hubble time (Hunter 1997; van Zee 2001).

17.5 Metallicity and Age

17.5.1 *Young Populations and Chemical Homogeneity*

The gas in Irrs and dIrrs is fairly well mixed, and mixing must proceed rapidly considering how homogeneous present-day H II region abundances at different locations within the same galaxy are. Nebular abundances of ionized gas trace the youngest populations and the chemical composition of the star-forming material. Why there is such a high degree of homogeneity is not fully understood, nor is it clear how the mixing proceeds; mechanisms may include winds and turbulence. Inhomogeneities are only expected to be detectable very shortly after the responsible stellar population formed (e.g., when pollution by Wolf-Rayet stars occurs; see Kobulnicky et al. 1997). Note that such global chemical homogeneity appears to be less pronounced in gas-deficient dSph galaxies, which seem to have experienced different star formation and chemical enrichment histories than dIrrs (see, e.g., Harbeck et al. 2001; Grebel et al. 2003).

If Irrs and dIrrs are chemically homogeneous, one would expect to measure comparable abundances in H II regions and young stars of a given dIrr, since both trace the same population. Due to their proximity, in the Magellanic Clouds supergiants, giants, and even massive main sequence stars can be analyzed with high-dispersion spectroscopy and individual element ratios can be measured. Indeed, in the Magellanic Clouds good agreement is found between nebular and stellar abundances (e.g., Hill, Andrievsky, & Spite 1995; Andrievsky et al. 2001). Star-to-star variations in the overall metallicity of young stars (B to K supergiants and B main sequence stars) are small (± 0.1 dex: Hill 1997; Luck et al. 1998; Venn 1999; Rolleston, Trundle, & Dufton 2002), and there is no evidence for a significant Population I metallicity spread in either Cloud. Also, the differences between the stellar abundances in young clusters and field stars are small (Gonzalez & Wallerstein 1999; Hill 1999; Korn et al. 2000, 2002; Rolleston et al. 2002). The mean metallicity of the young population in the LMC is [Fe/H] ≈ -0.3 dex and ~ -0.7 dex in the SMC.

Very good agreement between young stellar and nebular abundances is also found in the more distant dIrrs NGC 6822 (B supergiants: Muschielok et al. 1999; A supergiants: Venn et al. 2001, both yielding [Fe/H] $= -0.5$ dex), GR 8, and Sex A (Venn et al. 2004). The two blue supergiants analyzed in WLM, on the other hand, have clearly higher metallicities than found in its H II regions (Venn et al. 2003). The reasons for this discrepancy in WLM are still unknown.

17.5.2 Intermediate-age/Old Populations and Chemical Inhomogeneity

Whereas young populations in dIrrs tend to be fairly homogeneous, old and intermediate-age stellar populations show considerable metallicity spreads. In part this may be due to the large age range sampled here, and the difficulty of assigning ages to individual field stars. Metallicity spreads have mainly been derived based on the color width of the red giant branch in color-magnitude diagrams, or via metallicity-sensitive photometric systems (e.g., Cole, Smecker-Hane, & Gallagher 2000; Davidge 2003). These methods have the drawback that they are affected by the age-metallicity degeneracy. Near-infrared Ca II triplet spectroscopy is now increasingly being employed for general [Fe/H] derivations instead. In the LMC, red giants in different parts of the galaxy show significantly different mean abundances (Cole et al. 2000, 2004). Cole et al. conclude that in the LMC azimuthal metallicity variations may in part be due to different fractions of bar and disk stars sampled at different positions (with the bar stars being younger and more metal-rich). With regard to field populations, star clusters have the advantage of consisting of well-datable, single-age populations. Old globular clusters in the LMC may differ substantially in metallicity (Olszewski et al. 1991). Interestingly, there is also evidence for a radial abundance gradient in the LMC old cluster population, i.e., a trend for old clusters to be more metal-rich closer to the LMC's center (Da Costa 1999). In the SMC, there are indications that intermediate-age star clusters of a given age may occasionally differ by a few tenths of dex in [Fe/H] (Da Costa & Hatzidimitriou 1998; Da Costa 2002), which would indicate considerable chemical dif-

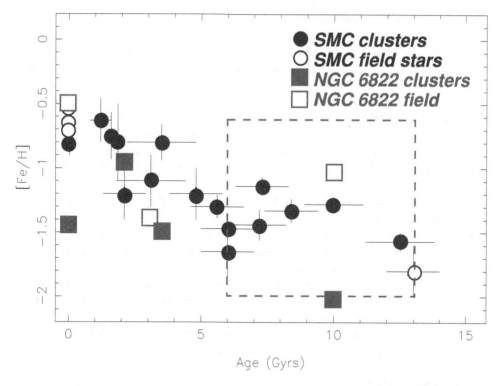

Fig. 17.3. Age versus metallicity for clusters and field stars in the SMC (circles) and in NGC 6822 (squares). The diagram for the SMC was adopted from Da Costa (2002) and comprises both spectroscopic and photometric abundances. The data points for NGC 6822 are based on a variety of different measurements and methods (clusters: Cohen & Blakeslee 1998; Chandar, Bianchi, & Ford 2000; Strader, Brodie, & Huchra 2003; field: Muschielok et al. 1999; Tolstoy et al. 2001; Venn et al. 2001) and should not be used to derive a quantitative age-metallicity relation. The solid-line box denoting the mean metallicity of NGC 6822's red giant field population is shown for an assumed age of 10 Gyr, while the much larger dashed box indicates the spread in metallicity and gives a rough idea of the possible age range.

ferences in the enrichment of their birth clouds—possibly due to infall. However, refined age determinations and more spectroscopic abundance determinations are needed to verify the SMC intermediate-age metallicity spread.

In NGC 6822 and IC 1613, abundance spreads among the field red giants have been confirmed spectroscopically (Tolstoy et al. 2001; Zucker & Wyder 2004); again, age uncertainties remain. Although the present-day field star metallicity in NGC 6822 lies between those of the LMC and the SMC, the cluster metallicities tend to lie below those of the SMC (Chandar, Bianchi, & Ford 2000; Strader et al. 2003; see Fig. 17.3). One needs to caution that the cluster measurements are based on a number of different studies and methods. Also, the present-day metallicity of NGC 6822 may have been enhanced by the recent interaction-triggered star formation episode (see §17.3.2).

17.5.3 Individual Element Abundances and Ratios

The [α/Fe] ratio in the Magellanic Clouds, NGC 6822, and WLM measured in the above studies is lower than the solar ratio. There are several possible reasons. This may be a consequence of the lower star formation rates in dIrrs or the possibility of a steeper initial mass function (see Tsujimoto et al. 1995; Pagel & Tautvaišienė 1998), i.e., a reduced contribution of Type II supernovae as compared to Type Ia supernovae (e.g., Hill et al. 2000; Smith et al. 2002), or the possibility of metal loss through selective winds (Pilyugin 1996). It seems that in the LMC, where α element abundances (most notably oxygen) were measured in clusters of different ages, the evolution of the [α/Fe] ratio was fairly flat with time (Hill et al. 2000 and Hill 2004). Field red giants also show reduced [O/Fe] values (Smith et al. 2002).

The r-process (traced by, e.g., Eu), which is the dominant process in massive stars, appears to prevail at low metallicities or the early stages of the LMC's evolution. The s-process (traced by, e.g., Ba and La), which takes place in cool intermediate-mass giants such as asymptotic giant branch stars, dominates at higher metallicities (Hill 2003). Nitrogen appears to be close to primary and may come mainly from nonmassive stars (see also Maeder, Grebel, & Mermilliod 1999). Stellar and gaseous nitrogen abundances in the Magellanic Clouds show considerable variations. Nitrogen enhancement in supergiants and old red giants may be due to mixing of CN-enhanced material to their surface during the first dredge-up (e.g., Russell & Dopita 1992; Venn 1999; Dufton et al. 2000; Smith et al. 2002; not observed, however, in LMC B main sequence stars; Korn et al. 2002).

17.5.4 Age-metallicity Relations

In order to derive a reliable, detailed, age-metallicity relation suitable to constrain the quantitative nature of the chemical evolution history of a galaxy, including the importance of infall, gas and metal loss, burstiness, etc., one would ideally want very well-resolved temporal sampling. This is not yet possible with the currently available, sparse data.

Present-day abundances are traced well by H II regions and massive stars. Progress is being made at intermediate and old ages using planetary nebulae and field red giants. Planetary nebulae have recently been used for an independent derivation of the age-metallicity evolution of the LMC at intermediate ages (Dopita et al. 1997). While extragalactic planetary nebulae cannot normally easily be age-dated, Dopita et al. extended their spectroscopic data set for the LMC to the ultraviolet to try to directly measure the flux from the central star and also used the size information for the nebulae. They were then able to not only derive abundances but also ages using full photoionization modeling, and found their results in good agreement with stellar absorption-line spectroscopy. In less massive dIrrs, the number of planetary nebulae tends to be small, making it more difficult to derive a well-sampled age-metallicity relation. In order to derive ages for individual field stars, one has to complement the spectroscopic abundances by photometric luminosities and colors and rely on isochrone models. Considerably more accurate information can presently be obtained when using star clusters as single-age, single-metallicity populations. Disadvantages of relying on star clusters are that one often only has very few such objects in a galaxy, and that their properties are not necessarily representative of the field populations.

In all Irrs and dIrrs studied in some detail to date, there is evidence for the expected increase in chemical enrichment with younger ages. The LMC's cluster age-metallicity relation clearly demonstrates this, although there is the famous cluster age gap in the age range from ~ 4 to ~ 9 Gyr (Da Costa 2002, and references therein). The SMC has the unique advantage among all Local Group dIrrs to have formed and preserved clusters throughout its lifetime. While spectroscopic abundance determinations and improved age determinations are still missing for many clusters, the SMC shows a very well-defined age-metallicity relation with what appears to be considerable metallicity scatter at a certain given age (see Fig. 17.3, adopted from Da Costa 2002, and discussion in the previous subsection). While the age-metallicity relation appears to be flatter than predicted by closed-box models in LMC and SMC, Da Costa (2002) notes that the presently available data do not yet permit one to distinguish between simple closed-box or leaky-box evolution models, bursty star formation histories, or infall.

Figure 17.3 also shows data points for clusters and field stars in NGC 6822, for which rough age estimates and spectroscopic abundance determinations are available from a variety of different sources. As noted by Chandar et al. (2000), the clusters generally seem to be more metal-poor than the SMC clusters and NGC 6822's field population. It is unclear whether these differences would be reduced if all metallicities were determined using the same method, but overall the graph seems to indicate a trend of increasing metallicity with decreasing age. Undoubtedly, these kinds of studies will be refined in the coming years.

17.6 Other Global Correlations

17.6.1 Gas and Environment

When plotting galaxy H I masses versus distance from the closest massive galaxy in the Local Group and its immediate surroundings, we see a tendency for H I masses to increase with galactocentric distance (Fig. 17.4; Grebel et al. 2003, and references therein). Only fairly large galaxies, such as the Magellanic Clouds, IC 10, and M33, with H I masses $\gg 10^7 \ M_\odot$, seem to be able to retain their gas reservoirs when closer than ~ 250 kpc to giants (Grebel et al. 2003). Note that weak lensing measurements and dynamical modeling indicate typical dark matter halo scales for massive galaxies of $260 h^{-1}$ kpc (e.g., McKay et al. 2002). The bulk of the Local Group dIrrs and transition-type galaxies are located beyond ~ 250 kpc from M31 and the Milky Way. At these distances these gas-rich galaxies seem to be less prone to galaxy harassment (i.e., in this case loss of gas through interactions with spirals), although the details will depend on their (yet unknown) orbital parameters (Grebel et al. 2003). A similar trend for dIrrs and dIrr/dSphs is seen in the Sculptor group (Skillman, Côté, & Miller 2003a).

Skillman, Côté, & Miller (2003b) investigated correlations between H I mass fraction and metallicity (oxygen abundance) for Local Group dIrrs and dIrrs in the Sculptor group of galaxies. They found that the Local Group dIrrs deviate from model curves expected for closed-box evolution, whereas the Sculptor group dIrrs follow these curves rather closely. Indeed the dIrrs that exhibit the strongest deviations are those with very low metallicity and gas content. Considering that the Sculptor group is, in contrast to the Local Group, a very loose and diffuse group or cloud (Karachentsev et al. 2003b), this may indicate that closed-box evolution is more likely to occur in low-density environments with little harassment.

17.6.2 Luminosity and Metallicity

Mean galaxy metallicity and mean galaxy luminosity are well correlated, as has been known for a long time. For dIrrs, one usually considers present-day oxygen abundances, and when comparing them to other galaxy types without ionized gas (such as dSphs),

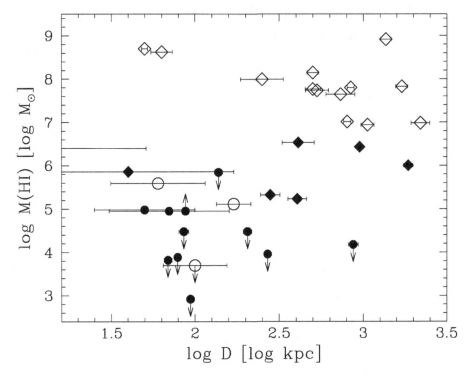

Fig. 17.4. Dwarf galaxy H I mass versus distance to the nearest massive galaxy. Filled circles stand for dwarf spheroidals (dSphs), open circles for dwarf ellipticals, open diamonds for dwarf irregulars (dIrrs), and filled diamonds for dIrr/dSph transition-type galaxies. Lower or upper H I mass limits are indicated by arrows. There is a general trend for the H I masses to increase with increasing distance from massive galaxies. DSphs lie typically below $10^5 \, M_\odot$ in H I mass limits, while potential transition-type galaxies have H I masses of $\sim 10^5$ to $10^7 \, M_\odot$. DIrr galaxies usually exceed $10^7 \, M_\odot$. (Figure from Grebel et al. 2003.)

stellar [Fe/H] values are converted into what is assumed to be the corresponding nebular abundance. This conversion comes with a number of uncertainties. Grebel et al. (2003) therefore used the stellar (red giant) metallicities of Local Group dwarf galaxies of all types to directly compare the properties of their old populations. A plot of V-band luminosity L_V versus $\langle$[Fe/H]$\rangle$ (Fig. 17.5, left panel) shows a clear trend of increasing luminosity with increasing mean red giant branch metallicity. However, different galaxy types (gas-rich dIrrs and gas-deficient dSphs) are offset from each other in that the dIrrs are more luminous at the same metallicity.

In other words, the dIrrs have too low a metallicity for their luminosity as compared to dSphs. *Thus, dSphs, most of which have been quiescent over at least the past few Gyr, must have experienced chemical enrichment faster and more efficiently than dIrrs, which continue to form stars until the present day* (Grebel et al. 2003). It is tempting to speculate that environment may once again have affected this in the sense of a denser environment leading to more vigorous early star formation rates.

When plotting the baryonic luminosity (Milgrom & Braun 1988; Matthews, van Driel, &

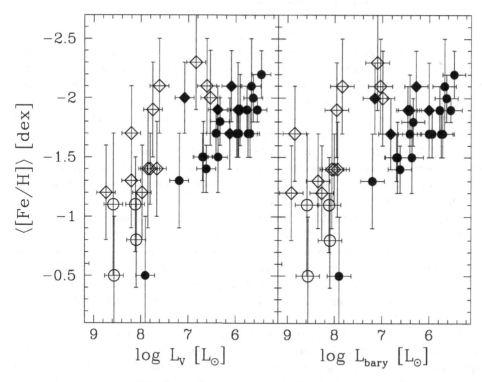

Fig. 17.5. *V*-band luminosity (*left panel*) and baryonic luminosity (*right panel*, corrected for baryon contribution of gas not yet turned into stars) versus mean metallicity of red giants. The symbols are the same as in Fig. 17.4. The error bars in metallicity indicate the metallicity spread in the old populations, not the uncertainty of the metallicity. dIrrs are more luminous at equal metallicity than dSphs. Or, in other words, dSphs are too metal rich for their low luminosity. However, several dIrr/dSph transition-type galaxies coincide with the dSph locus. These objects are indistinguishable from dSphs in all their properties except for gas content. (Figure from Grebel et al. 2003.)

Gallagher 1998) against metallicity (Fig. 17.5, right panel), the locus of the dSphs remains unchanged while the dIrrs move to higher luminosities as compared to the dSphs. Thus, if star formation in present-day dIrrs were terminated when all of their gas was converted into stars, then these fading dIrrs would be even further from the dSph luminosity-metallicity relation. For a discussion of the amount of fading, time scales, angular momentum loss, etc. required for converting a dIrr into a dSph, see Grebel et al. (2003). Here we simply want to emphasize that dIrrs follow a metallicity-luminosity relation that requires a different evolutionary path than in other types of dwarf galaxies. In particular, it seems that dIrrs are an intrinsically different type of galaxy than dSphs. We note in passing that for dIrrs not only do metallicities correlate well with luminosities, but also with surface brightness.

17.7 Summary

Irr and dIrr galaxies are usually gas-rich galaxies with ongoing or recent star formation. They are preferentially found in the outer regions of groups and clusters as well as in the field. Irrs and massive dIrrs exhibit solid body rotation, while low-mass dIrrs seem to be dominated by random motions. Spiral density waves are absent.

Irrs and dIrrs are often embedded in extended H I halos, which, in the absence of interactions, appear fairly regular. In low-mass dIrrs, the centroid of the H I distribution does not necessarily coincide with the optical center of the galaxy, and occasionally annular structures are seen. The neutral gas tends to be flocculent, dominated by shells and bubbles, and driven by the turbulent energy input from massive stars and supernovae. Molecular gas and dust form less easily and are more easily dissociated due to the high UV radiation field and fewer coolants in low-metallicity environments.

Irrs and dIrrs usually contain multiple distinct zones of concurrent star formation. Extended regions of active star formation tend to be long-lived and gradually migrate on time scales on a few tens to hundreds of Myr. Stochastic self-propagating star formation seems to be the main driver of star formation activity. There is no need for external triggering. In quiescently evolving dIrrs and/or dIrrs with slow or no rotation (usually the less massive dIrrs), the degree of central concentration of star formation is small, while the reverse trend is true for more massive and faster rotators. The formation of populous clusters seems to be preferred in more massive and/or interacting dIrrs. Generally, gas consumption is sufficiently low that star formation in Irrs and dIrrs may continue for another Hubble time. On global scales the star formation rate of Irrs and dIrrs is close to constant, with amplitude variations of factors of 2–3.

Old stellar populations are ubiquitous in all Irrs and dIrrs studied in detail so far, although their fractions vary widely. In contrast to the many scattered young OB associations and superassociations, older populations show a smooth and regular distribution that is much more extended than that of the young populations. Both young stellar populations and H II regions agree very well in their abundances, underlining the chemical homogeneity of Irrs and dIrrs. However, taken at face value, intermediate-age and old populations tend to exhibit considerably more scatter in their metallicity. There are indications that star clusters of the same age may differ by several tenths of dex in metallicity, although observational biases cannot yet be fully ruled out. Overall, Irrs and dIrrs follow the expected trend of increasing metal enrichment toward younger ages; the currently available data do not yet permit one to unambiguously distinguish between infall and leaky-box versus closed-box chemical evolution, nor to reliably evaluate the importance and impact of possible bursts.

Substantial progress is being made not only in spectroscopic measurements of stellar metallicities, but also in the determination of individual element abundances. The $[\alpha/Fe]$ ratios in Irrs and dIrrs, which tend to be lower than the solar ratio, and the lower effective yields may be interpreted as indicative of lower astration rates and a reduced contribution of Type II supernovae. Other interpretations (different initial mass functions, leaky-box chemical evolution with metal loss through selective winds) are being entertained as well.

Correlations between gas content and distance from massive galaxies as well as morphological segregation indicate that environment (in particular gas loss through ram pressure or tidal stripping; see also Parodi, Barazza, & Binggeli 2002; Lee, McCall, & Richer 2003 for the Virgo cluster) does have an impact on the evolution of Irrs and dIrrs. Irrs and dIrrs follow the well-known relation of increasing mean metallicity with increasing galaxy luminosity. The offset in this relation from the relation for dSphs, such that dIrrs are more luminous than dSphs at the same metallicity, indicates that the early chemical evolution in these two galaxy types proceeded differently, with dSphs becoming enriched more quickly.

Acknowledgements. I would like to thank the organizers, particularly Andy McWilliam, for their kind invitation to this very interesting Symposium, and for their patience while my paper was finished. I am grateful to Jay Gallagher for a critical reading of the text.

References

Andrievsky, S. M., Kovtyukh, V. V., Korotin, S. A., Spite, M., & Spite, F. 2001, A&A, 367, 605

Banas, K. R., Hughes, J. P., Bronfman, L., & Nyman, L.-A. 1997, ApJ, 480, 607

Battinelli, P., & Demers, S. 2000, AJ, 120, 1801

Becker, R., Henkel, C., Wilson, T. L., & Wouterloot, J. G. A. 1993, A&A, 268, 483

Bolatto, A. D., Jackson, J. M., Wilson, C. D., & Moriarty-Schieven, G. 2000, ApJ, 532, 909

Chandar, R., Bianchi, L., & Ford, H. C. 2000, AJ, 120, 3088

Cioni, M.-R. L., van der Marel, R. P., Loup, C., & Habing, H. J. 2000, A&A, 359, 601

Cohen, J. G., & Blakeslee, J. P. 1998, AJ, 115, 2356

Cole, A. A., et al. 1999, AJ, 118, 1657

Cole, A. A., Smecker-Hane, T. A., & Gallagher, J. S. 2000, AJ, 120, 1808

Cole, A. A., Smecker-Hane, T. A., Tolstoy, E., & Gallagher, J. S. 2004, in Carnegie Observatories Astrophysics Series, Vol. 4: Origin and Evolution of the Elements, ed. A. McWilliam & M. Rauch (Pasadena: Carnegie Observatories, http://www.ociw.edu/symposia/series/symposium4/proceedings.html)

Conselice, C. J., Gallagher, J. S., & Wyse, R. F. G. 2001, ApJ, 559, 791

Courteau, S., & van den Bergh, S. 1999, AJ, 118, 337

Da Costa, G. S. 1999, in IAU Symp. 190, New Views of the Magellanic Clouds, ed. Y.-H. Chu et al. (San Francisco, ASP), 397

——. 2002, in IAU Symp. 207, Extragalactic Star Clusters, ed. D. Geisler, E. K. Grebel, & D. Minniti (San Francisco: ASP), 83

Da Costa, G. S., & Hatzidimitriou, D. 1998, AJ, 115, 1934

Davidge, T. J. 2003, PASP, 115, 635

de Blok, W. J. G., & Walter, F. 2000, ApJ, 537, L95

——. 2003, MNRAS, 341, L39

Demers, S., & Battinelli, P. 1998, AJ, 115, 154

de Vaucouleurs, G. 1957, Leaflet of the ASP, 7, 329

Dickey, J. M., Mebold, U., Stanimirovic, S., & Staveley-Smith, L. 2000, ApJ, 536, 756

Dolphin, A. E., et al. 2002, AJ, 123, 3154

Dolphin, A. E., & Hunter, D. A. 1998, AJ, 116, 1275

Dohm-Palmer, R. C., et al. 1998, AJ, 116, 1227

Dohm-Palmer, R. C., Skillman, E. D., Mateo, M., Saha, A., Dolphin, A., Tolstoy, E., Gallagher, J. S., & Cole, A. A. 2002, AJ, 123, 813

Dufton, P. L., McErlean, N. D., Lennon, D. J., & Ryans, R. S. I. 2000, A&A, 353, 311

Dopita, M. A., et al. 1997, ApJ, 474, 188

Edgar, R. J., & Chevalier, R. A. 1986, ApJ, 310, L27

Elmegreen, B. G., Elmegreen, D. M., Salzer, J. J., & Mann, H. 1996, ApJ, 467, 579

Elmegreen, B. G., & Hunter, D. A. 2000, ApJ, 540, 814

Elmegreen, B. G., Kim, S., & Staveley-Smith, L. 2001, ApJ, 548, 749

Feitzinger, V., Glassgold, A. E., Gerola, H., & Seiden, P. E. 1981, A&A, 98, 371

Fukui, Y., et al., 1999, PASJ, 51, 745

Gardiner, L. T., & Noguchi, M. 1996, MNRAS, 278, 191

Gardiner, L. T., Sawa, T., & Fujimoto, M. 1994, MNRAS, 266, 567

Gardiner, L. T., Turfus, C., & Putman, M. E. 1998, ApJ, 507, L35

Garnett, D. R. 2004, in Carnegie Observatories Astrophysics Series, Vol. 4: Origin and Evolution of the Elements, ed. A. McWilliam & M. Rauch (Cambridge: Cambridge Univ. Press), in press

——. 2002, ApJ, 581, 1019

Gerola, H., & Seiden, P. E. 1978, ApJ, 223, 129

Gerola, H., Seiden, P. E., & Schulman, L. S. 1980, ApJ, 242, 517

Gibson, B. K., Giroux, M. L., Penton, S. V., Putman, M. E., Stocke, J. T., & Shull, J. M. 2000, AJ, 120, 1830

Girardi, L., Chiosi, C., Bertelli, G., & Bressan, A. 1995, A&A, 298, 87

Gonzalez, G., & Wallerstein, G. 1999, AJ, 117, 2286

Grebel, E. K. 1997, Reviews of Modern Astronomy, 10, 29

——. 1999, in IAU Symp. 192, The Stellar Content of the Local Group, ed. P. Whitelock & R. Cannon (Provo: ASP), 17

——. 2000, in Star Formation from the Small to the Large Scale, 33rd ESLAB Symp., SP-445, ed. F. Favata, A. A. Kaas, & A. Wilson (Noordwijk: ESA), 87

——. 2002a, in Gaseous Matter In Galaxies And Intergalactic Space, 17th IAP Colloq., ed. R. Ferlet et al. (Paris: Frontier Group), 171

——. 2002b, in IAU Symp. 207, Extragalactic Star Clusters, ed. D. Geisler, E. K. Grebel, & D. Minniti (San Francisco: ASP), 94

Grebel, E. K., & Brandner, W. 1998, in The Magellanic Clouds and Other Dwarf Galaxies, ed. T. Richtler & J. Braun (Aachen: Shaker Verlag), 151

Grebel, E. K., & Chu, Y. 2000, AJ, 119, 787

Grebel, E. K., Gallagher, J. S., & Harbeck, D. 2003, AJ, 125, 1926

Greggio, L., Marconi, G., Tosi, M., & Focardi, P. 1993, AJ, 105, 894

Groenewegen, M. A. T. 2002, in The Chemical Evolution of Dwarf Galaxies, Ringberg Workshop, ed. E. K. Grebel, astro-ph/0208449

Harbeck, D., et al. 2001, AJ, 122, 3092

Hatzidimitriou, D., Cannon, R. D., & Hawkins, M. R. S. 1993, MNRAS, 261, 873

Hill, V. 1997, A&A, 324, 435

——. 1999, A&A, 345, 430

——. 2004, in Carnegie Observatories Astrophysics Series, Vol. 4: Origin and Evolution of the Elements, ed. A. McWilliam & M. Rauch (Cambridge: Cambridge Univ. Press), in press

Hill, V., Andrievsky, S., & Spite, M. 1995, A&A, 293, 347

Hill, V., François, P., Spite, M., Primas, F., & Spite, F. 2000, A&A, 364, L19

Hodge, P. W. 1980, ApJ, 241, 125

Hodge, P. W., Dolphin, A. E., Smith, T. R., & Mateo, M. 1999, ApJ, 521, 577

Holtzman, J. A., Smith, G. H., & Grillmair, C. 2000, AJ, 120, 3060

Hoopes, C. G., Sembach, K. R., Howk, J. C., Savage, B. D., & Fullerton, A. W. 2002, ApJ, 569, 233

Hunter, D. A. 1997, PASP, 109, 937

Israel F. P. 1997, A&A, 317, 65

Israel, F. P., Bontekoe, T. R., & Kester, D. J. M. 1996, A&A, 308, 723

Israel, F. P., de Grauuw, Th., van de Stadt, H., & de Vries, C. P. 1986, ApJ, 303, 186

Johnson, J. A., Bolte, M., Stetson, P. B., Hesser, J. E., & Somerville, R. S. 1999, ApJ, 527, 199

Karachentsev, I. D., et al. 2002a, A&A, 385, 21

——. 2002b, A&A, 383, 125

——. 2002c, A&A, 389, 812

——. 2003a, A&A, 398, 467

——. 2003b, A&A, 404, 93

Kennicutt, R. C., Bresolin, F., Bomans, D. J., Bothun, G. D., & Thompson, I. B. 1995, AJ, 109, 594

Kim, S., Dopita, M. A., Staveley-Smith, L., & Bessell, M. S. 1999, AJ, 118, 2797

Kim, S., Staveley-Smith, L., Dopita, M. A., Freeman, K. C., Sault, R. J., Kesteven, M. J., & McConnell, D. 1998, ApJ, 503, 674

Kobulnicky, H. A., & Dickey, J. M. 1999, AJ, 117, 908

Kobulnicky, H. A., Skillman, E. D., Roy, J., Walsh, J. R., & Rosa, M. R. 1997, ApJ, 477, 679

Komiyama, Y., et al. 2003, ApJ, 590, L17

Koornneef, J. 1982, A&A, 107, 247

Korn, A. J., Becker, S. R., Gummersbach, C. A., & Wolf, B. 2000, A&A, 353, 655

Korn, A. J., Keller, S. C., Kaufer, A., Langer, N., Przybilla, N., Stahl, O., & Wolf, B. 2002, A&A, 385, 143

Kunkel, W. E., Demers, S., & Irwin, M. J. 2000, AJ, 119, 2789

Lee, H., McCall, M. L., & Richer, M. G. 2003, AJ, 125, 2975

Lehner, N., Sembach, K. R., Dufton, P. L., Rolleston, W. R. J., & Keenan, F. P. 2000, ApJ, 551, 781

Letarte, B., Demers, S., Battinelli, P., & Kunkel, W. E. 2002, AJ, 123, 832

Lin, D. N. C., Jones, B. F., & Klemola, A. R. 1995, ApJ, 439, 652

Lo, K. Y., Sargent, W. L. W., & Young, K. 1993, AJ, 106, 507

Lu, L., Sargent, W. L. W., Savage, B. D., Wakker, B. P., Sembach, K. R., & Oosterloo, T. A. 1998, AJ, 115, 162

Luck, R. E., Moffett, T. J., Barnes, T. G., & Gieren, W. P. 1998, AJ, 115, 605

Madden, S. C., Poglitsch, A., Geis, N., Stacey, G. J., & Townes, C. H. 1997, ApJ, 483, 200

Maeder, A., Grebel, E. K., & Mermilliod, J. 1999, A&A, 346, 459

Martínez-Delgado, D., Gallart, C., & Aparicio, A. 1999, AJ, 118, 862

Mastropietro, C., Moore, B., Mayer, L., Stadel, J., & Wadsley, J. 2004, in Satellite and Tidal Streams, ed. F. Prada, D. Martínez-Delgado, & T. Mahoney (San Francisco: ASP), in press (astro-ph/0309244)

Mateo, M. L. 1998, ARA&A, 36, 435

Matthews, L. D., van Driel, W., & Gallagher, J. S. 1998, AJ, 116, 2196

McCray, R., & Kafatos, M. 1987, ApJ, 317, 190

McKay, T. A., et al. 2002, ApJ, 571, L85

Meaburn, J. 1980, MNRAS, 192, 365

Milgrom, M., & Braun, E. 1988 ApJ, 334, 130

Minniti, D., & Zijlstra, A. A. 1996, ApJ, 467, L13

Minniti, D., Zijlstra, A. A., & Alonso, M. V. 1999, AJ, 117, 881

Murai, T., & Fujimoto, M. 1980, PASJ, 32, 581

Muschielok, B., et al. 1999, A&A, 352, L40

Olsen, K. A. G., Hodge, P. W., Mateo, M., Olszewski, E. W., Schommer, R. A., Suntzeff, N. B., & Walker, A. R. 1998, MNRAS, 300, 665

Olszewski, E. W., Schommer, R. A., Suntzeff, N. B., & Harris, H. C. 1991, AJ, 101, 515

Pagel, B. E. J., & Tautvaišienė, G. 1998, MNRAS, 299, 535

Parodi, B. R., Barazza, F. D., & Binggeli, B. 2002, A&A, 388, 29

Parodi, B. R., & Binggeli, B. 2003, A&A, 398, 501

Petitpas, G. R., & Wilson, C. D. 1998, ApJ, 496, 226

Pilyugin, L. S. 1996, A&A, 310, 751

Pilyugin, L. S., & Ferrini, F. 2000, A&A, 358, 72

Points, S. D., Chu, Y.-H., Kim, S., Smith, R. C., Snowden, S. L., Brandner, W., & Gruendl, R. 1999, ApJ, 518, 298

Putman, M. E., et al. 1998, Nature, 394, 752

Putman, M. E., Staveley-Smith, L., Freeman, K. C., Gibson, B. K., & Barnes, D. G. 2003, ApJ, 586, 170

Rejkuba, M., Minniti, D., Gregg, M. D., Zijlstra, A. A., Alonso, M. V., & Goudfrooij, P. 2000, AJ, 120, 801

Rhode, K. L., Salzer, J. J., Westphal, D. J., & Radice, L. A. 1999, AJ, 118, 323

Richer M. G., et al. 2001, A&A, 370, 34

Rolleston, W. R., Dufton, P. L., McErlean, N., & Venn, K. A. 1999, A&A, 348, 728

Rolleston, W. R. J., Trundle, C., & Dufton, P. L. 2002, A&A, 396, 53

Roye, E. W., & Hunter, D. A. 2000, AJ, 119, 1145

Rubio, M., Lequeux, J., & Boulanger, F. 1993, A&A, 271, 9

Russell, S. C., & Dopita, M. A. 1992, ApJ, 384, 508

Saitō, M., Sasaki, M., Ohta, K., & Yamada, T. 1992, PASJ, 44, 593

Sembach, K. R., Howk, J. C., Savage, B. D., & Shull, J. M. 2001, AJ, 121, 992

Skillman, E. D., Côté, S., & Miller, B. W. 2003a, AJ, 125, 593

——. 2003b, AJ, 125, 610

Smith, V. V., et al. 2002, AJ, 124, 3241

Stanimirovic, S., Staveley-Smith, L., Dickey, J. M., Sault, R. J., & Snowden, S. L. 1999, MNRAS, 302, 417

Stanimirovic, S., Staveley-Smith, L., van der Hulst, J. M., Bontekoe, T. R., Kester, D. J. M., & Jones, P. A. 2000, MNRAS, 315, 791

Staveley-Smith, L., Sault, R. J., Hatzidimitriou, D., Kesteven, M. J., & McConnell, D. 1997, MNRAS, 289, 255

St-Germain, J., Carignan, C., Côté, S., & Oosterloo, T. 1999, AJ, 118, 1235

Strader, J., Brodie, J. P., & Huchra, J. P. 2003, MNRAS, 339, 707

Tolstoy, E., Irwin, M. J., Cole, A. A., Pasquini, L., Gilmozzi, R., & Gallagher, J. S. 2001, MNRAS, 327, 918

Tomita, A., Ohta, K., & Saitō, M. 1993, PASJ, 45, 693

Tosi, M., Greggio, L., Marconi, G., & Focardi, P. 1991, AJ, 102, 951

Tsujimoto, T., Nomoto, K., Yoshii, Y., Hashimoto, M., Yanagida, S., & Thielemann, F.-K. 1995, MNRAS, 277, 945

Tumlinson, J., et al. 2002, ApJ, 566, 857

van den Bergh, S. 1960, ApJ, 131,215

——. 1966, AJ, 71, 922

——. 1999, A&ARv, 9, 273

——. 2000, The Galaxies of the Local Group (Cambridge: Cambridge Univ. Press)

van der Marel, R. P., Alves, D. R., Hardy, E., & Suntzeff, N. B. 2002, AJ, 124, 2639

Van Dyk, S. D., Puche, D., & Wong, T. 1998, AJ, 116, 2341

van Zee, L. 2001, AJ, 121, 2003

van Zee, L., Haynes, M. P., Salzer, J. J., & Broeils, A. H. 1997, AJ, 113, 1618

Venn, K. A. 1999, ApJ, 518, 405

Venn, K. A., et al. 2001, ApJ, 547, 765

Venn, K. A., Tolstoy, E., Kaufer, A., & Kudritzki, R. P. 2004, in Carnegie Observatories Astrophysics Series, Vol. 4: Origin and Evolution of the Elements, ed. A. McWilliam & M. Rauch (Pasadena: Carnegie Observatories, http://www.ociw.edu/symposia/series/symposium4/proceedings.html)

Venn, K. A., Tolstoy, E., Kaufer, A., Skillman, E. D., Clarkson, S. M., Smartt, S. J., Lennon, D. J., & Kudritzki, R. P. 2003, AJ, 126, 1326

Wakker, B., Howk, J. C., Chu, Y.-H., Bomans, D., & Points, S. 1998, ApJ, 499, L87

Walter, F., & Brinks, E. 1999, AJ, 118, 273

Weinberg, M. D. 2000, ApJ, 532, 922

Whiting, A. B., Hau, G. K. T., & Irwin, M. 1999, AJ, 118, 2767

Wilcots, E. M., & Miller, B. W. 1998, AJ, 116, 2363

Yamaguchi, R., Mizuno, N., Onishi, T., Mizuno, A., & Fukui, Y. 2001, PASJ, 53, 959

Yang, H., & Skillman, E. D. 1993, AJ, 106, 1448

Young L. M., & Lo K. Y. 1996, ApJ, 462, 203

——. 1997, ApJ, 490, 710

Young, L. M., van Zee, L., Lo, K. Y., Dohm-Palmer, R. C., & Beierle, M. E. 2003, ApJ, 592, 111

Zaritsky, D., Harris, J., Grebel, E. K., & Thompson, I. B. 2000, ApJ, 534, L53

Zucker, D. B., & Wyder, T. K. 2004, in Carnegie Observatories Astrophysics Series, Vol. 4: Origin and Evolution of the Elements, ed. A. McWilliam & M. Rauch (Pasadena: Carnegie Observatories, http://www.ociw.edu/symposia/series/symposium4/proceedings.html)

18

Chemical evolution of the old stellar populations of M31

R. MICHAEL RICH
Department of Physics and Astronomy, University of California, Los Angeles

Abstract

The stellar populations concept was developed based on Baade's studies of the central bulge of the Andromeda Galaxy, M31. Yet, the Andromeda system now appears to be anything but typical. Deep imaging reveals the outer halo to be a flattened system of complex tidal filaments roughly aligned with the disk major axis. Studies of the field remote from the nucleus find a metal-rich population that shows little variation in its mean abundance or distribution in projected locations ranging from 5 to 35 kpc from the nucleus. Every study to date confirms that the halo population is metal-rich, at $[Fe/H] \approx -0.5$, with a tail extending to lower metallicity. The halo within 20 kpc cannot be composed of disrupted dwarf spheroidal satellites of M31, which are very metal-poor and have well-populated blue horizontal branches. The stars in the interaction streams of M31 are more metal-poor, with a smaller abundance spread than the disk stars.

The age distribution of one deep halo field has been constrained from *HST* imaging, which shows that it has an old, metal-poor population, but also an equal population of 6–11 Gyr old, metal-rich stars. While the halo may be intermediate age, the bulge and a large fraction of the globular clusters appear to be old and similar to comparable populations in the Milky Way.

We show that M31 lies in the expected location on a plot of halo abundances as a function of total galaxy luminosity. While this relationship may be understood as implying that a metal-enriched wind controls chemical evolution, it is difficult to understand how a simple closed-box model with wind outflow could be consistent with the presence of a significant age range in the halo. The complexity present in the old populations of M31 stands to remind us how complicated the history of distant galaxies might be.

18.1   Introduction

It is indeed appropriate to consider the chemical evolution of M31 at this Carnegie Centennial Symposium. As is well known, Baade's investigations of M31 led directly to the development of the stellar populations paradigm (Baade 1944). Hubble's identification of a Cepheid variable in M31 certified its classification as an extragalactic system.

Baade might be surprised to know that while the stellar populations of M31 helped to develop the fundamental Population I and II paradigm, only within the last few years has M31 been revealed to be a system of remarkable complexity. M31 has a factor of 3 more globular clusters than the Milky Way, and the nature of the stellar halo is unclear. One sees a widespread metal-rich stellar population, but there is also clear evidence of tidal streams. The metal-rich population in the halo is intermediate age. Had this article been written a year earlier, it likely would have emphasized a relatively well-understood picture of chemical evolution driven by star formation and metal-enriched winds. This simplified picture is no longer appropriate for M31. The proximity of M31 permits studies of detail sufficient to uncover its peculiarities. Within the next decade, there is the potential for a wide-field imaging program in M31 from space, as well as the use of 10 m-class telescopes to measure detailed abundances of giants in the old populations. There is little doubt that more surprises are in store.

The impact of other Carnegie scientists on the subject of chemical evolution is well known, but bears mention because of the occasion of this meeting. The two dominant constructs for chemical evolution are the so-called ELS (Eggen, Lynden-Bell, & Sandage 1962) picture of prompt initial enrichment, violent relaxation, and an early burst of star formation and chemical evolution. The other construct is the so-called Searle-Zinn hypothesis, in which massive systems are built out of the merged remnants of low-mass systems (Searle & Zinn 1978). Sandage and Searle, both Carnegie astronomers, played a central role in the development of these ideas. Zinn was a Carnegie Fellow at the time of the Searle & Zinn paper. Searle & Sargent (1972) also wrote the landmark paper in which the simple model of chemical evolution is fully developed. Carnegie scientists are, in large part, responsible for developing the foundations of the subject area to which this Symposium is devoted.

In this review I am considering the old stellar populations of the bulge, disk, and stellar halo. I will not consider the abundances of the disk H II regions or young stars; for these and other topics that I do not address here, the reader may refer to Hodge's (1992) comprehensive volume.

18.2   Overview

As the nearest bright spiral galaxy, one might hope that M31 would be typical of its class. At 770 Kpc distant, M31 is near enough to enable a wide range of studies on its populations. The initial studies of its globular cluster system appeared to fulfill that promise. While more numerous than that of the Milky, the M31 clusters divide into a centrally concentrated red (metal-rich) and blue (old, metal-poor group), just as is the case for the Milky Way clusters. Imaging of ~ 20 M31 globular clusters using *HST* has turned up no clear signature of an intermediate-age population and will be addressed more in the next section. At the time of this writing, the main sequence turn-off has been reached in one M31 globular cluster (G312) and an age determination is underway (Brown et al. 2004).

Early investigations of the halo abundance distribution function (Durrell, Harris & Pritchet 1994; Rich et al. 1996a, b) find that the halo is remarkably metal-rich, about -0.6 dex, with an abundance distribution that roughly resembles that predicted from the "simple model" of chemical evolution (Searle & Sargent 1972). In the context of these observations, one could imagine that the halo and old globular clusters formed early, followed by a dissipative starburst giving rise to the inner disk and bulge.

The discovery of the giant stream (Ibata et al. 2001; Fig. 18.1) and its extent in the deep M31 halo might cause one to abandon hope for a simple chemical evolution model for the halo. The presence of intermediate-age, metal-rich stars in the M31 halo (discussed further in §18.4.2) shows that the old populations of M31 are complex, but that a description of their chemical evolution must defy any orderly or simple description.

The origin of the most distant populations in M31 remains unclear. In one remote field

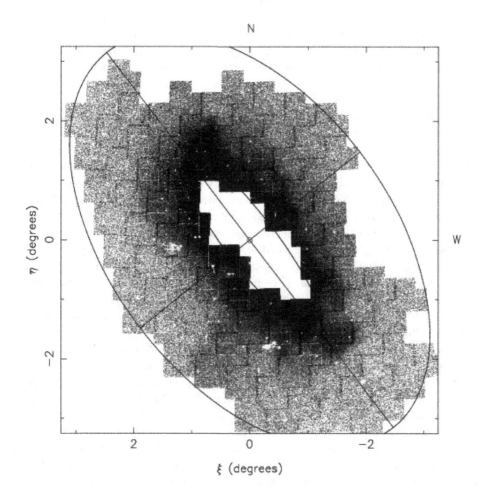

N

W

η (degrees)

ξ (degrees)

Fig. 18.1. This figure (Ferguson et al. 2002) shows the complex structure of the halo of M31, including the giant stream of Ibata et al. (2001) projecting on the lower left (SE direction). The complex structure along the SW major axis is near the globular cluster G1, although stars in that location do not share its radial velocity (§18.4.3 and Fig. 18.8; Reitzel, Guhathakurta, & Rich 2004). Projected onto the sky, the full extent of M31 and the Andromeda dwarfs spans roughly 10 degrees; were the eye sensitive to extreme low surface brightness, the halo of M31 would appear as large as the the Large Magellanic Cloud. Do these structures, the disk, and the smooth halo have differing star formation histories?

roughly 34 kpc distant on the major axis, the kinematics of the population appears consistent with a disk origin. On the other hand, the averaged star counts at large distances fit a smooth radial falloff. The question of whether the remote populations are dominated by the cumulative debris of dissolved stellar systems (Côté et al. 2000a) or by the early formation history of M31 must ultimately be settled by further investigations.

Our survey of M31 begins by considering the well-studied globular cluster system (§18.3). Following, §18.4 addresses the age, composition, and evolution of the halo. Section 18.4.2

presents the deepest *HST* imaging ever done in M31, while §18.4.3 analyzes a remote field on the major axis that appears to be dominated by the M31 disk. The dwarf satellites of M31 are discussed in §18.5, and §18.6 concerns the bulge of M31. In §18.7, we relate the M31 halo to the halo populations of other edge-on spiral galaxies. Section 18.8 summarizes this review and seeks a vision of chemical evolution for the old populations in M31.

18.3   Globular Clusters

The globular cluster system is the old stellar population of M31 that most closely resembles that of the Milky Way (Fig. 18.2). Based on *HST* color-magnitude diagrams reaching below the horizontal branch (Ajhar et al. 1996; Fusi Pecci et al. 1996; Rich et al. 1996b) the ~ 20 clusters with such deep color-magnitude diagrams appear to be as old as Milky Way globular clusters, with a similar primary strong correlation between horizontal branch morphology and metallicity. While an extended giant branch consisting of luminous carbon stars would likely have been detected in optical imaging and in the infrared (Stephens et al. 2001), no evidence for such an intermediate-age population is seen in resolved clusters. *HST* imaging of M31 globular clusters to date finds no evidence for extended giant branches, but the asymptotic giant branch (AGB) constrains less well the age of metal-rich populations (Guarnieri et al. 1998). Rich et al. (1996b) used WFPC2 to image the luminous globular cluster G1 at 1 μm, and infrared images have also been obtained (Stephens et al. 2001). If G1 had an extended AGB, we might (applying the fuel consumption theorem; Renzini & Buzzoni 1986) have seen upwards of 20 bright AGB stars.

Global studies of the cluster system, reaching back to Elson & Walterbos (1988), find no indication for a major subpopulation of intermediate-age or young clusters, such as is seen in the Magellanic Clouds. While the presence of a blue horizontal branch is *prima facie* evidence for an old stellar population, the discovery of candidate RR Lyrae variables in four M31 globular clusters with blue horizontal branches (Clementini et al. 2001) is reassuring.

A new approach by Jiang et al. (2003) fits multicolor photometry of the globular clusters to model spectral energy distributions. They confirm that a population of clusters as old as those in the Milky Way halo is present, but they also identify intermediate-age candidates. They argue further that the metal-rich clusters might be measurably younger than the metal-poor clusters. This approach is promising, but might be affected by the presence or absence of a blue horizontal branch, which is seen in some old, metal-rich clusters in the Milky Way (Rich et al. 1997). If infrared photometry and line strength measurements were to be added and included in the constraints, and the results were further calibrated using actual color-magnitude diagrams, this method might prove powerful in M31 and more distant galaxies.

As populations reach near 10 Gyr in age, relative age-related differences in their colors and spectra become subtle, and trace populations such as the blue horizontal branch or blue stragglers can affect the derived parameters. Beasley et al. (2004) fit Lick indices measured from high-quality spectra of 30 globular clusters using the population models of Thomas, Maraston, & Bender (2003). They find a subgroup of 13 metal-rich (> -1 dex) globular clusters consistent with 5 Gyr age. They confirm some of the Burstein et al. (1984) clusters with strong $H\beta$ but rely on other indices for their age constraint. However, their method gives an age of 7 ± 3 Gyr for the bulge of M31, and 3 ± 1.5 Gyr for M32, while the bright AGB stars required for such young populations are not present in the bulge of M31 (Stephens et al. (2003) or M32 (Davidge et al. 2000). Could some of the youth ascribed to these

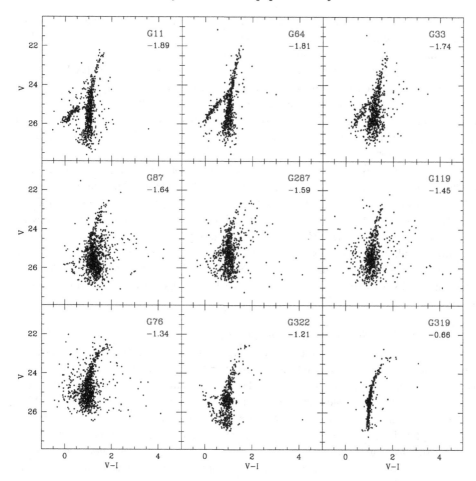

Fig. 18.2. Color-magnitude diagrams for globular clusters in M31 (Rich et al. 2004a) observed using *HST* and WFPC2. Notice that the agreement of horizontal branch morphology with metallicity follows the general trend established by the old globular clusters of the Milky Way halo. Only old stellar populations can produce the observed well-populated blue horizontal branches. The lack of luminous AGB stars and the good correlation between horizontal branch morphology and metallicity are both consistent with this sample being as old as the oldest Milky Way globular clusters. Cluster identifications from Sargent et al. (1977); metallicities from Perrett et al. (2002).

populations be due to the aforementioned blue horizontal branch problem, or related to some other difference in populations?

Turning now to abundances, the M31 cluster abundance distribution is bimodal (as is the case for the Milky Way and other extragalactic cluster populations), with a dominant metal-poor peak at −1.4 dex, and a metal-rich peak at −0.6 dex, which represents about 30% of the overall cluster population (Barmby et al. 2000).

Both metal-rich and metal-poor globular clusters show the clear signature of rotation; however, the metal-rich clusters have a steeper rotation curve and are also more centrally concentrated (Perrett et al. 2002). These observations would remain consistent with the picture in which the metal-rich clusters are more closely associated with the central bulge, while the metal-poor clusters are connected to the Population II halo.

New developments upset this satisfying picture. As is the case for the Milky Way, there appears to be kinematic subgroups of clusters (Saito & Iye 2000; Perrett et al. 2003; Morrison et al. 2004). This result is not surprising, given that seven dwarf spheroidal galaxies are already identified in the vicinity of M31, and the previously mentioned deep wide-field imaging of the outer halo shows that the galaxy is circled with debris trails extending on scales of tens of kiloparsecs.

But the new analysis of the Perrett et al. velocities would appear to contradict, or at least be inconsistent with, either picture. Morrison et al. (2004) argue that much of the M31 globular cluster population is associated with an extended thin disk. In contrast to the G dwarf problem of the Milky Way, they find the "thin disk" clusters to have as wide an abundance range ($-2 <$ [Fe/H] < 0) as is found anywhere else in M31. If this disk population is real, it poses a serious problem in chemical evolution. How does gas dissipate into a disk so rapidly as to not undergo general enrichment? It is difficult to imagine forming clusters at 1/100 solar abundance after the time scales and dissipation required for the thin disk to form.

18.3.1   *The Peculiar Spectra of M31 Globular Clusters*

In the 1980s, Faber acquired a very large sample of spectra at Lick observatory, using the then newly commissioned image-dissector scanner (Wampler 1977). It became possible to compare respectable samples of M31 globular clusters with their Galactic counterparts, and with the integrated light of galaxies. Burstein et al. (1984) find that the M31 globular clusters are distinguished spectroscopically from their Milky Way counterparts. Relative to globular clusters in the Milky Way, M31 globular clusters show enhanced CN, and metal-rich M31 globular clusters have stronger $H\beta$ than their Galactic counterparts; findings confirmed by Tripicco (1989). The $H\beta$ enhancement is often attributed to the M31 clusters being a few Gyr younger than the old, Galactic metal-rich clusters, but hot blue horizontal branch stars or blue stragglers may also be responsible.

Subsequent studies in the infrared (Davidge 1990) and in the ultraviolet (Ponder et al. 1998) confirm the CN enhancement. As the ultraviolet light arises mostly from the main sequence while the infrared light is dominated by giants, it becomes difficult to escape the hypothesis that nitrogen enhancement is responsible and caused by a valid difference in abundance. Challenges in both the ultraviolet and infrared observations keep the sample sizes small. Even so, the CN enhancement is confirmed in every case. The metal-rich clusters of the Milky Way bulge have strong CN relative to the bulge field population (Puzia et al. 2002). Chiappini, Romano, & Matteucci (2003) review the evolution of CNO elements in galaxies, and conclude that the evidence favors intermediate-mass stars ($4–9\ M_\odot$) as being the dominant source of nitrogen, not massive stars. One could argue that the low N/O ratios observed in a ultraviolet-selected sample of starburst galaxies (Contini et al. 2002) support this assertion, as do abundances in damped Lyα systems at high redshift (Lu, Sargent, & Barlow 1998). If the nitrogen excess is intrinsic, one may speculate that M31 globular clusters may have enriched more slowly, or that much of their star formation was extended on Gyr time scales (in sharp contrast to Galactic globular clusters, which have no measurable age dispersion). Perhaps the burst responsible for cluster formation in M31 may have been delayed by ~ 1 Gyr relative to other populations, a difference large enough to affect nucleosynthesis, yet small enough to be undetectable using the AGB, horizontal branch morphology, etc.

The emerging complexity of the M31 halo in both structure and, evidently, age, as well as the claims by Morrison et al. (2004) for a kinematic thin disk of globular clusters, would appear to be contradicted by the well-behaved color-magnitude diagrams of the clusters. On the other hand, the spectroscopic differences between the M31 clusters and those of the Galaxy, and between M31 and the integrated light of elliptical galaxies, argues for some different feature in their chemical evolution. Efforts to measure composition of these systems from high-resolution integrated spectra (Bernstein & McWilliam 2002) would be well worth pursuing.

18.4   The Stellar Halo Population

The perception of the M31 halo has changed fundamentally since the heroic imaging survey of Ibata et al. (2001) and Ferguson et al. (2002). Prior to that work, the dominant view of the halo largely remained that of de Vaucouleurs (1958)—that the halo is largely an extension of the central bulge (we return to this issue in §18.8, however). It would now appear that, instead, M31 is a confirmation of the hierarchical assembly of galaxies, and that the Searle-Zinn approach to considering chemical evolution in the halo is correct. In further support of this scenario is the ultra-deep halo imaging using the Advanced Camera for Surveys (ACS) on board *HST* (Brown et al. 2003), which evidently requires the presence of a 6–11 Gyr old, metal-rich population at 11 kpc (see §18.4.2).

Although the dramatic images of the outer halo of M31 show clearly the possibility of a major merger scenario, much remains to be explained that is simply not consistent with that picture. Interactions and their manifestations are interesting and often spectacular. In the Milky Way there is the Sagittarius dwarf spheroidal galaxy (Ibata, Gilmore, & Irwin 1995) and the tidally disrupted globular cluster globular cluster Palomar 5 (Rockosi et al. 2002). Yet, what fraction of the total halo mass is accounted for by such events? We start by considering the halo abundance distribution, distinctly more metal-rich than the Milky Way halo, yet remarkably uniform over a wide range of galactocentric distance in M31.

The pioneering work of Mould & Kristian (1986) was the first to show that the halo of M31 is composed of a metal-rich stellar population[*]. Pritchet & van den Bergh (1987) subsequently discovered RR Lyrae stars in a field 11 kpc distant from the nucleus on the SE minor axis; in that location they found also a metal-rich, globular cluster-like population with [Fe/H]≈ -1, and the first secure indication of a predominantly red horizontal branch (confirmed in all subsequent studies). Further work confirmed their early findings (e.g., Pritchet & van den Bergh 1988; Richer, Crabtree, & Pritchet 1990; Holland, Fahlman, & Richer 1996; Rich et al. 1996a; Durrell, Harris, & Pritchet 2001; Bellazzini et al. 2003). Studies of remote major axis fields (Rich et al. 1996b; Ferguson & Johnson 2001) also find a surprisingly metal-rich field population. Sarajedini & van Duyne (2001) argue that the metal-rich stars are related to a thick disk (§18.4.3 discusses the kinematics of these stars). Figure 18.3 shows how striking the high metallicity of the M31 halo is, from the new study of Bellazzini et al. (2003). We emphasize here that if the halo field population is younger

---

[*] Mould & Kristian's imaging of the halo of M31 was undertaken using the prime focus universal extragalactic instrument (PFUEI; Gunn & Westphal 1981) at the 200-inch telescope; the author of this article was present at the observing run.

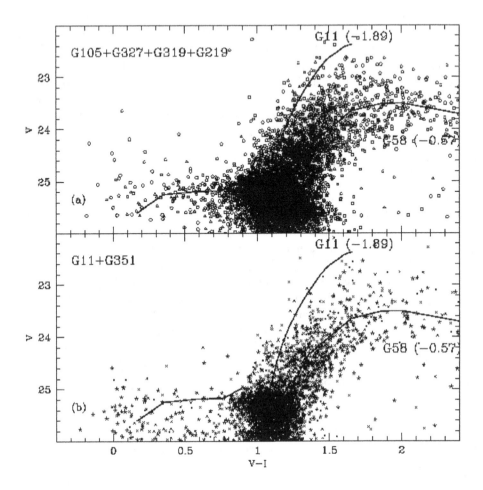

Fig. 18.3. The metal-rich halo of M31 (Bellazzini et al. 2003); photometry from WFC camera frames, where the PC is centered on the globular clusters listed in the two panels. Illustrated are the ridgelines of actual M31 globular clusters with indicated metallicities, as compared to the field halo giant branch. The horizontal branch is that of M68 at the distance of M31. G11 and G351 lie at projected distances > 20 kpc from the M31 nucleus. Halo fields show little change in their RGB color (abundance range) from 5 to 35 kpc from the nucleus. Note that the same analysis performed on NGC 147 (Han et al. 1997) finds no stars redward of the ridgeline defined by 47 Tuc.

than the globular clusters, the metallicity must be 0.1–0.3 dex *higher* than implied in Figure 18.3.

Durrell et al. (2001) find that a halo field 20 kpc distant on the minor axis has a peak

[Fe/H] = −0.8, a result that holds also for a field at 30 kpc (Durrell, Harris, & Pritchet 2004). The abundance distribution (peak at the abundance of 47 Tuc) resembles closely that found by Rich et al. (1996b) in a 34 kpc distant major axis field near G1. Durrell et al. (2001, 2004) determine photometric abundances and assume a globular cluster age population. The abundance distribution fits the standard simple model of chemical evolution, and they argue for a wind-driven chemical evolution model.

How widespread and uniform is the metal-rich population? At first glance, it might be a concern that the stars in the dwarf spheroidal galaxies of Andromeda range from −2 to −1.3, far more metal-poor than is found in the halo. However, (looking ahead to Fig. 18.9) the dwarfs are so distant from the metal-rich halo that they are far more likely to have evolved as self-contained systems, their chemical evolution being dominated by the simple model with metals removed via winds (Hartwick 1976). In contrast to the Milky Way dwarfs, these systems are so distant that it is difficult to consider them as belonging to the inner portions of the halo we address here. However, we can state confidently that accretion of systems similar in metallicity to the Andromeda dwarfs could not account for the bulk of the halo within 35 kpc, as the halo is nearly an order of magnitude too metal-rich. We will consider their chemical evolution in §18.5.

The recent study of Bellazzini et al. (2003) reports on the analysis of WFPC2 imagery of 16 fields ranging from 4.5 to 34 kpc from the nucleus (Fig. 18.4). While the WFPC2 field is tiny, the similarity of filters and exposure times, and clean detection of the red giant branch (RGB) permits us to explore abundance variations (from photometry of the RGB) as a function of position in the M31 halo. The imaging samples a wide range of environments in the galaxy, from fields very clearly superposed on the disk to those adjacent to globular clusters, 30 kpc or more in projected distance (including the G1 field, the most distant at 34 kpc projected distance). The principal result is that there is very little variation in the abundance distribution from field to field, with the exception that fields < 5 kpc from the nucleus reach higher metallicity. It would appear surprising that regardless of whether the field population is superposed on the bright disk and the color-magnitude diagram includes blue stars, or whether the field lies 20 kpc from the nucleus of M31, the color-magnitude diagram is largely the same: a red descending giant branch with a dominant red horizontal branch. However, all fields also have blue horizontal branch stars, which comprise about 15% of the population (from which they infer the fraction of metal-poor stars to be ∼15%).

One may ask (referring back to Fig. 18.1) whether any difference can be discerned as a function of environment: do stars in the disk, interaction streams, or undisturbed halo show any differences? Figure 18.5 shows that the disk is both more metal-rich and has a slightly wider abundance distribution than is found in the halo or the disturbed regions. It would appear that the material in these streamers is different from that in the disk of M31, making it less likely that the streams were formed as a consequence of some ancient interaction with the disk.

Metallicities for the field stars have been derived by interpolating between the giant branches of Galactic globular clusters; as mentioned earlier, if the field stars are systematically young, the abundances could be up to 0.3 dex higher. Regardless of whether the metallicities are derived from the Carretta & Gratton (1997) system or are based on the Zinn & West (1984) scale, the distribution has a peak near −0.5 dex with a tail toward the metal-poor end, in the reverse sense of the abundance distribution of the M31 globular cluster population.

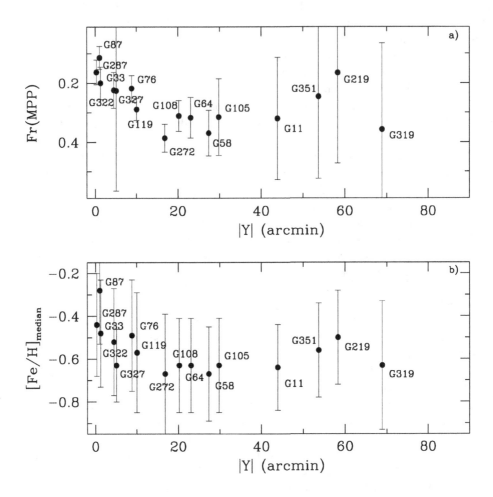

Fig. 18.4. Search for an abundance gradient in the halo of M31 (Bellazzini et al. 2003). The abundances are derived by interpolating between the RGBs of old (12–13 Gyr) Galactic globular clusters, and would be ~ 0.3 dex higher if the population were younger by 4 Gyr. *Upper panel* shows the fraction of stars with [Fe/H]< -0.8 as a function of absolute distance from the major axis in arcmin (note that 10 arcmin = 2.5 kpc). The *lower panel* plots the median abundance. The abundance distributions in M31 generally peak near -0.6 dex and have a tail toward low metallicity. There is little evidence for a gradient in the true halo, outside of 5 kpc, where the disk and inner bulge may contribute significantly. Notice how metal-rich the trend remains; this extends to 34 kpc, the most distant field measured (not shown on these plots), near the globular cluster G1. (Adapted from Bellazzini et al. 2003.)

18.4.1   *Spectroscopy in the Halo*

Red giant branch tip stars in M31 have $I > 20.5$ mag, faint enough that it is difficult to measure abundances, even with an 8–10 m-class telescope. The best choice for measurement of metallicity is the Ca infrared triplet, although that region of the spectrum suffers from strong night-sky OH lines, which play further havoc with velocity measurements.

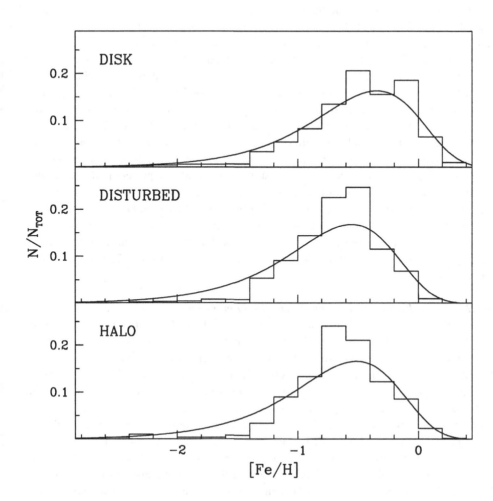

Fig. 18.5. Abundance distributions for giants in M31 fields (derived as in Fig. 18.4). The data are from Bellazzini et al. (2003); plot courtesy of L. Federici. This figure shows that it is unlikely that the interaction streams are comprised of material from the M31 disk. The sample is divided into fields dominated by the disk, interaction regions (see Fig. 18.1 and Ferguson et al. 2002), and the clearly undisturbed outer halo (see Bellazzini et al. 2003 for an illustration and a detailed description). For the disk, the median [Fe/H] = -0.46 ± 0.28 (N = 34091), while for the disturbed regions [Fe/H] = -0.64 ± 0.22 (N = 6220), and for the halo fields [Fe/H] = -0.62 ± 0.23 (N = 420). The solid curve is the fit to the simple model of chemical evolution, with the yield $\log(Z/Z_\odot)$ as a free parameter, normalized to unity. Fields superposed on the high-surface brightness disk appear to be more metal-rich and follow more closely the simple model than those in the halo or interaction streams, which resemble more closely each other.

Given the 500 km s^{-1} range of stellar velocities in M31, one or more lines of the triplet frequently coincide with night-sky emission; this compromises any spectroscopic abundance measure. Stars near zero radial velocity overlap with the Milky Way halo velocities. Reitzel

& Guhathakurta (2002) report radial velocities and abundances for 99 M31 halo candidates in a field 19 kpc from the nucleus on the SW minor axis. They find that nearly all of the stars in their sample are < −1 dex, whereas in a nearby field adjacent to the globular cluster G319, Bellazzini et al. (2003) find the peak of the (photometric) metallicity distribution to lie at −0.5 dex. If the population in this area is intermediate age, the derived metallicity would increase by 0.2–0.3 dex. As is illustrated in Figure 18.3, the RGB is observed to descend, or decline in luminosity at the RGB tip. Even though the bolometric luminosity of a red giant increases, when the metallicity exceeds ∼ −1 dex, molecular blanketing affects the V and, to a lesser extent, the I band. This effect can cause the tip of the RGB to become as faint in V as the red clump, and its widespread presence in the halo manifestly demonstrates that the halo population is metal-rich—or, at least, that TiO bands are present in the red giants, which could also be indicative of α element enhancement. Nonetheless, this observation is difficult to reconcile with the abundance distributions of Reitzel & Guhathakurta (2002), who find few stars more metal-rich than −1 dex. They argue that spectroscopic samples tend to successfully measure velocities and line strengths for only the brightest stars, and that these tip stars are metal-poor (the redder, more metal-rich stars have $V > 23$ mag in the M31 halo).

18.4.2   Age of Remote Fields from Deep HST Photometry

Recently, two *HST* imaging campaigns, in which this author is involved, have reached deeper into the halo population than prior studies. Brown et al. (2003) used a 120 *HST* orbit integration to solidly reach the old main sequence turn-off in a field 11 kpc from the nucleus on the SE minor axis (Fig. 18.6). In the first instance, some familiar features of the expected globular cluster-like population are present: the blue horizontal branch, a red clump, and a large turn-off population fainter than the horizontal branch by the canonical 3.5 mag. If the color-magnitude diagram were instead shallow and reached only just past the horizontal branch, it would be reasonable to conclude that the population in question has a component of old, metal-poor stars (based on the extended blue horizontal branch) and likely a fraction of old, metal-rich stars (based on the red clump and RGB). However, in that case we would be misled by the data, and this should be viewed as a cautionary example for those cases where the turn-off cannot be reached. In fact, there is evidently an age range in this population. To explore further the age, we examine the distribution of stars along a diagonal cut that crosses the subgiant branch (Fig. 18.7). It is not possible to fit the number distribution of stars along this cut, nor to achieve a fit to the general appearance of the color-magnitude diagram, using only old (globular cluster age) model populations. In order to avoid uncertainties with color transformations and from other effects, we use the identical filter/detector system to image well-studied Galactic globular clusters spanning a wide range in metallicity. The best fit to the subgiant branch retains a fraction of old (10–13 Gyr), metal-poor stars, but 55% of the population must be 6–11 Gyr old with [Fe/H]> −0.5 and have a continuous star formation history (roughly half of those stars are younger than 8 Gyr).

In a pioneering ground-based study, Pritchet & van den Bergh (1987) find 30 RR Lyraes in a field 9 kpc from the nucleus on the SE minor axis, 2.2 kpc closer than the deep *HST* imaging field discussed earlier. The high specific frequency of RR Lyrae led them to propose that the M31 halo has a well-developed horizontal branch; the specific frequency of RR Lyraes is higher than found in variable-rich Galactic globular clusters. If a majority of the

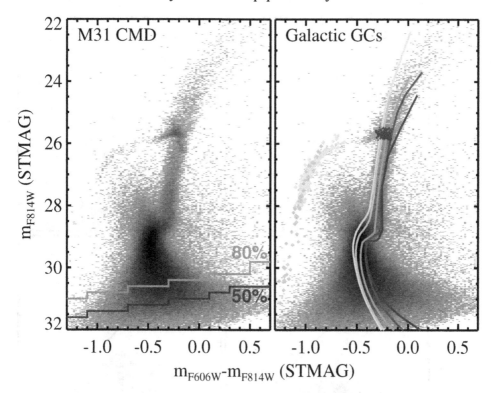

Fig. 18.6. Deep photometry in the M31 halo from the ACS imaging campaign of Brown et al. (2003) presented in the form of a Hess diagram (density plot). *Left panel:* The color-magnitude diagram derived from point-spread function fitting of the ACS images, with completeness limits indicated. Notice the blue horizontal branch (an indicator of great age) and that the old main sequence turn-off is clearly reached. *Right panel:* Ridgelines and horizontal branch loci of Galactic globular clusters, also observed with the same filter and camera as the halo field, are superimposed on the color-magnitude diagram. The clusters range from M92 ([Fe/H] = −2) to NGC 6528 ([Fe/H] = −0.2). The best available determinations in distance modulus and reddening are used to place the cluster loci, except for that of NGC 6528, which is force fit to the red horizontal branch of the M31 field. Note that while the horizontal branch loci of the Galactic clusters coincide with the data, *the subgiant branches of the more metal-rich (redder) clusters lie systematically fainter than the turn-off in the halo field.* (Adapted from Brown et al. 2003.)

more distant field at 11 kpc is composed of a metal-rich, intermediate-age population, then the abundance of RR Lyraes in the remaining old, metal-poor population is extraordinarily large, compared to Galactic globular clusters. The presence of the RR Lyraes requires that a substantial fraction of the population be old (globular cluster age). RR Lyraes are considered to indicate the presence of the oldest stellar populations, and it is difficult to find room for the evidently intermediate-age, metal-rich component.

18.4.3 *A Remote Field 34 kpc Distant on the Major Axis*
Although reaching only a fraction as deep as the halo field, a second deep field (and the second deepest analyzed to date) lies near the luminous globular cluster G1. Also nearby

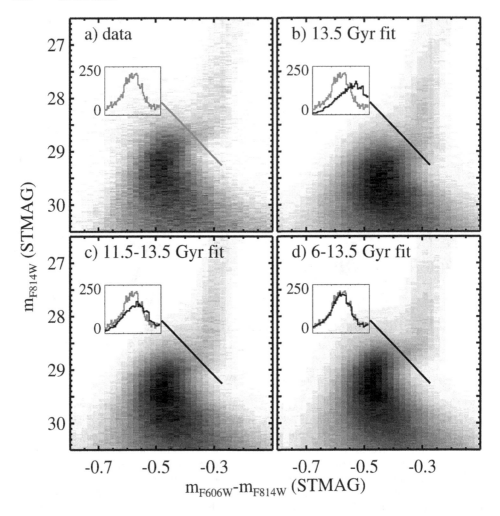

Fig. 18.7. Modeling of the turn-off population in M31; details given in Brown et al. (2003) and Brown (2004). These models are the basis for the argument that the M31 halo must contain a significant population of 6–11 Gyr old, metal-rich stars. The subgiant branch is reasonably complete and has the most straightforward sensitivity to age. (*a*) Histogram resulting from counts of stars along the illustrated line, for the data. (*b*) Fit to the color-magnitude diagram using only 13.5 Gyr stars with a wide range in metallicity. While able to fit the width of the RGB, the subgiant branch fit fails. (*c*) When the age range is broadened to 11.5–13.5 Gyr with a wide permitted abundance range, the fit still fails. (*d*) The fit to the age-sensitive subgiant branch succeeds when the model has 56% of the population with age 6–11 Gyr and −0.5 < [Fe/H] < 0.0, and retains a 30% component of 13 Gyr metal-poor stars that fit the blue horizontal branch. (Adapted from Brown et al. 2003.)

(Fig. 18.1) is a distinct plume in the irregular halo distribution, suggested by Ferguson et al. (2002) as possibly corresponding to the remnants of a low-luminosity galaxy that had G1 as its nucleus. Both Rich et al. (1996b) and Meylan et al. (2001) suggest that this massive ($\sim 10^7 M_\odot$) cluster could have once been the nucleus of a dwarf elliptical galaxy.

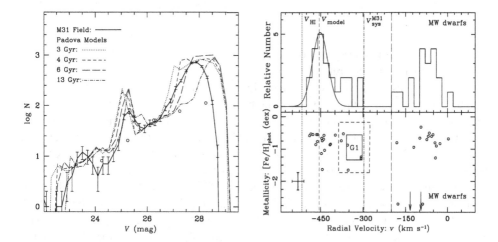

Fig. 18.8. *Left:* Deep luminosity function in the field of the M31 globular cluster G1, 34 kpc from the nucleus on the SW major axis (Rich et al. 2004b). Model luminosity functions constructed from the Padova isochrones show that a pure old globular cluster age population fails to fit the counts. The Padova luminosity functions include a completeness correction. The RGB slope indicates the field is metal-rich, at ~ -0.7 dex. *Right:* Radial velocities and photometric abundances for K giants in the same field near G1 (Reitzel et al. 2004). Half the stars are consistent with foreground Milky Way membership, while a handful are plausibly at the velocity of G1. The majority of stars lie near the velocity that the disk of M31 would have at that location, as inferred from the global velocity field of neutral hydrogen in M31. The velocity of the nearest detected neutral hydrogen, some 2 kpc distant, is indicated in the upper panel.

Support for this idea gained momentum both from the abundance spread found by both of the above studies, and by the strong hint of halo structure near the position of G1. Finally, the kinematic detection of a $10^4 M_\odot$ black hole in the core of G1 (Gebhardt, Rich, & Ho 2002) might appear most consistent with a galactic nucleus classification.

Fortunately, we have Keck spectroscopy in the field, and we are able to explore both the kinematics of the field as well as the deep photometry. Figure 18.8 shows our deep luminosity function in the vicinity of G1. Although the stellar population appears to be old in this vicinity, it is clearly not as old as a globular cluster age population. The presence of carbon stars some 32 kpc distant on the major axis (Brewer, Richer, & Crabtree 1995) would confirm the idea that the population in this field has a substantial disk component.

Keck radial velocities near G1 (Fig. 18.8) are a surprise. The majority of stars appear offset from the G1 radial velocity by > 100 km s^{-1}. We find that most of the stars in the field have velocities that would be consistent with the rotation of the disk, rather than with either the velocity of G1 or the systemic velocity of M31. While our velocities do not *prove* that the stars follow the kinematics of the disk, they do show that very few stars in this region are associated with G1. The star counts of Meylan et al. (2001) follow a King profile to a distance of 200″ from the center, but no hint of a deviation from a King profile (as might be expected if there is an associated dwarf spheroidal galaxy) is seen. It now appears as if the field stars have more to do with the fossil disk of M31 than with any remnant dwarf galaxy, or even the halo.

18.5   The Distant Companions

M31 has a growing retinue of companions, which include NGC 185, NGC 147, and the Andromeda dwarf spheroidal galaxies; Figure 18.9 shows that they all lie well outside even the outer portions of the surveyed stellar halo. The Andromeda dwarfs are significantly more metal-poor than the halo (Fig. 18.10), as determined from their blue RGBs, blue horizontal branches, and RR Lyrae stars (e.g., Mould & Kristian 1990; Da Costa, Armandroff, & Caldwell 2002; Pritzl et al. 2002). Spectroscopy of giants in And II (Côtè, Oke, & Cohen 1999) finds [Fe/H] = -1.47 ± 0.19 and a 0.35 dex spread. Even NGC 147, which is more metal-rich than the Andromeda dwarfs (Han et al. 1997), has a RGB that is clearly bounded on the red side by the locus of 47 Tuc, at -0.7 dex. The globular cluster systems of NGC 147, 185, and 205 are also metal-poor (Da Costa & Mould 1988). Their abundances range from -2 to -1.3 dex and, with the exception of Hubble V in NGC 205 (for which star formation is documented), the spectra are consistent with the clusters all being as old as their Galactic counterparts. The metallicity-luminosity and metallicity-surface brightness relationships for dwarf spheroidal galaxies have been considered in various papers. The plots given by Saviane, Held, & Piotto (1996) show that the Andromeda dwarfs, NGC 147, 185, and 205 fall on the same relationship as defined by Galactic dwarf spheroidal galaxies. The general trend and relatively low scatter suggest that in these systems the mean metallicity is regulated by the flow of metal-enriched winds out of the protogalaxy's potential well. More massive galaxies reach higher metallicities by retaining more gas. The chemical evolution for these galaxies would be described by Pagel & Patchett (1975), Hartwick (1976), and Mould (1984). The abundance distribution has the same shape as the simple model (broad abundance range), but the mean is lower because the wind-driven loss of metals reduces the effective yield. Dekel & Woo (2003) present a more current consideration of these issues in dwarf galaxies and are able to reproduce the observed metallicity-luminosity-surface brightness correlations as the result of supernova-driven winds bound by a dark matter potential, with metallicity regulated by the loss of gas.

The star formation history of the companions appears to be more complicated than that of the globular clusters. While it has been clearly demonstrated (e.g., NGC 205; Lee 1996) that the populations of the M31 companions have age ranges, RR Lyrae stars and blue horizontal branches have been found in every Andromeda dwarf for which the appropriate data have been taken (see Pritzl et al. 2002), as well as in NGC 147 (Saha, Hoessel, & Mossman 1990), NGC 185 (Saha & Hoessel 1990), and NGC 205 (Saha, Hoessel, & Krist 1992). The presence of RR Lyrae stars in these systems requires that they harbor at least a significant fraction of old and metal-poor stars, though Da Costa et al. (2002) suggest that the fraction of intermediate-age and young stars in these galaxies increases with distance from M31. Blitz & Robishaw (2000) found neutral hydrogen associated with And III and And V. Thilker et al. (2004; see Fig. 18.11) found numerous high-velocity clouds of neutral hydrogen in the vicinity of M31, which could provide fuel for ongoing star formation. The infall of gas has doubtless been important in chemical evolution and the building of the disk (Blitz et al. 1999).

Finally, there is the case of M32, a galaxy that has been suggested to have been tidally stripped from a formerly more massive system (Faber 1973; Choi, Guhathakurta, & Johnston

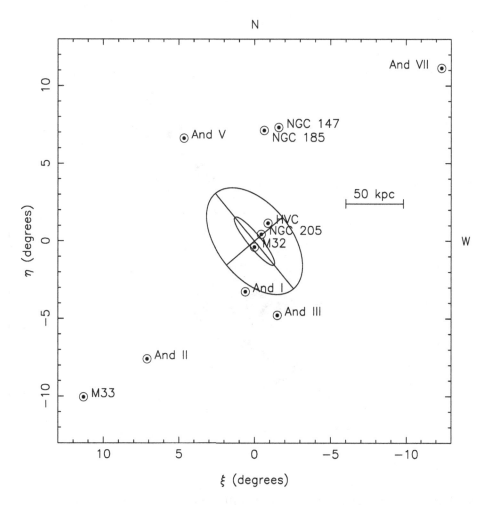

Fig. 18.9. Projected distribution of the satellite companions and Local Group galaxies near-est to M31 (from Ferguson et al. 2002). The figure also shows a massive cloud of neutral hydrogen (labeled HVC) detected by Davies (1975). The ellipse circumscribing M31 and the limit of the optical disk is the outer bound of the Ferguson et al. (2002) survey; the elliptical distribution of halo material lies within that bound. The great distance of the com-panions of M31 suggests why these systems appear to have evolved following wind-driven chemical evolution with some modification by gas infall (Blitz & Robishaw 2000 find H I in And III and And V). The M31 halo population remains metal-rich at 34 kpc on the major axis (Rich et al. 1996b) and to 30 kpc on the minor axis (Durrell et al. 2004), where both the mean metallicity near [Fe/H] = −0.7 and the dispersion are confirmed.

2002) and for which there is evidence of tidal tails. The stellar population in M32 is metal-rich; the RGB descends just as it does for the M31 metal-rich halo population, and the best-fit isochrone is at roughly −0.4 dex (Grillmair et al. 1996), while Trager et al. (2000) and Beasley et al. (2004) find solar [Fe/H] and [α/Fe]. Early studies found bright AGB stars in the infrared (Elston & Silva 1992; Freedman 1992). However, the infrared luminosity

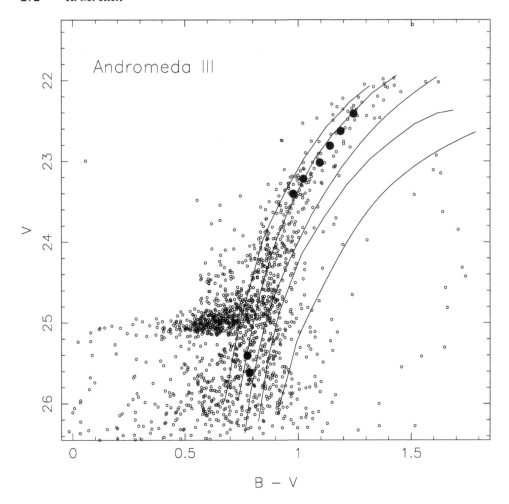

Fig. 18.10. Color-magnitude diagram of And III (Da Costa et al. 2002), which shows the metal-poor RGB and well-populated horizontal branch that typifies the dwarf spheroidal galaxies of M31. Variable stars are omitted in this plot. Empirical giant branches are from M68 ([Fe/H] = −2.09), M55 ([Fe/H] = −1.82), NGC 6752([Fe/H] = −1.54), NGC 362([Fe/H] = −1.28), and 47 Tuc ([Fe/H] = −0.71). The ridgeline of the typical M31 halo field is on average more metal-rich than that of 47 Tuc. The somewhat red distribution of the horizontal branch likely reflects that a significant fraction of the population is ∼ 10 Gyr old (i.e. intermediate age). Notice the wide spread in the RGB color, which is not typical for globular clusters.

function of Davidge et al. (2000) is not brighter than that in the M31 bulge (Stephens et al. 2003); yet, both the Trager and Beasley studies, mentioned above, fit M32 at 3 Gyr, significantly younger than M31. Discovery of RR Lyrae stars in M32 would confirm the presence of at least an underlying old population, and deep photometry might quantify the age spread; *HST* observations aimed at these programs have been obtained or are in progress.

Applied to the case of M31, it would appear that the halo field population within 20 kpc is not dominated by the debris of dwarf galaxies, unless they are the debris of a galaxy with

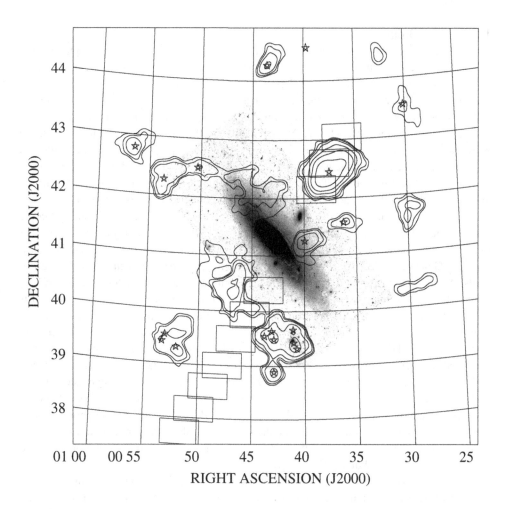

Fig. 18.11. High-velocity clouds of neutral hydrogen near M31, from the new imaging survey of Thilker et al. (2004). Emission from the M31 disk and NGC 205 has been masked and is not shown. A star symbol marks the position of each discrete cloud, and the series of rectangles shows the location of the stream (Ferguson et al. 2002). Contours are at 0.5, 1, 2, 10, and 20×10^{18} cm^{-2}. The large amounts of neutral hydrogen in the vicinity of M31 may have fueled the formation of intermediate-age populations that are detected in the outermost of the Andromeda dwarfs (even though their locations are well outside the boundaries of this plot).

$M_V < -18$ mag. Supporters of a "spaghetti" halo scenario for M31 might be disappointed, unless that spaghetti were to come from one plate, perhaps M32 or a galaxy as luminous as it. The halo of the Milky Way also does not appear to consist mostly of stars with chemistry resembling present-day dwarf spheroidal galaxies, as Shetrone, Côté, & Sargent (2001) and

Tolstoy et al. (2003) find that stars in the dwarf spheroidal galaxies are depressed relative to the halo in α elements.

18.6   The Bulge

"However, Morgan's spectra, taken at McDonald Observatory with a quartz spectrograph, showed that the cyanogen bands in the ultraviolet are very strong in the integrated spectrum—quite as strong as in normal giants" (Baade 1963; p. 254).

Morgan's early spectroscopy linked the nucleus of M31 to the giant ellipticals and gave strong indication of both large age and high metallicity. The pioneering photometry (U through K) of Sandage, Becklin, & Neugebauer (1969) showed the nuclear region to be as red as the most luminous Virgo ellipticals. Numerous spectroscopic studies followed and confirmed these fundamental results. Bica, Alloin, & Schmidt (1990) used a library of star clusters and modern CCD spectroscopy to argue that the nucleus and bulge are old and metal-rich, and that an abundance spread must be present (they also found a 10%–20% intermediate-age component). Davidge (1997) undertook a comprehensive spectroscopic assay of the bulge, finding strong CN lines (as in the M31 clusters) and spectral indices that place the bulge and clusters of M31 apart from the Milky Way globular clusters. Worthey, Faber, & González (1992) found a high Mg_2 Lick index, comparable with that in giant ellipticals. As is also the case for the giant elliptical population, the M31 nucleus has strong Mg_2 relative to iron, although Davidge (1997) argued that this declines toward the nucleus. (It is noteworthy that in the same plot of iron vs. Mg_2, M32 lies below the giant elliptical population on the solar composition line.) The population synthesis of Lick indices by Trager et al. (2000) gives [Fe/H] = 0.21 ± 0.05 and [E/H] = 0.19 ± 0.02, where [E/H] reflects elements produced in massive-star supernovae. Beasley et al. (2004) used new Keck/LRIS spectra and fit their indices to the Thomas et al. (2003) models to find the same α enhancement but higher [Fe/H] = 0.5 ± 0.1. Both studies, however, find a young age for the M31 bulge (6–7 Gyr), which is not consistent with the AGB properties described below. The photometry of red giants ~ 1 kpc from the nucleus is consistent with solar abundance with a dispersion (like the halo). The range likely reflects the simple model of chemical evolution, as is the case for the Milky Way bulge (Rich 1990).

Beasley (2004) models the abundance of the M31 and Galactic bulges using the Lick indices for Milky Way bulge fields from Puzia et al. (2002) and his own measurements in M31. The total metallicity [Z/H] = -0.10 ± 0.2 for the Galactic bulge, and [Z/H] = 0.48 ± 0.1 for the M31 bulge. The Type II supernova "α" enhancement is 0.3 ± 0.3 for the Galactic bulge, and 0.2 ± 0.1 for the Milky Way bulge. The mild α enhancement in the Milky Way bulge is seen by Maraston et al. (2003) and is found in the spectra of individual stars (McWilliam & Rich 1994).

For an intense starburst and rapid enrichment, chemical evolution models have long predicted enhanced α elements from the Type II supernova contribution (Matteucci, Donatella, & Molaro 1999). It is the intensity, not the absolute timing, of the starburst that controls the composition (i.e., the composition does not rule out an intermediate-age burst).

In the late 1980s the author noted that the brightest giants in the Galactic bulge are both luminous (reaching $M_K < -8$ mag) and rare (their lifetimes are $\sim 10^5$ yr). It occurred to the author that it might be feasible to image individual AGB stars in the bulge of M31 using ground-based infrared arrays. Rich et al. (1989) obtained optical spectra of individual, resolved red giants in M31, showing that they are ordinary red giants (not carbon stars).

Rich & Mould (1991) subsequently imaged a field in the M31 bulge roughly 500 pc from the nucleus, comparable to the distance of the well-studied Baade's Window field of the Galactic bulge from the Galactic nucleus. These observations created quite a stir at the time. Rich & Mould (1991) claimed that they might have detected more bright giants than are present in the Galactic bulge from the Frogel & Whitford (1987) luminosity function. Renzini (1998) simulated the tendency for crowding to produce spurious stars at the bright end. Higher-resolution imaging in the form of adaptive optics shows that M32 (Davidge et al. 2000) and the bulge of M31 (Davidge 2001) have AGB luminosity functions agreeing with that of the Milky Way bulge, hence consistent with an old age. The comprehensive *HST*/NICMOS study of the M31 bulge by Stephens et al. (2003) proves this convincingly. Their imaging finds that the RGB in M31 extends to $M_{bol} = -5$ mag, but that the shape of its luminosity function is similar to the Frogel & Whitford bulge luminosity function. However, Stephens et al. (2003) find no variation in the bright end of the luminosity function, regardless of whether disk- or bulge-dominated fields are considered. Frogel & Elias (1988) note that long-period variables in globular clusters reach $M_{bol} < -4.5$ mag, brighter than the He core flash limit. Guarnieri et al. (1998) find an AGB star at $M_{bol} = -5$ mag (long thought to be attained only by intermediate-age stars) in the old, metal-rich cluster NGC 6553. Present observations in the M31 bulge can only rule out a significant population of few Gyr old stars. Note that Rejkuba et al. (2004) find the brightest AGB stars in the halo of NGC 5128 to be 0.5 mag brighter than in the Milky Way bulge. The period distribution of the Miras in NGC 5128 is weighted more toward long periods (> 500 days) than in the Galactic bulge; even at high metallicity, there comes a point when intermediate age has an effect on the AGB. It is therefore reasonable to conclude that the similarity of the M31 bulge AGB luminosity function with that of the Milky Way bulge strongly argues for M31 being old, like the Milky Way bulge.

18.6.1 Abundance Spread and Gradient

Present observations are all consistent with the bulge of M31 having a population of *metal-rich* stars. The *HST*/WFPC2 imaging of Jablonka et al. (1999) resolves the red giants in three pure bulge fields 0.8 to 1.6 kpc from the nucleus [also finding no abundance or population gradient in that region, somewhat in contradiction to the $\Delta[\mathrm{Fe}/\mathrm{H}]/\Delta \log r = -0.5$ gradient from Davidge's (1997) spectra]. Figure 2 of Jablonka et al. shows a wide RGB similar to halo fields but extended toward the metal-rich side. However, the blue envelope is steep, as is seen in metal-poor globular clusters, suggesting that the metal-poor side is represented. Planetary nebulae near the bulge are metal-poor (Jacoby & Ciardullo 1999) but generally lie beyond 1 kpc. A new abundance distribution (derived from $V - K$ colors) for the Milky Way bulge (Zoccali et al. 2003) has fewer metal-poor stars than predicted by the simple model of chemical evolution. However, we have known since the work of Baade (1951) that the bulge has RR Lyrae stars, and hence at least some old, metal-poor population.

In the case of M31, it would be helpful to have higher spatial resolution optical photometry (optical colors are more sensitive to metals so the presence of metal-poor stars is more easily discerned) or spectroscopy of individual giants. The new Keck infrared spectrograph OSIRIS (Larkin et al. 2003), to be commissioned in 2004, might enable such studies. If the abundance range does lack the metal-poor tail, then the simple model cannot fit the abundance distribution. One might suspect that the first generation of long-lived stars in the protobulge was pre-enriched, perhaps by an earlier generation formed in a starburst so violent that low-mass stars were not formed.

18.7   The Halo of M31 and Extragalactic Halos

As the halo populations of more distant galaxies have come under study, it has become clearer that massive galaxies generally appear to have metal-rich halos. Just as is found in M31, the RGB is metal-rich in the halo of NGC 5128 (Harris, Harris, & Poole 1999), but there is more likelihood of an age range (Rejkuba et al. 2004). Evidence for a merger in NGC 5128 is not surprising, given the presence of many merger signatures in this system. In the case of M31, the presence of a significant intermediate-age population in the halo is more difficult to accept, even in the light of the apparent merger signatures in the Ferguson et al. (2002) study. Is the metallicity of halos controlled by the mass of the most significant merger, the depth of the potential well, or by the star formation history?

In order to address this question, the author has collaborated on an *HST* program to image fields in the halos of nearby spiral galaxies that are close to edge-on, thus having a geometry that permits study of the halo (GO-9086; PI H. Ferguson). We have imaged halo fields of galaxies covering a range in luminosity and have derived photometric metallicities for the stars by interpolating between empirical RGBs of Galactic globular clusters. This method has the potential to be in error if the populations are intermediate age, but the largest offset would be +0.3 dex, not large enough to affect the trend.

Figure 18.12 shows our trend of halo metallicity as a function of parent galaxy luminosity. It is interesting that bulges of galaxies show a similar relationship (Jablonka, Martin, & Arimoto 1996). It is difficult to imagine how such a relationship could result if halos are built, in general, out of the mergers of less luminous systems. Rather, the presence of such a trend suggests that the global potential well of the galaxy regulates the outflow of metals, which might be driven by a starburst (originally the chemical evolution model of Hartwick 1976). It is tempting to find support for such a scenario as well in the metal-rich intergalactic medium seen in X-ray studies of clusters of galaxies.

Perhaps wind regulation of the metallicity via the depth of the potential well is in fact the dominant physical mechanism at work. In that case, the *timing* of the starburst does not enter, only its intensity and duration. If further investigation finds punctuated starburst events in the halo, we could expect these events to be accompanied by α enhancement and winds. However, an intense starburst is required to establish a wind; if the star formation in the halo was continuous from 13 to 6 Gyr ago, that would be a problem for the wind scenario. In that case, one also expects the metal-rich stars to have a solar composition (enough time for Type Ia supernovae to contribute iron).

In seeking to explain the high metallicity of the bulge and halo relative to the Milky Way, one might hope to appeal to the greater mass of M31. But that argument would be wrong. If anything, the Milky Way is the more massive. Evans & Wilkinson (2000) find $12.3^{+18}_{-6} \times 10^{11} M_{\odot}$ for M31 and $19^{+36}_{-17} \times 10^{11} M_{\odot}$ for the Milky Way. In units of $10^{11} M_{\odot}$, Evans et al. (2000) find 7–10, while Côtè et al. (2000b) find a median mass of 7.9 ± 0.5. Gottesman, Hunter, & Boonyasait (2002) argue that the upper bound on the mass of M31 must be $6 \times 10^{11} M_{\odot}$, within a 350 kpc radius, less than that of the Milky Way. If we cannot appeal to the depth of the potential well to explain the difference, Hodge (1992) compares the total luminosity of M31 ($M_B = -21.1 \pm 0.4$ mag) with that of the Milky Way ($M_B = -20.5 \pm 0.5$ mag). Perhaps it is the mass in baryons, which controls the intensity of the

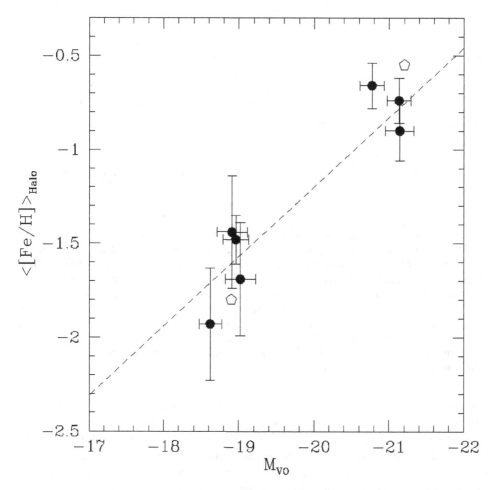

Fig. 18.12. Mean metallicity of the halo population at approximately 10 kpc projected distance from the nucleus, as a function of total luminosity of the parent galaxy. This plot is derived from measurements using WFPC2 onboard *HST* (see text; Ferguson et al. 2004). The pentagons indicate M33 (low metallicity) and M31, respectively. The trend suggests the presence of a luminosity-metallicity relationship for halo populations, as is found also in elliptical galaxies and bulges. The trend for more luminous systems to be more metal-rich would be consistent with the potential well regulating the loss of metals via enriched winds. A scenario in which halos are built from the merger of low-mass galaxies would not successfully produce such a relationship.

starburst, that drives the chemical evolution and may explain partially the high metallicity in the distant halo. The large study of Kauffmann et al. (2003) from the SDSS database shows that galaxies with stellar mass in excess of $3 \times 10^{10} M_{\odot}$ have old, metal-rich populations that are likely to have been in place at early times. The picture of the early wind enriching the M31 halo with metals is attractive, except for the Brown et al. (2004) finding of the age range. A consistent picture of chemical evolution remains beyond reach.

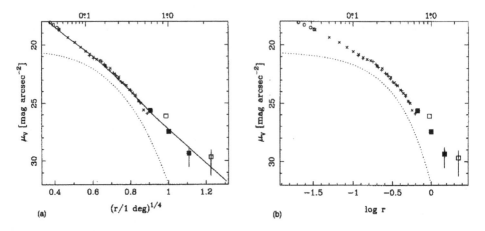

Fig. 18.13. The halo of M31 has a regular surface brightness profile on a scale of 10 kpc (0.75°), even if there are clear signs of a past merger (Pritchet & van den Bergh 1994). We also know that the abundance distribution from 20 to 30 kpc does not change (Bellazzini et al. 2003; Durrell et al. 2004). The *left panel* shows the radial profile versus $r^{1/4}$, hinting that the halo may be an extension of the bulge/spheroid. The *right panel* shows the profile vs. $\log r$. Plotted symbols are from the following sources: circles, Kent (1983); crosses, Walterbos & Kennicutt (1987); open and solid squares, Pritchet & van den Bergh (1994). The open squares are off-minor axis points, and provided the first hint that the halo is flattened. The dotted line shows the disk model of Walterbos & Kennicutt (1988), while the solid line is the sum of their disk plus spheroid model.

Figure 18.13 reminds us that earlier studies (Pritchet & van den Bergh 1994) find regularity on scales of 10–20 kpc in the surface brightness distribution. As we noted earlier, Durrell et al. (2001, 2004) find no change in the abundance distribution from 20–30 kpc. Are the giant stream and apparent flattened structure in Figure 18.1 (Ferguson et al. 2002) unduly focusing our attention to the merger picture as opposed to *in situ* starbursts? One might speculate that the structure formed from the metal-enriched gas that was expelled on Mpc scales from the galaxy by the bulge-forming primordial starburst (Adelberger et al. 2003) and then cooled (and probably self-enriched), forming the flattened halo. Large programs aimed at measuring the stellar dynamics of the M31 halo structures could lead toward some answers.

18.8   Summary

Since the days of Baade's intensive study of M31 from Mt. Wilson, the galaxy that provided the template for the stellar populations concept has proven to be incredibly complicated. The chemical evolution of the distant satellites of M31 appears to have been regulated by starburst-driven winds confined by a potential well. The globular clusters of M31 resemble in some broad way the system of the Milky Way, as opposed to that of, say, the Magellanic Clouds. The mean abundance of the M31 halo is broadly where one expects it should be for a galaxy of the luminosity of M31.

There are signs of great complexity in M31, however. A large fraction of the globular clusters may in fact be divided into kinematic subgroups, one of which may coincide with the thin disk. Although the color-magnitude diagrams of M31 globular clusters behave as expected, they evidently have a puzzling excess of nitrogen and are otherwise distinct in many line-strength trends from the well-studied Milky Way globular clusters. The halo has been revealed to have clear indications of interactions and mergers. Direct measurement of the old main sequence turn-off in a field 11 kpc from the nucleus, where one would expect a pure globular cluster-age population, instead can be understood only if half the stars are metal-rich and range in age from 6–11 Gyr.

In this modern day, the study of M31 once again has an important lesson to teach us, which is a caution against trying to draw general conclusions about the properties of galaxies far more distant, for which we have a great deal less information than is available for M31. After struggling to understand the wealth of information available presently on M31, I can only admire the prescient insight of Walter Baade, as I conclude with his words: "We must conclude, then, that in the central region of the Andromeda Nebula we have a metal poor Population II, which reaches -3^m for the brightest stars, and that underlying it there is a very much denser sheet of old stars, probably something like those in M67 or NGC 6752. We can be certain that these are enriched stars, because the cyanogen bands are strong, and so the metal/hydrogen ratio is very much closer to what we observe in the sun and in the present interstellar medium than to what is observed for Population II. And the process of enrichment has taken very little time. After the first generation of stars has been formed, we can hardly speak of a 'generation,' because the enrichment takes place so soon, and there is probably very little time difference. So the CN giants that contribute most of the light in the nuclear region of the Nebula must also be called old stars; they are not young."—Baade (1963, p. 256).

Acknowledgements. The author is grateful to the referee, Pat Durrell, for many valuable insights and suggestions and for his detailed reading of the manuscript. The author is also indebted to Carla Cacciari and Tom Brown for their detailed reading of an early draft. Additionally, Raja Guhathakurta and David Reitzel made many useful comments and suggestions. L. Federici prepared Figure 18.5 and provided a revised version of Figure 18.4; David Thilker provided Figure 18.11 in advance of publication. Mike Beasley kindly provided a comparison of the M31 and Milky Way bulge Lick indices using his models. The author thanks Tom Brown for Figures 18.6 and 18.7. The author acknowledges financial support from NSF grant AST-0307931.

References

Adelberger, K. L., Steidel, C. C., Shapley, A. E., & Pettini, M. 2003, ApJ, 594, 45
Ajhar, E. A., Grillmair, C. J., Lauer, T. R., Baum, W. A., Faber, S. M., Holtzman, J. A., Lynds, C. R., & O'Neil, E. J., Jr. 1996, AJ, 111, 1110
Baade, W. 1944, ApJ, 141, 45
——. 1951, Pub. Obs. Univ. Mich., 10, 7
——. 1963, in Evolution of Stars and Galaxies, ed. C. Payne Gaposchkin (Cambridge: Harvard Univ. Press), 263
Barmby, P., Huchra, J. P., Brodie, J., Forbes, D. A., Grillmair, C., & Schroder, L. 2000, AJ, 119, 727
Beasley, M. A. 2004, private communication
Beasley, M. A., Brodie, J. P., Forbes, D. A., Barmby, P., & Huchra, J. P. 2004, ApJ, submitted
Bellazzini, M., Cacciari, C., Federici, L., Fusi Pecci, F., & Rich, R. M. 2003, A&A, 405, 867
Bernstein, R. A., & McWilliam, A. 2002, in IAU Symp. 207, Extragalactic Star Clusters, ed. D. Geisler, E. K. Grebel, & D. Minniti (San Francisco: ASP), 739
Bica, E., Alloin, D., & Schmidt, H. R. 1990, A&A, 242, 241
Blitz, L., & Robishaw, T. 2000, ApJ, 541, 675

Blitz, L., Spergel, D. N., Teuben, P. J., Hartmann, D., & Burton, W. B. 1999, ApJ, 514, 818

Brewer, J. P., Richer, H. B., & Crabtree, D. R. 1995, AJ, 109, 2480

Brown, T. M. 2004, in The Local Group as an Astrophysical Laboratory, ed. M. Livio (Cambridge: Cambridge Univ. Press), in press (astro-ph/0308298)

Brown, T. M., Ferguson, H. C., Smith, E., Kimble, R. A., Sweigart, A. V., Renzini, A., Rich, R. M., & VandenBerg, D. A. 2003, ApJ, 592, L17

——. 2004, in preparation

Burstein, D., Faber, S. M., Gaskell, C. M., & Krumm, N. 1984, ApJ, 287, 586

Carretta, E., & Gratton, R. 1997, A&AS, 121, 95

Chiappini, C., Romano, D., & Matteucci, F. 2003, MNRAS, 339, 63

Choi, P. I., Guhathakurta, P., & Johnston, K. V. 2002, AJ, 124, 310

Clementini, G., Federici, L., Corsi, C., Cacciari, C., Bellazzini, M., & Smith, H. A. 2001, ApJ, 559, 109

Contini, T., Treyer, M., Sullivan, M., & Ellis, R. S. 2002, MNRAS, 329, 75

Côtè, P., Marzke, R. O., West, M. J., & Minniti, D. 2000a, ApJ, 533, 869

Côtè, P., Mateo, M., Sargent, W. L. W., & Olszewski, E. W. 2000b, ApJ, 537, L91

Côtè, P., Oke, J. B., & Cohen, J. G. 1999, AJ, 591, 850

Da Costa, G. S., Armandroff, T. E., & Caldwell, N. 2002, AJ, 124, 332

Da Costa, G. S., & Mould, J. R. 1988, ApJ, 334, 159

Davidge, T. J. 1990, ApJ, 351, L37

——. 1997, AJ, 113, 985

——. 2001, AJ, 122, 1386

Davidge, T. J., Rigaut, F., Chun, M., Brandner, W., Potter, D., Northcott, M., & Graves, J. E. 2000, ApJ, 545, L89

Davies, R. D. 1975, MNRAS, 170, 45

Dekel, A., & Woo, J. 2003, MNRAS, 344, 1131

de Vaucouleurs, G. 1958, ApJ, 128, 465

Durrell, P. R., Harris, W. E., & Pritchet, C. J. 1994, AJ, 108, 2114

——. 2001, AJ, 121, 2557

——. 2004, in preparation

Eggen, O. J., Lynden-Bell, D., & Sandage, A. R. 1962, ApJ, 136, 748

Elson, R. A. W., & Walterbos, R. A. M. 1988, ApJ, 333, 594

Elston, R., & Silva, D. R. 1992, AJ, 104, 1360

Evans, N. W., & Wilkinson, M. I. 2000, MNRAS, 316, 929

Evans, N. W., Wilkinson, M. E., Guhathakurta, P., Grebel, E. K., & Vogt, S. S. 2000, ApJ, 540, L9

Faber, S. M. 1973, ApJ, 179, 423

Ferguson, A. M. N., Irwin, M. J., Ibata, R. A., Lewis, G. F., & Tanvir, N. R. 2002, AJ, 124, 1452

Ferguson, A. M. H., & Johnson, R. A. 2001, ApJ, 559, L13

Ferguson, H. C., Rich, R. M., Mouhcine, M., Brown, T. M., & Smith, E. 2004, in preparation

Freedman, W. L. 1992, AJ, 104, 134

Frogel, J. A., & Elias, J. E. 1988, ApJ, 324, 823

Frogel, J. A., & Whitford, A. E. 1987, ApJ, 320, 199

Fusi Pecci, F., et al. 1996, AJ, 112, 1461

Gebhardt, K., Rich, R. M., & Ho, L. C. 2002, ApJ, 578, L41

Gottesman, S. T., Hunter, J. H., Jr., & Boonyasait, V. 2002, MNRAS, 337, 26

Grillmair, C. J., et al. 1996, AJ, 112, 1975

Guarnieri, M. D., Ortolani, S., Montegriffo, P., Renzini, A., Barbuy, B., Bica, E., & Moneti, A. 1998, A&A, 331, 70

Gunn, J. E., & Westphal, J. A. 1981, Proc. Soc. Photo-Opt. Instr. Eng., 290, 16

Han, M., Hoessel, J. G., Gallagher, J. S., III, Holtzman, J., & Stetson, P. B. 1997, AJ, 113, 1001

Harris, G. L. H., Harris, W. E., & Poole, G. B. 1999, AJ, 117, 855

Hartwick, F. D. A. 1976, ApJ, 209, 418

Hodge, P. 1992, The Andromeda Galaxy (Dordrecht: Kluwer)

Holland, S., Fahlman, G., & Richer, H. B. 1996, AJ, 112, 1035

Ibata, R. A., Gilmore, G., & Irwin, M. J. 1995, MNRAS, 277, 781

Ibata, R. A., Irwin, M. J., Lewis, G. F, Ferguson, A., & Tanvir, N. 2001, Nature, 412, 49

Jablonka, P., Bridges, T. J., Sarajedini, A., Meylan, G., Maeder, A., & Meynet, G. 1999, ApJ, 518, 627

Jablonka, P., Martin, P., & Arimoto, N. 1996, AJ, 112, 1415

Jacoby, G. H., & Ciardullo, R. 1999, ApJ, 515, 169

Jiang, L., Ma, J., Zhou, X., Chen, J., Wu, H., & Jiang, Z. 2003, AJ, 125, 727

Kauffmann, G., et al. 2003, MNRAS, 341, 54

Kent, S. M. 1983, ApJ, 266, 562

Larkin, J. E., et al. 2003, SPIE, 4841, 1600

Lee, M. G. 1996, AJ, 112, 1438

Lu, L., Sargent, W. L. W., & Barlow, T. A. 1998, AJ, 115, 55

Maraston, C., Greggio, L., Renzini, A., Ortolani, S., Saglia, R. P., Puzia, T. H., & Kissler-Patig, M. 2003, A&A, 400, 823

Matteucci, F., Donatella, R., & Molaro, P. 1999, A&A, 352, 117

McWilliam, A., & Rich, R. M. 1994, ApJS, 91, 7

Meylan, G., Sarajedini, A., Jablonka, P., Djorgovski, S. G., Bridges, T., & Rich, R. M. 2001, AJ, 122, 830

Morrison, H. L., Harding, P., Perrett, K., & Hurley-Keller, D. 2004, ApJ, submitted (astro-ph/0307302)

Mould, J. R. 1984, PASP, 96, 773

Mould, J. R., & Kristian, J. 1986, ApJ, 305, 591

——. 1990, ApJ, 354, 438

Pagel, B. E. J., & Patchett, R. 1975, MNRAS, 172, 13

Perrett, K. M., Bridges, T. J., Hanes, D. A., Irwin, M. J., Brodie, J. P., Carter, D., Huchra, J. P., & Watson, F. G. 2002, AJ, 123, 2490

Perrett, K. M., Stiff, D. A., Hanes, D. A., & Bridges, T. J. 2003, ApJ, 589, 790

Ponder, J. M., et al. 1998, AJ, 116, 2297

Pritchet, C. J., & van den Bergh, S. 1994, AJ, 107, 1730

——. 1987, ApJ, 316, 517

——. 1988, ApJ, 331, 135

lritzl, B. J., Armandroff, T. E., Jacoby, G. H., & Da Costa, G. S. 2002, AJ, 124, 1464

Puzia, T. H., Saglia, R. P., Kissler-Patig, M., Maraston, C., Greggio, L., Renzini, A., & Ortolani, S. 2002, A&A, 395, 45

Reitzel, D. B., & Guhathakurta, P. 2002, AJ, 201, 1411

Reitzel, D. B., Guhathakurta, P., & Rich, R. M. 2004, AJ, in press (astro-ph/0309295)

Rejkuba, M., Minniti, D., Silva, D., & Bedding, T. R. 2004, A&A, in press (astro-ph/0309795)

Renzini, A. 1998, AJ, 115, 2459

Renzini, A., & Buzzoni, A. 1986, in Spectral Evolution of Galaxies, ed. C. Chiosi & A. Renzini (Dordrecht: Reidel), 195

Rich, R. M. 1990, ApJ, 362, 604

Rich, R. M., et al. 1997 ApJ, 480, L35

Rich, R. M., Corsi, C., Cacciari, C., Federici, L., & Fusi Pecci, F., 2004a, in preparation

Rich, R. M., Mighell, K. J., Freedman, W. L., & Neill, J. D. 1996a, in Formation of the Galactic Halo, ed. H. Morrison & A. Sarajedini (San Francisco: ASP), 24

——. 1996b, AJ, 111, 768

Rich, R. M., & Mould, J. R. 1991, AJ, 101, 1286

Rich, R. M., Mould, J., Picard, A., Frogel, J. A., & Davies, R. 1989, ApJ, 341, L51

Rich, R. M., Reitzel, D. B., Guhathakurta, P., Gebhardt, K., & Ho, L. C. 2004b, AJ, in press (astro-ph/0309296)

Richer, H. B., Crabtree, D. R., & Pritchet, C. J. 1990, ApJ, 355, 448

Rockosi, C. M., et al. 2002, AJ, 124, 349

Saha, A., & Hoessel, J. G. 1990, AJ, 99, 97

Saha, A., Hoessel, J. G., & Krist, J. 1992, AJ, 103, 84

Saha, A., Hoessel, J. G., & Mossman, A. E. 1990, AJ, 100, 108

Saito, Y., & Iye, M. 2000, ApJ, 535, L95

Sandage, A., Becklin, E., & Neugebauer, G. 1969, ApJ, 157, 55

Sarajedini, A., & van Duyne, J. 2001, AJ, 122, 244

Sargent, W. L. W., Kowal, C. T., Harwick, F. D. A., & van den Bergh, S. 1977, AJ, 82, 947

Saviane, I., Held, E. V., & Piotto, G. P. 1996, A&A, 315, 40

Shetrone, M. D., Côtè, P., & Sargent, W. L. W. 2001, ApJ, 548, 592

Searle, L., & Sargent, W. L. W. 1972, ApJ, 173, 25

Searle, L., & Zinn, R. 1978, ApJ, 225, 357

Stephens, A. W., et al. 2001, AJ, 121, 2597

——. 2003, AJ, 125, 2473

Thilker, D. A., Braun, R., Walterbos, R. A., Corbelli, E., Lockman, F. J., Murphy, E., & Maddalena, R. 2004, Nature, submitted

Thomas, D., Maraston, C., & Bender, R. 2003, MNRAS, 339, 897

Tolstoy, E., Venn, K. A., Shetrone, M., Primas, F., Hill, V., Kaufer, A., & Szeifert, T. 2003, AJ, 125, 707

Trager, S., Faber, S. M., Worthey, G., & González, J. J. 2000, AJ, 119, 164

Tripicco, M. J. 1989, 97, 735

Walterbos, R. A. M., & Kennicutt, R. C., Jr. 1987, A&AS, 69, 311

——. 1988, A&A, 198, 61

Wampler, E. J. 1977, in IAU Collop. 40, Astronomical Applications of Image Detectors with Linear Response, ed. M. Duchesne & G. Leliever (Paris: Obs. Meudon), 14

Worthey, G., Faber, S. M., & González, J.J. 1992, ApJ, 398, 69

Zinn, R. J., & West, M. J. 1984, ApJS, 44, 45

Zoccali, M., et al. 2003, A&A, 399, 931

19

Stellar winds of hot massive stars nearby and beyond the Local Group

FABIO BRESOLIN and ROLF P. KUDRITZKI
Institute for Astronomy, University of Hawaii

Abstract

Photospheric radiation momentum is efficiently transferred by absorption through metal lines to the gaseous matter in the atmospheres of massive stars, sustaining strong winds and mass loss rates. Not only is this critical for the evolution of such stars, it also provides us with diagnostic UV/optical spectral lines for the determination of mass loss rates, chemical abundances, and absolute stellar luminosities. We review the mechanisms that render the wind parameters sensitive to the chemical composition, through the statistics of the wind-driving lines. Observational evidence in support of the radiation-driven wind theory is presented, with special regard to the wind momentum-luminosity relationship and the blue supergiant work carried out by our team in galaxies outside of the Local Group.

19.1 A Portrait of Mass Loss in Hot Stars

All stars in the upper Hertzsprung-Russell (H-R) diagram, above $\sim 10^4 L_\odot$, are affected by mass loss via stellar winds (Abbott 1979). Although this statement includes stars in different evolutionary stages [O and B main sequence stars, blue and red supergiants, Wolf-Rayet stars, luminous blue variables (LBVs)], this review will focus on a subset of the "hot" ($T_{eff} \gtrsim 8000$ K) objects, namely OB main sequence and blue supergiant stars. In the corresponding regime of luminosity and effective temperature the radiation field from the stellar photosphere is strong enough to be able to sustain powerful winds, by transfer of photon momentum to the metal ions present in the stellar outer layers. In spite of the low concentration of these ions, when compared to H and He, the momentum is efficiently distributed to the plasma by Coulomb interactions, so that the entire stellar envelope is being accelerated during this process (Springmann & Pauldrach 1992). It is intuitive that the chemical composition and overall metallicity of the stellar outer layers must play a crucial role in determining the strength of the winds and the rate of mass loss.

The basic description given above is the essence of the radiation-driven wind theory, developed during the course of the last 30 years in a number of landmark papers. Following the first suggestion by Lucy & Solomon (1970) concerning the absorption of radiation in the ultraviolet resonance lines of ions such as C IV, Si IV, and N V, recently detected at the time as P Cygni profiles from rocket experiments, the mechanism of wind acceleration by metal-line absorption has been developed further by Castor, Abbott, & Klein (1975) to include the approximate effect of a larger number of strong and weak metal lines. The theoretical framework has subsequently been perfected by the works of Abbott (1982) and Pauldrach, Puls, & Kudritzki (1986). Approximate analytical solutions for velocity fields and mass loss rates within this framework have been developed by Kudritzki et al. (1989). State-of-the-art modeling techniques attain nowadays a realistic description of radiation-driven winds, by solving the radiation transfer of millions of lines in an expanding atmosphere, producing synthetic spectra that match the observations of hot stars from the far-UV to the near-IR to a good degree of accuracy (Hillier & Miller 1998; Pauldrach, Hoffmann, & Lennon 2001; Puls et al. 2003).

19.1.1 Scale of the Phenomenon

Based mainly on IR or radio excess and Hα emission measurements, quite a broad range in the mass loss rate, $\dot{M}$, has been found among hot stars, scaling roughly with stellar luminosity as $L^{1.8}$ (Garmany & Conti 1984; de Jager, Nieuwenhuijzen, & van der Hucht 1988) during the main sequence and the supergiant phase (it is actually the modified wind momentum that scales with some power of the luminosity, as described later). Typical values up to a few times $10^{-6} M_\odot \, \mathrm{yr}^{-1}$ are found among such stars. Wolf-Rayet stars possess denser winds, with mass loss rates up to one order of magnitude higher than O stars, while even more extreme values, up to $10^{-3} M_\odot \, \mathrm{yr}^{-1}$, can be found among LBVs. For central stars of planetary nebulae, $\dot{M} \approx 10^{-7} - 10^{-9} M_\odot \, \mathrm{yr}^{-1}$, by virtue of their lower luminosities (see the recent review by Kudritzki & Puls 2000 for further information on stellar wind properties).

Evidently such rates of mass loss profoundly influence the evolution of massive stars (Chiosi & Maeder 1986). Moreover, winds from massive stars are relevant for the calculation of chemical yields and their inclusion in chemical evolution models of galaxies (Maeder 1992; Portinari, Chiosi, & Bressan 1998). Stellar rotation has also been shown to significantly affect the evolution of massive stars, in particular by enhancing mass loss rates and modifying the surface and wind chemical composition (Meynet & Maeder 2000). An example of recent evolutionary models for massive stars is shown in Figure 19.1, where the solar-metallicity tracks of Meynet & Maeder (2000) for an initial rotational velocity of 300 km s^{-1} are shown for $M = 40 M_\odot$ and $60 M_\odot$. Along the $40 M_\odot$ track the stellar mass and the N/C ratio are indicated at several stages of evolution.

19.2   Why Do Stellar Winds Depend on Metallicity?

Within the galaxies of the Local Group the metallicity varies by more than one order of magnitude, from metal-poor dwarf irregulars such as WLM to metal-rich giant spirals such as M31, offering the opportunity to probe the effects of chemical abundances on the character of the stellar winds. Expanded possibilities have appeared with the advent of 8–10 meter telescopes, which allow us to extend the spectroscopic study of individual stars even beyond the Local Group. Quantitative stellar spectroscopy in external galaxies is becoming important for a number of projects dealing with stellar abundances, the distance scale, the evolution of massive stars, the characterization of supernova progenitors, and the instabilities of massive stars in the very upper part of the H-R diagram, near the Humphreys-Davidson (1979) limit. The work carried out within our group is mostly concerned with the first two aspects. For the distance scale work, see the recent review by Bresolin (2004). Here we will succinctly illustrate how current observations of hot stars compare with the theoretical predictions concerning their stellar winds.

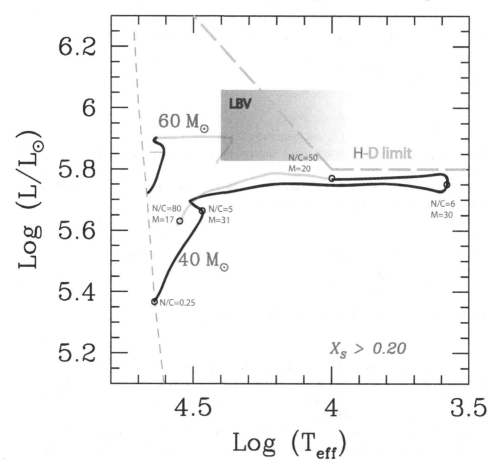

Fig. 19.1. Schematic H-R diagram showing the rotating ($V_i = 300 \,\mathrm{km\,s^{-1}}$) solar metallicity $40\,M_\odot$ and $60\,M_\odot$ stellar tracks from Meynet & Maeder (2000). For clarity the tracks are plotted up to the point at which the fractional hydrogen mass at the surface $X_S = 0.2$. The early WN phase is plotted in gray. The ZAMS location (short-dashed line) and the Humphreys-Davidson limit (long-dashed line) are shown, together with the typical location of LBV stars (shaded area). At several positions along the $40\,M_\odot$ track the stellar mass and the surface N/C ratio are reported.

19.2.1 *Theoretical Landscape*

As briefly mentioned in the previous section, the winds of hot massive stars are well described within the framework of the radiation-driven theory. For more details and references on the subject, the interested reader is referred to the reviews by Kudritzki (1998) and Kudritzki & Puls (2000).

Let us consider a hot and massive star radiating a total luminosity L out of its photosphere. The outer layers will be accelerated by radiation pressure through absorption in the metal lines. The maximum total momentum available from the radiation field is L/c, so that we can expect a dependence of the mechanical momentum flow:

$$\dot{M} v_\infty = f(L/c). \tag{19.1}$$

As a result of this transfer of momentum from the radiation to the ions, the wind is accelerated, starting from the photospheric layers, up to the asymptotic terminal velocity v_∞. The observed stellar wind velocity fields are customarily parameterized as a "β-law," given here in its simplest form:

$$v(r) = v_\infty \left(1 - \frac{R_\star}{r}\right)^\beta. \tag{19.2}$$

For the hot O stars, $\beta \simeq 0.8$, and therefore the ions are efficiently accelerated right above the photosphere, while for later-type A supergiants a more progressive acceleration takes place, corresponding to $\beta \simeq 2 - 4$.

When considering the hydrodynamics of a stationary line-driven wind, we start from the fundamental equations describing the conservation of mass

$$\dot{M} = 4\pi r^2 \rho v, \tag{19.3}$$

where $\rho = \rho(r)$ is the local density, together with the equation of motion

$$v\frac{dv}{dr} = -\frac{1}{\rho}\frac{dP_{gas}}{dr} - g + g_{rad}, \tag{19.4}$$

where the radiative acceleration g_{rad} can be expressed as the sum

$$g_{rad} = g_{rad}^{Th} + g_{rad}^{lines} + g_{rad}^{ff,bf}. \tag{19.5}$$

Of these three terms, only the second one (the acceleration due to line absorption) needs to be considered in more detail, the first one (Thomson scattering) will be included as a constant reduction of the gravity g, whereas the third one (free-free + bound-free transitions) is negligible in the winds of hot stars.

Since the work by Castor et al. (1975) and Pauldrach et al. (1986), the radiative term of the line acceleration is parameterized in terms of a *line force multiplier*, $M(t)$, such that

$$g_{rad}^{lines} = CF\, g_{rad}^{Th}\, M(t), \tag{19.6}$$

with CF being the correction factor that accounts for the finite extension of the stellar disk. The optical depth parameter t, multiplied by the dimensionless line strength k (i.e. the line opacity divided by the Thomson opacity $n_e \sigma_e$; see Kudritzki & Puls 2000), gives the optical depth in a given line transition in a supersonically expanding atmosphere:

$$\tau = kt \tag{19.7}$$

$$t = n_e \sigma_e \frac{v_{therm}}{dv/dr}. \tag{19.8}$$

In its simplest form (see Abbott 1982, Pauldrach et al. 1986, and Kudritzki 2002 for a more accurate description) the line force multiplier is then expressed as

$$M(t) \propto N_{eff}\, t^{-\alpha}, \tag{19.9}$$

thus depending on the effective number of metal lines driving the wind, N_{eff}, and on the optical depth through the line force parameter α. Solving the equation of motion (Eq. 19.4) with this parameterization of the line force, one obtains scaling relations for the mass loss rate and the terminal velocity (Kudritzki et al. 1989):

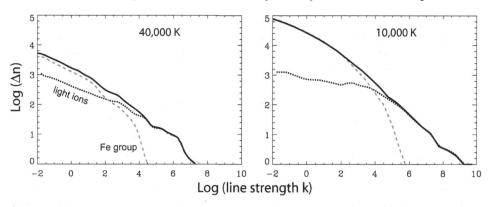

Fig. 19.2. Line distribution functions for $T_{eff} = 40{,}000$ K (*left*) and $T_{eff} = 10{,}000$ K (*right*) and solar composition, plotted separately for iron-group elements (dashed lines) and lighter ions (dotted lines). The full line represents the total distribution. (Adapted from Puls et al. 2000.)

$$\dot{M} \approx (N_{eff}L)^{1/\alpha} \left[M_\star (1 - \Gamma) \right]^{1-1/\alpha}$$

$$v_\infty \approx \frac{\alpha}{1 - \alpha} v_{esc}. \qquad (19.10)$$

Note that $M_\star(1-\Gamma)$ is the stellar *effective mass* (which accounts for the ratio Γ of Thomson scattering to gravitational acceleration).

The origin of the $M(t)$ expression used in the proportionality of Equation (19.9) lies in the statistics of the strengths of the wind-driving lines, which can be described by a simple *line strength distribution function*. It is found that, to a good approximation, the number of lines of a given strength k follows a power law (Fig. 19.2):

$$n(k)d(k) \propto k^{\alpha-2}dk \qquad (0 < \alpha < 1; \ 1 < k < k_{max}). \qquad (19.11)$$

The exponent α, which determines the slope of the distribution function, is mostly set by the laws of atomic physics (the distribution function of oscillator strengths), and is found to vary between 0.5 and 0.7 (for the hydrogen Lyman series one would obtain $\alpha = 2/3$). N_{eff} (see Eqs. 19.9 and 19.10) is related to the normalization of the line strength distribution function. Also note that a power law is only an approximation and that there is a slight curvature in the distribution function.

The line distribution function plays a primary role in the mechanisms governing radiation-driven winds (see Puls, Springmann, & Lennon 2000 for an in-depth study), and it is worthwhile spending a few additional words in relation to the type of ions that are most effective in driving the wind. It is clear that such ions will differ among stars of different chemical composition and/or effective temperatures, since the predominant ionization stages will change accordingly.

What kind of lines are driving the wind? The most prominent metal-line transitions are located in the far-UV and UV spectral ranges. As the emitted stellar flux peaks at lower frequencies with decreasing T_{eff}, so does the spectral range containing the lines contributing the most at driving the wind. The calculations by Abbott (1982) and Puls et al. (2000)

show that lines of high-ionization stages of O, N, P, and other heavy elements, located approximately between 800 and 1200 Å, are the dominant source of acceleration in the hottest (40,000 K) stars. At lower temperatures (10,000–20,000 K) lower-ionization species (such as Fe II, Fe III, Mg II, Ca II), having transitions at longer wavelengths, become predominant. The relative importance of CNO, Fe-group, and other elements changes with T_{eff} as well. For solar composition C, N, and O dominate over iron-group elements for $T_{\text{eff}} > 25,000$ K, and down to lower temperatures as the overall metallicity decreases (Vink, de Koter, & Lamers 2001). As the right panel in Figure 19.2 shows, at lower temperature the number of iron-group lines present, mostly Fe II–III, increases, resulting in a steepening of the line distribution function and a consequent decrease in α.

We can now answer our question regarding the origin of the metallicity dependence of the mass loss, based on the understanding of the line statistics briefly discussed so far. There are two main effects.

(1) To first order, the line strengths of the individual metal lines driving the wind are proportional to metallicity,

$$k \approx \frac{Z}{Z_\odot} k_\odot, \tag{19.12}$$

so that the distribution functions in Figure 19.2 shift horizontally in the plotted log–log plane. Consequently, the normalization over the range $1 < k < k_{\text{max}}$ changes with metallicity, affecting the effective number of lines driving the wind:

$$N_{\text{eff}} \propto Z^{1-\alpha}. \tag{19.13}$$

(2) There is a metallicity dependence of α, coming from the curvature of the line distribution function, which tends to become steeper at high line strengths (see Puls et al. 2000 for the details). It is important to note that the local acceleration in the wind comes mostly from lines with $\tau \simeq 1$. This means that strong winds are driven by lines with smaller line strengths, whereas weak winds rely on the acceleration of the fewer lines with large line strengths.

To conclude, the predicted scaling relations for the metallicity dependence of the wind parameters, valid approximately in the range $0.1 < Z/Z_\odot < 3$, are

$$\dot{M} \sim Z^{(1-\alpha)/\alpha} = Z^m, \tag{19.14}$$

with $m = 0.5 - 0.8$ for O- and B-type stars, and

$$v_\infty \sim Z^{0.13-0.15} \tag{19.15}$$

(Kudritzki, Pauldrach, & Puls 1987; Leitherer, Robert, & Drissen 1992). While the effects on the mass loss are mostly a direct result of (1) through Equations (19.10) and (19.13), the metallicity dependence of v_∞ is solely caused by (2) and Equation (19.10). At very low metallicities (below $10^{-2} Z_\odot$) these relations break down, and the depth dependence of the line force multipliers must be taken into account (Kudritzki 2002). We also conclude from Figure 19.2 that a change of the relative abundance ratio of α elements to Fe-group elements will have an additional effect.

19.3   Does Nature Conform to Theory?

The first obvious observational test of the effects of metallicity on the stellar winds of massive stars concerns the terminal velocities. These are easily measured from the P Cygni profiles of resonance lines in the UV, such as C IV $\lambda1550$ and Si IV $\lambda1400$. A clear decrease of v_∞ is found when comparing the terminal velocities measured in the Milky Way with those in the Magellanic Clouds, especially in the lower-metallicity SMC (Garmany & Conti 1985; Kudritzki & Puls 2000). Direct comparisons of O-star P Cygni profiles at high (Galactic) and low (SMC) metallicity can be found, for example, in Leitherer et al. (2001) and Heap, Hubeny, & Lanz (2001).

Concerning the mass loss rates, their derivation is more model dependent. The bulk of the measurements available for massive stars comes from Hα line profile fits. However, when correlating $\dot{M}$ with, for example, the luminosity L, a significant scatter is found, as a consequence of the dependence on the effective mass (see Eq. 19.10). This difficulty vanishes, at least theoretically, when considering the wind momentum through Equation (19.10), since $v_{\rm esc}$ depends on the square root of the effective gravitational potential:

$$\dot{M} v_\infty \propto \frac{(N_{\rm eff} L)^{1/\alpha}}{R_\star^{0.5}} \left[M_\star (1-\Gamma) \right]^{3/2 - 1/\alpha}. \tag{19.16}$$

The fortuitous nulling of the exponent for the effective mass (since $\alpha \simeq 2/3$ for hot stars) then relates the *modified wind momentum*

$$D_{\rm mom} = \dot{M} v_\infty (R_\star / R_\odot)^{0.5} \tag{19.17}$$

to the stellar luminosity in a simple wind momentum-luminosity relationship (WLR)

$$\log D_{\rm mom} = a + b \log \frac{L}{L_\odot} \tag{19.18}$$

(see Kudritzki 1988 or Kudritzki & Przybilla 2004 for a recent detailed derivation), which constitutes the basis of our program to use the winds of massive stars in external galaxies to determine their distances. [An independent method valid for blue supergiants, the flux-weighted gravity-luminosity relationship (FGLR) has been recently proposed by Kudritzki, Bresolin, & Przybilla (2003).]

From Equation (19.16), it is evident that a and $b = 1/\alpha$ in Equation (19.18), which are to be derived from observations for an empirical WLR, are a function of the predominant ionization stages in the stellar atmosphere (spectral type) and of the metallicity. The slope and zero-point of the WLR are, in fact, found to differ for O, B, and A supergiant stars by Kudritzki et al. (1999), as illustrated in Figure 19.3.

Combining the metallicity effect on $\dot{M}$ and v_∞ discussed in the previous section, the theoretical dependence of the modified wind momentum on metallicity becomes

$$D_{\rm mom} \sim Z^{0.6-0.8}. \tag{19.19}$$

Puls et al. (1996) and Herrero, Puls, & Najarro (2002) have shown that the Galactic O stars follow the WLR predicted by the radiation-driven theory rather convincingly. Data on B and A supergiants, as well as on lower-metallicity massive stars that would allow a test of Equation (19.19), remain scant. Puls et al. (1996) indeed found a decrease in the wind momentum for a small sample of O stars in the Magellanic Clouds compared to similar stars in the Milky Way, in rough agreement with the theoretical expectations. Vink et al. (2001)

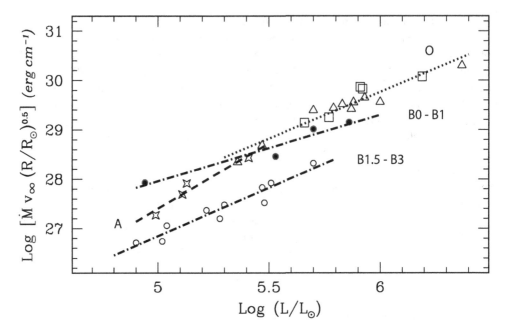

Fig. 19.3. Spectral type dependence of the WLR for Galactic supergiant stars. Different symbols are used for different spectral type ranges, and the corresponding linear regression lines are drawn. (Adapted from Kudritzki et al. 1999.)

have also found reasonable agreement between predicted and observed mass loss rates of O-type stars in the LMC and SMC, if the adopted metallicity is $0.8\,Z_\odot$ and $0.1\,Z_\odot$, respectively.

19.4   Looking Beyond the Local Group

Among the most exciting developments in stellar research in recent years has been the possibility of quantitative studies of individual stars in external galaxies located even beyond the Local Group, thanks to observations with 8 m-class telescopes. When working at $V \simeq 19-20$ mag or fainter in a galaxy several Mpc away, the natural targets become the visually brightest mid-B to early-A supergiants and hypergiants (with the occasional LBV). Reaching absolute magnitudes $M_V \simeq -8$ to -9, they can be studied spectroscopically out to $10-15$ Mpc. Such stars hold the promise of delivering important information on stellar abundance patterns, a welcome complement to H II region studies, and extragalactic distances (via the WLR and the FGLR) even at a moderate spectral resolution of $R \simeq 1000-2000$.

Pioneering results concerning quantitative stellar spectroscopy at such distances have been obtained recently within our collaboration, taking advantage of the multi-spectrum capabilities offered at the Very Large Telescope with the FORS instrument. Supergiant stars in NGC 3621, a late-type spiral with a Cepheid distance of 6.7 Mpc studied by Bresolin et al. (2001), remain the farthest *normal* stars for which information regarding the stellar wind and abundances have been obtained so far. The non-LTE spectral synthesis of two A-type supergiants in NGC 3621 reveals a sub-solar overall metallicity (Bresolin et al. 2001; Przy-

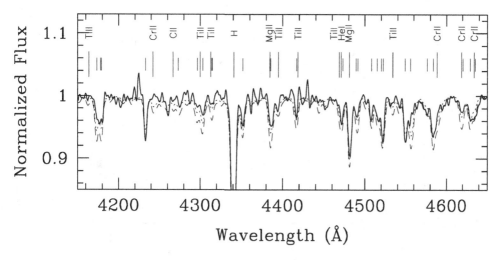

Fig. 19.4. Portion of the rectified blue VLT/FORS spectrum of an A0 Ia star in NGC 300. Principal spectral lines are identified at the top. Synthetic models are shown for $Z = Z_\odot$ (dashed line) and $Z = 0.5 Z_\odot$ (dotted line).

billa 2002) and indicates that the internal accuracy in abundance, even at the moderate FORS resolution, is ~ 0.2 dex.

Additional blue supergiants have been investigated spectroscopically in NGC 300, at a distance of 2 Mpc, by Bresolin et al. (2002a). This work is part of a larger project aimed at improving the accuracy of stellar candles used to measure distances of nearby galaxies. The study of the blue supergiants, in particular, will provide stellar chemical compositions needed to test the Cepheid period-luminosity relation at varying metallicities, and at the same time independent distances to the parent galaxies via the WLR and the FGLR, once calibrated in nearby galaxies (Bresolin 2004).

The abundance diagnostic power of the A supergiant VLT spectra in NGC 300 is illustrated in Figure 19.4, where $Z/Z_\odot \simeq 0.5$ is derived for an A0 Ia star from a comparison with synthetic spectra. A similar work has been carried out for early B-type supergiants from the same set of spectra by Urbaneja et al. (2003). This allows us a comparison between stellar and nebular abundances in NGC 300, using published results on the O/H abundance of H II regions (Deharveng et al. 1988) and adopting [O/H]=[M/H] for A stars, where M/H is the mean stellar metallicity (Fig. 19.5).

Only for very few external spiral galaxies has such a kind of comparison been carried out so far, namely the Local Group members M33 (Monteverde et al. 1997; Monteverde, Herrero, & Lennon 2000) and M31 (Smartt et al. 2001; Venn et al. 2001; Trundle et al. 2002; see also the contribution by Venn et al. 2004 on dwarf irregular galaxies at this conference). However, the importance of such comparisons cannot be overstated, because of the need for a check on the nebular abundances, which in the case of most spiral galaxies rely on empirical "strong line" methods, which can contain systematic errors on the estimated abundances amounting to factors of 2 or 3 (Kennicutt, Bresolin, & Garnett 2003). While the result for NGC 300 is still very preliminary, and based on a small number of stars, it suggests a rough agreement between stars and H II regions, even though the O/H scale for the latter is dependent upon which empirical calibration one chooses to adopt.

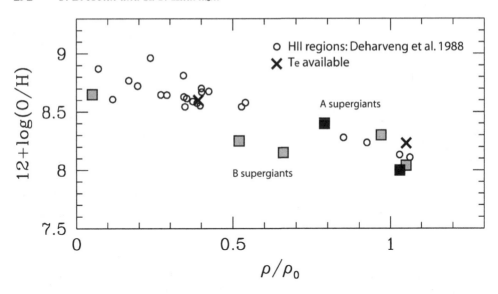

Fig. 19.5. Preliminary comparison of the nebular (circles) and stellar (squares) abundance gradients derived for NGC 300, expressed in terms of the fractional isophotal radius. The H II region abundances have been estimated from the semi-empirical $R_{23} = ([O\ II] + [O\ III])/H\beta$ calibration by Kobulnicky, Kennicutt, & Pizagno (1999), except for two regions (crosses), for which a direct abundance determination has been made possible.

19.4.1 Stellar Winds in Supergiants Outside the Local Group

High-resolution studies of individual blue supergiants outside of the Milky Way and the Magellanic Clouds have been made in M33 (McCarthy et al. 1995), M31 (McCarthy et al. 1997; Venn et al. 2000), NGC 6822 (Venn et al. 2001), and WLM (Venn et al. 2003). The winds of some of these stars have also been analyzed, and their strength roughly corresponds with the theoretical expectations, based on the measured luminosities and abundances. However, a systematic investigation of the wind properties of a sizable sample of blue supergiants in a single galaxy has yet to come. This is necessary for understanding the parameters affecting the strength of stellar winds, in particular their dependence on metallicity. The NGC 300 supergiant sample presented by Bresolin et al. (2002a), soon to be complemented by objects observed at the VLT in another Sculptor member, NGC 7793, gives us the chance to partly remedy the situation, although at such distances lower spectral resolutions must be used.

Six A supergiants in NGC 300 have been analyzed with the unblanketed version of the FASTWIND code described by Santolaya-Rey, Puls, & Herrero (1997) to measure mass loss rates from the Hα line profiles and gravities from Hγ. $\dot{M}$ was found to span the range $1.4 \times 10^{-8} - 2.3 \times 10^{-6} M_\odot$ yr^{-1}, with luminosities $\log(L/L_\odot) = 4.8 - 5.7$ ($M_V = -6.8$ to -8.6 mag).

The resulting WLR is displayed in Figure 19.6, where results for A supergiants in the Milky Way, M31 (Kudritzki et al. 1999), and NGC 3621 (Bresolin et al. 2001) are included. As can be seen, the $D_{mom} - Z$ scaling relation (§ 19.3) applied to the Galactic regression line for the mean metallicity of the NGC 300 stars ($Z \simeq 0.4 Z_\odot$, dotted line) provides a reasonable fit to the data. Some discrepancies are present, but our general conclusion is that theory and

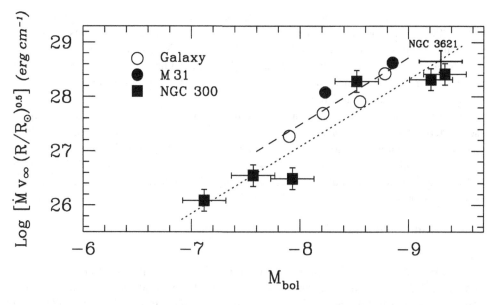

Fig. 19.6. The A-type supergiant WLR, including objects from the Milky Way, M31, NGC 300, and NGC 3621. The linear fit to the Galactic and M31 stars is shown by the dashed line. A theoretical scaling factor is applied to provide the expected relation at $Z/Z_\odot = 0.4$ (dotted line), the mean metallicity found for the NGC 300 and NGC 3621 stars included in the plot.

observations are in fair agreement. The analysis of a larger number of stars, including the effects of blanketing, is needed.

19.5 Multiwavelength Studies

The importance of the multiwavelength approach in the study of massive star winds and abundances has been clearly illustrated by Taresch et al. (1997), who analyzed far-UV (*ORFEUS*), UV (*IUE*) and optical spectra of the Galactic star HD 93129A (recently included in the newly defined O2 If* spectral class by Walborn et al. 2002). While the UV spectral range in general has been much more accessible than the far-UV, the amount of information regarding temperatures and abundances that can be collected from the UV alone is quite limited. Therefore, the iron abundance was estimated to be roughly solar from spectrum synthesis of the plethora of iron lines present in the UV (see Haser et al. 1998 for a similar approach applied to Magellanic Cloud O stars). This result was complemented by the measurement of optical lines to derive CNO abundances (finding a 2 × solar overabundance for N, and a ∼ 5 × depletion for C and O), and by the access to additional ionization stages of several elements (e.g., unsaturated lines of C III, N III–IV, O VI, S VI, P V), which become important in constraining the abundances and stellar effective temperatures, as well as the mass loss rates.

The effectiveness of this technique has been recently demonstrated by Crowther et al. (2002a), who analyzed *FUSE* far-UV wind-affected metal lines of four Magellanic Cloud O supergiants, together with *IUE* + *HST* UV and optical data. The far-UV spectra have been crucial in fixing the $T_{\rm eff}$'s, which resulted systematically 5000–7000 K lower than determined previously from unblanketed, plane-parallel models (see also Puls et al. 2003; Bianchi & Garcia 2002). Mass loss rates are also substantially revised downwards, as a consequence of lower luminosities. As in the case of HD 93129A studied by Taresch et al., strong N enrichment has been detected, together with modest C depletion, adding another piece of evidence for mixing processes affecting the surface chemical composition of evolved massive stars (see also Lennon, Dufton, & Crowley 2003).

As a final remark, it is worth emphasizing the importance of the modeling techniques used to derive the stellar parameters, currently relying on the full line blanketing and spherical extension treatment in non-LTE of the radiative transfer equation in the expanding atmospheres. Unified model atmosphere codes, such as CMFGEN (Hillier & Miller 1998) and WM-basic (Pauldrach et al. 2001), are now routinely employed to investigate the stellar properties of massive stars.

19.6   Additional Matters

19.6.1   Wolf-Rayet Stars

Wolf-Rayet stars are spectacular manifestations of the profound effects of mass loss on the evolution of massive stars. The removal of the outer layers of the stars progressively reveals, toward the late evolutionary stages, the products of H-burning (WN stars) and, subsequently, He-burning (WC stars). Because of the mechanism responsible for this effect—the transfer of photon momentum to the gas via metal-line absorption—the efficiency of this peeling process is highly dependent on metallicity. First of all, the minimum progenitor mass required for Wolf-Rayet star formation decreases with increasing metallicity (Maeder & Meynet 1994), ranging observationally from 20–25 $M_\odot$ in the Milky Way to 70 $M_\odot$ in the SMC (Massey, De Gioia-Eastwood, & Waterhouse 2000, 2001). Moreover, the WC/WN number ratio, which measures the efficiency of the evaporation process, increases from virtually zero at SMC abundance to almost one at super-solar abundance, as in the case of M31 (Massey 2003).

The high mass loss of Wolf-Rayet stars is manifested by the strong H, He, and metal emission lines present in their spectra. This facilitates their detection with narrow-band "on-off" imaging techniques and provides the spectroscopist with a means to probe detailed chemical abundance patterns in their outer layers, which invariably reflect their advanced evolutionary status. In this regard, besides quantitative work in the Milky Way, the Magellanic Clouds and a few Local Group galaxies (e.g., Herald, Hillier, & Schulte-Ladbeck 2001; Smartt et al. 2001; Crowther et al. 2002b), objects at much larger distances have started to be analyzed. Bresolin et al. (2002b) have presented chemical abundance patterns in a WN11 star they discovered in NGC 300. In this galaxy Schild et al. (2003) list nearly 60 Wolf-Rayet stars, based on a VLT imaging survey, and analyze the spectra of two WC stars. According to these authors there are at least a dozen Wolf-Rayet stars or Wolf-Rayet star candidates brighter than $V \simeq 19$ mag (the magnitude of the WN11 studied by Bresolin et al. 2002b) in this galaxy alone, opening up the possibility for quantitative studies of a large number of emission-line stars outside the Local Group.

19.6.2   Starbursts and the High-redshift Universe

It is of consolation to part of the hot-star community that the same kind of analyses and modeling used for nearby single stars can be successfully used for the understanding of stellar populations and chemical abundances in distant galaxies. Such is the case in the study of starburst events at distances where only the integrated light from a composite stellar population can be measured, where strong UV wind and photospheric lines can help constrain the properties of the emitting regions. Population synthesis models have become popular in order to disentangle several characterizing properties of the star-forming regions, such as the initial mass function, the star formation history, the age, and the chemical abundance. Recent examples of the application of this technique are given by González Delgado et al. (2002) for the metal-rich starburst galaxy NGC 3049, and Leitherer et al. (2001) for NGC 5253. In the latter case the main UV spectral features are reasonably well reproduced by synthetic models based on UV stellar libraries constructed from metal-poor Magellanic Cloud O stars, rather than Galactic ones. A similar result has been obtained by the same authors, as well as by Heap et al. (2001), for the well-known lensed galaxy MS1512–cB58. However, we have already run short of UV stellar templates, as we currently have no other metallicities represented in the synthetic models other than those of the Galactic and the Magellanic Cloud stars. Besides, B stars are included only for the Galactic case. The solution to these difficulties is represented by the use, instead of observed stellar spectra, of model UV spectra, where the metallicity can be varied at will. The reliability of the stellar models must, of course, first be tested against the existing templates. The outcome in the near future will be a better knowledge of the properties, in particular of the chemical abundances, of high-redshift star-forming regions and primordial galaxies.

References

Abbott, D. C 1979, in IAU Symp. 83, Mass Loss and Evolution of O-type Stars, ed. P. S. Conti & C. W. H. de Loore (Dordrecht: Reidel), 237
——. 1982, ApJ, 259, 282
Bianchi, L. & Garcia, M. 2002, ApJ, 581, 610
Bresolin, F. 2004, in Stellar Candles for the Extragalactic Distance Scale, ed. D. Alloin & W. Gieren (Berlin: Springer), in press
Bresolin, F., Gieren, W., Kudritzki, R. P., Pietrzynski, G., & Przybilla, N. 2002a, ApJ, 567, 277
Bresolin, F., Kudritzki, R. P., Mendez, R., & Przybilla, N. 2001, ApJ, 547, 123
Bresolin, F., Kudritzki, R. P., Najarro, F., Gieren, W., & Pietrzynski, G. 2002b, ApJ, 577, L107
Castor, J. I., Abbott, D. C., & Klein, R. I. 1975, ApJ, 195, 157
Chiosi, C., & Maeder, A. 1986, ARA&A, 24, 329
Crowther, P. A., Dessart, L., Hillier, D. J., Abbott, J. B., & Fullerton, A. W. 2002b, A&A, 392, 653
Crowther, P. A., Hillier, D. J., Evans, C. J., Fullerton, A. W., De Marco, O., & Willis, A. J. 2002a, ApJ, 579, 774
Deharveng, L., Caplan, J., Lequeux, J., Azzopardi, M., Breysacher, J., Tarenghi, M., & Westerlund, B. 1988, A&AS, 73, 407
de Jager, C., Nieuwenhuijzen, H., & vander Hucht, K. A. 1988, A&AS, 72, 259
Garmany, C. D., & Conti, P. S. 1984, ApJ, 284, 705
——. 1985, ApJ, 293, 407
González Delgado, R. M., Leitherer, C., Stasińska, G., & Heckman, T. M. 2002, ApJ, 580, 824
Haser, S. M., Pauldrach, A. W. A., Lennon, D. J., Kudritzki, R.-P., Lennon, M., Puls, J., & Voels, S. A. 1998, A&A, 330, 285
Heap, S. R., Hubeny, I., & Lanz, T. M. 2001, Ap&SSS, 277, 263
Herald, J. E., Hillier, D. J., & Schulte-Ladbek, R. E. 2001, ApJ, 548, 932
Herrero, A., Puls, J., & Najarro, F. 2002, A&A, 396, 949
Hillier, D. J., & Miller, D. L. 1998, ApJ, 496, 407
Humphreys, R. M., & Davidson, K. 1979, ApJ, 232, 409
Kennicutt, R. C., Bresolin, F., & Garnett, D. R. 2003, ApJ, 591, 801
Kobulnicky, H. A., Kennicutt, R. C., & Pizagno, J. 1999, ApJ, 514, 544
Kudritzki, R. P. 1998, in Stellar Astrophysics for the Local Group, ed. A. Aparicio, A. Herrero, & F. Sanchez (Cambridge: Cambridge Univ. Press), 149
——. 2002, ApJ, 577, 389

Kudritzki, R. P., Bresolin, F., & Przybilla, N. 2003, ApJ, 582, L83

Kudritzki, R. P., Pauldrach, A., & Puls, J. 1987, A&A, 173, 293

Kudritzki, R. P., Pauldrach, A., Puls, J., & Abbott, D. C. 1989, A&A, 219, 205

Kudritzki, R. P., & Przybilla, N. 2004, in Stellar Candles for the Extragalactic Distance Scale, ed. D. Alloin & W. Gieren (Berlin: Springer), in press

Kudritzki, R. P., & Puls, J. 2000, ARA&A, 38, 613

Kudritzki, R. P., Puls, J., Lennon, D. J., Venn, K. A., Reetz, J., Najarro, F., McCarthy, J. K., & Herrero, A. 1999, A&A, 350, 970

Leitherer, C., Leao, J. R. S., Heckman, T. M., Lennon, D. J., Pettini, M., & Robert, C. 2001, ApJ, 550, 724

Leitherer, C., Robert, C., & Drissen L. 1992, ApJ, 401, 596

Lennon, D. J., Dufton, P. L., & Crowley, C. 2003, A&A, 398, 455

Lucy, L. B., & Solomon, P. M. 1970, ApJ, 159, 879

Maeder, A. 1992, A&A, 264, 105

Maeder, A., & Meynet, G. 1994, A&A, 287, 803

McCarthy, J. K., Kudritzki, R. P., Lennon, D. J., Venn, K. A., & Puls, J. 1997, ApJ, 482, 757

McCarthy, J. K., Lennon, D. J., Venn, K. A., Kudritzki, R. P., Puls, J., & Najarro, F. 1995, ApJ, 455, L135

Massey, P. 2003, ARA&A, 41, 15

Massey, P., De Gioia-Eastwood, K., & Waterhouse, E. 2000, AJ, 119, 2214

——. 2001, AJ, 121, 1050

Meynet, G., & Maeder, A. 2000, A&A, 361, 101

Monteverde, M. I., Herrero, A., & Lennon, D. J. 2000, ApJ, 545, 813

Monteverde, M. I., Herrero, A., Lennon, D. J., & Kudritzki, R. P. 1997, ApJ, 474, L107

Pauldrach, A., Hoffmann, T. L., & Lennon, M. 2001, A&A, 375, 161

Pauldrach, A., Puls, J., & Kudritzki, R. P. 1986, A&A, 164, 86

Portinari, L., Chiosi, C., & Bressan, A. 1998, A&A, 334, 505

Przybilla, N. 2002, Ph.D. Thesis, Ludwig-Maximillians Universitaets

Puls, J., et al. 1996, A&A, 305, 171

Puls, J., Repolust, T., Hoffmann, T., Jokuthy, A., & Venero, R. 2003, in IAU Symp. 212, A Massive Star Odyssey, from Main Sequence to Supernova, ed. K. A. van der Hucht, A. Herrero, & C. Esteban (San Francisco: ASP), 61

Puls, J., Springmann, U., & Lennon, M. 2000, A&AS, 141, 23

Santolaya-Rey, A. E., Puls, J., & Herrero, A. 1997, A&A, 323, 488

Schild, H., Crowther, P. A., Abbott, J. B., & Schmutz, W. 2003, A&A, 397, 859

Smartt, S. J., Crowther, P. A., Dufton, P. L., Lennon, D. J., Kudritzki, R. P., Herrero, A., McCarthy, J. K., & Bresolin, F. 2001, MNRAS, 325, 257

Springmann, U., & Pauldrach, A. W. A. 1992, A&A, 262, 515

Taresch, G., et al. 1997, A&A, 321, 531

Trundle, C., Dufton, P. L., Lennon, D. J., Smartt, S. J., & Urbaneja, M. A. 2002, A&A, 395, 519

Urbaneja, M. A., Herrero, A., Bresolin, F., Kudritzki, R. P., Gieren, W., & Puls, J. 2003, ApJ, 584, L73

Venn, K. A., et al. 2001, ApJ, 547, 765

Venn, K. A., McCarthy, J. K., Lennon, D. J., Przybilla, N., Kudritzki, R. P., & Lemke, M. 2000, ApJ, 541, 610

Venn, K. A., Tolstoy, E., Kaufer, A., & Kudritzki, R. P. 2004, in Carnegie Observatories Astrophysics Series, Vol. 4: Origin and Evolution of the Elements, ed. A. McWilliam & M. Rauch (Pasadena: Carnegie Observatories, http://www.ociw.edu/symposia/series/symposium4/proceedings.html)

Venn, K. A., Tolstoy, E., Kaufer, A., Skillman, E. D., Clarkson, S. M., Smartt, S. J., Lennon, D. J., & Kudritzki, R. P. 2003, AJ, 126, 1326

Vink, J. S., de Koter, A., & Lamers, H. J. G. L. M. 2001, A&A, 369, 574

Walborn, N. R., Howarth, I. D., Lennon, D. J., Massey, P., Oey, M. S., Moffat, A. F. J., & Skalkowski, G. 2002, AJ, 123, 2754

20

Presolar stardust grains

DONALD D. CLAYTON[1] and LARRY R. NITTLER[2]
(1) Department of Physics and Astronomy, Clemson University
(2) Department of Terrestrial Magnetism, Carnegie Institution of Washington

Abstract
Meteorites and interplanetary dust particles contain presolar stardust grains: solid samples of stars that can be studied in the laboratory. We review the role of these grains for the sciences of nucleosynthesis, stellar evolution, grain condensation, and chemical and dynamic evolution of the Galaxy. We explain what stardust is, prediscovery ideas about it, how it is collected and recognized, and how evidence of thermal condensation and measured isotopic abundance ratios places individual grains into related families. Unique scientific information derives primarily from the high precision (in some cases <1%) of the measured isotopic ratios of large numbers of elements in single stardust grains. To clarify the new scientific scope, we focus on three of the stardust families: mainstream SiC grains from asymptotic giant branch carbon stars, oxide grains from oxygen-rich red giants, and carbide grains condensed within interiors of young expanding supernovae. Stardust science is just now reaching maturity and will remain an increasingly significant aspect of nucleosynthesis applications.

20.1 Introduction

Presolar stardust grains are solid samples of stars that can be studied in the laboratory. They condensed during cooling of gases in stellar outflows and explosions and became a part of the interstellar medium (ISM) from which our Solar System ultimately formed. The word "stardust" distinguishes such interstellar dust from the larger mass of interstellar dust that formed in other ways. The word "presolar" indicates that these grains formed in stars that evolved prior to the birth of the Sun. Presolar grains are overwhelmingly extracted from the most primitive meteorites, small asteroid fragments whose orbits have intersected the Earth, and from interplanetary dust particles, cometary and asteroidal particles collected at high altitudes in the atmosphere. Asteroids and comets both formed during the earliest stages of solar system formation. The stardust grains they contain must necessarily have formed from pre-existing stars, and each stardust grain is hence older than the Sun. Stardust grains that condensed in stars younger than the Sun certainly exist in the ISM today, but samples of them have not yet been collected. The current NASA STARDUST mission may return the first samples of younger stardust.

Presolar grains are recognized as such by their highly unusual isotopic compositions, in essentially every element that they contain, relative to all other materials available for laboratory study. Their isotopic variations are too large (>4 orders of magnitude in some cases) to be explained by chemical or physical fractionations. Rather, many of these variations

Table 20.1. *Types of presolar grains in meteorites and interplanetary dust particles (IDPs).*

Phase	Abundance (ppm)	Size	References
Diamond	1400	2 nm	[1]
SiC	14	0.1–20 μm	[1]
Graphite	10	1–20 μm	[1]
TiC, ZrC, MoC, RuC, FeC, Fe-Ni	(sub-grains in graphite)	5–220 nm	[2,3]
Silicon Nitride (Si_3N_4)	>0.002	~1 μm	[4]
Spinel ($MgAl_2O_4$)	1	0.1–3 μm	[4]
Corundum (Al_2O_3)	<0.1	0.1–3 μm	[5]
Hibonite ($CaAl_{12}O_{19}$)	0.002	~2 μm	[6]
Silicates	500 (in IDPs)	0.3–1 μm	[7]

Key to references: [1] Huss, Hutcheon, & Wasserburg 1997, [2] Bernatowicz et al. 1996, [3] Croat et al. 2004, [4] Nittler et al. 1995, [5] Zinner et al. 2004, [6] Choi, Wasserburg, & Huss 1999, [7] Messenger et al. 2003.

clearly point to nuclear reactions occurring in individual stars. Quite a large literature exists documenting the discovery of these grains and their astrophysical consequences. We cite several key papers as we proceed, but our goal is more to indicate the types of arguments and conclusions that they have uniquely brought to astrophysics. The state-of-the-art in the field has been reviewed recently by Zinner (1998) and Nittler (2003). Many details of the astrophysical implications of stardust may be found in the volume edited by Bernatowicz & Zinner (1997).

The known types of presolar stardust are listed along with their sizes and abundances in Table 20.1. Figure 20.1 shows scanning electron microscope images of two presolar stardust grains. Both examples are single crystals of silicon carbide (SiC), but their isotopic compositions (indicated in the caption) reveal that they formed in quite different stars. Figure 20.1a shows a so-called mainstream grain, and Figure 20.1b shows a so-called X grain. Both grains are a few μm in diameter, larger than is typical for both interstellar dust and meteoritic stardust, but amenable to isotopic analysis. New technologies are now allowing isotopic analysis of presolar stardust grains with more typical sizes (<500 nm). A 1μm SiC grain contains > 10^{11} atoms, sufficient for statistically precise measurements of isotopic ratios not only for C and Si but even for minor and trace elements within the SiC (such as N, Mg, Ca, Ti, and others in some cases). Such grains are also large enough for experimental determination of their mineralogy and microstructure. For example, most SiC stardust grains have been found to be single crystals with the cubic structure (β-SiC), rather than one of the hexagonal polymorphs (α-SiC) common in industrially produced SiC (Daulton et al. 2002). Crystalline SiC is expected to condense in a carbon-rich gas as it cools from high temperatures due to expansion, and β-SiC has, in fact, been identified in circumstellar envelopes of carbon stars (Speck, Hofmeister, & Barlow 1999). The extremely nonsolar isotopic compositions of the grains indicate that they completely condensed and cooled prior to any mixing with interstellar matter.

Note that the grains discussed here are not typical of the main mass of dust in the ISM.

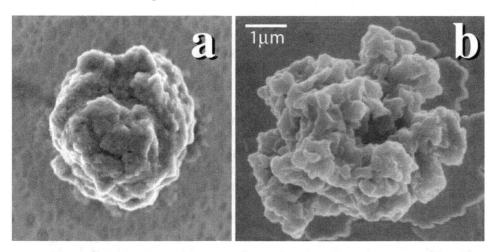

Fig. 20.1. Scanning electron microscope images of two presolar stardust grains of silicon carbide (SiC). (*a*) "Mainstream" grain with: $^{12}C/^{13}C = 60$ (solar ratio is 89), $^{14}N/^{15}N = 1{,}000$ (solar ratio is 272), $^{29,30}Si/^{28}Si = 1.05 \times$ solar, inferred $^{26}Al/^{27}Al \approx 10^{-3}$. This grain formed in a presolar C-rich AGB star. (*b*) "X" grain with: $^{12}C/^{13}C = 300$, $^{14}N/^{15}N = 60$, $^{29}Si/^{28}Si = 0.7 \times$ solar, $^{30}Si/^{28}Si = 0.5 \times$ solar, inferred $^{26}Al/^{27}Al \approx 0.3$. This grain formed in the ejecta of a presolar supernova.

Most dust mass is believed to consist of atoms and molecules accreted onto refractory grain cores when the diffuse ISM joins clouds (Draine 2003). Because the atoms and molecules that make up the grain mantles originated in diverse nucleosynthetic sites, most of the interstellar dust mass is probably isotopically homogeneous (~solar in the Sun's parent molecular cloud). It is evident that the stardust grains could not have formed by a process of heating of interstellar mantles accreted in molecular clouds, both because they are minerals (not amorphous solids) and because their isotopic ratios are too extreme and indicative of individual stars. For example, the highly nonsolar isotopic ratios of the SiC grains shown in Figure 20.1 are not expected in interstellar-grain mantles, but, based on both observations and theoretical models, these ratios speak strongly in favor of thermal condensation within asymptotic giant branch (AGB) C-star winds (mainstream grain) and supernova interiors (X grain).

20.1.1 Isolation and Analysis of Presolar Stardust

Presolar stardust comprises a small fraction of interstellar dust and of meteorites (Table 20.1). Meteoritic evidence is that stardust accounts for much less than 1% of refractory atoms in solar system solids, though how much preexisting dust was destroyed during solar system formation is unknown. Identifying this small fraction of stardust within a meteorite rock is not an easy matter. Most presolar grains have been identified in acid-resistant residues left over after most of the mass of the meteorite has been dissolved in strong acids. The chemical dissolution procedures were developed experimentally as a means of concentrating the unknown carriers of isotopically anomalous noble gases (Lewis & Anders 1983). Once these carrier concentrations were achieved, the residues were shown to contain many of the mineral phases listed in Table 20.1 (Amari, Lewis, & Anders 1994). These then were analyzed grain by grain for mineral structure, primarily by electron diffraction, and for isotopic ratios by secondary-ion mass spectrometry (SIMS). Such studies, which required breathtaking techniques for handling and analysis of microparticles, confirmed that individual stardust grains were the carriers of the noble gas anomalies and had isotopic variations in their major elements as well (Bernatowicz et al. 1987; Zinner, Tang, & Anders 1987). Identification of even rarer presolar phases (e.g., oxide minerals, Si_3N_4, and TiC) required development of automated particle analysis techniques (Nittler et al. 1995, 1997; Choi et al. 1998) and ultramicrotomy and transmission electron microscopy techniques (Bernatowicz et al. 1996). Recently, the development of new SIMS instrumentation has resulted in the discovery of presolar silicate stardust in interplanetary dust particles (Messenger et al. 2003). Such grains would have been destroyed in the prior acid treatments.

20.1.2   Astronomical Context

To glimpse the astrophysical potential of presolar stardust one can imagine a telescope capable of measuring isotope ratios to better than 1% precision in stars. Imagine what science one could do with such a telescope: nucleosynthesis in stars, stellar structure, and chemical evolution of the Galaxy. These are exactly the topics that the measured presolar grains clarify. But because the isotopic ratios are measured with considerably greater precision than is possible for astronomers, the questions that can be pursued are more precise and novel. The problem with this rosy picture is that each grain bears no label identifying its parent star! The parent star has long been gone, so at best one can identify the type of star and its evolutionary status. Knowledge of stars and their nucleosynthesis is required to identify the donor stars. Knowledge of condensation chemistry is also required. However, once the type of donor star is identified, the grain data can provide new knowledge about it. New science always follows from precise new information, and stardust is no exception to that rule.

20.1.3   Theoretical Precursors

The 1987 discovery of presolar grains from stars did not occur in a vacuum. In addition to a huge literature on interstellar dust there existed the following specific idea: dust condensed within stars and stellar outflows will record the isotopic signatures of that star and, as a natural component of the ISM, may be found in meteorites. Much of the observable expectation at the time lay in possible cosmic chemical memory (Clayton 1982) of anomalous ISM dust within solids that were grown from that dust in the solar system. Early suggestions were that the ^{16}O-richness of millimeter-sized solar system rocks called calcium-aluminum-rich inclusions resulted from their containing in their precursors dust that condensed in some undefined manner near supernovae (Clayton, Grossman, & Mayeda 1973) and that a ^{22}Ne-rich gas component within meteorites arose from the meteorites containing presolar dust rich in ^{22}Ne (Black 1972). These works did not suggest an origin for the isotopic anomalies in thermal condensation in stars, but they contributed to the emergence of the idea that isotopes might identify presolar material or rocks grown from it. What came as a surprise was the later isolation and characterization of several ISM stardust components themselves, apparently unchanged, within the meteorites.

In that early and fragmentary setting, the following ideas were set forth.

(1) Dust should condense thermally both in the interior of expanding supernovae and in red giant winds and be evidenced by huge and predictable isotopic anomalies (Clayton 1975, 1978).

(2) Supernova dust should have excess daughter abundances from extinct radioactivities (^{22}Na, ^{26}Al, ^{40}K, ^{41}Ca, ^{44}Ti, and ^{53}Mn) that were still alive when the grains condensed and stable isotope anomalies reflecting the compositions of the distinct supernova condensation zones.

(3) Differing grain phases would reflect differing supernova zone chemical compositions, such as sulfides replacing oxides in O-depleted inner supernova shells (Clayton & Ramadurai 1977).

(4) Stardust from red giant stars should contain *s*-process isotopes (Clayton & Ward 1978; Srinivasan & Anders 1978), with carbonaceous *s*-process carriers originating in AGB carbon stars (Clayton 1982, 1983; Swart et al. 1983).

(5) Dust should also condense in novae and be evidenced by extinct ^{26}Al and ^{22}Na, ^{30}Si-rich Si, ^{13}C-rich C, and ^{15}N-rich N (Clayton & Hoyle 1976).

(6) Finally, ^{13}C-rich carbon residues correlated with *s*-process Xe were attributed to stardust from a late-type red giant having C/O > 0.9 in order to condense carbon (Swart et al. 1983).

Once presolar grains were isolated in 1987, their isotopic compositions were investigated for every element measurable at the time. Many of the isotopic signatures were found to agree with the prior predictions. Moreover, these theoretical precursors, and even the controversy that they generated, provided fertile soil for quick interpretation of the first measurements of extreme isotopic anomalies in presolar grains after they were found in meteorites. The issues then became identification of the parent stars from the isotopic signatures of the presolar grains and the ways in which the precise isotopic ratios confronted stellar models.

20.1.4   *A Watershed of s/r Decomposition*

Many papers at this conference evaluate implications of astronomical observations of *s*-process and *r*-process elements, requiring that each heavy element be decomposed into contributions from those nucleosynthesis components. Based on such a decomposition, presolar stardust provided the first unambiguous demonstration that the *s*-process isotopes were synthesized in a special type of star and admixed into the ISM, including our solar system 4.6 Gyr ago. Quantitative decomposition of Solar heavy-element abundances into *s*-process and *r*-process contributions was first performed by Clayton et al. (1961), following their mathematical solution and evaluation of the *s*-process abundances. Their Table 4 listed N_r/N_s for each heavy element, and Clayton & Fowler (1961) provided the same for each of their isotopes. These first estimates were numerically rough because for N_r they utilized coarse, steady-state *r*-process calculations by B^{2}FH (Burbidge et al. 1957). An improved idea was introduced by Seeger, Fowler, & Clayton (1965), who introduced the exponential distribution of neutron fluences to calculate N_s and determined N_r by subtracting N_s from the total isotopic abundance. This improved procedure has become the method of choice, and was adopted by the Karlsruhe program of *s*-process studies, which has continued to present improved numerical evaluations of that same procedure (Käppeler et al. 1982; Käppeler 1999).

Nonetheless, the decomposition of solar abundances remained an act of theoretical faith until apparently pure *s*-process xenon gas was discovered in an acid-resistant residue of the Murchison carbonaceous meteorite (Srinivasan & Anders 1978). Their Figure 2 compared the isotopic composition of the Xe gas driven from that carbonaceous residue by high heating with the *s*-process decomposition of Xe calculated earlier by Clayton & Ward (1978). The stunning agreement shown in that figure was the first occasion on which essentially pure *s*-process isotopes could be measured experimentally. It advanced natural philosophy by

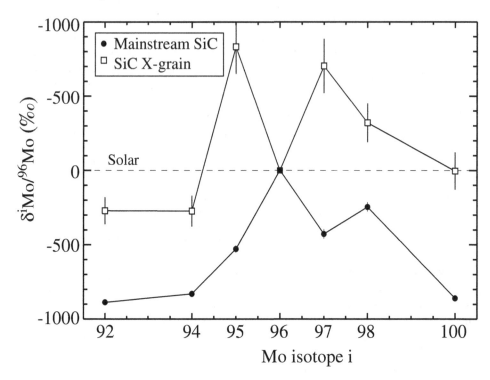

Fig. 20.2. Mo isotopic ratios measured in two presolar SiC grains. Ratios are expressed as δ-values, permil deviations from a terrestrial isotope standard: $\delta R = (R_{\text{meas}}/R_{\text{standard}} - 1) \times 10^3$. All isotopes in the mainstream grain are depleted relative to the *s*-process only ^{96}Mo. Thus, this grain clearly reveals an *s*-process signature, confirming an origin in a low-mass AGB star atmosphere. In contrast, the X grain (known to have formed in a supernova) is unusually enriched in ^{95}Mo and ^{97}Mo, indicating "neutron burst" nucleosynthesis (§20.5). Data from Nicolussi et al. (1998) and Pellin et al. (2000).

demonstrating that the *s*-process idea is correct, that it occurred in individual stars, and that the theoretical decomposition of the Solar mixture into *s* and *r* was justified. Clayton & Ward (1978) also argued that purely *s*-process isotopes could be sequestered in the ISM within red giant stardust, and they had emphasized Xe because other isotopic variations of Xe isotopes were already known; but, curiously, their prediction in 1975 was not accepted for publication until 1978 when the detection (Srinivasan & Anders 1978) supported its ideas. Building on this finding, Swart et al. (1983) showed that the *s*-process-rich residue was also very ^{13}C-rich, about twice the Solar complement, compared to ^{12}C, and suggested that dust from carbon stars carried both signatures into the meteoritic acid-resistant residue. An engaging historical account of this development can be found in Scientific American (Lewis & Anders 1983).

The demonstration that some unknown ^{13}C-rich chemical form of red giant stardust was present within the meteorites provided one of several motivations for further purification of that residue until the *s*-process carriers were identified as micron-sized grains of crystalline SiC (Bernatowicz et al. 1987), which were then shown to collectively carry *s*-process Xe (Lewis, Amari, & Anders 1994) and ^{13}C-rich carbon (Zinner et al. 1987). In this way the *s*-process pattern, and also ^{22}Ne-rich gas, contributed to the discovery of stardust and the isotopic astronomy that has been made possible by it. Newer high-sensitivity techniques in M. Pellin's laboratory at Argonne have been able to measure *s*-process isotopic patterns in individual presolar stardust grains (rather than collections of grains) for some refractory elements (Nicolussi et al. 1997, 1998; Pellin et al. 2000). Figure 20.2 shows their measurement of molybdenum isotopes in two SiC grains (Nicolussi et al. 1998). The "mainstream" grain contains essentially pure *s*-process Mo. As discussed in greater detail below, other isotopic data for the mainstream SiC stardust grains indicate an origin in AGB carbon-star atmospheres. Thus, the grains provide an unprecedented opportunity for confronting AGB star *s*-process nucleosynthesis models with high-precision observational data (Lugaro et al. 2003).

20.2   Classification of Isotopic Families of Stardust

For sake of illustration, suppose four elements can be measured isotopically (frequently true) in a single stardust grain. This provides at least one, and probably more, isotope ratio for each element. For example, in the most-studied stardust grain type, SiC (Fig. 20.1), elements routinely measured are Si, C, and N. The independent isotopic ratios that can be formed are Si (2), C (1), and N (1). Those measured ratios form the basis of a multidimensional space, 4-dimensional in this example, within which each grain assumes a unique position (a point). That space is mostly empty; but filled volumes enable identification of clustered groups and trends. Isotopically related grains form the basis of a classification system and enable science to be constructed. The significance of this classification system for stardust grains can be likened to the significance for stars of spectral types, the main sequence, and the Hertzsprung-Russell diagram.

The SiC classification is illustrated by Figure 20.3, which shows the ^{14}N/^{15}N vs. ^{12}C/^{13}C isotopic subspace. Differing symbols shown in these plots and associated labels (mainstream, A+B, X, Y, Z) have arisen historically (Hoppe et al. 1994) to label clusters or trends of individual SiC grains that are believed to be related. That the groupings are not arbitrary is indicated also by considering the Si-isotope space (Fig. 20.4). For example, the Y grains are defined on the basis of having ^{12}C/^{13}C > 100, but plot largely to the right of the slope 4/3 line formed by the mainstream grains on the Si 3-isotope plot. The Z grains, on the other hand have similar C isotopes to the mainstream, but quite different Si isotopes. Such considerations show the alphabet families to be less arbitrary than they might at first seem. In many SiC grains, it is also possible to measure more isotopic ratios than those of Si, C, and N (Hoppe et al. 1994, 1996; Hoppe & Ott 1997; Nittler & Alexander 2004). These include: ^{46,47,49,50}Ti/^{48}Ti; initial ^{26}Al/^{27}Al ratios inferred from excess radiogenic ^{26}Mg; Ca isotope ratios, enabling also the determination of initial ^{44}Ti/^{48}Ti ratios from radiogenic ^{44}Ca excesses and Ti/Ca ratios (Nittler et al., 1996); and, with special laser techniques, isotopic ratios of noble gases He and Ne (Nichols et al. 2004) and the heavy trace elements Ba, Fe, Sr, Mo, and Zr (Nicolussi et al. 1997; Pellin et al. 2000; Savina et al. 2003). Any of these help further constrain the donor star types for individual SiC grains.

Similar isotopic clustering has allowed subclassification of other stardust minerals as well, including graphite and oxides. In this paper we will discuss but a few presolar grain classes to illustrate main arenas of scientific impact, beginning with the mainstream SiC grains, which serve to illustrate how distinctive grain families are recognized and attributed to stellar sources. Space will prevent detailed discussion of the SiC grains of type Y (Amari et al.

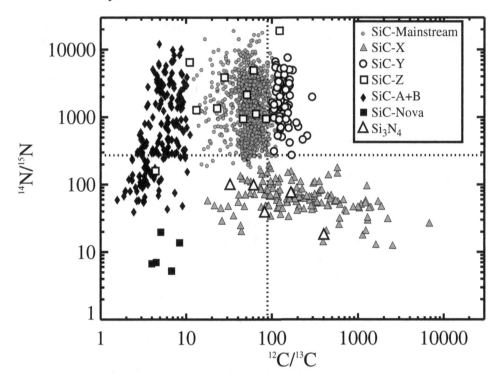

Fig. 20.3. C and N isotopic ratios measured in individual presolar SiC and Si$_3$N$_4$ grains (Data from: Hoppe et al. 1994, 2000; Nittler et al. 1995; Hoppe & Ott 1997; Huss et al. 1997). Dotted lines indicate the Solar isotope ratios here and in subsequent figures.

2001a), A+B (Amari et al. 2001b), and of nova grains (Amari et al. 2001c). Type X SiC will be discussed in §20.5. Also it must be said that the numbers of isotopic ratios that can be measured in stardust grains depends on the type of grain (Table 20.1), whose mineral structure and properties determine the abundances of trace elements that condensed sufficiently abundantly to yield isotopic measurements and grain size, which determines the total number of atoms available for measurement. Moreover, the SIMS technique is destructive, so that some grains can be consumed before all elements can be measured.

20.3   Mainstream SiC

The mainstream grains, which comprise ∼90% of the presolar SiC population (Figs. 20.3 and 20.4) were formed during mass loss from AGB carbon stars. This is attested to by: (1) their ^{12}C/^{13}C ratios distributed about the value 60, just as are the carbon stars (Smith & Lambert 1990); (2) large ^{14}N/^{15}N ratios owing to the dredge-up events that enrich the atmosphere in CN-burning products (Boothroyd & Sackmann 1999) along with the He-shell-burning products (Busso, Gallino, & Wasserburg 1999); (3) nearly pure *s*-process compositions of heavy elements (Srinivasan & Anders 1978; Prombo et al. 1993; Nicolussi et al. 1997), but including evidence of *s*-process branching unique to AGB stars (Gallino, Busso, & Lugaro 1997; Lugaro et al. 2003); (4) ^{26}Al/^{27}Al initial ratios up to ∼ 10^{-3} (Hoppe & Ott 1997) as expected for AGB stars (Forestini, Paulus, & Arnould 1991); and (5) no-

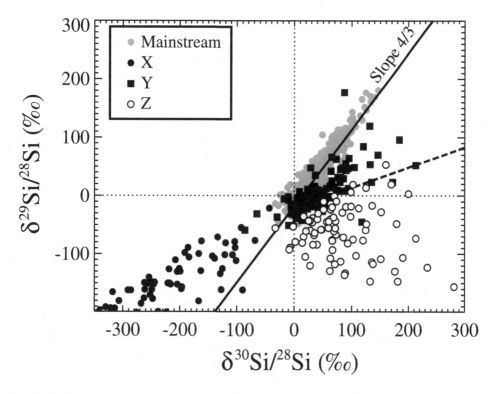

Fig. 20.4. Si isotopic ratios, expressed as δ-values, for presolar SiC grains of type mainstream, X, Y and Z (Data from: Hoppe et al. 1994, 2000; Hoppe & Ott 1997; Huss et al. 1997; Nittler & Alexander 2004). A+B grains have similar isotopic ratios to the mainstream, while the X grains extend to much more ^{28}Si-rich compositions than shown (lower left quadrant). Mainstream grains form an array with slope 1.3 (solid line). Long-dashed line indicates trajectory expected for dredge-up of He-shell material in an individual AGB star (see text).

ble gas isotopes reflecting a mixture between He-shell s-process and initial atmospheric gas (Lewis, Amari, & Anders 1990; Nichols et al. 2004), just as AGB models yield. Furthermore, an infrared feature associated with SiC has been observed around C stars (Speck, Barlow, & Skinner 1997), confirming the thermodynamic expectation that C-based solids (e.g., SiC) condense only in gas having C abundance comparable to or greater than O abundance (e.g., Lodders & Fegley 1995). Today it seems beyond doubt that AGB C stars are the birthplaces of mainstream SiC.

The good linear correlation of mainstream grain Si isotopes in Figure 20.4 surely attests to the mainstream being a physical family rather than an arbitrarily defined subgroup of stardust. However, a key observation is that the mainstream Si array has a much steeper slope on the Si 3-isotope plot and a larger range than expected from AGB star dredge-up processes (Gallino et al. 1994; Lugaro et al. 1999). Almost certainly, this array reflects a range of initial compositions in the Si isotopes of the stardust parent AGB stars. A related property of the mainstream grains is a strong correlation between Si and Ti isotopic ratios (Hoppe et al. 1994; Alexander & Nittler 1999), illustrated in Figure 20.5 for ^{29}Si/^{28}Si and

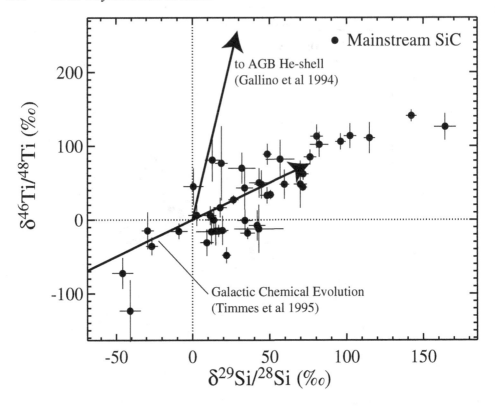

Fig. 20.5. Correlation of Si and Ti isotopic ratios in mainstream presolar SiC grains. The slope of the correlation is distinct from that expected for AGB star dredge-up, but compatible with models of Galactic chemical evolution. Data are from Hoppe et al. (1994) and Alexander & Nittler (1999).

^{46}Ti/^{48}Ti. Again, the observed correlation is primarily one of initial compositions for the AGB stars because the slope and extent of the line differ strongly from the expected shifts of these isotopes owing to *s*-process dredge-up in individual AGB stars.

20.3.1 New Astrophysics with Mainstream Grains

Several astrophysical issues are raised by the mainstream grains. These have led to new astrophysical knowledge. We illustrate only a few major issues here.

20.3.1.1 AGB Neutron Sources

Solar abundances already offered evidence of competition at branch points in the *s*-process path between neutron capture and temperature-dependent β decay rates (Ward, Newman, & Clayton 1976). The mainstream grains intensified this greatly since they are so close to pure *s*-process in their heavy elements. Their data revealed that no single temperature and neutron flux could simultaneously match the many branchings. A new physical model of the AGB *s*-process was developed in part because of this data-induced crisis. In that model, the ^{13}C$(\alpha,n)^{16}$O neutron source is the major source of the neutron fluence. The ^{13}C is produced by a pocket of mixing between envelope H and new ^{12}C after each thermal shell flash. The ^{13}C so produced burns, releasing its neutrons in hydrostatic equilibrium ($n_n = 10^7$ cm^{-3}, $kT = 8$ keV) during the interpulse phase of the AGB (Gallino et al. 1998; Arlandini et al. 1999). Subsequently during the He-shell flash, the ^{22}Ne source is activated to provide differing branch conditions ($n_n = 10^{10}$ cm^{-3}, $kT = 23$ keV) for a smaller neutron fluence, and this higher-T, higher-neutron-flux capture partially resets the branches established during the ^{13}C-burning interpulse. The resulting agreements not only brought many branch points into closer concordance but clearly established the ^{13}C neutron source as provider of the major fluence for the s-process. The new AGB model even provides an explanation for the bizarre observation that larger SiC grains have a greater s-process production ratio for ^{86}Kr/^{82}Kr than do smaller grains (Lewis et al. 1994), if higher-mass AGB stars (which have hotter ^{22}Ne flashes, larger n_n, and thus more ^{86}Kr) also produce the larger SiC grains. This two-neutron-source AGB model has not only been established by the SiC data and by models of AGB stars, it has also rejuvenated the science of accurate 10–30 keV neutron capture measurements for the s-process (Arlandini et al. 1999; Käppeler 1999).

20.3.1.2   *Presolar Metallicity Greater than Solar with Slope = 4/3*

Figures 20.4 and 20.5 show that most of the AGB stars donating mainstream grains had ^{29}Si/^{28}Si and ^{30}Si/^{28}Si ratios greater than solar (lying between about 95% and 120% of solar). This raises the question of how stars that were born and evolved prior to the birth of the Sun acquired such compositions. Clayton (1988) had suggested that the chemical evolution of the Galaxy would correlate ^{29}Si/^{28}Si and ^{30}Si/^{28}Si initial ratios within stars; but in a mixed-ISM, one-zone model that produces the Sun, presolar ratios should all be less than solar (Timmes & Clayton 1996). Furthermore, Timmes & Clayton (1996) showed that their deviations from solar should be equal when normalized to solar abundances (correlated along a line of slope 1), rather than distributed along a line of slope 4/3, as observed. Because the s-process in AGB stars enriches these ratios by only about 10 and 30 permil, respectively (Lugaro et al. 1999), much less than their observed spread along the mainstream line in Figure 20.4, the mainstream correlation is indeed a correlation in initial stellar abundances, as chemical evolution anticipates. But why the different slope and why are the mainstream grains isotopically heavier in Si than the Sun? This is a great and unexpected finding about the presolar Milky Way and/or the Sun. Attempts to resolve this crisis have taken four directions.

First, if the Sun (rather than the mainstream grains) is taken to be abnormal, then the heavy-isotope richness of presolar AGB stars can be regarded as a heavy-isotope deficiency of the Sun. Clayton & Timmes (1997) showed that it is hard to understand how the slope 4/3, which requires very different solar ^{29}Si/^{28}Si and ^{30}Si/^{28}Si isotope ratios than those of the evolving ISM, can, miraculously, also produce a Sun that lies almost on the mainstream line. Clayton & Timmes' Figure 4 shows that to do so requires a very large (about 30%) enrichment of ^{30}Si within each AGB star, much in excess of AGB model expectations and also apparently ruled out by C-isotopic ratios of AGB stars and mainstream grains (Alexander & Nittler 1999). Alternatively, it cannot be ruled out that the true ISM Si isotope evolution occurs along a slope 4/3 line. However, this would likely require that ^{30}Si is greatly overproduced relative to ^{29}Si in low-metallicity supernovae, and there is no hint of such behavior in current supernova calculations or reason to expect it from astrophysical considerations.

Second, Clayton (1997) proposed that new AGB stars born central to the solar orbit, where the Galactic metallicity gradient could result in stars having higher initial ^{29,30}Si/^{28}Si ratios, scatter outward from molecular clouds or other Galactic features to donate their SiC grains at the larger solar birth radius. Sellwood & Binney (2002) have given support to the general astronomical idea by calculating radial migration of low-mass stars as result of their "surfing" spiral density waves (rather than scattering from clouds). A semi-analytic model by Nittler & Alexander (1999) indicates that such outward orbital diffusion of stars would not necessarily result in the observed Si isotopic distribution of the SiC, but additional research is clearly needed.

Third, Lugaro et al. (1999) utilized the distinction (Timmes & Clayton 1996, Fig. 2) between low-mass and high-mass supernovae Si ejecta to construct an inhomogeneous chemical evolution model that produced stochastically a slope 4/3 in the Si three-isotope plot. Their model is certainly oversimplified and contains some arbitrary aspects that could invalidate it. Moreover, Nittler (2002) has shown that it does not reproduce the strong observed correlation between ^{46}Ti/^{48}Ti and ^{29}Si/^{28}Si ratios. The Si-Ti correlation does allow some of the spread in the data to be explained by such a model, but it seems unlikely that heterogeneous evolution can account for the entire mainstream Si distribution.

Fourth, the initial compositions of the AGB stars would have correlated isotopic compositions if they were varying mixtures of two distinct isotopic reservoirs. Clayton (2003) has advanced this as having been caused by a galactic merger about 6.5 Gyr ago between the gas of the Milky Way disk and that of a lower-metallicity satellite galaxy that therefore has lower ^{29}Si/^{28}Si and ^{30}Si/^{28}Si isotope ratios. The AGB stars appeared in a burst of star formation stimulated by the merger. The Sun, which formed about 2 Gyr later, had been enriched by supernova less massive than 25 $M_\odot$, whose ejecta do not increase and may even lower the Sun's ^{29,30}Si/^{28}Si ratios, allowing the Sun to remain at the lower extreme of the mainstream three-isotope Si plot (Fig. 20.4). This suggestion still needs to be rigorously investigated, but it appears promising to explain a range of observations.

20.4   Presolar Oxide Grains

Presolar oxide minerals (spinel, corundum, hibonite, TiO_2, and silicates) are more difficult to locate in meteorites than is C-rich dust due to a large background of isotopically solar dust formed in the solar system. Nonetheless, several hundred grains have now been found, mostly through automated techniques (e.g., Nittler et al. 1997; Choi et al. 1998). O isotope data are shown in Figure 20.6; the isotope ratios span several orders of magnitude and, as was seen above for SiC, tend to define distinct trends allowing for a classification system. Additional isotopic data for some grains, including inferred ^{26}Al/^{27}Al ratios and Ti and Mg isotopic ratios, have helped further constrain stellar origins of the grains. The majority of the grains (Groups 1 and 3, according to Nittler et al. 1997) have isotopic compositions in good agreement with observations of O-rich red giant and AGB stars (e.g., Harris & Lambert 1984) and with theoretical expectations for ^{17}O dredge-up in such stars (Boothroyd & Sackmann 1999). Comparison with the stellar models indicates that the grains formed from stars with a range of masses ($\sim 1.2-5M_\odot$) and with initial ^{18}O/^{16}O ratios probably reflecting Galactic chemical evolution. A supernova origin has been suggested for two grains, and the origin of the Group 4 grains remains enigmatic (Galactic chemical evolution and supernova mixing have been suggested). The ^{18}O-depleted Group 2 grains are discussed below.

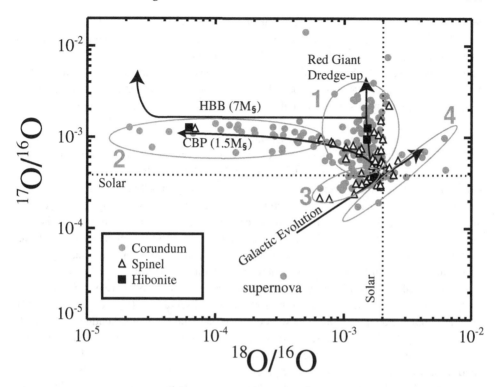

Fig. 20.6. O isotopic ratios measured in presolar oxide stardust. Grey ellipses indicate Group definitions of Nittler et al. (1997). Theoretical expectations for Galactic evolution (Timmes, Woosley, & Weaver 1995), red giant dredge-up (Boothroyd & Sackmann 1999), cool-bottom processing (CBP; Nollett, Busso, & Wasserburg 2003) and hot-bottom burning (HBB; Boothroyd, Sackmann, & Wasserburg 1995) are schematically shown. Data from Nittler (1997, and references therein), Choi et al. (1998, 1999), Nittler et al. (2003), and Zinner et al. (2004).

20.4.1 New Astrophysics with Presolar Oxide Grains

20.4.1.1 Extra Mixing in Red Giant and AGB Stars

Some 10% of presolar oxide grains (Group 2) have lower $^{18}O/^{16}O$ ratios and higher $^{26}Al/^{27}Al$ ratios than can be explained by standard stellar evolutionary models, indicating partial H-burning of the parent stars' envelopes. Two proposed mechanisms for this processing are hot-bottom burning (HBB) in intermediate-mass AGB stars (Boothroyd et al. 1995) and cool-bottom processing (CBP) in low-mass AGB stars (Wasserburg et al. 1995). CBP, also called "extra" or "deep" mixing, has previously been invoked to explain anomalously low $^{12}C/^{13}C$ ratios in low-mass red giants (Charbonnel 1994) as well as high Na and Al abundances in globular cluster giants (Shetrone 1996). Recent calculations by Nollett et al. (2003) strongly favor a CBP explanation for most of the Group 2 oxides and indicate that the grain compositions can constrain mixing rates and temperatures. Thus, the stardust grains reveal that CBP occurs in AGB stars as well as red giants, and it is hoped that the grain data will help identify the still unknown physical mechanism(s) driving the extra mixing. HBB is ruled out for most of the Group 2 oxide grains, because it can explain neither grains with intermediate $^{18}O/^{16}O$ ratios nor grains with $^{17}O/^{16}O$ ratios <0.001 (Boothroyd et al. 1995). However, Nittler et al. (2003) found an extreme Group 2 spinel grain with very high $^{25}Mg/^{24}Mg$ and $^{26}Mg/^{24}Mg$ ratios. The composition of this grain is most consistent with an origin in a $\sim 4-5\,M_\odot$ HBB AGB star. This result indicates that intermediate-mass stars were dust contributors to the protosolar cloud. Moreover, it emphasizes the importance of using multiple elements in individual grains to better constrain stellar origins.

20.4.1.2   Galactic Chemical Evolution and the Age of the Galaxy

The distribution of $^{18}O/^{16}O$ ratios observed in Group 1 and 3 oxide grains is in good agreement with expectations for Galactic chemical evolution, which predicts that $^{17}O/^{16}O$ and $^{18}O/^{16}O$ ratios should increase with metallicity (Clayton 1988; Nittler et al. 1997). This agreement indicates that the Galactic chemical evolution of O isotopes is reasonably well understood, and thus metallicities of parent stars can be inferred from grain O-isotope ratios and theoretical models. The grains can then be used to trace evolutionary histories of other isotope systems. For example, $^{25}Mg/^{24}Mg$ ratios measured in presolar spinel and corundum grains (Nittler et al. 2003) indicate that this ratio evolves much more slowly near solar metallicity than is predicted by numerical evolution models (e.g., Timmes et al. 1995). This result is in agreement with astronomical observations (Gay & Lambert 2000) and might indicate an important AGB star contribution to the Mg isotope budget at relatively low metallicity. Ti isotopes in a few corundum grains also seem to reflect Galactic chemical evolution (Choi et al. 1998; Hoppe et al. 2003). Finally, Nittler & Cowsik (1997) used the inferred masses and metallicities of the parent stars of Group 3 grains to put bounds on the age of the Milky Way. The age they derived (14 Gyr) has large systematic uncertainties, but is reasonably consistent with other estimates and was derived in a fundamentally new way.

20.5   Presolar Supernova Grains

Figures 20.3 and 20.4 show that SiC "X" grains form an isotopically distinct class, in many ways complementary to the mainstream. It is now well established that these grains, as well as a major fraction of presolar graphite grains and all known presolar Si_3N_4 grains, all condensed within the expanding ejecta of supernova explosions (see review by Amari & Zinner 1997). Although a special Type Ia supernova model was proposed to explain many of the observed isotopic signatures (Clayton et al. 1997), it now appears most likely that the grains formed in Type II events (SN II). Many of the observed isotope signatures (e.g., ^{28}Si and ^{15}N excesses, a wide range of $^{12}C/^{13}C$ ratios, and extremely high inferred $^{26}Al/^{27}Al$ ratios (up to 0.6) are qualitatively consistent with models of SN II nucleosynthesis. The smoking gun for a supernova origin (Clayton 1975) is the inferred presence of the extinct nuclides ^{44}Ti (Fig. 20.7) and ^{49}V, seen in the grains as ^{44}Ca and ^{49}Ti (Nittler et al. 1996; Hoppe & Besmehn 2002), since these short-lived nuclei are only synthesized within supernovae. Excesses in ^{22}Ne can in some cases also be attributed to supernova production of ^{22}Na (Nichols et al. 2004). The inferred $^{44}Ti/^{48}Ti$ ratios in ^{44}Ca-enriched grains are consistent with theoretical expectations for SN II. Moreover, they correlate with $^{29,30}Si/^{28}Si$ ratios (Fig. 20.7), consistent with the grains containing material from the innermost ^{28}Si-rich zones, where ^{44}Ti is synthesized during an α-rich freeze-out. Detailed discussions of the isotopic signatures of supernova grains can be found in Amari & Zinner (1997), Travaglio et al. (1999), and Hoppe et al. (2000). Of key importance is the observation that the grains carry the isotopic signatures of different mass zones of the parent supernovae. This indicates

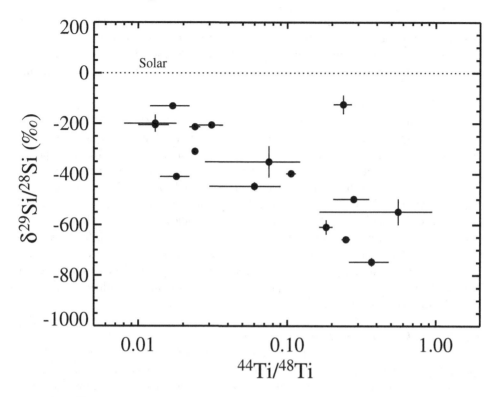

Fig. 20.7. δ-^{29}Si ratios plotted versus inferred ^{44}Ti/^{48}Ti ratios for presolar SiC X grains (Nittler et al. 1996; Hoppe et al. 2000; Besmehn & Hoppe 2004). The presence of extinct ^{44}Ti in the grains proves a supernova origin.

the need for selective mixing of material from different layers, with profound implications for processes of gas transport and dust nucleation and growth in SN II ejecta (see § 20.5.2).

20.5.1 New Nucleosynthesis Information from Supernova Grains

As we have already discussed for mainstream SiC grains and presolar oxides, the supernova grains provide new information about nucleosynthesis within their formation environments. We discuss two illustrative examples.

20.5.1.1 Nitrogen-15

Nitrogen-15 is produced in SN II models primarily by neutrino spallation from ^{16}O in inner shells (Woosley & Weaver 1995). Supernova presolar grains have large ^{15}N excesses coupled with isotopic signatures of H burning (very high ^{26}Al/^{27}Al ratios, low ^{12}C/^{13}C ratios in some grains). This has posed a significant problem for quantitatively explaining the grain data, since H-burning produces abundant ^{14}N-rich N. The measured ^{15}N/^{14}N ratios are higher than can be readily explained by SN II mixing models, if it is assumed that C > O for SiC or graphite condensation (Travaglio et al. 1999; Hoppe et al. 2000). Relaxing the C > O restriction might allow for SiC formation in the ^{15}N- and O-rich inner zones (see § 20.5.2), possibly removing the problem. Alternatively, recent observational evidence (Chin et al. 1999) supports much higher bulk ^{15}N production by SN II than is predicted by current models. Moreover, massive star models incorporating rotation indicate that abundant ^{15}N could be made hydrostatically when protons from the H shell get mixed into the He shell (Langer et al. 1998). These recent theoretical and observation results could help reduce the difficulty in explaining the grain data, if the ^{13}C and ^{15}N in the grains in fact originated in the He shell.

20.5.1.2   *Neutron-Burst Nucleosynthesis*

Pellin et al. (2000) and Davis et al. (2002) reported Sr, Zr, Mo, Ba, and Fe isotopic compositions for several SiC X grains. Figure 20.2 shows the Mo isotope pattern for one X grain. This ^{95}Mo- and ^{97}Mo-enriched pattern is clearly distinct, not only from the s-process signature of the plotted mainstream grain, but also from the composition expected for the r-process, associated with SN II. Similar results are found for the other studied elements. Meyer, Clayton, & The (2000) showed that the X grain Zr and Mo isotope signatures could be explained by a new kind of "neutron burst" nucleosynthesis, occurring when a large flux of neutrons are released as the shock wave passes through the He shell of the SN II. Although this process is but a minor contributor to bulk Mo isotopes in comparison with the s and r processes, unknown reasons apparently exist for its preferential enrichment of SiC X grains. This result clearly shows the power of presolar grains for probing conditions and processes in local regions of SN II ejecta.

20.5.2   *Condensation Problems within Supernovae*

The supernova interior offers a unique laboratory for condensation physics. It guarantees that chemistry must begin with gaseous atoms, with no trace of previous molecules or grains. The nucleosynthesis problems posed by isotopic ratios in supernova grains can not be decoupled from physical questions about their condensation. It is not physically clear whether the mixing apparently required by the isotope data represents molecular mixing in the young remnant or transport of a growing grain from one composition zone into another. Intimately related is an elemental composition question, namely, whether the bulk C abundance must exceed the O abundance to condense C-rich dust within supernovae. If the C/O ratio is less than unity, equilibrium condensation dictates that all carbon is locked up in the stable CO molecule, precluding condensation of SiC and graphite. Travaglio et al. (1999) performed SN II mixing models, assuming that the mixing is molecular, prior to condensation, and that only material with C > O could support graphite formation. In contrast, Clayton, Liu, & Dalgarno (1999) and Clayton, Deneault, & Meyer (2001) argued that radioactive destruction of CO molecules removes the C > O requirement. They advanced a kinetic theory of graphite growth and calculated its consequences in detail after advocating a specific nucleation model. They further argued that small graphite particles in a hot gas of C and O will associate with C faster than they will be oxidized by the more abundant O. Thus, even though oxidation would be the ultimate end, given adequate time, the expansion will terminate the chemistry after about two years, with large graphite grains remaining. This theory built on the finding that a large mass of CO was indeed destroyed in supernova 1987A (Liu & Dalgarno 1994, 1995). However, this theory is far from complete. For example, it is not known whether grains of TiC and Fe-Ni metal would condense with the graphite in this scenario, as required by observations of supernova graphites (Croat et al. 2004).

Similar but different questions surround the condensation of supernova SiC, the X grains of Figures 20.3 and 20.4. It seems plausible that radioactive liberation of free C atoms from CO molecules could also facilitate the condensation of SiC in O-rich gas. Although a kinetic route to SiC condensation has not been laid out, Deneault, Clayton, & Heger (2003) have formulated a physical description of the ejecta, which appears promising to explain many properties of SiC X grains. In their model, a reverse shock from increasing ρr^3 in the H envelope builds, after about a month, a dense shell in an inner layer of the supernova. Silicon-28 rich SiC condenses in this layer, allowed by the enhanced density and the radioactive dissociation of CO and SiO molecules. Mixing of a new type during condensation occurs if the reverse shock from the presupernova wind arrives at the condensation zone between six months to a year, because that shock slows the gas and forces the partially condensed SiC grains to propel forward through the decelerating gas into regions with different isotopic compositions. After $\sim 10^3$ yr, a third strong reverse shock from the ambient ISM propels SiC grains forward through overlying ejecta at high speed (typically 500 km s^{-1}) such that other atoms are implanted into the grains, perhaps accounting for some trace isotopes, as suggested for Mo and Fe (Clayton et al. 2002). Velocity-mixing instabilities prior to these reverse shocks will result in a spectrum of overlying column compositions so that grains from different supernovae (or different regions of a given supernova) could have a diversity of compositions.

Such physical modeling suggests that X grains provide new sampling techniques of the detailed physical structures of supernova ejecta, but it remains to be seen if the ideas will withstand the scrutiny of more detailed models. Especially threatening is the possibility of SiC destruction by sputtering, oxidation, or a shock that is too strong from the presupernova wind (Deneault et al. 2003). A key point becomes the spectrum of wind masses that accompany Type II supernovae, for those masses provide the mass of the shock-generating obstacle. Especially needed are 2-D and 3-D hydrodynamic calculations of the reverse shocks and of instability-induced velocity mixing by the primary outgoing shock and a detailed study of molecular mixing to ascertain the degree to which microscopic mixing can be called upon during the first year. Additionally, detailed microstructural investigations of supernova grains by analytical electron microscopy are likely to provide a great deal of information about their formation processes (e.g., Croat et al. 2004).

What now appears certain, despite these many open questions, is that supernova grains studied by isotopic analysis will provide, through details of condensation chemistry, a new sampling spectrum of young supernova interiors, just as have gamma-ray lines and hard X rays. The radioactivity that causes each type of sampling raises fundamental chemical questions as well.

20.6   Conclusions

Presolar stardust grains have been identifiable by cosmochemists because their several isotopic ratios are too unusual to be explained by any origin other than within ejected stellar matter prior to its mixing with the ISM. The high precision of this isotopic data, higher than traditionally obtainable by astronomical spectroscopy, supports detailed questioning about the stars that produced the grains. Stellar evolution theory, nucleosynthesis theory, and chemical evolution theory for the Galaxy must each be called upon to interpret the solid samples of the exploding or mass-losing stars. New insights into each have been demanded or suggested by the precise isotopic ratios of not only individual grains but of families of grains whose relatedness is established by iterative procedures. Grains grouped initially by purely isotopic criteria are examined with standard techniques of science: hypotheses about stars of origin are iterated with computed stellar models; hypotheses about trends within related grains are iterated with astrophysical models of trend-determining evolution; hypotheses about nuclear reactions in stars are iterated with the nuclear laboratories. By these processes of classification, posed hypotheses, and challenge by astrophysical theory, an increasingly refined scientific corpus has been established. This may be likened to the historic challenge of the spectral types of stars and of their positions and populations in the Hertzsprung-Russell diagram.

Stardust grains have also opened purely chemical frontiers of stellar physics. It is necessary to come to grips with details of the condensation processes and of physical aspects of stars that had not been previously demanded. How well mixed is the AGB dredge-up matter in the envelope? How is its wind initiated and does its stardust select overdense epochs of mass loss? How does the radioactive dissociation of the CO molecule alter chemical consequences of the C/O ratio? How do reverse shocks in supernovae enable supernova grains to grow and survive with the isotopes we find within them? How and when does supernova matter mix? These questions are but examples of many new cosmochemical frontiers that automatically open by the ability to study grains of stardust in detail in the laboratory.

Each presolar stardust grain is a measurement of some unknown star, a measurement undeveloped until cosmochemists isolate that grain and investigate it in the laboratory. Perhaps their laboratory techniques may justifiably be called new telescopes. They are the only telescopes capable of observing Galactic stars that died more than 5 Gyr ago. Understanding the early structure and chemical evolution of our Galaxy will increasingly rely on their high-precision message.

References

Alexander, C. M. O'D., & Nittler, L. R. 1999, ApJ, 519, 222

Amari, S., Gao, X., Nittler, L. R., Zinner, E., José, J., Hernanz, M., & Lewis, R. S. 2001c, ApJ, 551, 1065

Amari, S., Lewis, R. S., & Anders, E. 1994, Geochim. Cosmochim. Acta, 58, 459

Amari, S., Nittler, L. R., Zinner, E., Gallino, R., Lugaro, M., & Lewis, R. S. 2001a, ApJ, 546, 248

Amari, S., Nittler, L. R., Zinner, E., Lodders, K., & Lewis, R. S. 2001b, ApJ, 559, 463

Amari, S., & Zinner, E. 1997, in Astrophysical Implications of the Laboratory Study of Presolar Materials, ed. T. J. Bernatowicz & E. Zinner (New York: AIP), 287

Arlandini, C., Käppeler, F., Wisshak, K., Gallino, R., Lugaro, M., Busso, M., & Straniero, O. 1999, ApJ, 525, 886

Bernatowicz, T. J., Cowsik, R., Gibbons, P. C., Lodders, K., Fegley, Bruce, J., Amari, S., & Lewis, R. S. 1996, ApJ, 472, 760

Bernatowicz, T., Fraundorf, G., Tang, M., Anders, E., Wopenka, B., Zinner, E., & Fraundorf, P. 1987, Nature, 330, 728

Bernatowicz, T., & Zinner, E., ed. 1997, Astrophysical Implications of the Laboratory Study of Presolar Materials (New York: AIP)

Besmehn, A., & Hoppe, P. 2004, Geochim. Cosmochim. Acta, in press

Black, D. C. 1972, Geochim. Cosmochim. Acta, 36, 377

Boothroyd, A. I., & Sackmann, I.-J. 1999, ApJ, 510, 232

Boothroyd, A. I., Sackmann, I.-J., & Wasserburg, G. J. 1995, ApJ, 442, L21

Burbidge, E. M., Burbidge, G. R., Fowler, W. A., & Hoyle, F. 1957, Rev. Mod. Phys., 29, 547

Busso, M., Gallino, R., & Wasserburg, G. J. 1999, ARA&A, 37, 239

Charbonnel, C. 1994, A&AS, 282, 811

Chin, Y., Henkel, C., Langer, N., & Mauersberger, R. 1999, ApJ, 512, L143

Choi, B., Wasserburg, G. J., & Huss, G. R. 1999, ApJ, 522, L133

Choi, B.-G., Huss, G. R., Wasserburg, G. J., & Gallino, R. 1998, Science, 282, 1284

Clayton, D. D. 1975, Nature, 257, 36

——. 1978, Moon and Planets, 19, 109

——. 1982, J. Roy. Astr. Soc., 23, 174

——. 1983, ApJ, 271, L107

——. 1988, ApJ, 334, 191

——. 1997, ApJ, 484, L67

——. 2003, in Lun. Planet. Sci. Conf. XXXIV, No. 1059 (Houston: Lunar Planetary Inst.), (CD-ROM)

Clayton, D. D., Arnett, D., Kane, J., & Meyer, B. S. 1997, ApJ, 486, 824

Clayton, D. D., Deneault, E. A.-N., & Meyer, B. S. 2001, ApJ, 562, 480

Clayton, D. D., & Fowler, W. A. 1961, Ann. Phys., 16, 51

Clayton, D. D., Fowler, W. A., Hull, T. E., & Zimmerman, B. 1961, Ann. Phys., 12, 331

Clayton, D. D., & Hoyle, F. 1976, ApJ, 203, 490

Clayton, D. D., Liu, W., & Dalgarno, A. 1999, Science, 283, 1290

Clayton, D. D., Meyer, B. S., The, L., & El Eid, M. F. 2002, ApJ, 578, L83

Clayton, D. D., & Ramadurai, S. 1977, Nature, 265, 427

Clayton, D. D., & Timmes, F. X. 1997, ApJ, 483, 220

Clayton, D. D., & Ward, R. A. 1978, ApJ, 224, 1000

Clayton, R. N., Grossman, L., & Mayeda, T. K. 1973, Science, 182, 485

Croat, T. K., Bernatowicz, T., Amari, S., Messenger, S., & Stadermann, F. J. 2004, Geochim. Cosmochim. Acta, in press

Daulton, T. L., Bernatowicz, T. J., Lewis, R. S., Messenger, S., Stadermann, F. J., & Amari, S. 2002, Science, 296, 1852

Davis, A. M., Gallino, R., Lugaro, M., Tripa, C. E., Savina, M. R., Pellin, M. J., & Lewis, R. S. 2002, in Lun. Planet. Sci. Conf. XXXIII, No. 2018 (Houston: Lunar Planetary Inst.), (CD-ROM)

Deneault, E.-A., Clayton, D. D., & Heger, A. 2003, ApJ, 594, 312

Draine, B. T. 2003, ARA&A, 41, 241

Forestini, M., Paulus, G., & Arnould, M. 1991, A&AS, 252, 597

Gallino, R., Arlandini, C., Busso, M., Lugaro, M., Travaglio, C., Straniero, O., Chieffi, A., & Limongi, M. 1998, ApJ, 497, 388

Gallino, R., Busso, M., & Lugaro, M. 1997, in Astrophysical Implications of the Laboratory Study of Presolar Materials, ed. T. J. Bernatowicz & E. Zinner (New York: AIP), 115

Gallino, R., Raiteri, C. M., Busso, M., & Matteucci, F. 1994, ApJ, 430, 858

Gay, P. L., & Lambert, D. L. 2000, ApJ, 533, 260

Harris, M. J., & Lambert, D. L. 1984, ApJ, 285, 674

Hoppe, P., Amari, S., Zinner, E., Ireland, T., & Lewis, R. S. 1994, ApJ, 430, 870

Hoppe, P., & Besmehn, A. 2002, ApJ, 576, L69

Hoppe, P., Nittler, L. R., Mostefaoui, S., Alexander, C. M. O'D., & Marhas, K. K. 2003, in Lun. Planet. Sci. Conf. XXXIV, No. 1570 (Houston: Lunar Planetary Inst.), (CD-ROM)

Hoppe, P., & Ott, U. 1997, in Astrophysical Implications of the Laboratory Study of Presolar Materials, ed. T. J. Bernatowicz & E. Zinner (New York: AIP), 27

Hoppe, P., Strebel, R., Eberhardt, P., Amari, S., & Lewis, R. S. 1996, Geochim. Cosmochim. Acta, 60, 883

——. 2000, Meteoritics and Plan. Sci., 35, 1157

Huss, G. R., Hutcheon, I. D., & Wasserburg, G. J. 1997, Geochim. Cosmochim. Acta, 61, 5117

Käppeler, F. 1999, Prog. Nucl. Phys. 43, 419

Käppeler, F., Beer, H., Wisshak, K., Clayton, D. D., Macklin, R. L., & Ward, R. A. 1982, ApJ, 257, 821

Langer, N., Hefer, A., Woosley, S. E., & Herwig, F. 1998, in Nuclei in the Cosmos V, ed. N. Prantzos (Paris: Editions Frontiéres), 129

Lewis, R. S., Amari, S., & Anders, E. 1990, Nature, 348, 293

——. 1994, Geochim. Cosmochim. Acta, 58, 471

Lewis, R. S., & Anders, E. 1983, Scientific American, 249, 66

Liu, W., & Dalgarno, A. 1994, ApJ, 428, 769

——. 1995, ApJ, 454, 472

Lodders, K., & Fegley, B. 1995, Meteoritics, 30, 661

Lugaro, M., Davis, A. M., Gallino, R., Pellin, M. J., Straniero, O., & Käppeler, F. 2003, ApJ, 593, 486

Lugaro, M., Herwig, F., Lattanzio, J. C., Gallino, R., & Straniero, O. 2003, ApJ, 586, 1305

Lugaro, M., Zinner, E., Gallino, R., & Amari, S. 1999, ApJ, 527, 369

Messenger, S., Keller, L. P., Stadermann, F. J., Walker, R. M., & Zinner, E. 2003, Science, 300, 105

Meyer, B. S., Clayton, D. D., & The, L.-S. 2000, ApJ, 540, L49

Nichols, R. H., Jr., Kehm, K., Hoehenberg, C. M., Amari, S., & Lewis, R. S. 2004, Geochim. Cosmochim. Acta, in press

Nicolussi, G. K., Davis, A. M., Pellin, M. J., Lewis, R. S., Clayton, R. N., & Amari, S. 1997, Science, 277, 1281

Nicolussi, G. K., Pellin, M. J., Lewis, R. S., Davis, A. M., Amari, S., & Clayton, R. N. 1998, Geochim. Cosmochim. Acta, 62, 1093

Nittler, L. R. 1997, in Astrophysical Implications of the Laboratory Study of Presolar Materials, ed. T. J. Bernatowicz & E. Zinner (New York: AIP), 59

———. 2002, in Lun. Planet. Sci. Conf. XXXIII, No. 1650 (Houston: Lunar Planetary Inst.), (CD-ROM)

———. 2003, Earth Planet. Sci. Lett., 209, 259

Nittler, L. R., & Alexander, C. M. O'D. 1999, ApJ, 526, 249

———. 2004, Geochim. Cosmochim. Acta, in press

Nittler, L. R., et al. 1995, ApJ, 453, L25

Nittler, L. R., Alexander, C. M. O'D., Gao, X., Walker, R. M., & Zinner, E. 1997, ApJ, 483, 475

Nittler, L. R., Amari, S., Zinner, E., Woosley, S. E., & Lewis, R. S. 1996, ApJ, 462, L31

Nittler, L. R., & Cowsik, R. 1997, Phys. Rev. Lett., 78, 175

Nittler, L. R., Hoppe, P., Alexander, C. M. O'D., Busso, M., Gallino, R., Marhas, K. K., & Nollett, K. 2003, in Lun. Planet. Sci. Conf. XXXIV, No. 1703 (Houston: Lunar Planetary Inst.), (CD-ROM)

Nollett, K. M., Busso, M., & Wasserburg, G. J. 2003, ApJ, 582, 1036

Pellin, M. J., Calaway, W. F., Davis, A. M., Lewis, R. S., Amari, S., & Clayton, R. N. 2000, in Lun. Planet. Sci. Conf. XXXI, No. 1917 (Houston: Lunar Planetary Inst.), (CD-ROM)

Prombo, C. A., Podosek, F. A., Amari, S., & Lewis, R. S. 1993, ApJ, 410, 393

Savina, M. R., et al. 2003, Geochim. Cosmochim. Acta, 67, 3201

Seeger, P. A., Fowler, W. A., & Clayton, D. D. 1965, ApJS, 97, 121

Sellwood, J. A., & Binney, J. J. 2002, MNRAS, 336, 785

Shetrone, M. D. 1996, AJ, 112, 1517

Smith, V. V., & Lambert, D. L. 1990, ApJS, 72, 387

Speck, A. K., Barlow, M. J., & Skinner, C. J. 1997, MNRAS, 288, 431

Speck, A. K., Hofmeister, A. M., & Barlow, M. J. 1999, ApJ, 513, L87

Srinivasan, B., & Anders, E. 1978, Science, 201, 51

Swart, P. K., Grady, M. M., Pillinger, C. T., Lewis, R. S., & Anders, E. 1983, Science, 220, 406

Timmes, F. X., & Clayton, D. D. 1996, ApJ, 472, 723

Timmes, F. X., Woosley, S. E., & Weaver, T. A. 1995, ApJ, 98, 617

Travaglio, C., Gallino, R., Amari, S., Zinner, E., Woosley, S., & Lewis, R. S. 1999, ApJ, 510, 325

Ward, R. A., Newman, M. J., & Clayton, D. D. 1976, ApJS, 31, 33

Wasserburg, G. J., Gallino, R., Busso, M., Goswami, J. N., & Raiteri, C. M. 1995, ApJ, 440, L101

Woosley, S. E., & Weaver, T. A. 1995, ApJS, 101, 181

Zinner, E. 1998, ARE&EPS, 26, 147

Zinner, E., Amari, S., Guinness, R., Nguyen, A., Stadermann, F., Walker, R. M., & Lewis, R. S. 2004, Geochim. Cosmochim. Acta, in press

Zinner, E., Tang, M., & Anders, E. 1987, Nature, 330, 730

21

Interstellar dust

B. T. DRAINE
Princeton University Observatory

Abstract

In the interstellar medium of the Milky Way, certain elements (e.g., Mg, Si, Al, Ca, Ti, and Fe) reside predominantly in interstellar dust grains. These grains absorb, scatter, and emit electromagnetic radiation, heat the interstellar medium by photoelectric emission, play a role in the ionization balance of the gas, and catalyze the formation of molecules, particularly H_2. I review the state of our knowledge of the composition and sizes of interstellar grains, including what we can learn from spectral features, luminescence, scattering, infrared emission, and observed gas-phase depletions. The total grain volume in dust models that reproduce interstellar extinction is significantly greater than estimated from observed depletions.

Dust grains might reduce the gas-phase D/H ratio, providing an alternative mechanism to explain observed variations in the gas-phase D/H ratio in the local interstellar medium. Transport in dust grains could cause elemental abundances in newly formed stars to differ from interstellar abundances.

21.1   Introduction

Certain elements—Si and Fe being prime examples– are often extremely *depleted* from the interstellar gas (Jenkins 2004). The explanation for this depletion is that the atoms that are "missing" from the gas phase are located in solid particles—interstellar dust grains. The existence of this solid phase complicates efforts to determine elemental abundances, due to the difficult-to-determine abundance in grains.

The concentration of certain elements in dust grains also creates the possibility of selective transport of those elements through the gas, which could, at least in principle, lead to variations in elemental abundances from point to point in the interstellar medium, and possibly even differences between stellar and interstellar abundances. Dust grains might even deplete deuterium from the gas, providing an alternative explanation for observed variations in D/H ratios.

Observations of interstellar dust were recently reviewed by Whittet (2003) and Draine (2003a). The astrophysics of interstellar dust is a broad subject, encompassing dust grain formation and destruction, charging of dust grains, heating of interstellar gas by photoelectrons from dust, electron transfer from dust grains to metal ions, the optics of interstellar dust, optical luminescence and infrared emission from dust, chemistry on dust grains, and the dynamics of dust grains, including radiation-driven drift of grains relative to gas and alignment of dust by the magnetic field. Introductions to the astrophysics of dust can be found in Krügel (2003) and Draine (2003d).

21.2   Presolar Grains in Meteorites are Nonrepresentative

Genuine interstellar grains are found in meteorites (Clayton & Nittler 2004), but, because present search techniques rely on isotopic anomalies, the grains that are found are limited to those formed in stellar winds or ejecta. Theoretical studies of grain destruction by supernova-driven blastwaves lead to estimated grain lifetimes of only $\sim (2-3) \times 10^8$ yr (Draine & Salpeter 1979; Jones et al. 1994); this, together with the large depletions typically seen for elements like Si, implies that the bulk of interstellar grain material must be regrown in the interstellar medium (Draine & Salpeter 1979; Draine 1990).

Because the typical interstellar grains do not bear a distinctive isotopic signature, we do not know how to identify them in meteorites. We are therefore forced to study interstellar grains remotely.

21.3   Interstellar Reddening

Our knowledge of interstellar dust is largely derived from the interaction of dust particles with electromagnetic radiation: attenuation of starlight, scattering of light, and emission of infrared and far-infrared radiation.

The wavelength dependence of interstellar extinction tells us about both the size and composition of the grains. The extinction is best determined using the "pair method"— comparison of the fluxes from two stars with nearly identical spectroscopic features (and therefore photospheric temperature and gravity) but with one of the stars nearly unaffected by dust. With the assumption that the extinction goes to zero as wavelength $\lambda \rightarrow \infty$, it is possible to determine the extinction A_λ as a function of wavelength (see, e.g., Fitzpatrick & Massa 1990, and references therein).

The extinction A_λ is obviously proportional to the amount of dust, but A_λ/A_{λ_0}, the extinction normalized to some reference wavelength λ_0, characterizes the *kind* of dust present, and its size distribution. The quantity $R_V \equiv A_V/(A_B-A_V)$ characterizes the slope of the extinction curve between $V = 0.55\,\mu$m and $B = 0.44\,\mu$m; small values of R_V correspond to steep extinction curves.

In principle, the function A_λ/A_{λ_0} is unique to every sightline, but Cardelli, Clayton, & Mathis (1989) found that the observed A_λ/A_{λ_0} can be approximated by a one-parameter family of curves: $A_\lambda/A_{\lambda_0} = f(\lambda, R_V)$, where they chose $R_V \equiv A_V/(A_B-A_V)$ as the parameter because it varies significantly from one curve to another. Cardelli et al. obtained functional forms for $f(\lambda, R_V)$ that provided a good fit to observational data; Fitzpatrick (1999) revisited this question and, explicitly correcting for the finite width of photometric bands, obtained a slightly revised set of fitting functions. Figure 21.1 shows the Fitzpatrick (1999) fitting functions, using $I_C = 0.802\,\mu$m, the central wavelength of the Cousins I band, as the reference wavelength. The parameterization is shown for values of R_V ranging from 2.1 to 5.5, which spans the range of R_V values encountered on sightlines through diffuse clouds in the Milky Way. Also shown is an empirical fit to the extinction measured toward HD 210121, showing how an individual sightline can deviate from the one-parameter fitting function $f(\lambda, R_V)$.

Dust on sightlines with different values of R_V obviously must have either different compositions or different size distributions, or both. Also of interest is the total amount of dust per unit H. This requires measurement of the total H column density $N_H \equiv N(H)+2N(H_2)+$

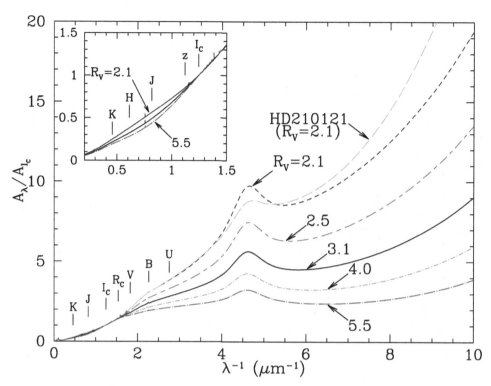

Fig. 21.1. Extinction normalized to Cousins *I*-band extinction for R_V values ranging from 2.1 to 5.5, using the Fitzpatrick (1999) parameterization, plus diffuse interstellar bands following Jenniskens & Desert (1994). Also shown is an improved fit to the extinction curve toward HD 210121, providing one example of how a sightline can deviate from the average behavior for the same value of R_V.

$N(H^+)$. On most sightlines the ionized hydrogen is a small correction; $N(H)$ and $N(H_2)$ can be measured using ultraviolet absorption lines.

Rachford et al. (2002) determined N_H to an estimated accuracy of better than a factor 1.5 on 14 sightlines. It appears that A_I/N_H is positively correlated with R_V, with

$$A_I/N_H \approx \left[2.96 - 3.55 \left(\frac{3.1}{R_V} - 1 \right) \right] \times 10^{-22} \text{cm}^2, \qquad (21.1)$$

providing an empirical fit (Draine 2003a). We can use the Fitzpatrick (1999) parameterization and Equation (21.1) to estimate A_λ/N_H for sightlines with different R_V. The results are shown in Figure 21.2—sightlines with larger R_V values appear to have larger values of A_λ/N_H for $\lambda^{-1} \lesssim 3\,\mu m^{-1}$, and decreased values for $\lambda^{-1} \gtrsim 4\,\mu m^{-1}$. This is interpreted as resulting from coagulation of a fraction of the smallest grains onto the larger grains; loss of small grains decreases the ultraviolet extinction, while adding mass to the larger grains increases the scattering at $\lambda \gtrsim 0.3\,\mu m$.

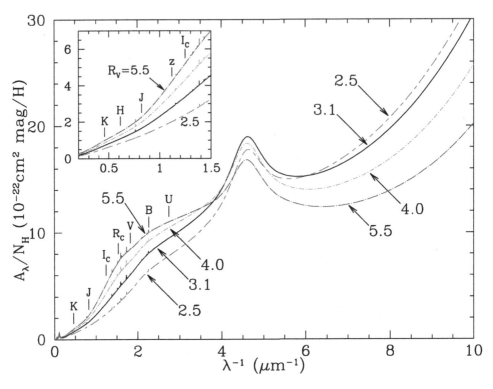

Fig. 21.2. Extinction per unit H column density, for different R_V. (Adapted from Draine 2003a.)

21.4   Spectroscopy of Dust in Extinction and Emission

21.4.1   *PAH Emission Features*

Interstellar dust glows in the infrared, with a significant fraction of the power radiated at $\lambda \lesssim 25\,\mu$m. In reflection nebulae the surface brightness is often high enough to permit spectroscopy of the emission; the 5–15 μm spectrum of NGC 7023 is shown in Figure 21.3. The spectrum has five very conspicuous emission peaks, at $\lambda = 12.7, 11.3, 8.6, 7.6,$ and 6.25 μm; there is an additional emission peak at 3.3 μm (not shown here), as well as weaker peaks at 12.0 and 13.6 μm.

These emission features are in striking agreement with the wavelengths of the major optically active vibrational modes for polycyclic aromatic hydrocarbon (PAH) molecules: the 5–15 μm features are labelled in Figure 21.3; the 3.3 μm feature (not shown) is the C-H stretching mode. The vibrational excitation results from single-photon heating (see §21.5). The strength of the observed emission requires that PAH molecules be a major component of interstellar dust. Modeling the observed emission in reflection nebulae indicates that the PAH species containing $\lesssim 10^3$ C atoms contain $\sim$40 ppm C/H—approximately 15% of $(\mathrm{C/H})_\odot = 246 \pm 23$ ppm (Allende Prieto, Lambert, & Asplund 2002).

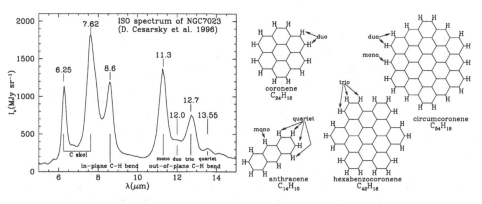

Fig. 21.3. PAH emission features in the 5–15 μm emission spectrum of the reflection nebula NGC 7023 (Cesarsky et al. 1996), and four PAH molecules, with examples of mono, duo, trio, and quartet H sites indicated.

21.4.2 2175 Å Feature: Graphitic C

By far the strongest feature in the extinction curve is the 2175 Å "bump" (see Figs. 21.1 and 21.2). Stecher & Donn (1965) pointed out that the observed feature coincided closely with the position and width of absorption expected from small spheres of graphite. Graphite consists of parallel sheets of graphene—two-dimensional hexagonal carbon lattices; adjacent graphene layers interact only through a weak van der Waals interaction.

The 2175 Å feature arises from a $\sigma \rightarrow \sigma^*$ excitation of a σ orbital in graphene; in small graphite spheres the feature strength corresponds to an oscillator strength per C of 0.16 (Draine 1989), so that the observed 2175 Å feature requires C/H $\approx$ 60 ppm $\approx 0.25(C/H)_\odot$ to account for the observed strength of the 2175 Å feature.

In a large PAH molecule, the interior C atoms have electronic orbitals closely resembling those in graphene, and one therefore expects that large PAH molecules will have a strong absorption feature peaking near 2175 Å, with an oscillator strength per C expected to be close to the value for graphite. It is of course interesting to note that the C/H in PAHs required to account for the 3–15 μm infrared emission is $\sim 2/3$ of that required to account for the 2175 Å bump. Since it would be entirely natural to have additional PAHs containing $> 10^3$ C atoms, it is plausible that the 2175 Å feature could be entirely due to PAHs, in which case the PAHs contain C/H $\approx$ 60 ppm.

The 2175 Å feature is suppressed in graphite grains with $a \gtrsim 0.02\,\mu$m because the grain becomes opaque throughout the 1800–2600 Å range. There can therefore be additional "aromatic" C within $a \gtrsim 0.02\,\mu$m grains.

21.4.3 10 and 18 μm Silicate Features

Interstellar dust has a strong absorption feature at 9.7 μm. While the precise composition and structure of the carrier remains uncertain, there is little doubt that the 9.7 μm feature is produced by the Si–O stretching mode in silicates. In the laboratory, crystalline silicates have multiple narrow features in their 10 μm spectra, while amorphous silicate material has a broad profile.

The interstellar 9.7 μm feature is seen in absorption on a number of sightlines. The observed profiles are broad and relatively featureless, indicative of amorphous silicate material. The observed strength of the absorption feature requires that much, perhaps most, of interstellar Si atoms reside in silicates. Li & Draine (2001a) estimated that at most 5% of interstellar silicate material was crystalline. Conceivably, a mixture of a large number of different crystalline minerals (Bowey & Adamson 2002) could blend together to produce the observed smooth profiles, although it seems unlikely that nature would have produced a blend of crystalline types with no fine structure evident in either absorption or emission, including emission in the far-infrared (Draine 2003a). The $9.7\,\mu$m extinction profile does not appear to be "universal"—sightlines through the diffuse interstellar medium show a *narrower* feature than sightlines through dense clouds (Roche & Aitken 1984; Bowey, Adamson, & Whittet 1998). Evidently the silicate material is altered in interstellar space.

Identification of the $9.7\,\mu$m feature as the Si–O stretching mode is confirmed by the presence of a broad feature centered at $\sim 18\,\mu$m (McCarthy et al. 1980; Smith et al. 2000) that is interpreted as the O–Si–O bending mode in silicates.

Spectral features of crystalline silicates are seen in some circumstellar disks (Artymowicz 2000; Waelkens, Malfait, & Waters 2000) and some comets (Hanner 1999), but even in these objects only a minority of the silicate material is crystalline (e.g., Bouwman et al. 2001).

21.4.4   *3.4 μm C-H Stretch: Aliphatic Hydrocarbons*

Sightlines with sufficient obscuration reveal a broad absorption feature at $3.4\,\mu$m that is identified as the C–H stretching mode in aliphatic (i.e., chain-like) hydrocarbons. Unfortunately, it has not proved possible to identify the specific aliphatic hydrocarbon material, and the band strength of this mode varies significantly from one aliphatic material to another. Sandford et al. (1991) suggest that the $3.4\,\mu$m feature is due to short saturated aliphatic chains incorporating C/H $\approx$ 11 ppm, while Duley et al. (1998) attribute the feature to hydrogenated amorphous carbon (HAC) material containing C/H $\approx$ 85 ppm.

21.4.5   *Diffuse Interstellar Bands*

In addition to narrow absorption features identified as atoms, ions, and small molecules, the observed extinction includes a large number of broader features—known as the "diffuse interstellar bands" (DIBs). The first DIB was recognized over 80 years ago (Heger 1922), and shown to be interstellar 70 years ago (Merrill 1934). A recent survey by Jenniskens & Desert (1994) lists over 154 "certain" DIBs, with another 52 "probable" features. Amazingly, not a single one has yet been positively identified!

High resolution spectroscopy of the 5797 Å and 6614 Å DIBs reveals fine structure that is consistent with rotational bands in a molecule with tens of atoms (Kerr et al. 1996, 1998), and there is tantalizing evidence that DIBs at 9577 Å and 9632 Å may be due to C_{60}^+ (Foing & Ehrenfreund 1994; Galazutdinov et al. 2000; but see also Jenniskens et al. 1997 and Moutou et al. 1999.)

It appears likely that some or all of the DIBs are due to absorption in large molecules or ultrasmall grains. As noted above, a large population of PAHs is required to account for the observed infrared emission, and it is reasonable to suppose that these PAHs may be responsible for many of the DIBs. What is needed now is laboratory gas-phase absorption spectra for comparison with observed DIBs. Until we have precise wavelengths (and band profiles) from gas-phase measurements, secure identification of DIBs will remain problematic.

21.4.6 Extended Red Emission

Interstellar dust grains luminesce in the far-red, a phenomenon referred to as the "extended red emission" (ERE). The highest signal-to-noise ratio observations are in reflection nebulae, where a broad, featureless emission band is observed to peak at wavelengths 6100 Å $\lesssim \lambda_p \lesssim$ 8200 Å, with a FWHM in the range 600–1000 Å (Witt & Schild 1985; Witt & Boroson 1990). The ERE has also been seen in H II regions (Darbon et al. 2000), planetary nebulae (Furton & Witt 1990), and the diffuse interstellar medium of our Galaxy (Gordon, Witt, & Friedmann 1998; Szomoru & Guhathakurta 1998).

The ERE is photoluminescence: absorption of a starlight photon raises the grain to an excited state from which it decays by spontaneous emission of a lower-energy photon. The reported detection of ERE from the diffuse interstellar medium (Gordon et al. 1998; Szomoru & Guhathakurta 1998) appears to require that the interstellar dust mixture have an overall photoconversion efficiency of order ∼10% for photons shortward of ∼5000 Å. If the overall efficiency is ∼10%, the ERE carrier itself must contribute a significant fraction of the overall absorption by interstellar dust at $\lambda \lesssim$ 5000 Å.

Candidate ERE carriers that have been proposed include PAHs (d'Hendecourt et al. 1986) and silicon nanoparticles (Ledoux et al. 1998; Witt, Gordon, & Furton 1998; Smith & Witt 2002). While some PAHs are known to luminesce, attribution of the ERE to PAHs is difficult because of nondetection of PAH emission from some regions where ERE is seen (Sivan & Perrin 1993; Darbon et al. 2000) and nondetection of ERE in some reflection nebulae with PAH emission (Darbon, Perrin, & Sivan 1999). Oxide-coated silicon nanoparticles appear to be ruled out by nondetection of infrared emission at ∼ 20 μm (Li & Draine 2002a). The search to identify the ERE carrier continues.

21.4.7 X-ray Absorption Edges

Dust grains become nearly transparent at X-ray energies, so that the measured X-ray absorption is sensitive to *all* of the atoms, not just those in the gas phase. Consider, for example, K shell absorption by an oxygen atom, where photoabsorption excites one of the $1s$ electrons to a higher (initially vacant) energy level. Transitions to a bound state ($2p$, $3p$, $4p$, ...) produce a series of absorption lines, and transitions to unbound "free" (i.e., "photoelectron") states result in a continuum beginning at the "absorption edge." If the atom is in a solid, the absorption spectrum is modified because the available bound states and free electron states are modified by the presence of other nearby atoms. High-resolution X-ray absorption spectroscopy of interstellar matter could thereby identify the chemical form in which elements are bound in dust grains (Forrey, Woo, & Cho 1998, and references therein).

Figure 21.4 shows the structure expected for scattering, absorption, and extinction near the major X-ray absorption edges in a grain model composed of sp^2-bonded carbon grains plus amorphous $MgFeSiO_4$ grains (Draine 2003c). Spectroscopy with $\lesssim$ 1 eV energy resolution near these absorption edges for both interstellar sightlines and laboratory samples can test this model for the composition of interstellar dust. The *Chandra X-ray Observatory* has measured the wavelength-dependent extinction near the absorption edges of O (Paerels et al. 2001; Takei et al. 2002) and O, Mg, Si, and Fe (Schulz et al. 2002), but it has not yet proved possible to identify the chemical form in which the solid-phase Mg, Si, Fe, and O reside.

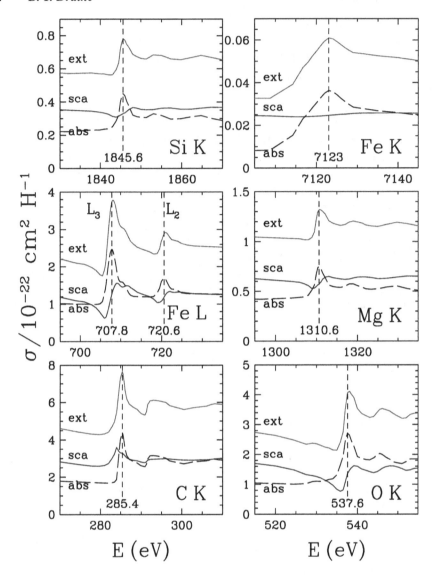

Fig. 21.4. Scattering and absorption cross-sections per H nucleon near principal X-ray absorption edges, for an interstellar dust model consisting of graphite and amorphous silicate grains. (From Draine 2003c.)

21.5   Infrared Emission Spectrum of Interstellar Dust

Interstellar dust grains are heated primarily by absorption of starlight photons. A small fraction of the absorbed starlight energy goes into luminescence or ejection of a photoelectron, but the major part of the absorbed starlight energy goes into heating (i.e., vibrationally exciting) the interstellar grain material. Figure 21.5 shows grain temperature vs. time simulated for four sizes of carbonaceous grains exposed to the average interstellar radiation field. For grain radii $a \gtrsim 100$ Å, individual photon absorptions are relatively fre-

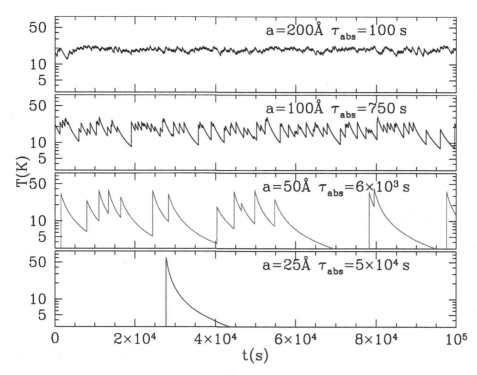

Fig. 21.5. A day in the life of four carbonaceous grains, exposed to the average starlight background. τ_{abs} is the mean time between photon absorptions.

quent, and the grain heat capacity is large enough that the temperature excursions following individual photon absorptions are relatively small; it is reasonable to approximate the grain temperature as approximately constant in time. For $a \lesssim 50$ Å grains, however, the grain is able to cool appreciably in the time between photon absorptions; as a result, individual photon absorption events raise the grain temperature to well above the mean value. To calculate the time-averaged infrared emission spectrum for these grains, one requires the temperature distribution function (see, e.g., Draine & Li 2001).

Figure 21.6 shows the average emission spectrum of interstellar dust, based on observations of the far-infrared emission at high galactic latitudes, plus observations of a section of the Galactic plane where the surface brightness is high enough to permit spectroscopy by the *IRTS* satellite (Onaka et al. 1996; Tanaka et al. 1996).

The similarity of the 5–15 μm spectrum with that of reflection nebulae is evident (see Fig. 21.3). Approximately 21% of the total power is radiated between 3 and 12 μm, with another $\sim 14\%$ between 12 and 50 μm. This emission is from dust grains that are so small that single-photon heating (see Fig. 21.5) is important. The remaining $\sim 65\%$ of the power is radiated in the far-infrared, with λI_λ peaking at ~ 130 μm. At far-infrared wavelengths, the grain opacity varies as $\sim \lambda^{-2}$, and $\lambda I_\lambda \propto \lambda^{-6}/(e^{hc/\lambda kT_d} - 1)$ peaks at $\lambda = hc/5.985kT_d = 134(18\,\mathrm{K}/T_d)$ μm. The emission spectrum for 18 K dust with opacity $\propto \lambda^{-2}$ shown in Figure 21.6 provides a good fit to the observed spectrum for $\lambda \geq 80$ μm, but falls far below the observed emission at $\lambda \leq 50$ μm. From Figure 21.6 it is apparent that $\sim 60\%$ of the radiated

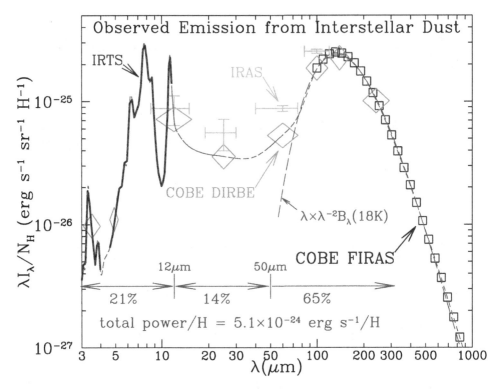

Fig. 21.6. Observed emission spectrum of diffuse interstellar dust in the Milky Way. Crosses: *IRAS* (Boulanger & Perault 1988); squares: *COBE*-FIRAS (Finkbeiner, Davis, & Schlegel 1999); diamonds: *COBE*-DIRBE (Arendt et al. 1998); heavy curve for 3–4.5 μm and 5 – 11.5 μm: *IRTS* (Onaka et al. 1996; Tanaka et al. 1996). The total power, $\sim 5.1 \times 10^{-24}\,\mathrm{erg\,s^{-1}}$/H, is estimated from the interpolated broken line.

power appears to originate from grains that are sufficiently large (radii $a \gtrsim 100$ Å) so that individual photon absorption events do not appreciably raise the grain temperature.

21.6   Dust Grain Size Distribution

A physical grain model of PAHs, carbonaceous grains, and amorphous silicate grains has been constructed to reproduce observations of interstellar extinction and infrared emission (Li & Draine 2001a; Weingartner & Draine 2001a). The material dielectric functions and heat capacities, and absorption cross-sections for the PAHs, are consistent with laboratory data and physics; once the size distribution is specified, the properties of the grain model (scattering, extinction, infrared emission) can be calculated.

The grain size distribution for average diffuse clouds ($R_V \approx 3.1$) in our region of the Milky Way is shown in Figure 21.7. The mass distribution must peak near $\sim 0.3\,\mu$m in order to reproduce the observed extinction near visual wavelengths. The peak in the PAH distribution near $0.0005\,\mu$m is required for single-photon heating to reproduce the observed 3–12 μm emission, but the secondary peak near $0.005\,\mu$m in Figure 21.7 may be an artifact of the fitting procedure.

Milky Way extinction curves with different R_V values, or extinction curves for the LMC

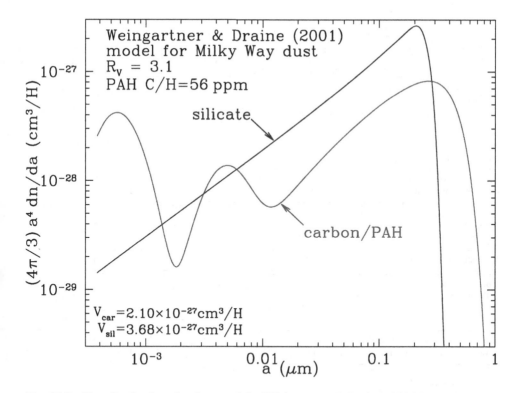

Fig. 21.7. Size distributions for dust model of Weingartner & Draine (2001a).

and SMC, can be reproduced by varying the size distribution. The grain model appears to be consistent with the observed scattering properties of interstellar dust in the optical and ultraviolet (Draine 2003b) and at X-ray energies (Draine 2003c).

The grain model also reproduces the observed infrared emission in the diffuse interstellar medium of the Milky Way (Li & Draine 2001a), reflection nebulae (Li & Draine 2002b), and the SMC Bar (Li & Draine 2002c).

21.7   Dust Abundances vs. Depletion Patterns

Jenkins (2004) finds that observed depletions on many sightlines suggest two types of grain material: a "core" material that is returned to the gas phase only in very high velocity shock waves, and a "mantle" material that is more easily stripped. For low velocity gas, variations in depletion from one sightline to another are interpreted as due to invariant grain cores plus varying amounts of mantle material.

What do Jenkins' abundances of elements in the cores and mantles suggest as regards the volumes of cores and mantles? Table 21.1 shows one possible mix of minerals that could reproduce the overall elemental compositions deduced by Jenkins for sightlines with "depletion multiplier" $F_* = 1$, representative of "cool disk" material. The aims of this exercise are to estimate (1) how much oxygen is likely to be in the grain material and (2) the volumes of core and mantle material that would be consistent with the depletion patterns found by Jenkins, for comparison with the grain volume required by physical dust models. The compositions in Table 21.1 are purely illustrative and should *not* be taken to be realistic.

Table 21.1. *Jenkins (2004) Grain Composition: One Illustrative Possibility*

Material	C [a]	O [a]	Mg [a]	Si [a]	Al [a]	Ca [a]	Fe [a]	Ni [a]	ρ [b]	V [c]
Grain Cores										
C,PAH,HAC,...	71	-	-	-	-	-	-	-	2.2	6.5
$MgFeSiO_4$ olivine	-	52	13	13	-	-	13	-	3.8	9.8
$CaMgSiO_4$ monticellite	-	8	2	2	-	2	-	-	3.2	1.6
Fe_2O_3 hematite	-	18	-	-	-	-	12	-	5.3	3.0
Al_2O_3 corundum	-	4.5	-	-	3	-	-	-	4.02	0.6
Ni_2O_3 dinickel trioxide	-	2.4	-	-	-	-	-	1.6	4.84	0.5
Illustrative Core Total	71	85	15	15	3	2	25	1.6	3.5	22.1
Observed Core Total [d]	71^{+61}_{-71}	53^{+49}_{-53}	15	14	3.0	2.2	25	1.6		
Grain Mantles										
C,PAH,HAC,...	35	-	-	-	-	-	-	-	2.2	3.2
$Mg_{0.9}Fe_{0.1}SiO_3$ pyroxene	-	57	17	19	-	-	2	-	3.3	9.9
Illustrative Mantle Total	35	57	17	19	-	-	2	-	3.5	13.1
Cores + Mantles										
C,PAH,HAC,...	106	-	-	-	-	-	-	-	2.2	9.7 [e]
silicates	-	117	32	34	-	2	15	-	3.5	21.4 [e]
other	-	24	-	-	3	-	12	1.6	5.2	4.0
Illustrative Core + Mantle Total	106	142	32	34	3	2	27	1.6	3.5	35.2 [e]
Observed Core + Mantle Total [d]	106^{+16}_{-20}	134^{+22}_{-23}	32	33	3.0	2.2	28	1.8		

[a] Atomic abundance (ppm) per total H.

[b] Solid density ($g\,cm^{-3}$).

[c] Grain volume per total H ($10^{-28}\,cm^3$).

[d] From Jenkins (2004). Quoted uncertainties do not include uncertainties in assumed total abundances.

[e] Models that reproduce the observed interstellar extinction per H require a greater volume of grain material than provided by the depletions found by Jenkins (2004). A model based on carbonaceous grains plus silicate grains (see Draine 2003a) has $V = 37 \times 10^{-28}\,cm^3/H$ for silicate grains and $V = 21 \times 10^{-28}\,cm^3/H$ for carbonaceous grains (see text).

In Table 21.1 the overall composition of the core material is consistent with Jenkins' results. The amounts of Mg, Fe, and Si in the mantle require $\sim$57 ppm O if combined into pyroxene—consistent with the O/H = 81^{+54}_{-58} ppm that Jenkins assigns to the mantle.

Studies of total depletions infer the amount of grain material based on an assumed value for the total abundance of each element; Jenkins adopted current estimates for solar abundances. It is important to remember that estimates of solar abundances have varied considerably over time; the review by Anders & Grevesse (1989) had C/H= 363 ± 35 ppm and O/H = 851 ± 72 ppm, whereas the most recent redeterminations find C/H= 246 ± 23 ppm (Allende Prieto, Lambert, & Asplund 2001) and O/H = 490 ± 47 ppm (Allende Prieto et al. 2002)—C/H has gone down by a factor 1.5, and O/H by a factor 1.7! Tomorrow's "solar abundance" values may differ from today's. Furthermore, interstellar abundances may not be equal to solar, or even to the abundances in young stars, as discussed in §21.10 below.

The total grain volume indicated by Jenkins' abundance estimates is $35 \times 10^{-28}\,cm^3/H$, only $\sim$60% of the total grain volume for the physical dust model of Weingartner & Draine (2001a). The dust modeling has approximated the dust as solid spheres, and it is expected that nonspherical porous grains would allow the extinction to be reproduced with a slightly reduced total solid volume, but it is not clear that this would reduce the required solid volume by 40%.

One is compelled to consider seriously the possibility that interstellar abundances of the depleted elements—especially C—may exceed solar abundances. This will be discussed further in §21.10.

21.8 Time Scale for Depletion

Lifetimes of grains in H I clouds against destruction in supernova blast waves are only $\sim 3 \times 10^8$ yr (e.g, Draine & Salpeter 1979; Jones et al. 1994). How then can we understand the very small gas-phase abundances routinely found for elements like Si? The kinetics of depletion have been studied by Weingartner & Draine (1999), who show that the population of very small grains is capable of accreting metal ions rapidly enough to achieve the required depletions, provided that interstellar material is rapidly cycled between dense clouds and diffuse clouds.

21.9 Local Variations in D/H: Does Dust Play a Role?

The evidence concerning primordial abundances of D, He, and Li appear to be consistent with D/H $\approx 28 \pm 4$ ppm produced by nucleosynthesis in the early Universe (O'Meara et al. 2001; Kirkman et al. 2004). Observations of D/H in the Milky Way have been reviewed by Linsky (2004). The local interstellar medium has a weighted mean D/H $= 15.2 \pm 0.8$ ppm (1-σ uncertainties) within ~ 180 pc of the Sun (Moos et al. 2002). This value could be consistent with "astration" if $\sim 50\%$ of the H atoms now in the interstellar medium have previously been in stars that burnt D to ^{3}He. Observations appear to find spatial variations in the D/H ratio in the interstellar medium within ~ 500 pc, with values ranging from $7.4^{+1.9}_{-1.3}$ ppm toward δ Orionis (Jenkins et al. 1999) to $21.8^{+3.6}_{-3.1}$ ppm toward γ^2 Vel (Sonneborn et al. 2000). On longer sightlines in the Galactic disk, Hoopes et al. (2003) find D/H $= 7.8^{+2.6}_{-1.3}$ ppm toward HD 191877 ($d = 2200 \pm 550$ pc) and $8.5^{+1.7}_{-1.2}$ ppm toward HD 195965 ($d = 800 \pm 200$ pc). These variations in D/H are usually interpreted as indicating variations in "astration," with as much as $\sim 75\%$ of the D on the sightline to δ Ori having been "burnt," vs. only $\sim 20\%$ of the D toward γ^2Vel. Such large variations in astration between regions situated just a few hundred pc apart would be surprising, since the interstellar medium appears to be sufficiently well mixed that large local variations in the abundances of elements like N or O are not seen outside of recognizable stellar ejecta such as planetary nebulae or supernova remnants.

Jura (1982) pointed out that interstellar grains could conceivably sequester a significant amount of D. Could the missing deuterium conceivably be in dust grains?

Let us suppose that dust grains contain 200 ppm C relative to total H (as in the dust model of Weingartner & Draine 2001a) with ~ 60 ppm in PAHs containing $N_C \lesssim 10^4$ atoms.

The solid carbon will be hydrogenated to some degree. The most highly pericondensed PAH molecules (coronene $C_{24}H_{12}$, circumcoronene $C_{54}H_{18}$, dicircumcoronene $C_{96}H_{24}'$) have H/C $= \sqrt{6/N_C}$, where N_C is the number of C atoms; other PAHs have higher H/C ratios for a given N_C. Let us suppose that the overall carbon grain material—including small PAHs and larger carbonaceous grains—has H/C $= 0.25$.

The carbonaceous grain population would then contain ~ 50 ppm of hydrogen. If $\sim 20\%$ of the hydrogen in the carbonaceous grains was deuterium, the deuterium in the grains would then be $(D)_{grain}/(H)_{total} \approx 10$ ppm. If the total D/H $= 20$ ppm, this would reduce the gas-phase D/H to 10 ppm, comparable to the value observed toward δ Ori.

Is it conceivable the D/H ratio in dust grains could be $\sim 10^4$ times higher than the overall D/H ratio? Some interplanetary dust particles have D/H as high as 0.0017 (Messenger & Walker 1997; Keller, Messenger, & Bradley 2000), although this factor $\sim$85 enrichment (relative to D/H = 20 ppm) is still 2 orders of magnitude less than what is required to significantly affect the gas-phase D/H value. Extreme D enrichments are seen in some interstellar molecules—D_2CO/H_2CO ratios in the range of 0.01–0.1 are seen (Ceccarelli et al. 2001; Bacmann et al. 2003), and attributed to chemistry on cold grain surfaces in dense clouds.

Could such extreme enrichments occur in the diffuse interstellar medium? The thermodynamics is favorable. The H or D would be bound to the carbon via a C-H bond. The C-H bond—with a bond strength $\sim 3.5\,\mathrm{eV}$—has a stretching mode at $\lambda_{CH} = 3.3\,\mu$m, while the C-D bond, with a larger reduced mass, has its stretching mode at $\lambda_{CD} \approx \sqrt{2}\lambda_{CH} \approx 4.67\,\mu$m. Because of the difference in zero-point energy, the difference in binding energies is

$$\Delta E_{CD-CH} = \frac{hc}{2}\sum_{j=1}^{3}(\lambda_{CH,j}^{-1} - \lambda_{CD,j}^{-1}) \approx \frac{hc}{2}\left(1 - \frac{1}{\sqrt{2}}\right)\sum_{j=1}^{3}\lambda_{CH,j}^{-1} = 0.092\,\mathrm{eV}\,, \quad (21.2)$$

where the sum is over the stretching, in-plane bending, and out-of-plane bending modes, with $\lambda_{CH} = 3.3$, 8.6, and 11.3 μm. This exceeds the difference $\Delta E_{HD-H_2} = 0.035\,\mathrm{eV}$ in binding energy between HD and H_2. It is therefore energetically favored for impinging D atoms to displace bound H atoms via reactions of the form (Bauschlicher 1998)

$$C_lD_mH_n^+ + H \quad \rightarrow \quad C_lD_mH_{n+1}^+ \quad\quad\quad\quad\quad\quad\quad\quad\quad\quad (21.3)$$

$$C_lD_mH_{n+1}^+ + D \quad \rightarrow \quad C_lD_{m+1}H_{n-1}^+ + H_2 \quad \text{branching fraction } f_1 \quad (21.4)$$

$$\text{versus} \quad\quad\quad\quad \rightarrow \quad C_lD_mH_n^+ + HD \quad \text{branching fraction } f_2\,. \quad (21.5)$$

The branching ratio $f_1/f_2 \approx \exp[(\Delta E_{CD-CH} - \Delta E_{HD-H_2})/kT_d] > 10^4$ if $T_d \lesssim 70\,\mathrm{K}$. If no other reactions affect the grain hydrogenation, then the grains would gradually become D enriched.

However, the interstellar medium is far from LTE—the hydrogen is atomic (rather than molecular) and, indeed, partially ionized because of the presence of ultraviolet photons, X-rays, and cosmic rays. It is not yet clear whether the mixture of non-LTE reactions will allow the grains to become deuterated to a level approaching D/H $\approx 1/4$, but it seems possible that this may occur. Deuterated PAHs would radiate in the C-D stretching and bending modes at ~ 4.67, 12.2, and 16.0 μm. The 12.2 μm emission will be confused with C-H out-of-plane bending emission (see Fig. 21.3), but the other two modes should be searched for.

Even if extreme D-enrichment of carbonaceous grains is possible, it will take time to develop. Meanwhile, the gas in which the grain is found may undergo a high-velocity shock, with grain destruction by a combination of sputtering and ion field emission in the high-temperature postshock gas. D incorporated into dust grains would be released and returned to the gas phase if those dust grains are destroyed. PAHs, in particular, would be expected to be easily sputtered in shock-heated gas; destruction by ion field emission would be expected to be even more rapid. Thus, if D is depleted into dust grains, we would expect to see the gas-phase D/H to be larger in recently shocked regions. This could explain the large D/H value observed by Sonneborn et al. (2000) toward γ^2 Vel.

It should be noted that significant depletion of D from the gas can only occur if there is sufficient carbonaceous grain material to retain the D. This can occur if gas-phase abundances are approximately solar (with $\sim$200 ppm C in dust) but would not be possible for abundances significantly below solar (e.g., the abundances in the LMC and SMC). The factor-of-two variations in D/H seen in the local interstellar medium would *not* be possible in gas with metallicities characteristic of the LMC, SMC, or high-velocity clouds such as "Complex C" (see Jenkins 2004).

21.10 Transport of Elements in Dust Grains

Elements like Mg, Si, Al, Ca, Ti, Fe, and Ni are concentrated in interstellar dust grains. Since dust grains can move through the gas, transport in dust grains could produce local variations in elemental abundances.

21.10.1 Anisotropic Starlight

Starlight is typically anisotropic, as can be confirmed by viewing the night sky on a clear night. The anisotropy is a function of wavelength, with larger anisotropies at shorter wavelengths because ultraviolet is (1) more strongly attenuated by dust grains and (2) originates in a smaller number of short-lived stars that are clustered. At the location of the Sun, Weingartner & Draine (2001b) found starlight anistropies ranging from 3% at 5500 Å to 21% at 1565 Å.

The dust grains are charged, and fairly well coupled to the magnetic field lines. Let θ_B be the angle between the starlight anisotropy direction and the magnetic field. The drift tends to be approximately parallel to the magnetic field direction, with a magnitude $v_d \approx v_{d0} \cos \theta_B$; for a starlight anisotropy of 10%, $v_{d0} \approx 0.5$ km s^{-1} in the "warm neutral medium," and 0.03 km s^{-1} in the "cold neutral medium" (Weingartner & Draine 2001b).

If the radiation anisotropy is stable for $\sim 10^7$ yr, a dust grain in the cold neutral medium would be driven $0.3 \cos \theta_B$ pc from the gas element in which it was originally located.

If the magnetic field is nonuniform, spatial variations in the drift velocity can then lead to variations in the dust/gas ratio. Indeed, if there are "valleys" in the magnetic field (with respect to the direction of radiation anisotropy), the dust grains would tend to be concentrated there. Radiation pressure acting on the concentrated dust grains would in fact cause such perturbations in the magnetic field to be unstable.

Dust impact detectors on the *Ulysses* and *Galileo* spacecraft measure the flux and mass distribution of interstellar grains entering the solar system, finding a much higher flux of very large ($a \gtrsim 0.5\,\mu$m) dust grains than would be expected for the average interstellar grain size distribution (Frisch et al. 1999). The inferred local dust size distribution is very difficult to reconcile with the average interstellar grain population (Weingartner & Draine 2001a), but we must keep in mind that the dust grains entering the solar system over a time scale of a few years are sampling a tiny region of the interstellar medium of order a few AU in size. It is possible that the solar system just happens to be passing through a region in the local interstellar cloud that has been enhanced in the abundance of large grains due to radiation-driven grain drift.

21.10.2 Star-forming Regions

Various processes could alter the gas-to-dust ratio in the material forming a star:

(1) Motion of dust through gas can result from gravitational sedimentation of dust grains in star-forming clouds (Flannery & Krook 1978). This could lead to enhanced abundances in stars of those elements that are depleted into dust in star-forming clouds.

(2) Star formation is accompanied by ambipolar diffusion of magnetic field out of the contracting gas cloud. Charged dust grains, while not perfectly coupled to the magnetic field, will tend to drift outward, resulting in reduction of the abundances of the depleted elements in the star-forming core (Ciolek & Mouschovias 1996).

(3) Gravitational sedimentation in an accretion disk can concentrate dust at the midplane. Gammie (1996) has suggested "layered accretion," in which viscous stresses are effective along the surface layers of the disk, but the midplane is a quiescent "dead zone" as far as the magnetorotational instability is concerned. This could suppress stellar abundances of depleted elements.

(4) In thick disks, small bodies orbit more rapidly than the gas, with gas drag leading to radial infall of these bodies. This could enhance stellar abundances of depleted species.

(5) Before accretion terminates, a massive star may attain a pre-main sequence luminosity high enough for radiation pressure to drive a drift of dust grains away from it. This would suppress stellar abundances of depleted elements.

Given these competing mechanisms, the net effect could be of either sign, but, as pointed out by Snow (2000), it would not be suprising if stars had abundances that differed from the overall abundances of the interstellar medium out of which they formed, and it would also not be surprising if the resulting abundances depended on stellar mass. In this connection it is interesting to note that the study by Sofia & Meyer (2001) finds that abundances of Mg and Si in B stars appear to be significantly below the abundances of these elements in young F and G disk stars, or in the Sun; they conclude that the abundances in B stars do not provide a good representation of interstellar abundances.

21.11   Ion Recombination on Dust Grains

As discussed by Jenkins (2004), determinations of elemental abundances from interstellar absorption-line observations may require estimation of the fraction in unobserved ionization states, such as Na II, K II, or Ca III. Such estimates require knowledge of the rates for ionization and recombination processes. Neutral or negatively charged dust grains provide a pathway for ion neutralization that for some cases can be faster than ordinary radiative recombination with free electrons (Weingartner & Draine 2002).

21.12   Summary

This Symposium has been concerned with the origin and evolution of the elements. The main points of the present paper are as follows:

(1) The chemical composition of interstellar dust remains uncertain. PAH molecules are present, and a substantial fraction of the grain mass most likely consists of amorphous silicate, but precise compositional information still eludes us.

(2) The grain size distribution is strongly constrained by observations of extinction, scaterring, and infrared emission, and extends from grains containing just tens of atoms to grains with radii $a \gtrsim 0.3 \, \mu$m.

(3) Dust grains will drift through interstellar gas, and this transport process could produce local variations in the dust/gas ratio.

(4) The dust mass required to account for interstellar extinction using homogeneous spherical grains exceeds that inferred from depletion studies (Jenkins 2004) by a factor ~ 1.5.

(5) Dust grains could possibly deplete a significant fraction of interstellar D. This mechanism

should be considered as a possible explanation for observed variations in the interstellar D/H ratio.

Acknowledgements. I thank Ed Jenkins and Todd Tripp for many valuable discussions, and Robert Lupton for availability of the SM software package. This research was supported in part by NSF grant AST-9988126.

References

Allende Prieto, C., Lambert, D. L., & Asplund, M. 2001 ApJ, 556, L63
——. 2002, ApJ, 573, L137
Anders, E. A., & Grevesse, N. 1989, Geochim. Cosmochim. Acta, 53, 197
Arendt, R. G., et al. 1998, ApJ, 508, 74
Artymowicz, P. 2000, Space Sci. Rev., 92, 69
Bacmann, A., Lefloch, B., Ceccarelli, C., Steinacker, J., Castets, A., & Loinard, L. 2003, ApJ, 585, L55
Bauschlicher, C. W. 1998, ApJ, 509, L125
Boulanger, F., & Perault, M. 1988, ApJ, 330, 964
Bouwman, J., Meeus, G., de Koter, A., Hony, S., Dominik, C., & Waters, L. B. F. M. 2001, A&A, 375, 950
Bowey, J. E., & Adamson, A. J. 2002, MNRAS, 334, 94
Bowey, J. E., Adamson, A. J, & Whittet, D. C. B. 1998, MNRAS, 298, 131
Ceccarelli, C., Loinard, L., Castets, A., Tielens, A. G. G. M., Caux, E., Lefloch, B., & Vastel, C. 2001, A&A, 373, 998
Cardelli, J. A., Clayton, G. C., & Mathis, J. S. 1989, ApJ, 345, 245
Cesarsky, D., Lequeux, J., Abergel, A., Perault, M., Palazzi, E., Madden, S., & Tran, D. 1996, A&A, 315, L305
Ciolek, G. E., & Mouschovias, T. C. 1996, ApJ, 468, 749
Clayton, D. D., & Nittler, L. R. 2004, in Carnegie Observatories Astrophysics Series, Vol. 4: Origin and Evolution of the Elements, ed. A. McWilliam & M. Rauch (Cambridge: Cambridge Univ. Press), in press
Darbon, S., Perrin, J. M., & Sivan, J. P. 1999, A&A, 348, 990
Darbon, S., Zavagno, A., Perrin, J. M., Savine, C., Ducci, V., & Sivan, J. P. 2000, A&A, 364, 723
d'Hendecourt, L. B., Léger, A., Olofson, G., & Schmidt, W. 1986, A&A, 170, 91
Draine, B. T. 1989, in IAU Symp. 135, Interstellar Dust, ed. L. J. Allamandola & A. G. G. M. Tielens (Dordrecht: Kluwer), 313
——. 1990, in Evolution of the Interstellar Medium, ed. L. Blitz (San Francisco: ASP), 193
——. 2003a, ARA&A, 41, 241
——. 2003b, ApJ, 598, 1017
——. 2003c, ApJ, 598, 1026
——. 2003d, in The Cold Universe, Saas-Fee Advanced Course 32, ed. D. Pfenniger (Berlin: Springer-Verlag), 000 (astro-ph/0304489)
Draine, B. T., & Li, A. 2001, ApJ, 551, 809
Draine, B. T., & Salpeter, E. E. 1979, ApJ, 231, 438
Duley, W. W., Scott, A. D., Seahra, S., & Dadswell, G. 1998, ApJ, 571, L117
Finkbeiner, D. P., Davis, M., & Schlegel, D. J. 1999, ApJ, 524, 867
Fitzpatrick, E. L. 1999, PASP, 111, 63
Fitzpatrick, E. L., & Massa, D. 1990, ApJS, 72, 163
Flannery, B. P., & Krook, M. 1978, ApJ, 223, 447
Foing, B. H., & Ehrenfreund, P. 1994, Nature, 369, 296
Forrey, R. C., Woo, J. W., & Cho, K. 1998, ApJ, 505, 236
Frisch, P. C., et al. 1999, ApJ, 525, 492
Furton, D. G., & Witt, A. N. 1990, ApJ, 364, L45
Galazutdinov, G. A., Krelowski, J., Musaev, F. A., Ehrenfreund, P., & Foing, B. H. 2000, MNRAS, 317, 750
Gammie, C. F. 1996, ApJ, 457, 355
Gordon, K. D., Witt, A. N., & Friedmann, B. C. 1998, ApJ, 498, 522
Hanner, M. 1999, Space Sci. Rev., 90, 99
Heger, M. L. 1922, Lick Obs. Bull. , 10, 146
Hoopes, C. G., Sembach, K. R., Hébrard, G., Moos, H. W., & Knauth, D. C. 2003, ApJ, 586, 1094

Jenkins, E. B. 2004, in Carnegie Observatories Astrophysics Series, Vol. 4: Origin and Evolution of the Elements, ed. A. McWilliam & M. Rauch (Cambridge: Cambridge Univ. Press), in press

Jenkins, E. B., Tripp, T. M., Wozniak, P. R., Sofia, U. J., & Sonneborn, G. 1999, ApJ, 520, 182

Jenniskens, P., & Desert, F. X. 1994, A&AS, 106, 39

Jenniskens, P., Mulas, G., Porceddu, I., & Benvenuti, P. 1997, A&A, 327, 337

Jones, A. P., Tielens, A. G. G. M., Hollenbach, D. J., & McKee, C. F. 1994, ApJ, 433, 797

Jura, M. 1982, in Advances in Ultraviolet Astronomy, ed. Y. Kondo, (NASA CP-2238), 54

Keller, L. P., Messenger, S., & Bradley, J. P. 2000, J. Geophys. Res. , 105, 10397

Kerr, T. H., Hibbins, R. E., Fossey, S. J., Miles, J. R., & Sarre, P. J. 1998, ApJ, 495, 941

Kerr, T. H., Hibbins, R. E., Miles, J. R., Fossey, S. J., Sommerville, W. B., & Sarre, P. J. 1996, MNRAS, 283, 1104

Kirkman, D., Tytler, D., Suzuki, N., O'Meara, J. M., & Lubin, D. 2004, ApJS, submitted (astro-ph/0302006)

Krügel, E. (2003), The Physics of Interstellar Dust (Bristol: Inst. of Physics)

Ledoux, G., et al. 1998, A&A, 333, L39

Li, A., & Draine, B. T. 2001a, ApJ, 550, L213

——. 2001b, ApJ, 554, 778

——. 2002a, ApJ, 564, 803

——. 2002b, ApJ, 569, 232

——. 2002c, ApJ, 576, 762

Linsky, J. L. 2004, in Carnegie Observatories Astrophysics Series, Vol. 4: Origin and Evolution of the Elements, ed. A. McWilliam & M. Rauch (Pasadena: Carnegie Observatories, http://www.ociw.edu/ociw/symposia/series/symposium4/proceedings.html)

McCarthy, J. F., Forrest, W. J., Briotta, D. A., & Houck, J. R. 1980, ApJ, 242, 965

Merrill, P. W. 1934, PASP, 46, 206

Messenger, S., & Walker, R. M. 1997, in Astrophysical Implications of the Laboratory Study of Presolar Materials, ed. T. J. Bernatowitcz & E. K. Zinner (New York: AIP), 545

Moos H. W., et al. 2002, ApJS, 140, 3

Moutou, C., Sellgren, K., Verstraete, L., & Léger, A. 1999, A&A, 347, 949

O'Meara, J. M., Tytler, D., Kirkman, D., Suzuki, N., Prochaska, J. X., Lubin, D., & Wolfe, A. M. 2001, ApJ, 552, 718

Onaka, T., Yamamura, I., Tanabe, T., Roellig, T. L., & Yuen, L. 1996, Proc. Astr. Soc. Japan, 48, L59

Paerels, F., et al. 2001, ApJ, 546, 338

Rachford, B. L., et al. 2002, ApJ, 577, 221

Roche, P. F., & Aitken, D. K. 1984, MNRAS, 208, 481

Sandford, S. A., Allamandola, L. J., Tielens, A. G. G. M., Sellgren, K., Tapia, M., & Pendleton, Y. 1991, ApJ, 371, 607

Schulz, N. S., Cui, W., Canizares, C. R., Marshall, H. L., Lee, J. C., Miller, J. M., & Lewin, W. H. G. 2002, ApJ, 556, 1141

Sivan, J. P., & Perrin, J. M. 1993, ApJ, 404, 258

Smith, C. H., Wright, C. M., Aitken, D. K., Roche, P. F., & Hough, J. H. 2000, MNRAS, 312, 327

Smith, T. L., & Witt, A. N. 2002, ApJ, 565, 304

Snow, T. P. 2000, J. Geophys. Res. , 105, 10239

Sofia, U. J., & Meyer, D. M. 2001, ApJ, 554, L221 (erratum: 558, L147)

Sonneborn, G., Tripp, T. M., Ferlet, R., Jenkins, E. B., Sofia, U. J., Vidal-Madjar, A., & Wozniak, P. R. 2000, ApJ, 545, 277

Stecher, T. P., & Donn, B. 1965, ApJ, 142, 1681

Szomoru, A., & Guhathakurta, P. 1998, ApJ, 494, L93

Takei, Y., Fujimoto, R., Mitsuda, K., & Onaka, T. 2002, ApJ, 581, 307

Tanaka, M., Matsumoto, T., Murakami, H., Kawada, M., Noda, M., & Matsuura, S. 1996, Proc. Astr. Soc. Japan, 48, L53

Waelkens, C., Malfait, K., & Waters, L. B. F. M. 2000, in IAU Symp. 197, Astrochemistry, ed. Y. C. Minh & E. F. van Dishoeck (San Francisco: ASP), 435

Weingartner, J. C., & Draine, B. T. 1999, ApJ, 517, 292

——. 2001a, ApJ, 548, 296

——. 2001b, ApJ, 553, 581

——. 2002, ApJ, 563, 842

Whittet, D. C. B. 2003, Dust in the Galactic Environment, 2nd Ed. (Bristol: IOP)

Witt, A. N., & Boroson, T. A. 1990, ApJ, 355, 182

Witt, A. N., Gordon, K. L., & Furton, D. G. 1998, ApJ, 501, L111
Witt, A. N., & Schild, R. E. 1985, ApJ, 294, 225

22

Interstellar atomic abundances

EDWARD B. JENKINS
Princeton University Observatory

Abstract
A broad array of interstellar absorption features that appear in the ultraviolet spectra of bright sources allows us to measure the abundances and ionization states of many important heavy elements that exist as free atoms in the interstellar medium. By comparing these abundances with reference values in the Sun, we find that some elements have abundances relative to hydrogen that are approximately consistent with their respective solar values, while others are depleted by factors that range from a few up to around 1000. These depletions are caused by the atoms condensing into solid form onto dust grains. Their strengths are governed by the volatility of compounds that are produced, together with the densities and velocities of the gas clouds. We may characterize the depletion trends in terms of a limited set of parameters; ones derived here are based on measurements of 15 elements toward 144 stars with known values of $N(\text{H I})$ and $N(\text{H}_2)$. In turn, these parameters may be applied to studies of the production, destruction, and composition of the dust grains. The interpretations must be done with care, however, since in some cases deviations from the classical assumptions about missing atoms in unseen ionization stages can create significant errors. Our experience with the disk of our Galaxy offers important lessons for properly unravelling results for more distant systems, such as high- and intermediate-velocity clouds in the Galactic halo, material in the Magellanic Stream and Magellanic Clouds, and otherwise invisible gas systems at large redshifts (detected by absorption features in quasar spectra). Inferences about the total (gas plus dust) abundances of such systems offer meaningful information on their origins and/or chemical evolution.

22.1 Introduction

For those who study the interstellar medium (ISM), the manifold states that atoms can assume in space are both a benefit and a hindrance. On the favorable side, the apportionments of atoms in various ionization stages and the fractions that become bound in molecular compounds (including solids) reveal the outcomes of fundamental processes that arise from various physical and chemical influences, and these, in turn, disclose the nature of nearby objects or conditions that created these circumstances. Conversely, if it turns out that we have a poor understanding of the fractions of atoms that exist in unseen states, we are hampered in our ability to gauge the relative total abundances of different elements. These two considerations underscore the importance of our mastering the principles that govern how the atoms are subdivided into different states within environments that range from the local part of our Galaxy to the most distant gas systems that we can detect in the Universe. A major theme in this article will be to cover the underlying principles of this topic.

Viewing atomic species by their absorption features in the spectra of background continuum sources, such as stars or quasars, offers the simplicity of not having to understand the details of how the atoms are excited, as we must do if we are observing emission lines arising from collisional excitation or recombination. Moreover, the radiative transfer of absorption features in the optical and ultraviolet regions is straightforward, since the attenuation of light follows a simple exponential absorption law. Studies of absorption lines have revealed a broad assortment of findings on the state and composition of interstellar gases, and we review here many of these conclusions.

At this point, it is appropriate to digress and offer a brief historical perspective. The papers in this volume arose from one of four symposia that celebrated the founding of the Carnegie Observatories exactly 100 years ago. We may reflect on the fact that some two years after that event, Hartmann (1904) reported that stationary Ca II lines in the spectrum of the spectroscopic binary δ Ori arose from the intervening medium, rather than from the atmospheres of the stars. As with many important discoveries, doubts about the interpretation remained (Young 1922), but eventually the interstellar nature of these features was firmly established (Plaskett & Pearce 1933). There soon followed large surveys of interstellar lines, with Carnegie astronomers playing an important role in the process (Merrill et al. 1937; Adams 1949). Nowadays, modern echelle spectrographs with very high resolving powers show exquisite recordings of the complex superpositions of narrow velocity features, and these data allow one to study either weak features or those arising from very small parcels of gas along the line of sight to the target stars (e.g., Blades, Wynne-Jones, & Wayte 1980; Welsh, Vedder, & Vallerga 1990; Welty, Hobbs, & Kulkarni 1994; Barlow et al. 1995; Crane, Lambert, & Sheffer, 1995; Welty, Morton, & Hobbs 1996; Crawford 2001; Welty & Hobbs 2001; Knauth, Federman, & Lambert 2003; Welty, Hobbs, & Morton 2003).

Against the backdrop of almost 100 years of studies of the interstellar medium using absorption features in the visible part of the spectrum, we celebrate yet another anniversary. Some 30 years ago the first papers describing results from the *Copernicus* satellite (Rogerson et al. 1973a), launched in 1972, demonstrated the extraordinary utility of using the ultraviolet part of the spectrum to study a vast array of features arising from important interstellar atomic constituents (Morton et al. 1973; Rogerson et al. 1973b), along with our first insights on the excitation and widespread presence of molecular hydrogen (Spitzer et al. 1973; Spitzer & Cochran 1973). Spitzer & Jenkins (1975) reviewed many of the principal conclusions that came from this initial venture into the ultraviolet.

Additional inferences on the nature of the ISM arose from the *International Ultraviolet Explorer (IUE)* (Kondo et al. 1987), which bridged the *Copernicus* era and that of the *Hubble Space Telescope (HST)*, launched in 1990. *HST* still functions as a premier facility for gathering useful observations of interstellar absorption lines through the use of a new instrument, the Space Telescope Imaging Spectrograph (STIS) that was installed in 1997 (Kimble et al. 1998; Woodgate et al. 1998). A comprehensive review of the findings on interstellar abundances gathered from *HST* up to 1996 [using a first-generation spectrograph, the Goddard High-Resolution Spectrograph (GHRS)], has been written by Savage & Sembach (1996). Finally, the *Far Ultraviolet Spectroscopic Explorer (FUSE)* (Moos et al. 2000) offers an opportunity to revisit the spectral region $912-1185$ Å with high sensitivity, a band that has been mostly inaccessible since the days of *Copernicus*.

In parallel with investigations of the Galactic ISM using observatories in Earth orbit, large-aperture telescopes on the ground have allowed us to study absorptions by the same UV transitions arising from gaseous material at very large redshifts. Much of the progress in understanding the abundances in these distant systems has drawn upon lessons learned from studies of Galactic material, where we have better insights on the fundamental abundance patterns (from the Sun and stars) and the physical conditions. Recent reviews by Pettini (2003) and Calura et al. (2004) summarize the current state of many of the conclusions on highly redshifted systems that have arisen from such studies.

In the sections that follow, we will explore how the apparent abundances of elements are altered by the effects of ionization and depletion onto dust grains. We bypass discussions of the techniques employed to study the absorption features. The reader is referred to a number of papers that cover some key points on the methodology of this type of research (Cowie & Songaila 1986; Jenkins 1986, 1996; Savage & Sembach 1991; Sembach & Savage 1992), plus a list of possible pitfalls in its conduct (Jenkins 1987).

22.2   Ionization Corrections

Of all the species that have absorption features in the visible part of the spectrum, Ti II is the only one that represents an element in its favored stage of ionization within H I regions. The remaining species, Li I, Ca I, Ca II, Na I, and K I all have an ionization potential less than that of H I, with the consequence that they are mostly ionized to higher stages by UV starlight photons that can easily penetrate the regions. One may attempt to derive the gas-phase abundance of an element X from solutions of the ionization equilibrium equation involving the element's ionization rate $\Gamma(X)$ from starlight photons, balanced against its rate of recombination governed by the rate coefficients $\alpha(X, e)$ with free electrons and $\alpha(X, g)$ with negatively charged or neutral, very small dust grains [at a rate that is normalized to the hydrogen density $n(H)^*$] (Weingartner & Draine 2001). That is, in principle the balance

$$n(X^0)\Gamma(X) = n(X^+)[\alpha(X, e)n(e) + \alpha(X, g)n(H)] \tag{22.1}$$

may be solved to yield a total abundance of X if the parameters are well defined. If the local values of $\Gamma(X)$ and the electron and hydrogen densities $n(e)$ and $n(H)$ are poorly known, it is still possible to take Equation 22.1 for element X and divide it by that for another element (Y) to derive a relative abundance of the two (X/Y). While this may seem to be an attractive solution for studying abundance trends, recent comparisons made by Welty, Hobbs & Morton (2003) indicate that different elements showing features in the ultraviolet show mutually inconsistent values for the ratios of adjacent ionization stages when Equation 22.1 is invoked in the respective cases. These disparities may arise from either errors in the atomic constants or, perhaps in some cases, from rapid changes in conditions that prevent the equilibria from being fully established. Until these effects are better understood, comparing abundances from minor ionization stages is probably a risky undertaking.

For the major stages of ionization of different elements in H I regions, i.e., the lowest stages with ionization potentials greater than 13.6 eV, it is generally assumed that nearly all of the atoms are in that stage. This is probably a trustworthy notion for individual regions with hydrogen shielding depths $N(H\ I) > 10^{19.5}\ cm^{-2}$, but below this column density

---

* The rate coefficient $\alpha(X, g)$ also depends on many incidental factors, such as the character and concentration of dust grains, the local density of starlight photons, and the local electron density.

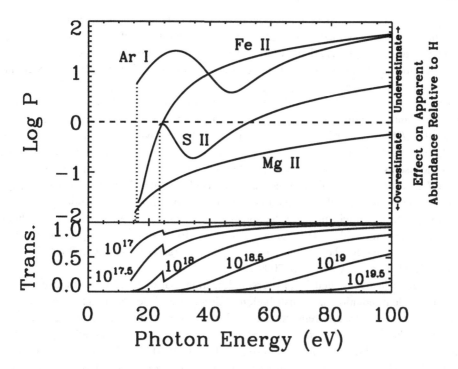

Fig. 22.1. *Top section:* Relative ease of photoionizing Ar, S, Fe, and Mg to ionization stages above the preferred ones (Ar I, S II, Fe II, and Mg II) relative to that of partially ionizing H, expressed in the form of the logarithm of P defined in Eq. 22.3 as a function of the photon energy, under the simplified condition that the radiation field is purely monoenergetic. The amounts by which the derived abundances relative relative to that of H are either underestimated ($\log P > 0$) or overestimated ($\log P < 0$) are defined by Eq. 22.2. Curves for additional elements are shown by Sofia & Jenkins (1998). *Bottom section:* The transmission of photoionizing radiation through different shielding column densities of hydrogen, with the curves labeled according to different values of $\log N(\mathrm{H\,I})$.

photons with energies $< 100\,\mathrm{eV}$ can penetrate the region, as illustrated in the bottom section of Figure 22.1. Once this happens, some fractions of both the element in question and the accompanying hydrogen can be elevated to higher stages of ionization, but by different amounts. It follows that a comparison of an element's most abundant stage to that of H I could lead to a misleading abundance ratio, which differs from the true one by a logarithmic difference

$$\log \zeta(X) - \log \zeta(\mathrm{H}) = \log \left[\frac{1 + \frac{n(e)}{n(\mathrm{H})}}{1 + P(X)\frac{n(e)}{n(\mathrm{H})}} \right], \tag{22.2}$$

where

$$\zeta = (\text{preferred stage})/(\text{total}) \quad \text{and} \quad P(X) = \left[\frac{\Gamma(X)\alpha(\mathrm{H})}{\Gamma(\mathrm{H})\alpha(X)} \right]. \tag{22.3}$$

These equations are discussed in more detail by Sofia & Jenkins (1998)*, who also present their more complex modified forms that account for the additional effects of charge exchange with hydrogen. Alterations arising from charge exchange can be very important for certain elements, such as N and O, which have large reaction rate coefficients (Field & Steigman 1971; Butler & Dalgarno 1979).

The curves for $\log P$ for several elements are shown in Figure 22.1. This figure illustrates that under conditions of partial ionization, $N(\text{Ar I})/N(\text{H I})$ underestimates the true value of Ar/H because Ar is more easily ionized than H, while the converse is true for $N(\text{Mg II})/N(\text{H I})$. At first glance, S might appear to be more immune to ionization corrections because its $\log P$ is not far from zero. However, the second ionization potential of S is 23.4 eV, i.e., far above the 13.6 eV needed to ionize hydrogen. Thus, there is some danger that significant amounts of S II could arise from a fully ionized H II region in front of a target star, and this might increase the apparent abundance of S II over that which should have been identified with an H I region under study. To illustrate how important this effect can be, we note that for the extreme case of β CMa the ratio of foreground S II to H I is 10 times solar (Jenkins, Gry, & Dupin 2000).

The picture just presented is an oversimplification, but one that is instructive. In real life, the incident radiation has a distribution of fluxes over different energies, and it is scattered and attenuated by different amounts as deeper layers of a cloud are penetrated. What we view is the composite absorption produced by these layers. One popular approach for calculating these effects is to use the program CLOUDY (Ferland et al. 1998), which is designed to evaluate the various ion fractions at different locations within a cloud that is irradiated from the outside. This program has been made generally available to the public (http://www.nublado.org) and is continuously updated. Another complication is that in some circumstances anomalously large temperatures created by shock heating may lead to collisional ionization, which can elevate even further the populations of high-ionization states (Trapero et al. 1996).

Useful general discussions of the effects that can arise from a mixture of ionized and neutral regions are given by Sembach et al. (2000). It is clear that they are especially important in low-density gases within about 100 pc of the Sun (Jenkins et al. 2000; Jenkins, Gry, & Dupin 2000; Gry & Jenkins 2001). Howk & Sembach (1999) point out the dangers inherent in deriving abundances in damped Lyα systems, and they back up their cautions by noting that the velocity profiles of Al III are very similar to those from species that are expected to arise from H I regions. The notion that such systems may have a considerable amount of partially ionized gas is reinforced by the findings of Vladilo et al. (2003) who noted that Ar I seems to be deficient relative to other α-capture elements that are not expected to be appreciably depleted onto dust grains (see §22.3 below). By definition, damped Lyα systems have column densities $\log N(\text{H I}) > 20.3$, well in excess of the limit $\log N(\text{H I}) = 19.5$ for shielding out photons with $E \lesssim 100\,\text{eV}$. Perhaps these systems are either (1) subjected to appreciable fluxes with $E > 100\,\text{eV}$, (2) are made up of many thinner regions that are each separately exposed to the ionizing radiation, or (3) have strong sources of internal ionization arising from rapid star formation.

---

* Sofia & Jenkins refer to ζ as δ, but δ is not adopted here because δ has often been used in much of the literature to denote a depletion factor, as expressed here in Eq. 22.4.

22.3   Depletions of Free Atoms onto Dust Grains

22.3.1   *Basic Premise*

The fact that appreciable fractions of some elements in the ISM have condensed into solid form is supported by a number of independent observations. First, starlight is absorbed and scattered by dust particles or very large molecules—the opacity takes the form of continuous attenuation in the visible and UV, together with discrete absorption bands (2200 Å feature, diffuse interstellar bands in the visible, and infrared absorption features). When compared with the amount of hydrogen present, the magnitude of the attenuation indicates that the proportions of some atoms in solid form are large (see the review by Draine 2004). Second, it is clear from circumstellar emissions in the infrared that many types of evolved stars cast off their dusty envelopes into the ISM. While such dust probably does not account for most of the material that is in solid form within the general ISM, it probably provides nucleation sites for further growth in dense clouds. Finally, we may compare interstellar atomic abundances with solar (or stellar) abundances in the Galactic disk and surmise that the inferred depletions measure the fractional amounts of material hidden in some kind of solid form (or within large molecules that are difficult to identify spectroscopically). Specifically, we may define a depletion of free atoms using the usual bracket notation,

$$[X/\mathrm{H}] = \log(N(X)/N(\mathrm{H})) - \log(X/\mathrm{H})_{\odot} \tag{22.4}$$

and the amount of an element locked up in solids, relative to H, is

$$(X/\mathrm{H})_{\mathrm{dust}} = (X/\mathrm{H})_{\odot}(1 - 10^{[X/\mathrm{H}]}) , \tag{22.5}$$

assuming that solar abundances are good reference values that truly reflect the total abundances (see the second footnote in §22.3.3). Much of the discussion that follows in this paper concentrates on the findings on depletions, ones that are strikingly evident from the UV absorption data, but which have also been evident even from the visible absorption lines of Ti II (Wallerstein & Goldsmith 1974; Stokes & Hobbs 1976; Stokes 1978) and the features of Ca II (Snow, Timothy, & Seab 1983; Crinklaw, Federman, & Joseph 1994) (notwithstanding the inherent difficulties in dealing with the uncertainties of the ionization equilibrium of the latter; see Eq. 22.1).

22.3.2   *General Properties of the Depletion Trends*

To obtain a better understanding of the underlying causes of the depletions of free atoms in space and the various factors that are influential, it is important to identify systematic effects that occur from one element to the next and from one environmental factor to the next. Three important effects have been observed:

(1) Elements that reside within refractory compounds are more severely depleted than ones that mostly condense into volatile ones. Field (1974) was the first to identify this effect by comparing the depletions of different elements toward ζ Oph with their condensation temperatures, defined according to the temperature at which some fixed proportion of an element in a gaseous cosmic mixture freezes out into some solid compound(s). The interpretation offered by Field was that in cooling and expanding mass loss outflows from stars, the dense, inner zones of the flows gave rise to nearly complete depletions of elements making up the refractory compounds, whereas the outer, less dense zones that were cool enough to freeze

out more volatile substances could not fully complete the process. While this is an attractive concept, we must also acknowledge that condensation temperatures are an indication of general strengths of chemical bonds and may thus also apply to the balance of creation and destruction of dust grains in the general ISM (Cardelli 1994). A more modern presentation of the correlations of depletions with condensation temperature (again for ζ Oph) is shown in Figure 4 of the review article by Savage & Sembach (1996).

(2) The strengths of depletions decrease for gas moving at large velocities relative to the undisturbed material in a given region. This effect was first pointed out by Routly & Spitzer (1952) who compared lines of Ca II to those of Na I, and it has been substantiated in greater detail by Siluk & Silk (1974), Vallerga et al. (1993), and Sembach & Danks (1994). Misgivings that this effect could arise from changes in ionization have been overcome by the ultraviolet data (Jenkins, Silk, & Wallerstein 1976; Shull, York, & Hobbs 1977), where several ionization stages for certain elements could be observed. The interpretation of this effect is that gas at high velocities was recently subjected to a shock that was strong enough to partly (or completely) destroy the grains (Barlow 1978a, b; Draine & Salpeter 1979; Jones et al. 1994; Tielens et al. 1994).

(3) The strengths of depletions scale in proportion to the average density of gas (i.e., N(H) divided by distance) along a line of sight (Savage & Bohlin 1979; Harris, Gry, & Bromage 1984; Spitzer 1985; Jenkins, Savage, & Spitzer 1986; Jenkins 1987; Crinklaw et al. 1994). Such a measure is a very crude indication of conditions, since there is no way to discriminate between very compact clouds with high densities and a small filling factor from far more extended regions having only moderately high densities but large filling factors. Nevertheless, the trend supports the propositions that the growth of dust grains is markedly enhanced in dense regions and/or that within these regions the destructive effects of shocks are muted. Successful alternative ways to gauge the predisposition for elements to deplete include the fraction of hydrogen in molecular form (Cardelli 1994) and, toward very distant sources at high Galactic latitudes, simply N(H I) (Wakker & Mathis 2000).

22.3.3   *A Reinvestigation of the Depletion Trends*

Seven years ago, Savage & Sembach (1996) summarized the interstellar abundances derived from observations using the GHRS instrument aboard *HST*. Since that time, the continued success of *HST* (with the STIS instrument that replaced GHRS) and emergence of *FUSE* have enabled a significant number of important new surveys of interstellar abundances to be carried out. For this reason, it is useful to revisit the general trends, now that a larger suite of results, both new and old, can be compiled and analyzed. Unlike previous studies that directly compared the strengths of depletions of specific elements with tangible properties of the lines of sight (e.g., Jenkins 1987 or Cardelli 1994), we will initially ignore these external factors and instead characterize depletions simply to arise as the result of some measure of each element's general propensity to deplete, which works in combination with an index F_* for the overall strength of depletions (applied to all elements) along the line of sight to a star. This method has the advantage that links in the depletions from element to element are evaluated directly. As a result, they are not weakened by the inevitable inaccuracies in the relationships between individual depletions and some independently measured quantity (either average density or the fraction of H_2). After values of F_* have been defined for many different stars, we can then probe how they correlate with the average density (or any other interesting attribute for a line of sight). When we investigate

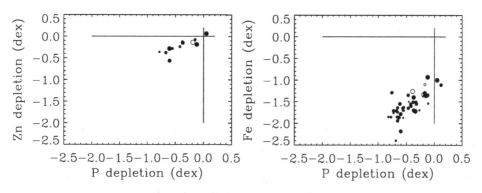

Fig. 22.2. Two comparisons of depletions of pairs of elements along different lines of sight: *Left panel:* Zn vs. *P* and *Right panel:* Fe vs. *P*. Points with large diameters represent determinations with small errors; open circles apply to cases where $\log N(\mathrm{H}) < 19.5$, where ionization effects may be important (see §22.2).

how depletions change under different conditions, we enjoy the benefit of being able to draw upon a wide selection of examples and not just those for a single element.

For the study presented here, abundances toward 144 stars were identified from over 70 investigations reported in the literature (far too numerous to cite here). The targets were limited to those for which we can extract reliable measures of hydrogen in the form of neutral atoms and H_2, derived from measurements of the Lyα damping profiles and the H_2 Lyman and Werner bands.* Since it is usually impossible (or very difficult) to identify the amounts of hydrogen associated with individual velocity components along a line of sight (without applying simplifying assumptions arising from "undepleted" species), only total column densities over all components were considered. Exceptions to this principle could be made for a few cases where one component strongly dominated the total.

We begin with the simplest assumption that from one line of sight to the next, the depletions of different elements scale with each other in some fixed proportion, as is illustrated for Zn and P shown in the left-hand panel of Figure 22.2. However, this simple prediction does not hold for all elements, as one can see for the case of Fe and P in the right-hand panel of this figure. This behavior seems likely to be explained by the concept that typical dust grains have a mantle that is easily created and destroyed in the ISM, and that this mantle surrounds a relatively indestructible core that survives all but the most energetic (and very rare) shock events (Spitzer 1985; Jenkins, Savage, & Spitzer 1986; Joseph 1988; Spitzer & Fitzpatrick 1993). We are thus forced to consider at least three parameters for a general law that characterizes the depletion of an element X,

$$[X/\mathrm{H}] = A_X F_* + A_{0,X}, \tag{22.6}$$

where A_X represents the relative ease with which an element's individual depletion changes as the general line-of-sight depletion multiplier F_* varies, and $A_{0,X}$ represents an offset attributable to special compounds within the most rugged portions of the grains. For the study

---

* In some instances, definite determinations for $N(H_2)$ were not available. However, when upper limits for $N(H_2)$ were known to be smaller than $0.02N(\mathrm{H\,I})$, we could disregard the effect of H_2 in increasing the value of $N(H_{\mathrm{total}}) = N(\mathrm{H\,I}) + 2N(H_2)$. Also, based on general findings for H_2 (Savage et al. 1977; Rachford et al. 2002), it is safe to ignore H_2 if the star's $B-V$ color excess is less than 0.1, or (equivalently) $\log N(\mathrm{H\,I}) < 20.3$.

Table 22.1. *Best-fit Depletion Constants*

Elem. X	Abund$_\odot$ (H = 12)	A_X ($\pm 1\sigma$ error)	$A_{0,X}$ ($\pm 1\sigma$ error)	χ^2	d.f.	Probability of a Worse Fit
C	8.39	-0.097 ± 0.208	-0.148 ± 0.189	2.2	6	0.899
N	7.93	-0.060 ± 0.075	-0.080 ± 0.043	27.9	15	0.022
O	8.69	-0.089 ± 0.067	-0.050 ± 0.052	38.7	28	0.086
Mg	7.54	-0.861 ± 0.046	-0.248 ± 0.026	42.9	62	0.969
Si	7.54	-1.076 ± 0.122	-0.223 ± 0.033	5.0	7	0.660
P	5.57	-0.967 ± 0.055	0.088 ± 0.028	31.9	52	0.987
Cl	5.27	-0.950 ± 0.108	0.300 ± 0.068	41.3	35	0.215
Ti	4.93	-2.226 ± 0.108	-0.844 ± 0.064	30.5	32	0.543
Cr	5.68	-1.373 ± 0.066	-0.854 ± 0.036	16.8	12	0.158
Mn	5.53	-0.685 ± 0.045	-0.774 ± 0.025	52.7	59	0.704
Fe	7.45	-1.198 ± 0.045	-0.950 ± 0.023	59.7	54	0.275
Ni	6.25	-1.440 ± 0.088	-0.917 ± 0.046	31.0	9	0.000
Cu	4.27	-0.408 ± 0.223	-0.974 ± 0.204	7.7	9	0.561
Zn	4.65	-0.522 ± 0.083	0.042 ± 0.053	12.6	12	0.399
Kr	3.23	-0.178 ± 0.134	-0.116 ± 0.104	3.1	9	0.961

presented here, the normalization of F_* was defined to make a value of 1.0 equal to the depletions on line of sight toward ζ Oph, a star that shows for one of its velocity components a significant amount of depletion. It is also a case that has probably received the most attention of any star for depletions of a broad range of different elements [see Table 5 of Savage & Sembach (1996) for a summary].

Weighted least-squares determinations for the constants A_X and $A_{0,X}$ associated with all of the elements covered in this study† are given in Table 22.1. Figure 22.3 shows plots of the observed depletions as a function of F_*; in each case a dashed line is drawn to show the relationship expressed by Equation 22.6 for the appropriate values of A_X and $A_{0,X}$ shown in the table. Within the uncertainties of A_X, the elements C, N, O, and Kr all have slopes that are consistent with zero, in accord with the findings of Cardelli et al. (1996), Meyer, Cardelli, & Sofia (1997), Meyer, Jura, & Cardelli (1998) and Cartledge, Meyer, & Lauroesch (2004). Table 22.1 also indicates the values of χ^2 for the fits, the associated degrees of freedom (d.f. equal to the number of cases minus 2), and the probability of obtaining a worse value of χ^2 assuming that the errors are accurate and a linear fit of the depletions [X/H] to F_* is a correct model. Values of these probabilities that are unreasonably close to 0 or 1 probably reflect unrealistically large or small errors assigned by the investigators. An exception is Cl, which seems to exhibit a genuinely poor linear relationship.

---

† Initially, the elements S, Ar, and Ge were also considered, but there were too few cases that met the acceptance criteria to produce meaningful determinations of the depletion parameters. Reference abundances are shown in column 2 of Table 22.1 and were taken from the compilation by Anders & Grevesse (1989), except for the revised solar abundances of C (Allende Prieto, Lambert, & Asplund 2002), N, Mg, Si, Fe (Holweger 2001), and O (Allende Prieto, Lambert, & Asplund 2001). The revised solar abundances appear to have resolved some outstanding discrepancies that appeared to arise between the B-star abundances and the solar ones (Sofia & Meyer 2001), which led to large ambiguities in [X/H] and (X/H)$_{\text{dust}}$ (Snow & Witt 1996).

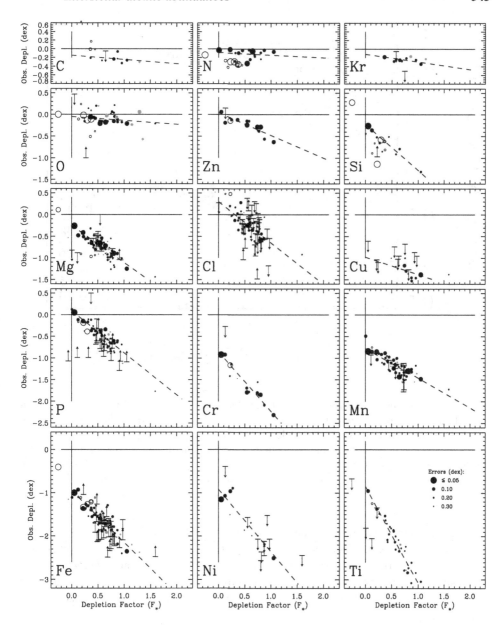

Fig. 22.3. Observed element depletions as a function of the line-of-sight depletion multiplier F_*. In each case, the dashed line represents the linear fit defined by Eq. 22.6 with the constants A_X and $A_{0,X}$ listed in Table 22.1. Cases excluded in the best-fit determinations of these constants include upper and lower limits (arrows) and those with $\log N(H_{total}) < 19.5$ (open circles). Errors in the observations are designated by the sizes of the symbols (smaller symbols represent less certain results), according to the legend in the lower right panel. Cases for which fewer than three other elements were available to establish a measure of F_* are shown in gray rather than black. The open circles with negative F_* correspond to β CMa, which has an extraordinarily large fraction of ionized gas in front of it (Gry, York, & Vidal-Madjar 1985; Jenkins, Gry, & Dupin 2000).

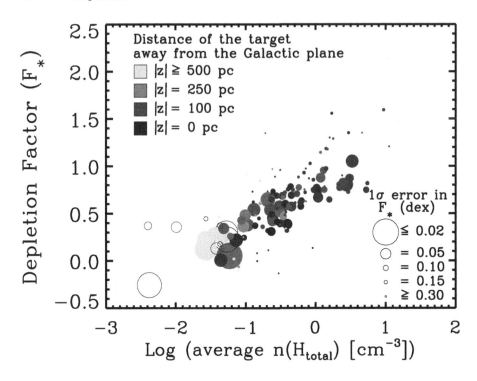

Fig. 22.4. Relationship between the depletion factors F_* toward different stars and their respective average hydrogen densities along the lines of sight. Large symbols are more reliable than small ones, as indicated by the legend in the lower right-hand corner, either because more elements were sampled or the errors in the column densities were smaller. Open circles denote cases where ionization corrections might be important because $\log N(\text{H I}) < 19.5$. The darknesses of the filled symbols indicate the target star's distance above or below the Galactic plane.

Figure 22.4 shows how well the values of F_* correlate with the average hydrogen densities along the lines of sight. There is a clear trend between the two, which substantiates the statements made earlier in §22.3.2 (point nr. 3). The lack of any apparent separation between the darker and lighter shaded points indicates that for this sample, there is no clear effect that distinguishes paths toward stars in the lower halo from those well within the Galactic plane (but see §22.5.1).

Savage & Sembach (1996) summarized depletions of the elements Mg, Si, S, Mn, Cr, Fe, and Ni in terms of broad categories that were labeled "warm disk" and "cool disk," in addition to identifying circumstances that included some gas arising from the Galactic halo (see their Table 6). As a point of reference, if one takes their listed depletions and evaluates the corresponding values of F_* for all of the elements*, the warm disk gas is equivalent to $F_* = 0.21$ while the cool disk material is equivalent to $F_* = 0.99$. The rms dispersion of

---

* For Mg, one should raise the Mg abundances listed by Savage & Sembach by 0.29 dex, in recognition of the improved f-values of the 1240 Å doublet determined by Fitzpatrick (1997). Likewise, abundances of Ni should be increased by 0.27 dex (Fedchak & Lawler 1999)

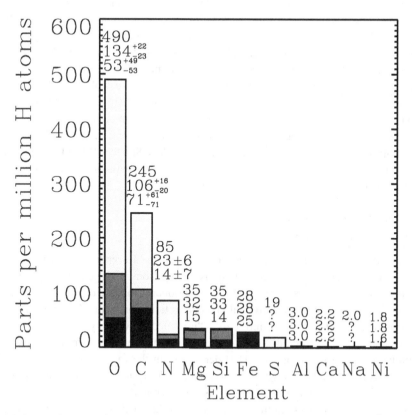

Fig. 22.5. Depictions of elements having solar abundances greater than 10^{-6} times that of H (excluding H and the noble gases He and Ar), with the horizontal positions arranged according to the ranks in total abundances shown by the heights of the bars. The interior of each bar is broken down by the element's fraction in gaseous form (upper, clear portion) and solid form $(X/H)_{dust}$ evaluated from Eqs. 22.5 and 22.6 according to whether $F_* = 1.0$ (gray plus black lower portions) or $F_* = 0.0$ (only the black portion). Numbers over each bar state the values (and errors, if more than 3 ppm) corresponding to the tops of the black, gray and clear zones.

these two numbers from one element to the next is 0.05. From strong deficiencies of Al III compared to S III along certain lines of sight, Howk & Savage (1999) concluded that dust grains reside within ionized regions as well.

22.4   Insights on the Composition of Interstellar Dust

As an aid to help us visualize the composition of dust grains, Figure 22.5 indicates what fractions of the more abundant elements exist in the simplest characterizations of (1) resilient grain cores depicted by the lower, black portions of the bars, (2) the outer, somewhat interchangeable forms (sometimes in mantles and other times as free atoms) shown by the gray extensions, and (3) atoms perpetually in free gaseous form shown by the clear portions that extend to the top of each bar (with a length equal to the total abundance). The divisions between the three zones are based on evaluations of Equation 22.6 for $F_* = 1.0$

(tops of the gray zones) and $F_* = 0.0$ (the boundary between the gray and black zones). The presumption here is that these two values of F_* offer good representations of the quantities of material found in grains that are, respectively, either stripped or well nourished. Note from a previous statement in §22.3.3 that it is not clear from the observations that C, N, and O show significant changes in depletion as F_* progresses from one extreme to the other, so in truth the differences between the gray and black zones are not well established. Two elements shown in the figure were not studied here, primarily because there were problems with the ionization effects discussed in §22.2: Na is seen only as Na I (an unfavored stage of ionization) in the visible, and significant contributions of S II may arise from H II regions. Sofia et al. (1994), Spitzer & Fitzpatrick (1993), and Fitzpatrick & Spitzer (1994) have considered the possible mix of compounds in grains that were consistent with the observed depletions; the current state of this topic is reviewed by Draine (2004).

22.5   Outside the Galactic Plane

22.5.1   *Lower Halo*

In order to study the composition of gases in the lower halo, it is necessary to eliminate the inevitable foreground contributions arising from material in the Galactic disk. This can be accomplished by selecting velocity components that are displaced from that of the local gas by virtue of either differential Galactic rotation or special kinematics (usually infall, as with material that is often referred to as "intermediate velocity clouds"*). However, in this circumstance it is difficult to obtain useful information on the amount of hydrogen that accompanies this gas. That is, N(H I) derived from the Lyα damping wings applies to all of the hydrogen present, regardless of velocity. Measures of 21-cm emission show hydrogen as function of velocity, but the results are often not very satisfactory because the radio beam covers a solid angle in the sky that is much larger than a typical cloud size and thus may detect material that is not on top of the source viewed in the UV. Also, some of the signal may arise from hydrogen beyond the target if it is a halo star and not a distant QSO or AGN.

One method of overcoming the lack of good information on hydrogen is to use some other element as a proxy, preferably one that is known to be very lightly depleted. For instance, Sembach & Savage (1996) summarized the abundances of various elements relative to the corresponding amounts of Zn in halo gas components along various sightlines. Those findings are reproduced in Table 22.2 in the form of a value followed by the observed dispersion of results from one halo star to the next. For comparison, the table also shows similar ratios for disk gas at two extremes for the depletion index, $F_* = 0.0$ and 1.0.

We must be cautious when interpreting the data shown in Table 22.2. The column densities of Zn II typically indicate that $\log N$(H I) ≈ 19.5, which means that some of the gas could be partially photoionized, and this in turn would indicate that the column density comparisons may be subject to the ionization corrections discussed in §22.2. Nevertheless, even with the possibility that the apparent abundances may be perturbed in this manner, we can still formulate some useful conclusions. For instance, we note from Figure 22.1 that Fe II is more easily ionized to higher stages than Mg II, regardless of the characteristic photon energy. However, according to the entries in Table 22.2, $[\text{Fe/Zn}]_{\text{halo}} > [\text{Fe/Zn}]_{\text{disk}}$, while $[\text{Mg/Zn}]_{\text{halo}} \approx [\text{Mg/Zn}]_{\text{disk}}$ for $F_* = 0.0$. It is therefore safe to say that Fe is liberated from

---

* Not to be confused with the high velocity clouds discussed in §22.5.2.

Table 22.2. *Abundances Compared to Zn for Halo and Disk Gas*

Ratio	Halo Value	rms Dispersion	Disk with $F_* = 0.0$	Disk with $F_* = 1.0$
[Mg/Zn]*	−0.23	< 0.19	−0.29	−0.63
[Si/Zn]	−0.26	0.14	−0.27	−0.82
[S/Zn]	−0.05	0.14	...	...
[Cr/Zn]	−0.51	0.13	−0.90	−1.75
[Mn/Zn]	−0.61	0.09	−0.82	−0.98
[Fe/Zn]	−0.64	0.04	−0.99	−1.67
[Ni/Zn]*	−0.56	0.07	−0.96	−1.88

*As noted in an earlier footnote, the originally quoted abundances of Mg and Ni have been increased by 0.29 and 0.27 dex, respectively.

dust grains more easily in the halo than in the disk, regardless of the strength of the ionization (or the value of F_* adopted for the disk gas). Evidently, in the halo the amount of residual iron left in grain cores is not as large as in the most lightly depleted gas in the disk, i.e., $[Fe/H]_{halo} > A_{0,Fe} = -0.95$. This effect might arise from the condition that much of the gas in the halo was heavily shocked as it was ejected from the disk and thus the grains were more completely destroyed, or, alternatively, that perhaps ejecta from Type Ia supernovae in the halo have had a chance to enrich the Fe in a time that is shorter than the circulation time between the disk and halo (Jenkins & Wallerstein 1996).

22.5.2 High-velocity Clouds

We now move on to examine the abundances in distinct cloud complexes at high Galactic latitudes that are moving at extraordinarily large velocities (Oort 1966; Mathewson, Cleary, & Murray 1974), i.e., with kinematics clearly not associated with differential Galactic rotation. A current and very comprehensive review of this material has been presented by Wakker (2001). Here, our objective is not to assume some reference abundances and then study the condensation into dust grains, but instead, we seek to use abundances as a tool to gain a better understanding about the origin and history of the material. However, we must still apply the concepts that we have considered earlier and account for the alterations caused by ionization effects and the removal of atoms as they condense into dust grains.

For sightlines that penetrate through well-known high-velocity cloud complexes called Complex C and the Magellanic Stream, Table 22.3 lists the observed deficiencies of the dominant ion X_i of element X in H I regions with respect to neutral hydrogen,

$$[X_i/\text{H I}] = \log\{N(X_i)/N(\text{H I})\} - \log(X/\text{H})_\odot, \qquad (22.7)$$

from which one may define a true depletion of the element

$$[X/\text{H}] = [X_i/\text{H I}] + \log\{f(\text{H I})/f(X_i)\} \qquad (22.8)$$

if the ion ratio $f(\text{H I})/f(X_i)$ can be calculated. Tripp et al. (2003) argue that for the Complex C gas in front of 3C 351 the corrections of Equation 22.8 are substantial and are most likely to arise from collisional ionization. The last row of Table 22.3 shows the element

Table 22.3. *Observed Abundance Ratios in the Magellanic Stream (MS) and Complex C (C)*

Object	[N I/ H I]	[O I/ H I]	[S II/ H I]	[Si II/ H I]	[Fe II/ H I]	[Ar I/ H I]	log N(H I)	Refs.
Fairall 9 (MS)	...	...	−0.6	> −1.2	...	...	19.97	[1,2]
NGC 3783 (MS)	> −1.9	...	−0.6	−0.8	−1.5	< −0.2	19.90	[3,4]
Mrk 279 (C)	< −1.2	−0.5	−0.3	−0.4	−0.8	< −0.9	19.49	[5,6]
Mrk 290 (C)	...	...	−1.1	> −1.6	...	...	19.96	[5,7]
Mrk 817 (C)	< −1.4	−0.5	−0.4	−0.5	−0.6	< −1.1	19.48	[5,6]
Mrk 876 (C)	−1.4	...	...	−0.2	−0.4	< −0.9	19.37	[5,6,8]
PG 1259+593 (C)	−1.8	−0.9	−0.8	−0.8	−1.0	−1.24	19.92	[6,9]
3C 351 (C)	< −1.1	−0.7	< 0.1	−0.4	−0.3	...	18.62	[10]
ioniz. corr.	**[N/H]**	**[O/H]**	**[S/H]**	**[Si/H]**	**[Fe/H]**	**[Ar/H]**		
3C 351 (C)	**< −1.4**	**−0.8**	**< −0.3**	**−0.7**	**−0.5**	...		[10]

Key to references: [1] Gibson et al. 2000, [2] Lu, Savage, & Sembach 1994, [3] Lu et al. 1998, [4] Sembach et al. 2001, [5] Gibson et al. 2001, [6] Collins, Shull, & Giroux 2003, [7] Wakker et al. 1999, [8] Murphy et al. 2000, [9] Richter et al. 2001b, [10] Tripp et al. 2003. The values of [O I/H I] listed by Collins et al. (2003) have been adjusted to a reference solar abundance of 8.69 given by Allende Prieto et al. (2001).

abundances after these corrections were made. One can argue that the other four directions through Complex C are probably less susceptible to ionization effects, particularly from photoionization, because their high-velocity gas components have much larger values of N(H I) than that toward 3C 351. While this may be true, the fact that [Ar I/H I] < [O I/H I] for Mrk 279, Mrk 817, and PG 1259+593 warns us that ionization corrections for other elements, while probably small, are not completely inconsequential. (Of all the elements, Ar is the most responsive one to the effects arising from partial photoionization.)

The Complex C determinations shown in Table 22.3 indicate that the generally very mildly depleted species N and O (see §22.3.3) are substantially below their corresponding solar abundances. The findings for O are particularly secure since their corrections arising from Equation 22.8 are very small (e.g., 0.03 dex* for 3C 351), principally because the ionization of O is strongly coupled to that of H through charge exchange reactions that have large rate constants (see §22.2). The fact that N seems even more deficient than O [even after applying the corrections, which amount to < 0.2 dex according to Tripp et al. (2003)] indicates that the gas in Complex C has not only a low metallicity, but it has not undergone multiple generations of secondary element processing that are characteristic of the disk of our Galaxy (Henry, Edmunds, & Köppen 2000). These two considerations disfavor the proposition that Complex C arose from a recent ejection of gas from the Galactic disk, even after some heavy dilution from more pristine material in the Galactic halo or Local Group. Finally, the fact that [Fe/H] is not substantially less than the values for other elements indicates that the fraction of elements locked into grains is much smaller than for gas within the Galactic disk. This finding is consistent with an apparent lack of H_2 for material arising from Complex C in front of PG 1259+593 (Richter et al. 2001a), since H_2 is produced primarily on the surfaces of grains. A question may arise about a possible enhancement of Fe over α-capture elements, but Tripp et al. caution that currently the errors in both the measurements and correction factors are too large to make a sound evaluation of this proposition.

---

* In Table 22.3 the correction appears to be 0.1 dex, but this is due to roundoff error.

The picture for the Magellanic Stream is not as complete as that for Complex C. Nevertheless, the pattern appears to be very similar, except for evidence that Fe is deficient toward NGC 3783, which in turn suggests that some dust is present [further support for the presence of dust in the Magellanic Stream is suggested by the presence of H_2 (Richter et al. 2001a; Sembach et al. 2001)]. The evidence presented in Table 22.3 is consistent with the picture that the Magellanic Stream is material that has been tidally stripped from the Magellanic Clouds.

22.6 Application to Observations of Abundances Elsewhere

The topics discussed in the preceding sections all lead to interesting insights on the nature of processes that govern the apparent abundances of elements in the gas phase within and near our own Galaxy. A new and exciting frontier is the exploration of abundances in very distant systems (Prochaska, Howk, & Wolfe 2003; Prochaska et al. 2003a, b), once again through the observations of absorption lines against continuum sources (QSOs), but ones at very large redshifts. This is a productive way to understand the chemical state of systems that are otherwise invisible, owing to their great distances. The lessons we have learned from studying the interstellar material in an environment that we know well (our Galaxy) can be applied elsewhere. For instance, we may start with the assumption that the patterns of dust depletion are the same everywhere, regardless of the overall level of metallicity. At the very least, this assumption has been shown to hold true for Galaxy and the LMC by Vladilo (2002). Provided one has measurements of two or more elements with very different indices A_X and $A_{0,X}$, but with somewhat similar production origins and responses to ionization, it should be possible to solve simultaneously the different forms of Equation 22.6 to arrive at a representative F_*. Once this has been done, the F_* and observations of all of the elements may once again be substituted into Equation 22.6 to derive the absolute abundances. These, in turn, lead to our understanding of the chemical histories of such systems, along with key properties of their internal environments.

Acknowledgements. The preparation of this paper was supported by NASA contract NAS5-30110. The author thanks B. T. Draine, J. X. Prochaska, T. M. Tripp, B. D. Savage, and D. Welty for helpful comments on early drafts of this paper.

References

Adams, W. S. 1949, ApJ, 109, 354

Allende Prieto, C., Lambert, D. L., & Asplund, M. 2001, ApJ, 556, L63

——. 2002, ApJ, 573, L137

Anders, E., & Grevesse, N. 1989, Geochim. Cos. Acta, 53, 197

Barlow, M. J. 1978a, MNRAS, 183, 367

——. 1978b, MNRAS, 183, 397

Barlow, M. J., Crawford, I. A., Diego, F., Dryburgh, M., Fish, A. C., Howarth, I. D., Spyromilo, J., & Walker, D. D. 1995, MNRAS, 272, 333

Blades, J. C., Wynne-Jones, I., & Wayte, R. C. 1980, MNRAS, 193, 849

Butler, S. E., & Dalgarno, A. 1979, ApJ, 234, 765

Calura, F., Matteucci, F., Dessauges-Zavadsky, M., D'Odorico, S., Prochaska, J. X., & Vladilo, G. 2004, in Carnegie Observatories Astrophysics Series, Vol. 4: Origin and Evolution of the Elements, ed. A. McWilliam & M. Rauch (Pasadena: Carnegie Observatories, http://www.ociw.edu/symposia/series/symposium4/proceedings.html)

Cardelli, J. A. 1994, Science, 265, 209

Cardelli, J. A., Meyer, D. M., Jura, M., & Savage, B. D. 1996, ApJ, 467, 334

Cartledge, S. I. B., Meyer, D. M., & Lauroesch, J. T. 2004, ApJ, in press (astro-ph/0307182)

Collins, J. A., Shull, J. M., & Giroux, M. L. 2003, ApJ, 585, 336

Cowie, L. L., & Songaila, A. 1986, ARA&A, 24, 499

Crane, P., Lambert, D. L., & Sheffer, Y. 1995, ApJS, 99, 107

Crawford, I. A. 2001, MNRAS, 328, 1115

Crinklaw, G., Federman, S. R., & Joseph, C. L. 1994, ApJ, 424, 748

Draine, B. T. 2004, in Carnegie Observatories Astrophysics Series, Vol. 4: Origin and Evolution of the Elements, ed. A. McWilliam & M. Rauch (Cambridge: Cambridge Univ. Press), in press

Draine, B. T., & Salpeter, E. E. 1979, ApJ, 231, 77

Fedchak, J. A., & Lawler, J. E. 1999, ApJ, 523, 734

Ferland, G. J., Korista, K. T., Verner, D. A., Ferguson, J. W., Kingdon, J. B., & Verner, E. M. 1998, PASP, 110, 761

Field, G. B. 1974, ApJ, 187, 453

Field, G. B., & Steigman, G. 1971, ApJ, 166, 59

Fitzpatrick, E. L. 1997, ApJ, 482, L199

Fitzpatrick, E. L., & Spitzer, L. 1994, ApJ, 427, 232

Gibson, B. K., Giroux, M. L., Penton, S. V., Putman, M. E., Stocke, J. T., & Shull, J. M. 2000, AJ, 120, 1830

Gibson, B. K., Giroux, M. L., Penton, S. V., Stocke, J. T., Shull, J. M., & Tumlinson, J. 2001, AJ, 122, 3280

Gry, C., & Jenkins, E. B. 2001, A&A, 367, 617

Gry, C., York, D. G., & Vidal-Madjar, A. 1985, ApJ, 296, 593

Harris, A. W., Gry, C., & Bromage, G. E. 1984, ApJ, 284, 157

Hartmann, J. 1904, ApJ, 19, 268

Henry, R. B. C., Edmunds, M. G., & Köppen, J. 2000, ApJ, 541, 660

Holweger, H. 2001, in Solar and Galactic Composition, A Joint SOHO/ACE Workshop, ed. R. F. Wimmer-Schweingruber (New York: AIP), 23

Howk, J. C., & Savage, B. D. 1999, ApJ, 517, 746

Howk, J. C., & Sembach, K. R. 1999, ApJ, 523, L141

Jenkins, E. B. 1986, ApJ, 304, 739

——. 1987, in Interstellar Processes, ed. D. J. Hollenbach & H. A. Thronson Jr. (Dordrecht: Reidel), 533

——. 1996, ApJ, 471, 292

Jenkins, E. B., et al. 2000, ApJ, 538, L81

Jenkins, E. B., Gry, C., & Dupin, O. 2000, A&A, 354, 253

Jenkins, E. B., Savage, B. D., & Spitzer, L., Jr. 1986, ApJ, 301, 355

Jenkins, E. B., Silk, J., & Wallerstein, G. 1976, ApJS, 32, 681

Jenkins, E. B., & Wallerstein, G. 1996, ApJ, 462, 758

Jones, A. P., Tielens, A. G. G. M., Hollenbach, D. J., & McKee, C. F. 1994, ApJ, 433, 797

Joseph, C. L. 1988, ApJ, 335, 157

Kimble, R. A., et al. 1998, ApJ, 492, L83

Knauth, D. C., Federman, S. R., & Lambert, D. L. 2003, ApJ, 586, 268

Kondo, Y., Wamsteker, W., Boggess, A., Grewing, M., de Jager, C., Lane, A. L., Linsky, J. L., & Wilson, R. 1987, Exploring the Universe with the IUE Satellite (Dordrecht: Reidel)

Lu, L., Savage, B. D., & Sembach, K. R. 1994, ApJ, 437, L119

Lu, L., Savage, B. D., Sembach, K. R., Wakker, B., Sargent, W. L. W., & Oosterloo, T. A. 1998, AJ, 115, 162

Mathewson, D. S., Cleary, M. N., & Murray, J. D. 1974, ApJ, 190, 291

Merrill, P. W., Sanford, R. F., Wilson, O. C., & Burwell, C. G. 1937, ApJ, 86, 274

Meyer, D. M., Cardelli, J. A., & Sofia, U. J. 1997, ApJ, 490, L103

Meyer, D. M., Jura, M., & Cardelli, J. A. 1998, ApJ, 493, 222

Moos, H. W., et al. 2000, ApJ, 538, L1

Morton, D. C., Drake, J. F., Jenkins, E. B., Rogerson, J. B., Spitzer, L., & York, D. G. 1973, ApJ, 181, L103

Murphy, E. M., et al. 2000, ApJ, 538, L35

Oort, J. H. 1966, BAIN, 18, 421

Pettini, M. 2003, in Cosmochemistry: The Melting Pot of Elements (Cambridge: Cambridge Univ. Press), in press (astro-ph/0303272)

Plaskett, J. S., & Pearce, J. A. 1933, Pub. Dominion Ap. Obs., 5, 167

Prochaska, J. X., Howk, J. C., & Wolfe, A. M. 2003, Nature, 423, 57

Prochaska, J. X., Gawiser, E., Wolfe, A. M., Castro, S., & Djorgovski, S. G. 2003a, ApJ, 595, L9

Prochaska, J. X., Gawiser, E., Wolfe, A. M., Cooke, J., & Gelino, D. 2003b, ApJS, 147, 227

Rachford, B., et al. 2002, ApJ, 577, 221

Richter, P., Sembach, K. R., Wakker, B. P., & Savage, B. D. 2001a, ApJ, 562, L181

Richter, P., Sembach, K. R., Wakker, B. P., Savage, B. D., Tripp, T. M., Murphy, E. M., Kalberla, P. M. W., & Jenkins, E. B. 2001b, ApJ, 559, 318

Rogerson, J. B., Spitzer, L., Drake, J. F., Dressler, K., Jenkins, E. B., Morton, D. C., & York, D. G. 1973a, ApJ, 181, L97

Rogerson, J. B., York, D. G., Drake, J. F., Jenkins, E. B., Morton, D. C., & Spitzer, L. 1973b, ApJ, 181, L110

Routly, P. M., & Spitzer, L. 1952, ApJ, 115, 227

Savage, B. D., & Bohlin, R. C. 1979, ApJ, 229, 136

Savage, B. D., Bohlin, R. C., Drake, J. F., & Budich, W. 1977, ApJ, 216, 291

Savage, B. D., & Sembach, K. R. 1991, ApJ, 379, 245

——. 1996, ARA&A, 34, 279

Sembach, K. R., & Danks, A. C. 1994, A&A, 289, 539

Sembach, K. R., Howk, J. C., Ryans, R. S. I., & Keenan, F. P. 2000, ApJ, 528, 310

Sembach, K. R., Howk, J. C., Savage, B. D., & Shull, J. M. 2001, AJ, 121, 992

Sembach, K. R., & Savage, B. D. 1992, ApJS, 83, 147

——. 1996, ApJ, 457, 211

Shull, J. M., York, D. G., & Hobbs, L. M. 1977, ApJ, 211, L139

Siluk, R. S., & Silk, J. 1974, ApJ, 192, 51

Snow, T. P., Timothy, J. G., & Seab, C. G. 1983, ApJ, 265, L67

Snow, T. P., & Witt, A. N. 1996, ApJ, 468, L65

Sofia, U. J., Cardelli, J. A., & Savage, B. D. 1994, ApJ, 430, 650

Sofia, U. J., & Jenkins, E. B. 1998, ApJ, 499, 951

Sofia, U. J., & Meyer, D. M. 2001, ApJ, 554, L221

Spitzer, L. 1985, ApJ, 290, L21

Spitzer, L., & Cochran, W. D. 1973, ApJ, 186, L23

Spitzer, L., Drake, J. F., Jenkins, E. B., Morton, D. C., Rogerson, J. B., & York, D. G. 1973, ApJ, 181, L116

Spitzer, L., & Fitzpatrick, E. L. 1993, ApJ, 409, 299

Spitzer, L., & Jenkins, E. B. 1975, ARA&A, 13, 133

Stokes, G. M. 1978, ApJS, 36, 115

Stokes, G. M., & Hobbs, L. M. 1976, ApJ, 208, L95

Tielens, A. G. G. M., McKee, C. F., Seab, C. G., & Hollenbach, D. J. 1994, ApJ, 431, 321

Trapero, J., Welty, D. E., Hobbs, L. M., Lauroesch, J. T., Morton, D. C., Spitzer, L., & York, D. G. 1996, ApJ, 468, 290

Tripp, T. M., et al. 2003, AJ, 125, 3122

Vallerga, J. V., Vedder, P. W., Craig, N., & Welsh, B. Y. 1993, ApJ, 411, 729

Vladilo, G. 2002, ApJ, 569, 295

Vladilo, G., Centurión, M., D'Odorico, V., & Péroux, C. 2003, A&A, 402, 487

Wakker, B. P. 2001, ApJS, 136, 463

Wakker, B. P., et al. 1999, Nature, 402, 388

Wakker, B. P., & Mathis, J. S. 2000, ApJ, 544, L107

Wallerstein, G., & Goldsmith, D. 1974, ApJ, 187, 237

Weingartner, J. C., & Draine, B. T. 2001, ApJ, 563, 842

Welsh, B. Y., Vedder, P. W., & Vallerga, J. V. 1990, ApJ, 358, 473

Welty, D. E., & Hobbs, L. M. 2001, ApJS, 133, 345

Welty, D. E., Hobbs, L. M., & Kulkarni, V. P. 1994, ApJ, 436, 152

Welty, D. E., Hobbs, L. M., & Morton, D. C. 2003, ApJS, 147, 61

Welty, D. E., Morton, D. C., & Hobbs, L. M. 1996, ApJS, 106, 533

Woodgate, B. E., et al. 1998, PASP, 110, 1183

Young, R. K. 1922, Pub. Dominion Ap. Obs., 1, 219

23

Molecules in the interstellar medium

TOMMY WIKLIND

ESA Space Telescope Division, Space Telescope Science Institute

Abstract

The molecular component of the interstellar medium contains more than 120 molecular species and represent the coldest, densest, and most obscured part of the interstellar gas. Molecules are the only probes of the physical and chemical conditions in these obscured regions and allow us to study the earliest stages of star formation. Molecules and the molecular chemistry also strongly influence the physical conditions, mainly by acting as efficient cooling agents. In this overview, the main characteristics of the molecular component of the interstellar medium is reviewed. Some outstanding problems are discussed, together with a summary of challenges for the future.

23.1   Introduction

The molecular component of the interstellar medium (ISM) represents the coldest and densest part of the ISM. Molecular gas is distributed in discrete clouds, which can be self-gravitating and directly involved in the formation of new generations of stars. These cold and dense regions can only be probed at far-infrared and longer wavelengths. Photons from molecular transitions can reach us from the densest and most opaque parts and can be used to trace the physical structure of the clouds, as well as allowing a detailed study of the dynamics of the stellar formation processes. Molecular transitions can also be used to derive the magnetic field strength and the cosmic ray ionization rate.

Interstellar molecules are the result of dynamic chemical processing in the ISM itself. The conditions found in the ISM are difficult or impossible to achieve in terrestrial laboratories, and the molecular ISM is therefore a unique laboratory where exotic reactions can be studied. Indeed, several molecular species were first detected in interstellar space before they were synthesized in laboratories on Earth. The interstellar molecules are diagnostics of the chemical evolution, and their abundances reflect the raw material that will ultimately be part of the formation of protoplanetary disks and planetary systems. The molecules are, however, not just passive probes. They are important cooling agents for the gas and directly influence the hydrodynamical gas evolution.

23.1.1   Historical Background

The existence of molecules in interstellar space has been known for almost 70 years. The first detection of the molecules in the ISM was done through observations of both narrow and broad absorption lines seen toward bright stars (Merrill 1934, 1936). It was quickly realized that these lines did not originate from some unknown atomic species in the atmosphere of the stars. The first suggestion that these lines were of molecular origin was made by Russell (1935). Swings & Rosenfeld (1937) found a theoretical match with electronic transitions of a few diatomic molecules. This suggestion was put on firm grounds through the observational work of McKellar (1940), who positively identified several lines as due to the molecules CH^+, CH, and CN. Meanwhile, more diffuse lines had been detected by Beals & Blanchet (1938). In a provocative paper by Swings & Öhman (1939), these lines were tentatively identified with small crystals or amorphous aggregates of light elements, such as solid hydrogen, oxygen, or ice of CO_2.

By the early 1940s a few simple diatomic molecules were known to exist in interstellar space, their number densities had been estimated with a surprising accuracy, and their excitation temperatures were known (e.g., McKellar 1940; Adams 1941). In fact, through observations of electronic transitions of adjacent rotational levels of CN, it was established that the "effective temperature of interstellar space" was around 3 K. These early observers of the molecular component of the ISM had, unwittingly, made the first measurement of the temperature of the cosmic microwave background radiation, not to be discovered for another 25 years.

After this initial flurry of observational and theoretical work, not much was done concerning the molecular component of the ISM until interstellar OH was discovered in 1963 (Weinreb et al. 1963). This time the observations were done using the new technique of radio astronomy. Just as the early optical observations, the radio OH line was initially seen in absorption, but was soon found to also exhibit strong maser emission. It was still believed that only simple diatomic molecules could exist in the harsh environment of interstellar space. During the following 10 years, however, there were radio detections of more complex molecules, such as ammonia NH_3 (Cheung et al. 1969) and formaldehyde H_2CO (Snyder et al. 1969). In 1973 the abundant CO molecule was discovered (Wilson, Jefferts, & Penzias 1970), as well as several radio lines of CH (Rydbeck, Elldér, & Irwine 1973).

With the introduction of telescope and receiver systems capable of observing in the 100–200 GHz frequency range (corresponding to wavelengths of 3–1 mm), many new molecules, with increasingly complex structure, were detected in a rapid pace. The reason for this rapid development is the richness of molecular rotational transitions with energies corresponding to this wavelength range, coupled with the cold excitation conditions of interstellar space as depicted by McKellar and Adams 40 years earlier.

23.1.2 Present Status

Today there are more than 120 molecular species known to exist in interstellar space (see Table 23.1), and the list continues to grow as instruments become more sensitive and as studies in terrestrial laboratories allow more lines to be identified in already obtained spectra. Several of the more prominent molecule-rich hot cores have been part of so-called spectral scans (e.g., Nummelin et al. 1998). The complexity of the detected molecules has grown from simple diatomic molecules to those containing up to 13 atoms. Although the heaviest molecules observed to date consist of long carbon chains, some molecules have a more intricate structure. One example of the latter is the recently detected glycolaldehyde CH_2OHCHO, an interstellar monosaccharide sugar compound (Hollis, Lovas, & Jewell 2000). Some molecular species remain only tentatively detected, although they are believed to exist in the molecular ISM. An example of these species is glycine, the simplest amino acid (NH_2CH_2COOH). Several searches for it have been done. The main difficulty in detecting glycine is the large number of allowed transitions and the line confusion that limits the observed spectral sensitivity. Some light hydrides also remain undetected, mainly because their ground transition fall at wavelengths inaccessible to existing telescopes and instruments. The LiH molecule has, however, been tentatively detected at a redshift $z = 0.68$ (Combes & Wiklind 1998), by virtue of being redshifted into an observable wavelength regime.

The field of molecular astrophysics has grown to involve and impact almost all branches of astrophysics. Astrochemistry tells about how molecules are formed and which processes are important in different environments. Molecular emission and absorption lines can be used to determine the detailed physical conditions of the interstellar gas in which the molecules reside. In this case it is necessary to follow the transfer of radiation in great detail. The cooling provided by the molecular gas is of paramount importance in the formation of stars.

The molecules in the ISM are thus not just a probe of the conditions, but take an active part in regulating the temperature of the gas. In order to understand complex processes such as star formation, it is therefore necessary to study the interplay between molecules, its chemistry, the transfer of radiation, and the effect the molecules have on the dynamical process itself. The simultaneous solution of molecular astrochemistry, transfer of radiation, and hydrodynamics has not yet been achieved, but when it is done it is likely to give us a good understanding of how stars are formed.

23.2   Components of the Molecular ISM

23.2.1   *Molecular Species*

Astrochemists usually divide the molecular species found in the ISM into organic and inorganic molecules. The organic ones are simply those containing carbon, while the inorganic do not. The vast majority of detected molecules in the ISM are organic (see Table 23.1). The organic molecules can be divided into simple diatomic molecules, such as CO, CH, CS, etc., and more complex hydrocarbons, such as methanol (CH_3OH), ethyl alcohol (CH_3CH_2OH), etc.

Most of the interstellar molecules are linear in shape: examples of linear molecules are CO, CH_3OH, CH_3CH_2OH, and HC_9N. A few species of ring molecules have now also been detected: C_3H, C_3H_2, C_2H_4O, and SiC_2. There are also some molecules that have only been seen in circumstellar envelopes and are therefore not part of the ISM proper: examples of these are MgCN, AlF, and KCl. On the other hand, many of the molecules found in the ISM are also found in circumstellar envelopes.

In general, molecules with a simple structure have fewer symmetries and therefore a smaller number of transitions than more complex molecules. Molecules with several symmetries, and a correspondingly large number of transitions can be difficult to identify since the line strength in any particular line will be low. Also, these complex molecules with many low-intensity lines contribute significantly to confusion. This is especially true in molecular clouds with rich chemistries.

There are three main types of transitions between different energy levels of a molecule: (1) electronic transitions with $\Delta E \approx$ a few eV, corresponding to UV/optical wavelengths, (2) vibrational transitions with $\Delta E \approx 0.1 - 0.01$ eV, corresponding to near-infrared/infrared

Table 23.1. *Known Interstellar and Circumstellar Molecules*

Hydrogen compounds					
H_2	H_3^+				

Hydrogen and carbon compounds					
CH	CH^+	C_2	CH_2	C_2H	C_3
CH_3	C_2H_2	C_3H	c-C_3H	CH_4	c-C_3H_2
H_2CCC	C_4H	C_5	C_2H_4	C_5H	H_2C_4
CH_3C_2H	C_6H	H_2C_6	C_7H	CH_3C_4H	C_8H

Hydrogen, carbon, and oxygen compounds					
OH	CO	CO^+	H_2O	HCO	HCO^+
HOC^+	C_2O	CO_2	H_3O^+	$HOCO^+$	H_2CO
C_3O	CH_2CH	HCOOH	H_2COH^+	CH_3OH	HC_2CHO
C_5O	CH_3CH_2OH	c-C_2H_4O	CH_3OCHO	CH_2OHCHO	CH_3COOH
CH_3OCH_3	CH_3CH_3OH	$(CH_3)_2CO$			

Hydrogen, carbon, and nitrogen compounds					
NH	CN	NH_2	HCN	HNC	N_2H^+
NH_3	$HCNH^+$	H_2CN	HCCN	C_3N	CH_2CN
CH_2NH	HC_2CN	HC_2NC	NH_2CN	C_3NH	CH_3CN
CH_3NC	HC_3NH^+	C_5N	CH_3NH_2	CH_2CHCN	HC_5N
CH_3C_3N	CH_3CH_2CN	HC_7N	CH_3C_5N	HC_9N	$HC_{11}N$

Sulphur containing compounds					
SH	CS	SO	SO^+	NS	SiS
H_2S	C_2S	SO_2	OCS	HCS^+	H_2CS
HNCS	C_3S	CH_3SH	C_5S		

Other compounds					
SiH	SiC	SiN	SiO	HCl	NaCl
AlCl	KCl	HF	AlF	CP	PN
c-SiC_2	NaCN	MgCN	MgNC	$HSiC_2$	SiC_3
SiH_4	SiC_4	NO	HNO	N_2O	HNCO
NH_2CHO					

Molecules preceded by "*c–*" are ring-shaped.

wavelengths, and (3) rotational transitions with $\Delta E \approx 10^{-3}$ eV, corresponding to mm/submm wavelengths. There are also other, less frequent modes, such as bending transitions.

The bulk of the molecular gas is characterized by very low kinetic temperatures, 5–100 K. Since the molecular gas is shielded from UV radiation by dust grains and, for the most abundant molecules, through self-shielding, the main excitation mechanism is collisions with H_2. With the low kinetic temperatures, usually only the rotational levels will be populated. Vibrational and electronic transitions can occur in warm molecular gas, at the periphery of molecular clouds, and in photo-dissociation regions (see § 23.2.2).

23.2.2 *Molecular Clouds*

Already the first observers of interstellar molecules realized that they occurred in discrete clouds rather than being spread over large volumes as in the case for warm ionized gas (e.g., Adams 1948). Clouds containing molecules are divided into different types, on a somewhat *ad hoc* basis. There are five main types of molecular clouds: diffuse, translucent, dark, giant molecular cloud (GMC), and hot cores. Their properties are listed in Table 23.1. In addition, molecules can be found in photo-dissociation regions or photon-dominated regions (PDRs). PDRs exhibit a relatively rich chemistry, driven by the photon field. In

Table 23.2. *Types of Molecular Clouds*

Type of Cloud	A_V (mag)	n (cm^{-3})	T_k (K)	Contents
Diffuse	<1	10–100	50–100	simple molecules, atoms, ions
Translucent	1–5	100–10^3	20–50	ions, more complex molecules
Dark	>5	> 10^3	10	hydrocarbons, few ions
GMC	large	∼ 10^3	20–40	mix of different types of clouds
Hot core	large	> 10^4	40–100	rich chemistry

principle, any of the clouds listed in Table 23.1 becomes a PDR if it is exposed to a high flux of UV photons. The chemistry and physical conditions will, however, change as the photo-processes start to dominate the chemistry. The sizes of molecular clouds can vary from less than 1 parsec for the smallest cores to 100 pc for the largest GMCs. There are indications that very small-scale structure (∼10 AU) exists in the molecular ISM (e.g., Rollinde et al. 2003).

In reality, it is rare to encounter a single isolated molecular cloud that can be classified according to the above types. In most cases, the molecular gas exist in a complicated mixture of different physical conditions. A part of a cloud complex can be a typical dark cloud, while the surroundings are more characteristic of translucent gas. Also, the topology of molecular clouds is in most cases complicated, often showing a filamentary structure. Work by Falgarone and collaborators has shown that over several orders of magnitude, molecular gas can have a self-similar structure (Puget & Falgarone 1990; Falgarone et al. 1998). The inhomogeneous structure of the molecular gas can have profound effects on the chemistry and the excitation conditions of the molecules, as UV photons can penetrate deeper into the cloud complex. This will be discussed further in § 23.4.

23.3   Molecular ISM Chemistry

When the first molecules were discovered in 1934, it was not known whether they were formed *in situ* or originated from stellar atmospheres. It was known that molecules, such as TiO, exist in cool stars, and with a sufficiently strong stellar wind, the ISM could in principle be enriched by molecules. Today we know that the vast majority of all molecules originate in the ISM. The number of molecules coming from circumstellar envelopes is very small indeed. The contribution of dust grains from circumstellar envelopes and Type II supernovae, however, is significant and also necessary in order to start the production of ISM molecules.

There are two types of chemical processes driving the chemistry in the ISM: gas-phase reactions and grain-phase reactions. The grain-phase reactions are important, but not yet fully understood. They involve surface migration of the atoms and molecules, sticking co-efficients and desorption efficiencies. Gas-phase reactions, on the other hand, are quite well known and can successfully be modeled. However, as will be seen in § 23.4, a successful modeling of the chemistry of molecular gas needs both gas- and grain-phase reactions to be able to accurately predict molecular abundances.

In principle, there are only nine processes with which molecular bonds can be formed, rearranged, or destroyed (see Table 23.3). Which ones dominate in any given situation de-

Table 23.3. *Molecular Bond Processes*

radiative association	$A + B$	$\rightarrow$	$AB + h\nu$	
associative attachment	$A^- + B$	$\rightarrow$	$AB + e$	bond formation
grain surface formation	$A + B{:}g$	$\rightarrow$	$AB + g$	
neutral–neutral	$A + BC$	$\rightarrow$	$AB + C$	
ion–molecule	$A^+ + BC$	$\rightarrow$	$AB^+ + C$	bond rearrangement
charge transfer	$A^+ + BC$	$\rightarrow$	$A + BC^+$	
photodissociation	$AB + h\nu$	$\rightarrow$	$A + B$	
dissociative recombination	$AB^+ + e$	$\rightarrow$	$A + B$	bond destruction
collisional dissociation	$AB + M$	$\rightarrow$	$A + B + M$	

pends on the type of clouds where the chemistry is taking place (e.g., the flux of UV photons, density of ions and electrons, etc.). On the other hand, there are up to 400 different molecular species that needs to be considered when modeling the chemistry, and the total number of reactions can be around 4000 (e.g., Millar, Farquhar, & Willacy 1997; van Dishoeck 1998). Each reaction proceeds at a rate that can be a function of temperature. A reaction is governed by a rate equation, which defines the abundance n of a given molecule (e.g., Duley & Williams 1984). The result is a chemical reaction network consisting of 4000 coupled nonlinear differential equations. Equilibrium chemical models are used to study the steady-state chemical properties of molecular gas and the influence of varying degrees of UV extinction. Full time-dependent models have become more common, and can allow for self-consistent solutions involving changing physical parameters such as density and temperature (e.g., Bergin & Langer 1997). The time-dependent models are important since the chemical time scales often are of the same order as the dynamical time scales for molecular gas. A goal with astrochemistry is to use certain key molecules as diagnostic of the temporal evolutionary state of a molecular gas entity: a star-forming molecular cloud, a protostellar core, or a protoplanetary disk (e.g., van Dishoeck & Blake 1998; Doty et al. 2002).

There are now also models that incorporate gas-grain chemistry where molecules are accreted, formed, and desorbed from grain surfaces (e.g., d'Hendecourt, Allamandola, & Greenberg 1985; Hasegawa & Herbst 1993a, b; Roberts & Herbst 2002). Since it is not easy to reproduce the interstellar conditions in terrestrial laboratories, progress in understanding the grain surface chemistry and its connection to the gas phase has been slow. As will be seen in § 23.4, the grain chemistry is most likely very important for a complete understanding of astrochemistry, as well as for understanding the stellar formation process.

23.3.1 *Molecular Absorption*

The very first molecular observations were done using absorption lines, both at optical and radio wavelengths. The use of absorption lines is, however, difficult at radio and millimeter wavelengths due to the fact that the necessary background continuum sources generally have an angular extent much smaller than the telescope beam. It was only with the introduction of millimeter interferometry techniques that molecular absorption-line studies could be done in earnest.

The main difference between absorption- and emission-line observations at millimeter wavelengths is that while molecular emission lines need a relatively dense and warm gas in order to produce an observable signal (the observed signal is approximately proportional to the temperature characterizing the level population), absorption-line measurements are most sensitive to the cold and diffuse molecular component, where the observed signal $\propto (\mu_0/T)^2$, where μ_0 is the permanent electric dipole moment of the molecule (e.g., Wiklind & Combes 1997b). The dependence of the opacity on the permanent dipole moment implies that less abundant molecules can be as easily detectable as more abundant ones, if their permanent dipole moment is larger. For instance, the HCO^+/CO abundance ratio is approximately 5×10^{-4}, yet HCO^+ can be more easily detected in absorption than CO.

Using the IRAM Plateau de Bure millimeter-wave interferometer, the molecular abundances in diffuse molecular gas in the Milky Way galaxy has been extensively explored using the technique of absorption lines (e.g., Lucas & Liszt 1996, 2002; Liszt & Lucas 1998, 2002). Molecular absorption lines have also been used to probe the detailed properties of molecular gas in galaxies up to redshifts of $z \approx 0.9$ (Wiklind & Combes 1994, 1999; Combes & Wiklind 1997, 1999; Menten, Carilli, & Reid 1999). The results from these studies show that molecular abundances in Galactic diffuse clouds are similar to that of dark clouds. Furthermore, for both nearby and distant systems, some molecules have much higher abundances than what is suggested by chemical models. For instance, HCO^+ is found to be up to 100 times more abundant in diffuse molecular gas than indicated in models. This clearly shows that conventional models of gas-phase chemistry in low-density and moderate-extinction regions cannot account for all the processes taking place.

23.4   The Missing Ingredients

Gas-phase chemical models, especially the ion-molecule reaction scheme, have been very successful in explaining the abundances of simple hydrocarbons in diffuse and translucent clouds, the high abundances of protonated ions such as HCO^+ and N_2H^+, seen in emission, as well as the high abundances of unsaturated and metastable molecules in dark clouds. The models also explain the high isotopic fractionation found for many species, in particular the deuterium fractionation (e.g., Millar, Bennett, & Herbst 1989; Millar 2002). However, as mentioned in § 23.3.1, absorption-line measurements indicate abundances for some molecular species that are higher than what the chemical models suggest.

There are also a few molecules that are important ingredients in the chemical modeling, as well as important for the constraining the physical properties of the gas itself, which have yet to be detected. Among these are water vapor, H_2O, and molecular oxygen, O_2. Chemical models predict high abundances of these species in dark molecular clouds, but they have remained impossible or difficult to observe due to their presence in the terrestrial atmosphere.

23.4.1   Molecular Oxygen

Oxygen is the third most abundant element in the Universe and is an important part of the interstellar molecular chemistry. The simplest oxygen-bearing molecule is molecular oxygen, O_2, which is believed to be an important cooling agent for molecular gas (e.g., Goldsmith & Langer 1978) and thus important for the formation of stars.

The formation and destruction pathways for O_2 are fairly simple and well known (e.g., Black & Smith 1984; Bergman 1995). In dark quiescent clouds, the O_2 abundance should reach $\sim 20\% - 40\%$ of that of CO. Despite its predicted high abundance, O_2 remains undetected in the ISM. The transitions of O_2 are mostly obscured by the Earth's atmosphere.

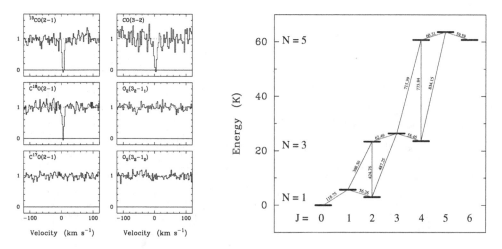

Fig. 23.1. *Left:* Molecular absorption lines of rotational transitions seen toward B0218+357 at $z = 0.685$ (Wiklind & Combes 1995). The left three panels show absorption of $^{13}CO(2-1)$, $C^{18}O(2-1)$, and a nondetection of $C^{17}O(2-1)$. The right hand panels show $^{12}CO(2-1)$ on the top and nondetections of the $N_J = 3_2 - 1_1$ and $N_J = 3_2 = 1_2$ transitions of O_2 (Combes & Wiklind 1995; Combes, Wiklind, & Nakai 1997). *Right:* Energy level diagram for molecular oxygen, O_2. The transition frequencies in GHz are shown, as well as the energy levels in K (E/k).

Early upper limits to its abundance were therefore obtained through observations of isotope variants, $^{16}O^{18}O$ (Goldsmith et al. 1985; Liszt & vanden Bout 1985; Pagani, Langer, & Castets 1993; Maréchal et al. 1997a). Observations of the main isotope from redshifted objects have been obtained through emission studies (Liszt 1985, 1992; Goldsmith & Young 1989) as well as absorption-line studies (Combes & Wiklind 1995; Combes, Wiklind & Nakai 1997).

The best upper limit to the O_2 abundance was obtained through molecular absorption lines (see also § 23.3.1) in the $z = 0.68$ intervening system toward the background QSO B0218+357 (Combes & Wiklind 1995; Combes et al. 1997). Additional observations of ^{12}CO, ^{13}CO, ^{18}CO, and a nondetection of ^{17}CO led to an upper limit of $N(O_2)/N(CO) < 2 \times 10^{-3}$ (Fig. 23.1). The theoretical value of the abundance ratio O_2/CO is quite insensitive to the density and extinction and is ~ 0.4 (Maréchal, Viala, & Banayoun 1997b). Adopting a typical Galactic CO/H_2 abundance ratio of 5×10^{-4}, the observed upper limit corresponds to an O_2 abundance a few times $< 10^{-7}$. This is almost 3 orders of magnitude lower than predicted ISM abundances and was a first indication that the chemistry in the ISM may have surprises in store.

Recently, two satellites have provided new data on the abundance of interstellar O_2: *SWAS* (the *Submillimeter Wavelength Astronomy Satellite*) launched in 1998 (Melnick et al. 2000), and *Odin* launched in 2001 (Hjalmarson et al. 2003). Both satellites carry a small submillimeter telescope above the obscuring terrestrial atmosphere. While both *Odin* and *SWAS* could observe the higher transition, $N_J = 3_3 - 1_2$ (see Fig. 23.1), *Odin* also carried a detector for the ground transition at 118.75 GHz, $N_J = 1_1 - 1_0$, allowing for a better sensitivity. Neither satellite detected the O_2 molecule, despite heroic efforts (Goldsmith et al. 2000; Pagani

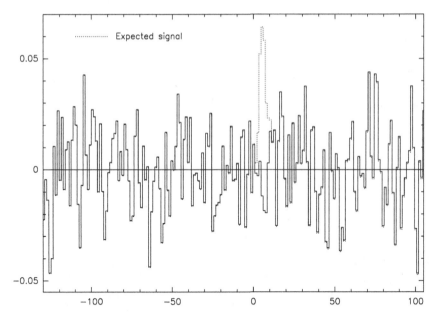

Fig. 23.2. *Odin* spectra of the ground transition of O_2 toward ρ Oph (Pagani et al. 2003). The dotted line indicates a previously reported tentative O_2 detection by *SWAS*. The detection is not confirmed by the more sensitive *Odin* data.

et al. 2003). The best 3 σ upper limits for the $N(O_2)/N(H_2)$ ratio are $< 2.6 \times 10^{-7}$ for *SWAS* (Goldsmith et al. 2000) and $< 1.0 \times 10^{-7}$ for *Odin* (Pagani et al. 2003). The value expected from chemical models of dark clouds is $N(O_2)/N(H_2) = (2-5) \times 10^{-4}$. A tentative detection of O_2 emission reported by *SWAS* (Goldsmith et al. 2002) in an outflow region of the ρ Oph cloud was not confirmed with *Odin* data (Pagani et al. 2003), despite a better noise level (Fig. 23.2).

23.4.2   *Reasons for O_2 Nondetections.*

There are several possible explanations to the low O_2 abundances found by *SWAS*, *Odin*, and through molecular absorption lines. Not all of the explanations give results that are consistent with other abundances, notably the low water vapor abundances also found by *SWAS* and *Odin*.

High C/O ratio. The simplest way to decrease the O_2 abundance is to increase the C/O abundance ratio from its nominal cosmic abundance value of 0.45. An increase in C/O by a factor ~ 2 decreases the O_2 abundance by a factor 150 (Maréchal et al. 1997b). This is accompanied by a similarly large decrease in the H_2O abundance as well as in the OH abundance. The reason for the O_2 abundance dependence on the C/O ratio is simply that if C/O<1 there are oxygen atoms left when all the gas-phase carbon has been bound up in CO, leaving enough gas-phase oxygen to build other molecules. However, even in the densest and most opaque clouds, the atomic oxygen remains more abundant than O_2 and represents between 20% and 30% of the available gas-phase oxygen (Maréchal et al. 1997b).

Clumpy cloud structure. Molecular oxygen is easily photo-dissociated and therefore not expected to be abundant in diffuse gas or in PDRs. Spaans & van Dishoeck (2001) computed models of the thermal and chemical balance of molecular gas with a clumpy cloud structure and exposed to UV radiation, where the abundances of O_2 and H_2O were computed for various density distributions, radiation field strengths, and geometries. They found that an inhomogeneous density distribution lowers the column density of both H_2O and O_2 compared to the homogeneous case by more than an order of magnitude for the same A_V. In particular O_2 is sensitive to the penetrating UV radiation, and the destruction of O_2 remains dominated by photodissociation to large extinction as the clumpy structure allows UV photons to penetrate to larger depths.

Dust coagulation. Casu, Cecchi-Pestellini, & Aiello (2001) argue that dust coagulation in dense clouds can alter the amount of UV photons that penetrate deep into the cloud interior. This model agree with some observational evidence that the UV-to-visual extinction is smaller in dense clouds than in more diffuse clouds. This could affect molecules like O_2 and H_2O, which are susceptible to photodissociation, while a more abundant molecule like CO would be protected through self-shielding. This model predicts that the main carrier of oxygen in dense molecular clouds should be atomic oxygen, while CO is the main carrier of gas-phase carbon. Another interesting, and testable, effect is an increase of the HOC^+/HCO^+ ratio by a factor ~ 100.

Diffusive mixing and turbulence. Turbulence is the dominant component of support and dynamics of most molecular clouds. It may also play a role in the mixing of material. If this is the case, chemical abundances will be altered compared to static models if the transport time scale between zones is less than the chemical time scale. Willacy, Langer, & Allen (2002) couple diffusion and gas-grain interaction (freeze-out) as well as photodissociation processes. They find that diffusive processes significantly alter the predicted chemical abundances. Increases in the abundances of C I and C II and most carbon-bearing species are seen together with a reduction of O_2 and H_2O. Willacy et al. (2002) find that the H I/H_2 ratio should be an observable measure of the mixing.

Grain freeze-out. Roberts & Herbst (2002) have recently presented new gas-grain chemical models for cold cloud cores, which simultaneously follows the chemistry that occurs in the gas and on grain surfaces. These models give results that strongly depart from pure gas-phase models at times greater than 10^5 years. The gas-phase and grain-surface chemistries are linked through accretion and desorption processes, based on new experimental results on the surface mobility of hydrogen atoms on olivine and amorphous carbon. Two sets of gas-grain models are considered, differing in the mobility of various atoms. Even though both accretion and desorption mechanisms act in these models, a steady-state chemistry is never reached. The abundances of many gas-phase species, including CO, H_2O, and O_2, rise until somewhere between $10^5 - 10^6$ years, at which time freeze-out begins to dominate and their abundances decline. With such behavior, molecules such as O_2, that takes a significantly long time to reach a high abundance in gas-phase models, are never able to reach this abundance in gas-grain models.

Although very promising, the gas-grain models do not optimally match all the observed constraints of well-known dark clouds at a single time. Pure gas-phase models with C/O=1 are able to match the observations fairly accurate at a single time. Explanations could be (1) differential desorption rates and (2) different chemical evolutionary times within a given cloud. There are some observational evidence for chemical variation across the dark cloud TMC-1 (e.g., Saito et al. 2002).

23.5   Molecules Far Away

Observations of the molecular gas component in distant galaxies enable us to study the conditions of star formation and the chemical evolution of objects at early epochs. Molecular gas has been detected in galaxies up to redshift $z = 4.7$ (Ohta et al. 1996; Omont et al. 1996), corresponding to a look-back time of almost 90% of the age of the Universe.

There are three sources of information about the molecular gas at high redshift: (1) H_2 in damped Lyα systems, (2) CO emission from far-infrared luminous galaxies and AGNs, and (3) molecular absorption-line systems at intermediate redshifts. Furthermore, molecules are believed to play an important role in formation of the very first generation of stars. This is due to the presence of H_2, HD, and LiH molecules in the primordial gas (e.g., Stancil, Lepp, & Dalgarno 1996, 1998; Galli & Palla 2002). Although the abundances of these primordial molecules are minute, they act as cooling agents of the primordial gas, and thereby allow the formation of stellar objects less massive than what is the case for a pure atomic gas.

Molecular hydrogen has been detected in a few damped Lyα systems (e.g., Petitjean, Srianand, & Ledoux 2000, 2002; Ledoux, Srianand, & Petitjean 2002), but with very low column densities, $N_{H_2} \leq 10^{18}$ cm^{-2}. The H_2 seen in these system appears to originate in a warm ($\sim$3000 K) diffuse gas. This gas is not typical for the normal molecular ISM component found in galaxies.

Carbon monoxide, CO, has been detected in 15 galaxies above $z = 1$ (see Wiklind & Alloin 2002 for a recent compilation). The detected systems are all far-infrared luminous galaxies with elevated star formation rates. Most of the CO-detected objects are gravitationally lensed. The CO lines give information about the global properties of the molecular ISM, such as the overall mass and the average excitation conditions. It can also be used to derive the dynamical mass. If the commonly used conversion factor between CO integrated intensity and H_2 column density is used, many of these distant systems have a mass that is dominated by the molecular gas component. Due to the extreme weakness of the observed signal, in almost all cases only the CO lines are observable with present-day instrumentation. Apart from the global parameters, little can be learned about the chemical evolutionary state of these distant galaxies.

A few molecular absorption systems have been detected at redshifts $z = 0.25 - 0.89$ (e.g., Wiklind & Combes 1994; Combes & Wiklind 1999; Menten et al. 1999). These rare systems, where the line of sight to a background radio-loud QSO penetrates the inner region of an intervening galaxy, gives an unprecedented view of the molecular gas component in distant galaxies (see Fig. 23.3). Although only four such systems have been detected to date, 22 different molecular species have been observed, including several isotopic variants, and most molecules have been observed in at least two different transitions. These data give detailed information about the chemistry, the abundances, and the physical conditions in the molecular gas in these distant galaxies. Comparison with similar data from Galactic absorption-line studies and those of a few nearby galaxies shows no significant differences between the molecular gas properties (Fig. 23.4). This suggests that the star-forming conditions in distant galaxies are essentially the same as we find in local systems.

23.6   Challenges for the Future

Astrochemistry remains very much an active research area. This brief overview of molecules in the ISM can only point to some of the unsolved problems and their impact on

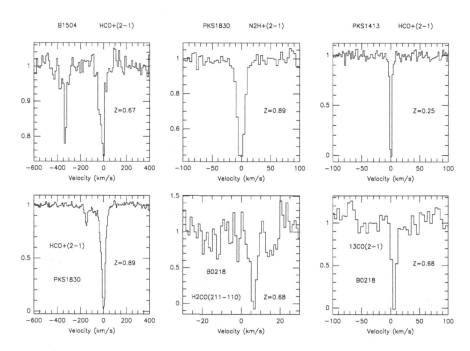

Fig. 23.3. Several different molecular transitions seen in absorption toward the four known molecular absorption-line systems at intermediate redshifts. (Adapted from Combes & Wiklind 1998.)

our understanding of the physical and chemical processes related to the formation of stars and planetary systems. A non-exhaustive list of challenges for the future includes:

- A better physical understanding of the combined gas-phase and grain-phase chemistry in dark clouds, as well as the formation of molecular ices. This can possibly lead to an explanation of the low observed abundances of molecular oxygen and water vapor in dark clouds. It may also explain the elevated abundances of molecular species found through absorption-line studies of diffuse and translucent molecular gas.

- Molecular absorption-line observations of diffuse and translucent molecular clouds in the Milky Way show that the chemistry is very similar to that of dark clouds. This suggests that the interior of dark clouds may not be as "dark" as previously thought. The inhomogeneous structure of molecular gas needs to be investigated further. The possibility of very small-scale structure, as shown by recent reports, needs to be studied as well.

- In order to understand the physics of stellar formation it is necessary to construct self-consistent models combining gas- and grain-phase chemistry, hydrodynamical models of the protostellar cloud, and the full transfer of radiation. Small steps toward such models have been taken, but the fully consistent model is yet to be constructed.

- Observations of molecules at high redshift allow not just a study of the global properties, but also of the detailed physical and chemical evolutionary status of the gas. Since the molecular gas is directly involved in the formation of new stars, these data give infor-

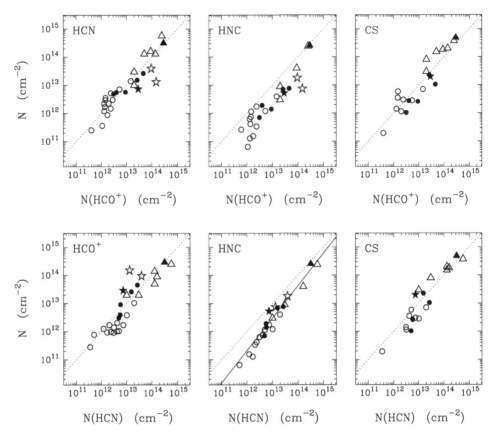

Fig. 23.4. The *top* panels show the column density of HCO$^+$ plotted versus column densities of HCN, HNC, and CS. The *bottom* panels show column density of HCN plotted versus column densities of HCO$^+$, HNC, and CS. Open circles represent Galactic diffuse clouds (Lucas & Liszt 1996), filled circles represent data from Cen A (Wiklind & Combes 1997a), open triangles absorption data toward Sgr B2 (Greaves & Nyman 1996), filled star absorption data for PKS 1413+135 at $z = 0.25$ (Wiklind & Combes 1997b), open stars B31504+377 at $z = 0.67$ (Wiklind & Combes 1996b), and the filled triangle PKS 1830–211 at $z = 0.89$ (Wiklind & Combes 1996a, 1998). The dashed line is a one-to-one correspondence between the column densities and not a fit to the data.

mation on which processes are at work in these earlier epochs and if the star formation proceeds in the same manner as in nearby systems.

The future looks promising indeed for observations of molecules in the ISM, both in the Milky Way and in very distant galaxies. The Atacama Large Millimeter Array (ALMA) will increase the sensitivity at millimeter and submillimeter wavelengths by a factor 10–100 (depending on the wavelength regime). Paired with a similarly large increase in the angular resolving power, it will be able to do very sensitive observations of molecules on small spatial scales. This is likely to give insights into how protostars form and evolve, as well as allow studies of protoplanetary disks. The increased sensitivity will allow detailed studies of molecular absorption-line systems. An example of what can be achieved with ALMA

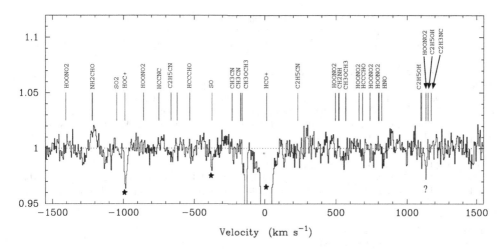

Fig. 23.5. Coadded observations from a monitoring project of the $HCO^+(2\text{-}1)$ absorption line at $z = 0.89$ toward PKS 1830–211. The two deep absorption lines correspond to two lines of $HCO^+(2\text{-}1)$. In addition a weak line of $HOC^+(2\text{-}1)$ can be seen, as well as SO. This spectrum represents several tens of hours of total observing time with the 15 m SEST (Wiklind & Combes, unpublished). With ALMA a similar spectrum can be obtained in a matter of minutes.

in a few minutes of observing time, but which requires several days with present-day instruments, is shown in Figure 23.5. Here coadded spectra from a monitoring program of $HCO^+(2\text{-}1)$ absorption at $z = 0.89$ toward PKS 1830–211 reveal the presence of the isomer $HOC^+(2\text{-}1)$ as well as a line of sulphur monoxide, SO. The HOC^+ molecule has an abundance that is $\sim 2.6 \times 10^{-4}$ that of HCO^+, similar to values found in our Galaxy. Assuming Galactic abundance ratios for CO/H_2, the HOC^+/H_2 abundance ratio for the $z = 0.89$ system is $\sim 10^{-10}$.

On a shorter time scale, the new Atacama PAthfinder EXperiment (APEX) will put a single dish submillimeter telescope on Chanjantor at an altitude of 5000 m. This will allow observations up to the THz region (300 μm) and make it possible to observe for the first time the ground transitions of many light hydrides believed to exist and play an important role in the molecular chemistry.

References

Adams, W. S. 1941, PASP, 53, 73
———. 1948, PASP, 60, 174
Beals, C. S., & Blanchet, G. H., 1938, MNRAS, 98, 398
Bergin, E. A., & Langer, W. D. 1997, ApJ, 486, 316
Bergman, P. 1995, ApJ, 445, L167
Black, J. H., & Smith, P. L. 1984, ApJ, 277, 562
Casu, S., Cecchi-Pestellini, C.-C., & Aiello, S. 2001, MNRAS, 325, 826
Cheung, A. C., et al. 1969, Nature, 221, 626
Combes, F., & Wiklind, T. 1995, A&A, 303, L61
———. 1996, in Cold Gas at High Redshift, ed. M. Bremer, P. van der Werf, & C. Carilli (Dordrecht: Kluwer), 215
———. 1997, ApJ, 486, L79
———. 1998a, A&A, 334, L81

——. 1998b, in 13th IAP Colloquium: Structure and Evolution of the IGM from QSO Absorption Line Systems, ed. P. Petitjean & S. Charlot (Paris: Editions Frontières), 317

——. 1999, in Highly Redshifted Radio Lines, ed. C. L. Carilli, K. M. Menten, & G. I. Langston (San Francisco: ASP), 210

Combes, F., Wiklind, T., & Nakai, N. 1997, A&A, 327, L17

d'Hendecourt, L., Allamandola, L. J., & Greenberg, J. M. 1985, A&A, 152, 130

Doty, S. D., van Dishoeck, E. F., van der Tak, F. F. S., & Boonman, A. M. S. 2002, A&A, 389, 446

Duley, W. W., & Williams, D. A. 1984, Interstellar Chemistry (London: Academic Press)

Falgarone, E., Panis, J.-F., Heithausen, A., Pérault, M., Stutzki, J., Puget, J.-L., & Bensch, F. 1998, A&A, 331, 669

Galli, D., & Palla, F. 2002, P&SS, 50, 1197

Goldsmith, P. F., et al. 2000, ApJ, 539, 537

Goldsmith, P. F., & Langer, W. D. 1978, ApJ, 222, 881

Goldsmith, P. F., Li, D., Melnick, G. J., Tolls, V., Howe, J. E., Snell, R. L., & Neufeld, D. A. 2002, ApJ, 576, 814

Goldsmith, P. F., Snell, R. L., Erickson, N. R., Dickman, R. L., Schloerb, F. P., & Irvine, W. M. 1985, ApJ, 289, 613

Goldsmith, P. F, & Young, J. Y. 1989, ApJ, 341, 718

Greaves, J. S., & Nyman, L.-Å 1996, A&A, 305, 950

Hasegawa, T. I., & Herbst E. 1993a, MNRAS, 261, 83

——. 1993b, MNRAS, 263, 589

Hjalmarson, Å, et al. 2003, A&A, 402, L39

Hollis, J. M., Lovas, F. J., & Jewell, P. R. 2000, ApJ, 540, L107

Ledoux, C., Srianand, R., & Petitjean, P. 2002, A&A, 392, 781

Liszt, H. S. 1985, ApJ, 298, 281

Liszt, H. S., & Lucas, R. 1998, A&A, 339, 561

——. 2002, A&A, 391, 693

Liszt, H. S., & vanden Bout, P. A. 1985, ApJ, 291, 178

Lucas, R., & Liszt, H. S. 1996, A&A, 307, 237

——. 2002, A&A, 384, 1054

Maréchal, P., Pagani, L., Langer, W. D., & Castets, A. 1997a, A&A, 318, 252

Maréchal, P., Viala, Y. P., & Benayoun, J. J. 1997b, A&A, 324, 221

McKellar, A. 1940, PASP, 52, 187

Melnick, G. J., et al. 2000, ApJ, 539, L77

Menten, K. M., Carilli, C. L., & Reid, M. J. 1999, in Highly Redshifted Radio Lines, ed. C. L. Carilli, K. M. Menten, & G. I. Langston (San Francisco: ASP), 218

Merrill, P. W. 1934, PASP, 46, 206

——. 1936, ApJ, 83, 126

Millar, T. J. 2002, P&SS, 50, 1189

Millar, T. J., Bennett, A., & Herbst, E. 1989, ApJ, 340, 906

Millar, T. J., Farquhar, Y., & Willacy, K. 1997, A&AS, 121, 139

Nummelin, A., Bergman, P., Hjalmarson, A., Friberg, P., Irvine, W. M., Millar, T. J., Ohishi, M., & Saito, S. 1998, A&AS, 117, 427

Ohta, K., Yamada, T., Nakanishi, K., Khono, K., Akiyama, M., & Kawabe, R. 1996, Nature, 382, 426

Omont, A., Petitjean, P., Guilloteau, S., McMahon, R. G., Solomon, P. M., & Pecontal, E. 1996, Nature, 382, 428

Pagani, L., et al. 2003, A&A, 402, L77

Pagani, L., Langer, W. D., & Castets, A. 1993, A&A, 274, L13

Petitjean, P., Srianand, R., & Ledoux, C. 2000, A&A, 364, 26

——. 2002, MNRAS, 332, 383

Puget, J.-L., & Falgarone, E. 1990, in Submillimeter Astronomy, ed. G. D. Watts & A. S. Webster (Dordrecht: Kluwer), 3

Roberts, H., & Herbst, E. 2002, A&A, 395, 233

Rollinde, E., Boissé, P., Federman, S. R., & Pan, K. 2003, A&A, 401, 215

Russell, H. N. 1935, MNRAS, 95, 635

Rydbeck, O. E. H., Elldér, J., & Irwine, W. M. 1973, Nature, 246, 466

Saito, S., Aikawa, Y., Herbst, E., Ohishi, M., Hirota, T., Yamamoto, S., & Kaifu, N. 2002, ApJ, 569, 836

Snyder, L. E., Buhl, D., Zuckerman, B., & Palmer, P. 1969, Phys. Rev. Lett., 22, 679

Spaans, M., & van Dishoeck, E. F. 2001, ApJ, 548, L217

Stancil, P. C., Lepp, S., & Dalgarno, A. 1996, ApJ, 458, 1996

——. 1998, ApJ, 509, 1

Swings, P., & Öman, Y. 1939, The Observatory, 62, 150

Swings, P., & Rosenfeld, L. 1937, ApJ, 86, 483

van Dishoeck, E. F. 1998, in The Molecular Astrophysics of Stars and Galaxies, ed. T. W. Hartquist & D. A. Williams (Oxford: Clarendon Press), 53

van Dishoeck, E. F., & Blake, G. A. 1998, ARA&A, 36, 317

Weinreb, S., Barrett, A. H., Meeks, M. L., & Henry, J. C. 1963, Nature, 200, 829

Whittet, D. C. B., et al. 1996, A&A, 315, L357

Wiklind, T., & Alloin, D. 2002, in Gravitational Lensing: An Astrophysical Tool, ed. F. Courbin & D. Minniti (Berlin: Springer), 124

Wiklind, T., & Combes, F. 1994, A&A, 286, L9

——. 1995, A&A, 299, 382

——. 1996a, Nature, 379, 139

——. 1996b, A&A, 315, 86

——. 1997a, A&A, 324, 51

——. 1997b, A&A, 328, 48

——. 1998, ApJ, 500, 129

——. 1999, in Highly Redshifted Radio Lines, ed. C. L. Carilli, K. M. Menten, & G. I. Langston (San Francisco: ASP), 202

Willacy, K., Langer, W. D., & Allen, M. 2002, ApJ, 573, L119

Wilson, R. W., Jefferts, K. B., & Penzias, A. A. 1970, ApJ, 161, L453

24

Metal ejection by Galactic winds

CRYSTAL L. MARTIN
University of California, Santa Barbara

Abstract
The development of high-resolution X-ray spectral imaging has made it possible to constrain the metal content of hot galactic winds and to explore the relationship between cold, warm, and hot gas in galactic outflows. I review the empirically derived limits on metal ejection and mass loss rates and discuss their variation with galactic mass. I use these results to normalize standard leaky-box models of chemical evolution and examine the galactic chemical properties that they predict. The amount of intergalactic metal enrichment by starburst winds is also discussed.

24.1 Introduction

In a little over a decade, our knowledge of galactic winds has grown from a theoretical parameter to a phenomenon studied at nearly every electromagnetic frequency. This widespread effort is fueled by the increasing recognition of the large number of high-redshift galaxies with massive outflows (e.g., Pettini et al. 2002; Adelberger et al. 2003), and the relevance of winds for the properties of galaxies and intergalactic gas.

This review concentrates on measurements of the metal content and mass flow rate in winds, with an emphasis on nearby galaxies. I begin with a simplified theoretical description of the coupling between supernovae and the interstellar medium (ISM), which illustrates the multi-phase nature of winds. I then describe the X-ray observations and analysis techniques that constrain the mass and metallicity of the hot wind. I summarize the variation in X-ray properties with the gravitational potential depth and draw attention to the biased view provided by observations of emission processes. I then review the known absorption-line signatures of winds in some detail and speculate about the connections to some classes of intervening absorption-line systems. Finally, I combine the information afforded by these complementary techniques to empirically constrain the parameters in leaky-box models of galactic chemical evolution. The emphasis is on the starburst epoch throughout, although I illustrate the impact of burst duty cycle on the time-integrated metal loss and intergalactic medium (IGM) enrichment.

24.2 Chemodynamical Modeling of Galactic Outflows

Galactic-scale outflows are driven by a large number of massive stars, so estimates of the mechanical power and metal injection rate are generally averaged over the stellar population. In this section I discuss properties of starburst synthesis models critical to galactic wind dynamics and enrichment. Metal yields from models of individual supernovae are then reviewed. Finally, I illustrate how the hydrodynamic interaction of the stellar ejecta with the ISM mixes the two fluids.

24.2.1 Energy and Metal Production by Core-collapse Supernovae

24.2.1.1 Synthesis Models

Synthesis models use a grid of stellar evolution tracks, an initial stellar mass function, and a particular star formation history to derive the effective temperature and bolometric luminosity of each star as a function of time. Observable properties are then defined by assigning a stellar atmosphere model to each star and adding all the stellar templates together. The basic trends in mass and energy production by a starburst are illustrated in Figure 24.1, which is based on the Starburst 99 synthesis models (Leitherer et al. 1999). Several properties of this mechanical feedback affect the development of galactic winds.

First, the time scales for stellar winds and supernovae differ. Stars more massive than about 20 $M_\odot$—e.g., O stars, early-B stars, and Wolf-Rayet stars—generate powerful stellar winds (Leitherer, Robert, & Drissen 1992); but all stars with masses exceeding $\sim 8\, M_\odot$ are thought to produce core-collapse supernovae (Type II) of similar energy. Following an instantaneous burst of star formation, the mass and energy deposition from supernovae will overtake the contribution from stellar winds after only a few million years. The least massive Type II supernovae explode after ~ 40 Myr.

Second, the mechanical power from supernovae is roughly constant in time. This result reflects the near-perfect cancellation between the larger number of low-mass progenitors and their longer lifetimes. The number of supernova progenitors per unit mass increases according to the Salpeter mass function, $dN \propto M^{-2.35}$, but the evolutionary rate $dM/d\tau$ slows down according to the stellar mass–lifetime relation, $\tau(M) \propto M^{-1.2}$ (Schaerer et al. 1993).

A modest burst of $10^6\, M_\odot$ generates $\sim 10^4$ supernova explosions in 40 Myr. In a starburst, these supernova are correlated spatially as well as temporally. The high pressure of the remnants drives a shock front into the surrounding interstellar gas. In a medium of uniform density, the dynamical evolution of the swept-up shell of interstellar matter is described by (Weaver et al. 1977),

$$R(t) = 104\, L_{40}^{1/5} n_0^{-1/5} t_6^{3/5}\ \text{pc} \tag{24.1}$$

and

$$v(t) = 61\, L_{40}^{1/5} n_0^{-1/5} t_6^{-2/5}\ \text{km s}^{-1}, \tag{24.2}$$

where the mechanical power and age are in units of 10^{40} erg s^{-1} and Myr, respectively. Since the supershell velocity is only a weak power of the starburst strength, the dynamical time scale of a galactic outflow is determined largely by the scale height of the gaseous disk, H. The time scale for a galactic wind to develop, $t_d \approx 8 \left(\frac{H}{500\ \text{pc}} \right) \left(\frac{61\ \text{km s}^{-1}}{v} \right)$ Myr, is longer than the lifetimes of O stars and Wolf-Rayet stars. Hence, supernovae dominate the returned energy and matter by the time a galactic outflow develops.

Finally, it should be noted that the adopted mass loss rates in synthesis models such as Starburst 99 are generally not taken directly from the stellar evolutionary models, which predict considerly higher mass loss rates than are measured. The large mass loss parameter in the stellar models probably compensates for other physical processes yet to be included

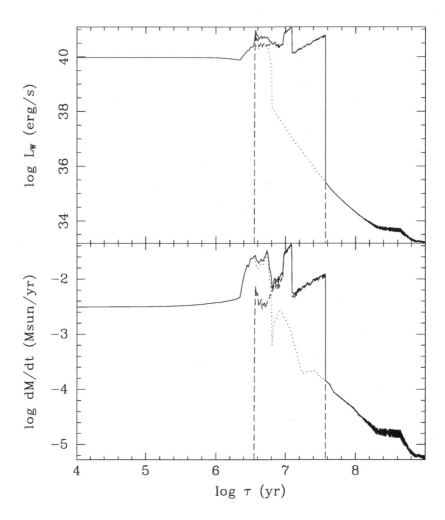

Fig. 24.1. Population synthesis models from Starburst 99 (Leitherer et al. 1999). *Top:* Production of mechanical power by a coeval starburst population. The contribution of stellar winds (dotted line) is important before a few Myr, but supernovae (dashed line) dominate at later times. *Bottom:* Mass of material returned to the ISM by stars. The curves are normalized to 10^6 $M_\odot$ of stars formed in an instantaneous burst.

in the modeling. A combination of empirical results and theoretical predictions is thought to provide a more accurate description of stellar mass loss. This prescription is sensitive to stellar metallicity in two ways (e.g., Leitherer et al. 1992). First, the ratio of Wolf-Rayet stars to O stars is metallicity dependent, and this is believed to affect the early enrichment. Second, the stellar winds grow stronger (weaker) with increasing (decreasing) metallicity because the momentum of the radiation couples to the atmosphere via metal lines. However, since the supernova properties are relatively insensitive to metallicty, neither the mass loss nor the mechanical power is sensitive to metallicity on the time scales of most galactic-scale outflows.

24.2.1.2 Supernova Yields

To determine the metal production rate, the population synthesis models must include a grid of supernova yields. I will use the models of Woosley & Weaver (1995) to illustrate some general properties. Stars in the 11–$40\,M_\odot$ range produce most of the intermediate-mass elements. The 8–$10\,M_\odot$ stars synthesize relatively little of these elements, so the low-mass cut-off for Type II supernovae has little impact on the ejected mass. Since increasing the minimum mass from $8\,M_\odot$ to $10\,M_\odot$ reduces the number of supernovae, the low-mass cut-off does have some affect on the mechanical feedback.

The high-mass cut-off for supernovae, in contrast to the low-mass cut-off, does affect the yield but has little impact on *the number of supernovae*. In the Woosley & Weaver (1995) models, stars heavier than $40\,M_\odot$ undergo considerable reimplosion of heavy elements, reducing their yield. The mass above which stars stop contributing to the yield is not firmly established, however. In Figure 24.2, I have weighted the Woosley & Weaver (1995) yields by a Salpeter initial mass function to illustrate the O production in various stellar mass intervals. The mass of O produced in each stellar mass interval is given per unit initial mass of 0.1–$100\,M_\odot$ stars formed. The O production peaks around $30\,M_\odot$ because more massive stars retain a large fraction of their synthesized metals.

Integrating over stellar mass, the total oxygen yield is 0.0061 for the $0.1\,Z_\odot$ models and increases to 0.0065 for the solar-metallicity models. The ^{16}O isotope comprises about half of the heavy element yield from massive stars. Since ^{16}O is produced by He burning and Ne burning, its yield is not sensitive to the metallicity of the stellar model. In contrast, the yields of ^{17}O and ^{18}O scale roughly linearly with the initial metallicity of the star. (The ^{17}O is synthesized by the CNO tri-cycle during H shell burning, and α-capture on ^{14}N during He shell burning produces ^{18}O.) Since the abundance of ^{16}O is 3 to 5 orders of magnitude higher than that of ^{17}O or ^{18}O, the total mass of oxygen produced depends little on metallicity. The metallicity dependence of the O yield is illustrated in Figure 24.2. The same models give an Fe yield of 2.041×10^{-5} for the $0.1\,Z_\odot$ models.

Yield calculations vary by factors of a few among authors. The adopted $^{12}CO(\alpha,\gamma)^{16}O$ reaction rate is a critical parameter that accounts for some, but not all, of the differences. The amount of reimplosion also has an important impact on the yield through the contribution from the highest-mass stars. The models plotted use relatively low explosion energy and give a conservative O yield. The discretization into a finte number of explosion models also reduced the accuracy of the final yield of the population. The enrichment from a coeval burst takes place over a 40 Myr period, and many starbursts are observed at a slightly younger age before we can expect this enrichment to have been completed.

24.2.2 Mass Loading

The metallicity of the stellar ejecta does not provide much direct information on the chemical abundances in galactic outflows. The strong hydrodynamical interaction between the supernova fluid and the ISM generates a complex multi-phase flow that partially mixes the two fluids.

Numerical simulations have been instrumental in developing a basic understanding of the outflow dynamics (e.g., Suchkov et al. 1994; Suchkov et al. 1996; MacLow & Ferrara

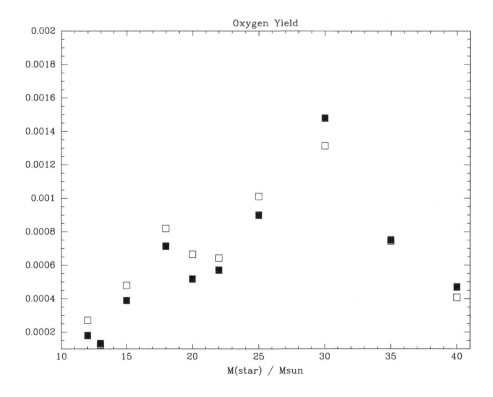

Fig. 24.2. Mass of O produced per unit mass of stars. Solid and open symbols represent $0.1\,Z_\odot$ and $1.0\,Z_\odot$ models for the initial stellar metallicity. The low explosion energy yields from Woosley & Weaver (1995) were weighted by a Salpeter initial mass function extending to $0.1\,M_\odot$. The most massive stars are rare, and they also retain a larger fraction of the synthesized oxygen. Note that stars less massive than $10\,M_\odot$ make no contribution to the total yield in this plot, so it is trivial to scale the results to other low-mass cut-offs. The total yield is 0.0061 (0.0065) per unit mass of 0.1 to $100\,M_\odot$ stars formed.

1999; Strickland & Stevens 2000). They show the collision of the wind fluid with the ISM generates shock fronts in the disk gas and the halo gas. If the outflows are fast enough, the shocked disk and halo material emits thermal X-rays. The brightness of this shocked interstellar gas can exceed that of the X-ray emission from the supernova ejecta (Suchkov et al. 1996; Strickland & Stevens 2000). The emission from the supernova fluid, usually called the *free wind*, depends on the degree of mass loading.

Mass loading takes place as the supernova ejecta punch through the galactic disk along the galaxy's minor axis. The shear at the interface between the free wind and the dense disk gas creates vortices that draw disk clouds into the free wind. The subsequent evaporation of the interstellar clouds in the hot wind fluid mixes gas with the galactic abundance pattern into the highly enriched free wind. The interstellar gas advected into these vortices is whisked into relatively dense filamentary structures. The metallicity of these cold filaments (or warm filaments if photoionized by the starburst) is that of the ambient interstellar gas. The mass of evaporated cloud material and the metallicity of the supernova ejecta determine the net metallicity of the free wind.

The X-ray observations of starburst winds empirically constrain the degree of mass loading. The X-ray emissivity increases quadratically with the gas density, so mass loading brightens outflows relative to non-loaded winds. Comparisons of data and models suggest the mass entrainment rate in M82 is 3 to 6 times larger than the mass deposition rate due to supernova (Suchkov et al. 1996). Similarly, in the dwarf starburst galaxy NGC 1569, the mass loading factor is about 9 times the mass of stellar ejecta (Martin, Kobulnicky, & Heckman 2002). *An importatnt consequence of mass loading is simply that the supernova metals are highly diluted by mixing with lower-metallicity gas.*

Figure 24.3 illustrates the dilution process for a range of supernova yields. The mass loading factor on the abscissa, χ, is defined to be the ratio of the mass of entrained ISM to the mass of supernova ejecta. Adding the entrained ISM metals to the metals from supernovae, it is straightforward to show that the wind metallicity becomes

$$Z = (Z_{SN} + \chi Z_{ISM})/(1+\chi), \tag{24.3}$$

where Z_{SN} and Z_{ISM} are the initial metallicities of the two fluids being mixed. The same reasoning shows the abundance ratio of α elements to iron in the wind will be,

$$Z_O/Z_{Fe} = (Z_{O,SN} + \chi Z_{O,ISM})/(Z_{Fe,SN} + \chi Z_{Fe,ISM}). \tag{24.4}$$

As more ISM is mixed into the free wind (see Fig. 24.3), the metallicity of the latter approaches that of the ambient ISM. The inflection point is determined by the ratio of initial supernova metallicity to ISM metallicity and moves to higher χ values in low-metallicity galaxies. Low-metallicity galaxies therefore present the most favorable environment for detecting supernova enrichment in galactic outflows.

24.2.3 *Feedback from Type Ia Supernovae*

The contribution of Type Ia supernovae to metal and energy injection is generally ignored in starburst models due to the longer evolutionary time scale of their progenitors relative to Type II supernovae. The abundance of Fe in the intracluster medium clearly indicates, however, that a significant fraction of the metals liberated by Type Ia supernovae reach the intracluster medium (Renzini 2004, and references therein). Late winds, driven by Type Ia supernova on time scales $\gtrsim 10^9$ yr, have been advocated to explain this enrichment and the mass-metallicity relation (Mathews & Baker 1972; Ciotti et al. 1991). The empirical evidence for winds, however, is mostly restricted to the early winds in starburst galaxies. The strength and ubiquity of late-type winds need to be established. Some models include Type Ia supernova progenitors as young as 100 Myr (e.g., Nomoto et al. 2003, and references therein), which, if shown to be correct, could change the way we model metal and energy feedback.

24.3 Observational Diagnostics

One of the primary challenges in describing the feedback process is understanding the mass and energy transfer between hot, warm, and cold gas in the outflows. Kinematics, composition, density, and temperatue should be measured for each component, which is not currently possible. I describe here the main techniques that are being used to measure a subset of these quantities. The enormous temperature range requires observations from X-ray to radio wavelengths. The data suggest the various components of the outflow undergo strong hydrodynamical and thermal interactions.

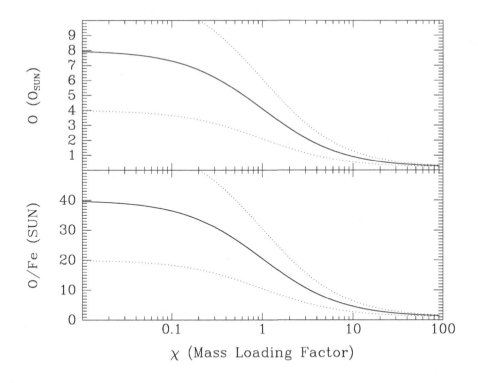

Fig. 24.3. Dilution of wind metallicity by mass entrainment. The initial mass function-averaged metallicity of supernova ejecta is ~ 8 times solar for oxygen and ~ 0.2 times solar for Fe (model A runs of Woosley & Weaver 1995 at $0.1\,Z_\odot$). The oxygen and iron yields of massive stars are only known to an accuracy of a factor of 2 to 3, so the dotted curves illustrate this range. The galactic ISM is assumed to have $0.25\,Z_\odot$ metallicity appropriate for a dwarf galaxy.

24.3.1   Optical Emission Lines

Starburst galaxies generally present moderately bright optical emission-line filaments that extend up to ~ 10 kpc out of the galactic disk (e.g., Heckman, Armus, & Miley 1990; Marlowe et al. 1995; Martin 1999). One can gain some appreciation for the excitation mechanism by considering the potential power sources. In a steady state situation, i.e. continuous star formation at a constant rate, the luminosity of the ultraviolet continuum (shortward of the ionization potential of H) is about 7 times larger than the mechanical power supplied by stellar winds and supernovae. (In a single, instantaneous burst, the ionizing luminosity only exceeds the mechanical power up to an age of 6.3 Myr when the O stars die off.) Hence, it is no surprise that the optical recombination line luminosity and the collisionally excited line luminosity are found to be dominated by emission from photoionized clouds (e.g., Lehnert & Heckman 1996; Martin 1997; Miller & Veilleux 2003).

Despite the dominance of the emission from photoionized clouds on a global scale, strong evidence for the presense of shocks is seen several kpc from the starburst region in the more extended filaments. The line ratios observed along the outflows evolve toward shock-like line ratios with increasing distance from the starburst region. Figure 2 of Martin (1997) shows several examples in dwarf starburst galaxies. Line ratios become more "shock-like" (high ratios of [N II] $\lambda6583$/Hα, [S II] $\lambda\lambda6716$, 6731/Hα, and [O II] $\lambda6300$/Hα) further out along the minor axis. Hybrid models that combine photoionization, shocks, and/or emission from turbulent mixing layers match the data better than photoionization alone. The rough constraints available on the chemical abundances in the dwarf starburst filaments are consistent with them being composed of disk material (Martin 1996, 1997).

The dynamics of the optical filaments were established in the seminal paper by Heckman et al. (1990). They showed that the network of Hα-emitting filaments extending along the minor axis of starburst galaxies presents double-peaked emission profiles characteristic of a moving shell. The pressure gradients indicated the shells were driven by an extended power source, and the starbursts were found to be strong enough to explain the observed shell velocities of 100–300 km s^{-1}. Subsequent studies found scaled-down versions of these flows in dwarf galaxies (Marlowe et al. 1995; Martin 1998).

For starbursts of all sizes, the estimated mass flux in the warm ionized filaments is similar to the star formation rate (SFR) (Martin 1999). This assertion rests on a volume filling factor derived from the assumption that the thermal pressure in the filaments is similar to that of the X-ray emitting region (Martin 1999). The projected velocities of the warm shells are not much larger than the escape speed, however, so much of the warm gas probably remains bound to the parent halo. The appropriate picture here may be closer to a fountain than an escaping wind. Indeed, the disks of particularly active spiral galaxies like NGC 891 and NGC 4631 also heat and transport warm gas into the halo at rates similar to the SFR per unit area.

24.3.2 X-ray Emission

The absence of local enrichment in H II regions suggests that most of the metals from core-collapse supernovae are injected directly into the hot ISM (Kobulnicky & Skillman 1996, 1997; Martin 1996). Measurements of chemical abundances from X-ray spectrocopy are therefore of great interest. In principle, abundance determinations for a collisionally ionized plasma are easier than in the nebular case in which photoionization must be included in the model. Under coronal conditions, the ionization structure is completely determined by the electron temperature, which is fixed by processes like hydrodynamic shocks.

In practice, metal abundances derived from thermal X-ray spectra are often unreliable. The observations have several biases. First, until recently, the spectral apertures were large compared to the thickness of superbubble shells. The X-ray emission within an aperture was the sum of emission from multiple components—i.e. shocked shells, mixing layers, hot bubbles, and point sources—all weighted disproportionately by the densest regions. Second, the observations were not very sensitive to gas cooler than 10^6 K or hotter than $10^{7.5}$ K. Third, due to a degeneracy between high-metallicity, two-temperature models and low-metallicity, single-temperature models, metallicities fitted to the data tended to be pushed lower than their true value if the temperature range was not fully represented (Buote & Canizares 1994). Fourth, the omission of intrinsic absorption from the spectral models increased the apparent strength of Mg and Si lines relative to Fe L lines. Unless lower-energy lines from O were detected, statistically acceptable fits could be obtained with an anomalously high α-element-to-Fe abundance ratio (e.g., Dahlem, Weaver, & Heckman 1998; Weaver, Heckman, & Dahlem 2000). Generally a volume filling factor of unity and solar abundances were simply assumed, leaving significant uncertainties about the mass and composition of the hot wind.

The *Chandra* X-ray Observatory has revolutionized our understanding of the origins of X-ray emission in starbursts. The arcsecond spatial resolution of the ACIS imaging spectrometer reveals that the brightest X-ray emitting gas lies adjacent to the warm, optical filaments. For example, in the prototypical starburst NGC 253, the X-ray emission was resolved into filaments out to 9 kpc above the disk (Strickland et al. 2000, 2002). Associations between the X-ray emission and the warm, ionized filaments have also been found in NGC 3079 (Cecil, Bland-Hawthorn, & Veilleux 2002) and in NGC 1482 and NGC 6240 (Veilleux et al. 2003). The correlation of soft X-ray emission with $H\alpha$ filaments suggests the brightest X-ray emission comes from the boundary layer between the optical filaments and a hotter, undetected free wind. The extraordinary sensitivity of *XMM-Newton* has very recently detected X-ray emission over a larger region, albeit with a high degree of structure, out to 14 kpc from the disk of M82 (Stevens, Read, & Bravo-Guerrero 2003).

An ultra-deep *Chandra* exposure of perhaps the closest starburst galaxy, NGC 1569, is perhaps the best available for constraining abundances (Martin et al. 2002). The metallicity of the ISM in this galaxy is only 0.25 $Z_\odot$, so the contrast with supernova-enriched gas is larger than it is in more luminous starbursts. Figure 24.4 shows the ACIS spectrum of NGC 1569. At such low spectral resolution, there is no single spectral model that uniquely fits this complex spectrum. The high-resolution imaging data can break the model degeneracy, however, when combined with data obtained in optical and radio bands. For example, the X-ray color gradients reveal intrinsic absorption that is tightly correlated with the location of the 21-cm emission from the cold disk gas. The relatively hard point source component is resolved and fitted independently. The X-ray colors show there are at least two thermal emission components, and that they have different mean absorbing columns. The strong spectral lines require an α-element abundance of at least 0.25 $Z_\odot$ to fit the spectrum. However, the wind metallicity can be increased arbitrarily above this level provided the emission measure, which is proportional to the continuum normalization, is decreased proportionately. Using population synthesis models, normalized to the starburst in NGC 1569, to estimate the mass of supernova ejecta, which is a lower limit on the wind mass (i.e. no mass loading), the upper limit on the wind metallicity is about twice solar. The best fit implies α-element abundances in the wind of roughly solar, a mass loading factor ~ 9, and $Z_\alpha/Z_{Fe} \approx 3$. In NGC 1569, most of the metals in the outflow come from the supernova ejecta, not the entrained ISM, even though the ISM supplies the majority of the mass in the outflow. It appears likely that the wind carries most of the metals synthesized in the recent burst. The fraction of the total stellar mass formed in such bursts is not yet known.

24.3.3 *Resonance Absorption Lines*

Starburst outflows transport cold gas and dust in addition to hot gas and warm gas. When a suitably bright background light source is observed, the kinematics of the cold gas can be probed in absorption. This method probes lower column densities than the best 21-cm emission observations. The ultraviolet bandpass is preferred for absorption-line studies owing to the wide range of ionization states and elements with ultraviolet transitions. The much higher sensitivity of ground-based telescopes means in practice that large samples of galaxies have only been observed in optical lines, particularly the Na I $\lambda\lambda5896$, 5890 doublet. Hydrogen is neutral in regions where Na I is the dominant ionization state, so

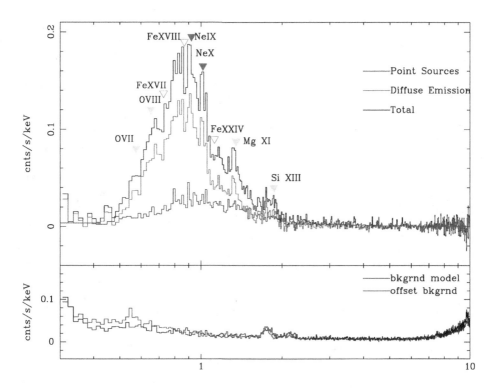

Fig. 24.4. Integrated spectrum of NGC 1569 (*top curve*). The contribution from point sources (*lower curve*) is harder than that from diffuse gas (*middle curve*). Bottom panel shows the folded X-ray background. See discussion in Martin et al. (2002), from which this figure was taken.

the Na absorption probes the cold component of the outflow, complementing the optical emission-line and X-ray emission studies.

Most starburst galaxies show Na I absorption that is blueshifted relative to the systemic velocity (Heckman et al. 2000; Rupke, Veilleux, & Sanders 2002; Martin & Armus 2004; Schwartz & Martin 2004). The absorption is measured against star clusters in the nucleus of the host galaxy, so there is no ambiguity about the direction of the flow. Figure 24.5 summarizes the expansion velocities measured at maximum optical depth for these data sets. Taken together, there is a clear trend for the velocities to increase with increasing burst strength. The envelope is steeper than the $V \propto L_{\rm Bol}^{1/4}$ relation expected for a flow driven by a starburst of maximum surface brightness (Meurer et al. 1997; Heckman et al. 2000).

It is interesting to compare the mass flux of the cold flow to that in the warm and hot gas phases. An equivalent hydrogen column is obtained from assumptions about the Na abundance and ionization state. The inferred mass flux is particulary sensitive, however, to the size of the region the sightline represents. Heckman et al. (2000) measured an extent of 1 to 10 kpc along their slits for the absorbing region, and they argued the cold clouds carried a mass flux similar to the SFR. Rupke et al. (2002) found the mass flux in the ultraluminous starbursts appears to be about an order of magnitude less than the SFR. Of those dwarf

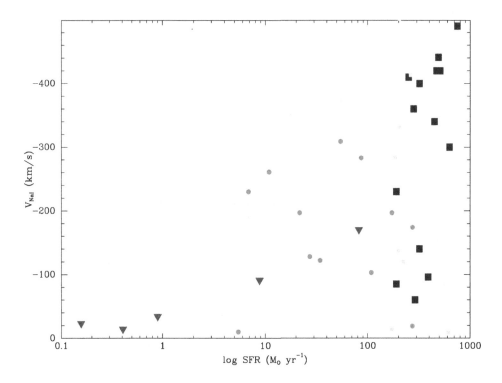

Fig. 24.5. Expansion velocity of the minimum in the Na I absorption-line profile. Filled circles represent luminous infrared galaxies from Heckman et al. (2000). Ultraluminous infrared galaxies are shown as large squares (Martin & Armus 2004) and small squares (Rupke et al. 2002). Dwarf starbursts from Schwartz & Martin (2004) are denoted by triangles. The maximum outflow velocities are larger in the more powerful starbursts.

galaxies that show interstellar Na I absorption, the area of the outflow would need to be at least 1 kpc to bring the flux of cold material up to that of the SFR (Schwartz & Martin 2004). The neutral clouds may carry a significant fraction of the mass in the flow.

24.4   Escape to the IGM?

While hot, warm, and cold components of the outflow clearly leave the starburst region, reaching heights several kpc above the disk, the amount of mass escaping to the IGM is not yet firmly established. Here I summarize the measurements by phase. For comparison, recall that the escape velocities are at least $1.4 V_c$, where V_c is the rotation speed of the disk. Simple models for an isothermal halo raise the escape velocity to 3.4 times V_c (for a halo extending 100 times further than the starburst region).

The projected expansion velocities measured in the Hα filaments reach 120 km s^{-1} in NGC 1569, whose $V_c \approx 30$ km s^{-1}, and hundreds of km s^{-1} in larger galaxies. The fraction of the warm gas mass that has been accelerated past the escape velocity cannot be measured directly due to the uncertain details of the geometry. However, most studies have concluded that the fraction of warm gas escaping is less than the total warm gas mass—often much less.

The Na I absorption studies sample the mass distribution in the wind as a function of velocity more fairly than the emission-line studies, since they measure the gas column directly and are not biased toward the densest regions. The terminal velocities in the neutral flow are much larger than the velocities of maximum optical depth shown in Figure 24.5. They reach 700 km s^{-1} in the ultraluminous starbursts (Martin & Armus 2004). These systems are mergers of L^* galaxies, however, which have rotation speeds $\sim$ 200 to 300 km s^{-1} . These terminal velocities, although large, are actually similar to the escape velocity. It is intriguing that the measured terminal velocities in the dwarf galaxies are not only lower, but once again approximately equal to the escape speed. Efforts should be made to understand why this conspiracy between the terminal velocity of the cold clouds and the escape velocity arises.

The soft X-ray emission from the wind traces a mixture of free wind and entrained material and provides the best direct detection to date of material that appears likely to escape from the galaxy and enrich the IGM. Figure 24.6 summarizes the X-ray temperatures of the diffuse emission from starburst galaxies. The temperatures exhibit no dependence on the depth of the gravitational potential. The simple interpretation is that the hot component escapes from the dwarf galaxies but is marginally bound to large galaxies (e.g., Martin 1999).

Two factors make the interpretation of Figure 24.6 more complicated. First, the emission from boundary/mixing layers is the easiest to detect; and observations are likely failing to detect the hottest component of the wind in massive starbursts (Strickland & Stevens 2000). However, upper limits on the mass carried by this hotter component indicate it does carry the bulk of the mass; it may carry significant energy (Strickland et al. 2002). Second, large gaseous halos can significantly increase the energy requirements for expelling newly processed matter (Silich & Tenorio-Tagle 2001). The dwarf irregular galaxy NGC 4449 is situated in such a halo, and the powerful outflow will probably generate little mass loss (Summers et al. 2003). The tidal interactions in the M81 group apparently affect the fate of the outflow from the compact dwarf, NGC 3077 (Ott, Martin, & Walter 2003). Yet in some dwarf galaxies like NGC 1569 (Martin et al. 2002) or Mrk 33 (Summers, Stevens, & Strickland 2001), the outflows appear destined to transport metals into the IGM.

24.5   Impact on Galactic Chemical Properties

It has been known since the early work on the stellar metallicity distribution in the solar neighborhood that the chemical evolution of the Galaxy did not proceed as a closed system. The simplicity of a closed-box model sometimes lends useful insight, however. In particular, the addition of outflows to the closed-box model, i.e. the *leaky-box* model, provides some insight into the impact of galactic winds on chemical evolution.

24.5.1   Properties of a Leaky Box

The properties of the leaky-box model were worked out by Larson (1974) and further generalized by Edmunds (1990). I will attempt to constrain the parameters of this model using the observations described in this review. Following the notation introduced by Binney & Merrifield (1998), the key parameter is the escape efficiency—call it C—that ties the rate of mass outflow to the SFR such that

$$\frac{dM_T}{dt} = -C\frac{dM_*}{dt}. \tag{24.5}$$

It is straightforward to show, using mass conservation, that the gas-phase metallicity rises as

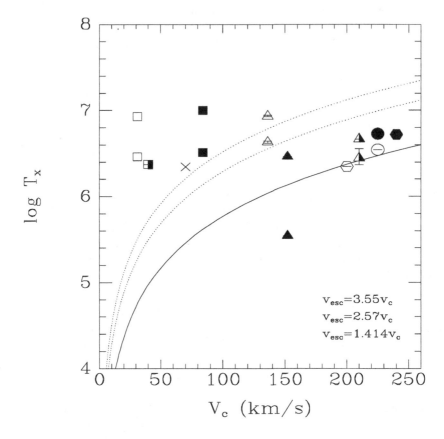

Fig. 24.6. Temperature of the diffuse X-ray emission vs. galactic rotation speed. From left to right, the galaxies are NGC 1569 (Martin et al. 2002), NGC 3077 (Ott et al. 2003), He 2-10 (Kobulnicky & Martin 2004), NGC 4449 (Summers et al. 2003), M82 (Stevens et al. 2003), NGC 4631 (Dahlem et al. 1998), NGC 3628 (Dahlem et al. 1998), NGC 253 (Strickland et al. 2002), NGC 3079 (Dahlem et al. 1998), NGC 891 (Bregman & Houck 1997), and NGC 2146 (Dahlem et al. 1998).

$$Z = \frac{-p}{1+C} \ln\left(1 - (1+C)\frac{M_*}{M_T(0)}\right), \tag{24.6}$$

which is equivalent to reducing the stellar yield, p, by $(1+C)^{-1}$. Measurement of the effective yields in dwarfs are below the solar yield, consistent with the outflow scenario, but have large uncertainties due to the difficulty of measuring gas mass fraction (e.g., Garnett 2000).

The stellar metallicity distribution for various values of the escape efficiency is plotted in Figure 24.7. With outflow, only a fraction $(1+C)^{-1}$ of the initial mass, $M_t(0)$, is turned into stars. A larger fraction of the stars have low metallicity when strong outflows are included. In constrast, accretion of gas and the subsequent formation of additional stars from enriched gas produce a paucity of low-metallicity stars. Measurement of the stellar metallicity distribution may ultimately prove or refute the importance of outflows in dwarf galaxies.

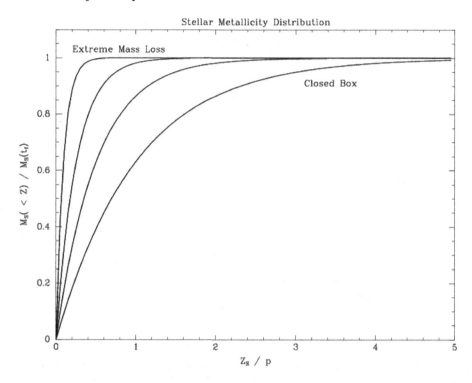

Fig. 24.7. Stellar mass distribution normalized by the total number of stars formed. Curves show values of $C = 0, 1, 3, 10$, where C is the ratio of the mass loss rate to the SFR. Note that the total mass of stars is lower in the models with mass loss.

To date, only a few stars in nearby dwarf galaxies have measured metallicities, so we can only make comparisons to the average stellar metallicity. In the leaky-box model, the average stellar metallicity reaches the effective yield, $\langle Z_* \rangle = p/(1+C)$, once gas consumption is complete. Dwarf irregular galaxies have large gas mass fractions, so the final stellar metallicity distribution is not established. The largely gas-free dwarf spheroidals, on the other hand, have reached the terminus of their chemical evolution. The measured [Fe/H] in Local Group dSphs are typically ~ -1.5, with some estimates reaching -2.0 (Mateo 1998). Outflow models would require an escape efficiency $C = 10$ to 100 to produce such low mean metallicities. Type II supernova ejecta that escapes from the galaxy cannot be incorporated into the next stellar generation of stars, so the stellar [α/Fe] ratio will be low in subsequent stellar generations. Recent measurements of the stellar [α/Fe] in dwarf spheroidals come in almost as low as the Galactic disk (Tolstoy et al. 2003), although those authors interpret the result as evidence for a truncated stellar mass function (no stars more massive than $12\ M_\odot$). In nearby starburst galaxies, the estimated outflow rates are 1 to 10 times the SFR, which is not much less than the escape efficiencies required to explain the low metallicities in the dwarf spheroidal galaxies. The higher escape efficiencies in low-mass galaxies will produce a strong correlation between mass and metallicity.

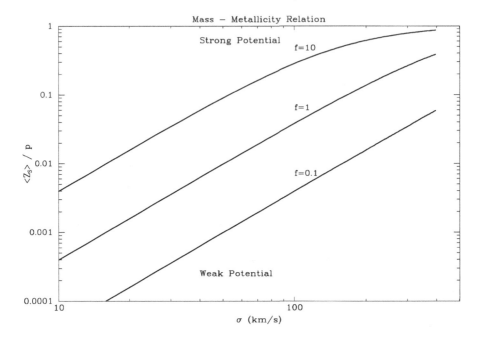

Fig. 24.8. Mass-metallicity relation for different dependences of the outflow efficiency on the potential depth. The parameter f describes the shape of the potential, such that $\phi \approx -f\sigma_*^2$.

24.5.2   *The Mass-metallicity Relation among Galaxies*

Gas removal requires several criteria to be met. In addition to acquiring enough energy to escape from the gravitational potential, the hot gas must not radiate all of its energy or expend its energy pushing aside a massive gaseous halo.

It is easy to illustrate how the potential depth affects the escape efficiency. Let $\epsilon_{SN} = 3.75 \times 10^{15}$ erg g^{-1} be the supernova energy per unit mass of 0.1–$100\ M_\odot$ stars and assume a fraction, f_{rad}, of the supernova energy is radiated away. The mass of gas that can be lifted out of the galaxy, dM_g/dt, is

$$(1 - f_{rad})\epsilon_{SN}dM_*/dt = |\phi dM_g/dt|, \qquad (24.7)$$

where the gravitational potential ϕ is proportional to the square of the stellar velocity dispersion— i.e. $\phi = -f\sigma_*^2$. It follows that the escape efficiency will not exceed

$$C = \left\{ \frac{dM_g/dt}{dM_*/dt} \right\} \approx \epsilon_{SN}(1 - f_{rad})/(f\sigma_*^2). \qquad (24.8)$$

The leaky-box model therefore predicts that the mean stellar metallicity grows with the potential depth as

$$\langle Z_* \rangle = \frac{p}{1+C} = \frac{p}{1+x/\sigma_*^2}, \qquad (24.9)$$

where $x \equiv \epsilon_{SN}(1 - f_{rad})/f$.

Figure 24.8 illustrates the mass-metallicity relation for various feedback parameters in the leaky-box model with $f_{rad} = 20\%$. Clearly, when σ_*^2 is small compared to ϵ_{SN}, the metallicity grows linearly with the square of the stellar velocity dispersion. Ignoring variations in mass-to-light ratio with mass, an equivalent expression is $\log \langle Z_* \rangle / p \approx 0.26 M_V + \text{constant}$, where M_V is the absolute V-band magnitude. The slope agrees very well with the relation fitted to stellar abundance measurements in dwarf spheroidals and transition objects (e.g., Mateo 1998; Fig. 7).

The agreement in slope between theory and observation has been noted previously (Larson 1974; Dekel & Silk 1986). The new aspect here is the connection to the empirical value of the feedback parameter. For example, taking the gravitational potential to be proportional to the stellar velocity dispersion and a proportionality constant $f \approx 3$, which is appropriate for an isothermal sphere, the feedback parameter C decreases from values of roughly 1000 in the smallest dwarf galaxies with $\sigma = 10$ km s^{-1} to unity in a giant galaxy with $\sigma_* = 300$ km s^{-1}. These values of C are higher than the measured range. Bringing the escape efficiency down to more reasonable values, like 0.3 and 300 in the high- and low-mass halos, respectively, requires a stronger potential with $f \approx 10$. The model with $f = 10$ in Figure 24.8 turns over very near the break at $3 \times 10^{10} \, M_\odot$ seen by Tremonti et al. (2004).

Outflows may not be a unique explanation for the strong correlation between metallicity and luminosity among dwarf galaxies. If winds are the solution, then dwarf galaxies form a significant fraction of their stars in a starburst mode. The average SFR in a dSph, $\sim 10^{-4} \, M_\odot$ yr^{-1}, is not sufficient to sustain a wind from even a dwarf galaxy. However, if the duty cycle of the star formation episodes is 1% or less, then outflows become plausible. Although the dSphs clearly contain stars with a wide range of ages, it is not yet clear that the star formation histories have been resolved on a fine enough time scale to make this distinction.

24.6   Summary: Global Impact of Metal Loss via Galactic Winds

Metal ejection is now directly observed from some nearby galaxies. The importance of outflows for galaxy scaling relations and the enrichment of the IGM is still debated, however. Nearly all starburst galaxies present winds, but the fraction of stellar mass formed in starbursts is not yet known. Several factors suggest winds played a prominent role in the past.

Winds are ubiquitous in Lyman-break-selected galaxies at redshift 3–4, and these systems appear to account for at least half of the star formation at that epoch. Yet a careful census of the metals in stars and gas at redshift 2.5 recovers only 10%–15% of the metals produced by galaxies up until that time (Pettini 2003). These observations are easily reconciled if a large fraction of the metals were ejected by winds and remain in a hot, and therefore hard to detect, phase. The relatively high metallicity of the intracluster medium also suggests galaxies lose more metals than they retain (Renzini 2004).

Relic winds would be detectable in absorption long after they have faded from prominence in emission. Establishing an association with some class(es) of quasar absorption-line systems would greatly improve our understanding of the spatial extent and ubiquity of winds at high redshift. One interesting, though controversial, hypothesis is that strong Mg II systems are associated with superwinds (Bond et al. 2001). This class of absorption-line system evolves away from redshift 2 to the present, in parallel to the starburst population, and has about the right number of systems per unit redshift to be associated with starburst galaxies.

The irony of this situation is that, despite the larger escape efficiency in dwarf galaxies, the stars formed in these starburst events reside in massive galaxies today rather than in dwarf galaxies. It remains possible that winds are relatively rare in dwarf galaxies today. The star formation properties of complete samples of dwarfs selected on the basis of H I mass should resolve the role of dwarf starburst winds at the present epoch.

Acknowledgements. The author thanks Henry "Chip" Kobulnicky for discussions about the chemical evolution in galaxies. This work was supported by grants from the David and Lucile Packard Foundation and the Alfred P. Sloan Foundation.

References

Adelberger, K. L., Steidel, C. C., Shapley, A. E., & Pettini, M. 2003, ApJ, 584, 45
Binney, J., & Merrifield, M. 1998, Galactic Astronomy (Princeton: Princeton Unvi. Press)
Bond, N. A., Churchill, C. W., Charlton, J. C., & Vogt, S. S. 2001, ApJ, 562, 641
Bregman, J. N., & Houck, J. C. 1997, ApJ, 485, 159
Buote, D. A., & Canizares, C. R. 1994, ApJ, 427, 86
Cecil, G., Bland-Hawthorn, J., & Veilleux, S. 2002, ApJ, 576, 745
Ciotti, L., D'Ercole, A., Pellegrini, S., & Renzini, A. 1991, ApJ, 376, 380
Dahlem, M., Weaver, K. A., & Heckman, T. M. 1998, ApJS, 118, 401
Edmunds, M. G. 1990, MNRAS, 246, 678
Garnett, D. R. 2002, ApJ, 581, 1019
Heckman, T. M., Armus, L., & Miley, G. K. 1990, ApJS, 74, 833
Heckman, T. M., Lehnert, M. D., Strickland, D. K., & Armus, L. 2000, ApJS, 129, 493
Kobulnicky, H. A., & Martin, C. L. 2004, in preparation
Kobulnicky, H. A., & Skillman, E. D. 1996, ApJ, 471, 211
——. 1997, ApJ, 489, 636
Larson, R. B. 1974, MNRAS, 166, 585
Lehnert, M. D., & Heckman, T. M. 1996, ApJ, 462, 651
Leitherer, C., et al. 1999, ApJS, 123, 3
Leitherer, C., Robert, C., & Drissen, L. 1992, ApJ, 401, 596
MacLow, M.-M., & Ferrara, A. 1999, ApJ, 513, 142
Marlowe, A. T., Heckman, T. M., Wyse, R. F. G., & Schommer, R. 1995, ApJ, 438, 563
Martin, C. L. 1996, ApJ, 465, 680
——. 1997, ApJ, 491, 561
——. 1998, ApJ, 506, 222
——. 1999, ApJ, 513, 156
Martin, C. L., & Armus, L. 2004, in preparation
Martin, C. L., Kobulnicky, H. A., & Heckman, T. M. 2002, ApJ, 574, 663
Mateo, M. 1998, ARA&A, 36, 435
Mathews, W. G., & Baker, J. C. 1971, ApJ, 170, 241
Meurer, G., et al. 1997, AJ, 114, 54
Miller, S. T., & Veilleux, S. 2003, ApJ, 592, 79
Nomoto, K., Uenishi, T., Kobayashi, C., Umeda, H., Ohkubo, T., Hachisu, I., & Kato, M. 2003, in From Twilight to Highlight: The Physics of Supernovae, ed. W. Hillebrandt & B. Leibundgut (Berlin: Springer), 115
Ott, J., Martin, C. L., & Walter, F. 2003, ApJ, 594, 776
Pettini, M. 2003, in Cosmochemistry: The Melting Pot of Elements (Cambridge: Cambridge Univ. Press), in press (astro-ph/0303272)
Pettini, M., Rix, S. A., Steidel, C. C., Adelberger, K. L., Hunt, M. P., & Shapley, A. E. 2002, ApJ, 569, 742
Renzini, A. 2004, in Carnegie Observatories Astrophysics Series, Vol. 4: Origin and Evolution of the Elements, ed. A. McWilliam & M. Rauch (Cambridge: Cambridge Univ. Press), in press
Rupke, D. S., Veilleux, S., & Sanders, D. B. 2002, ApJ, 570, 588
Schaerer, D., Charbonnel, C., Meynet, G., Maeder, A., & Schaller, G. 1993, A&AS, 102, 339
Schwartz, C., & Martin, C. L. 2004, ApJ, submitted
Silich, S., & Tenorio-Tagle, G. 2001, ApJ, 552, 91
Stevens, I. R., Read, A. M., & Bravo-Guerrero, J. 2003, MNRAS, 343, L47
Strickland, D. K., Heckman, T. M., Weaver, K. A., & Dahlem, M. 2000, AJ, 120, 2965
Strickland, D. K., Heckman, T. M., Weaver, K. A., Hoopes, C. G., & Dahlem, M. 2002, ApJ, 568, 689

Strickland, D. K., & Stevens, I. R. 2000, MNRAS, 314, 511

Suchkov, A. A., Balsara, D. S., Heckman, T. M., & Leitherer, C. 1994, ApJ, 463, 528

Suchkov, A. A., Berman, V. G., Heckman, T. M., & Balsara, D. S. 1996, ApJ, 463, 528

Summers, L. K., Stevens, I. R., & Strickland, D. K. 2001, MNRAS, 327, 385

Summers, L. K., Stevens, I. R., Strickland, D. K., & Heckman, T. M. 2003, MNRAS, 342, 690

Tolstoy, E., Venn, K. A., Shetrone, M., Primas, F., Hill, V., Kaufer, A., & Szeifert, T. 2003, AJ, 125, 707

Tremonti, C., et al. 2004, in preparation

Veilleux, S., Shopbell, P. L., Rupke, D. S., Bland-Hawthorn, J., & Cecil, G. 2003, AJ, 126, 2185

Weaver, K. A., Heckman, T. M., & Dahlem, M. 2000, ApJ, 534, 684

Weaver, R., McCray, R., Castor, J., Shapiro, P., & Moore, R. 1977, ApJ, 218, 377

Woosley, S. E., & Weaver, T. A. 1995, ApJS, 101, 181

25

Abundances from the integrated
light of globular clusters and galaxies

SCOTT C. TRAGER
Kapteyn Astronomical Institute, Groningen, The Netherlands

Abstract

It is currently impossible to determine the abundances and ages of the stellar populations of distant, dense stellar systems star by star. Therefore, methods to analyze the composite light of stellar systems are required. I review the modeling and analysis of integrated spectra of the stellar populations of individual globular clusters, globular cluster systems, early-type galaxies, and the bulges of spiral galaxies, with a focus on their abundances and abundance ratios. I conclude with a list of continuing difficulties in the modeling that complicate the interpretation of integrated spectra, as well as a look ahead to new methods and new observations.

25.1 Introduction

The reviews presented in this volume demonstrate that the nucleosynthetic history of the Milky Way and its satellites can now be probed in exquisite detail, revealing a wealth of information about the processes by which our own Galaxy and its neighbors have formed. Such star-by-star analysis of the abundances of distant globular clusters and distant and/or dense galaxies is, however, currently beyond the reach of current telescopes. Although this situation might change with the advent of overwhelmingly large telescopes of the 30–100 m class with high-precision adaptive optics systems, during the 20th century and at the beginning of the 21st, we have been limited to studying the *integrated* stellar populations of distant globular clusters and galaxies.

In this review, I will discuss spectroscopic techniques for determining abundances from the integrated light of galaxies and globular clusters. I will begin with a discussion of the ingredients of the models required to interpret the integrated-light spectra, the calibration of those models, and a few of their pitfalls. I will then discuss the abundances determined from integrated-light spectroscopy of globular clusters in our Galaxy, the Local Group, and the globular cluster systems of early-type galaxies. I will then consider the abundances of nearby early-type galaxies and the bulges of early-type spirals determined from integrated-light spectroscopy, including recent results on abundance anomalies in these systems. Although I will focus on deriving abundances from optical spectra, these results necessarily depend on the *ages* of the systems (through the age-metallicity degeneracy; §25.2.1), and so I must briefly discuss the ages of these systems. Finally, I will summarize the current state of the observations, the current problems in understanding stellar populations from integrated-light spectroscopy, and current and future directions in this subject.

Perusal of this outline reveals that there are a number of topics I will not discuss, due either to space constraints or to difficulties in interpretation. I will not be able to discuss the fascinating history of this subject in detail. I will not discuss purely photometric methods of determining abundances, such as attempts to derive abundances of globular clusters and early-type galaxies from broad-band colors, as the interpretation of broad-band optical colors is seriously compromised by the age-metallicity degeneracy. I will not have space to discuss abundances derived from wavebands other than the optical; there have been excellent recent reviews and studies of the stellar populations of globular clusters and early-type galaxies determined from, for example, UV spectroscopy (O'Connell 1999; Dorman, O'Connell, & Rood 2003; Peterson et al. 2003, 2004). Last, but not least, I will not discuss stellar abundances of irregular or dwarf galaxies themselves (apart from M32), as this subject is ably covered elsewhere in this volume, and I will touch only lightly on abundances derived from the integrated *stellar* light of bulges of early-type spiral galaxies. Garnett (2004) thoroughly and cogently discusses the *gas-phase* abundances of such galaxies.

25.2 Stellar Population Modeling of Integrated-light Spectra

The interpretation of the integrated light of globular clusters and galaxies requires astrophysically constrained models. In this section, I outline the basic purpose and ingredients of such models and the steps required to calibrate them.

25.2.1 The Basic Problem

O'Connell (1986) gives a clear physical explanation of the *age-metallicity degeneracy* that plagues the study of the integrated light of stellar populations. Because the emergent flux from a stellar atmosphere is a superposition of spectra that are characterized by different temperatures weighted by a function of opacity, increases in opacity in a stellar atmosphere due to increased metallicity are equivalent to a decreased mean temperature for the emergent flux. Therefore, a metal-rich population can be simulated by decreasing the mean temperature—that is, decreasing the age—of a more metal-poor population (Fig. 25.1).

The key breakthrough in the analysis of integrated stellar populations was made by Rabin (1980, 1982) and Gunn, Stryker, & Tinsley (1981), who noticed independently that the strengths of the Balmer lines of hydrogens derived from evolutionary population synthesis models allow an accurate age of a single-burst stellar population to be measured. Rabin expressed this idea in a powerful graphical form: the "hydrogen-metals diagnostic diagram," in which the strength of Balmer lines are plotted as a function of a metal line (in this case, Ca II K) for the integrated spectra and stellar population models. Using this diagram, he could break the age-metallicity degeneracy. Applying this method to clusters in the Magellanic Clouds, Rabin confirmed that the Searle, Wilkinson & Bagnuolo (1980; hereafter SWB) ranking of those clusters was indeed an age ranking.

Worthey (1994) and Worthey & Ottaviani (1997) quantified the age-metallicity degeneracy for a large number of absorption-line strengths and broad-band colors. Metal lines (like Mgb and $\langle$Fe$\rangle$) and broad-band colors are slightly more sensitive to metallicity than age, in the ratio $\partial \log t / \partial \log Z = 3/2$ (Worthey 1994; Fig. 25.1). This means that a factor of 3 change in age looks like a factor of 2 change in metallicity for a simple stellar population. Clearly, age-sensitive indices are desired—lines that preferentially measure the (luminosity-weighted) temperature of the MSTO. Following Rabin (1980, 1982), Worthey (1994) showed that $\partial \log t / \partial \log Z \approx 2/3$ for Hβ (and ~ 1 for Hδ and Hγ; Worthey & Ot-

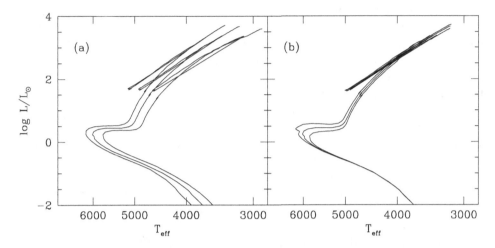

Fig. 25.1. The age-metallicity degeneracy. (*a*) Varying metallicity at fixed age: 12 Gyr old Padova (Girardi et al. 2000) isochrones with metallicities of [Fe/H] = −0.68, −0.38, 0 (left to right). (*b*) Varying age at fixed metallicity: [Fe/H] = −0.38 Padova isochrones with ages of *t* = 7, 10, 14 Gyr (left to right). Note that in this view of the age-metallicity degeneracy, the main sequence turnoff (MSTO) varies in temperature roughly equally in each case, but the red giant branch (RGB) varies little with age at fixed metallicity. To break the degeneracy, one needs to measure the temperature of both the MSTO and the RGB to decouple the age and metallicity.

taviani 1997). Thus, a Balmer-line index *combined* with a metal-line index *breaks* the age-metallicity degeneracy.

25.2.2   *Ingredients*

The first attempt to synthesize the spectral lines and colors of a galaxy appears to have been made by Whipple (1935). Whipple developed population synthesis, in which arbitrary permutations of stellar types are combined to make a synthetic galaxy spectrum and colors; this method was highly influential on many later workers (e.g., de Vaucouleurs & de Vaucouleurs 1959; Spinrad 1962a,b; Spinrad & Taylor 1969, 1971; Lasker 1970; Faber 1972; O'Connell 1976; Williams 1976). Tinsley (1968, 1972; Tinsley & Gunn 1976) invented the method that has generally superseded population synthesis: *evolutionary* population synthesis. In this method, populations are modeled from the starting point of an isochrone and a luminosity function and then matched to the observations, instead of attempting to extract a color-magnitude diagram (CMD) and luminosity function from the observations. Tinsley's method has become the basis for modern stellar population models (e.g., Bruzual 1983; Buzzoni 1989; Charlot & Bruzual 1991; Bruzual & Charlot 1993; Worthey 1994; Bressan, Chiosi, & Fagotto 1994; Maraston 1998).

To build a model of a *simple stellar population* (SSP)—that is, a single-age, single-metallicity stellar population—that can be used to analyze the integrated spectrum of a given object, four major ingredients are required.

Isochrones An isochrone set that is best calibrated against the populations of interest (young, intermediate-aged, or old; metal poor or metal rich) is needed as the basic astrophysical constraint. Popular choices for the intermediate-aged and old populations considered here are currently the Padova set (Girardi et al. 2000; Salasnich et al. 2000), the set from Cassisi, Castellani, & Castellani (1997) and Bono et al. (1997), and the set of Salaris & Weiss (1998) and more recent extensions (see, e.g., Schiavon et al. 2002b).

Initial mass function An IMF populates the isochrones as needed. Typically, a Salpeter (1955) IMF is chosen, although other choices are possible (Vazdekis et al. 1996).

Stellar fluxes While it may be preferable to have a set of observed stellar fluxes covering a comprehensive range in temperature, gravity, and metallicity (and even $[\alpha/\text{Fe}]$), such a set is not currently available (although see Vazdekis 1999 for another approach and Le Borgne et al. 2003 for a hint of what is to come). We therefore are reliant on the theoretical flux library of Kurucz (1993), which covers a broad range of parameter space, or modifications of that library to attempt to bring its colors into better agreement with real stars (Lejeune, Cuisinier, & Buser 1997, 1998; Westera et al. 2002).

Stellar spectra or absorption-line strengths Because we require the strengths of Balmer-line and metal-line indices to break the age-metallicity degeneracy, a library of either stellar spectra (e.g., Jones 1998) or stellar absorption-line strengths (e.g., Worthey et al. 1994) is required. Given the difficulties in matching the Lick/IDS system precisely (Worthey & Ottaviani 1997; Trager, Faber, & Dressler 2004), we would like to find new libraries for population syntheses. Such libraries are on their way (Peletier, Rose, Schiavon, and Worthey, private communications).

Once these ingredients are available, one can construct a stellar population model, as shown in Figure 25.2. However, such models clearly have a problem: ages and metallicities derived from different pairs of Balmer- and metal-line indices are different for giant elliptical and many S0 galaxies. These discrepancies result from the nonsolar abundance ratios of these objects (Faber & Jackson 1976; O'Connell 1976; Peterson 1976; Peletier 1989; Worthey, Faber, & González 1992; Greggio 1997; Worthey 1998; Trager et al. 2000a). Although we refer to these nonsolar abundance ratios by the shorthand phrase "α-enhancements," we stress that the proper interpretation of these discrepancies is actually as "Fe-deficiencies," because the "α-elements" O and Mg dominate the metallicity determinations (Greggio 1997; Trager et al. 2000a). Correcting for $[\alpha/\text{Fe}] \neq 0$ (or more generally, for $[X_i/\text{Fe}] \neq 0$ for any element X_i; Thomas, Maraston & Bender 2003a, hereafter TMB03) requires either an empirical stellar library with the needed enhancements or stellar atmosphere models in which synthetic spectra with the needed enhancements can be generated, as well as isochrones with $[\alpha/\text{Fe}] \neq 0$, such as those available in the Padova and Salaris & Weiss sets (cf. Trager et al. 2000a; but see Maraston & Thomas 2003). The empirical approach was attempted first by Weiss, Peletier, & Matteucci (1995), who coupled fitting functions for Mg_2 and $\langle\text{Fe}\rangle$ from the Galactic bulge giants of Rich (1988) with α-enhanced isochrones; unfortunately, the Rich library was not complete enough for a reasonable coverage of parameter space and, more importantly, did not include the crucial $\text{H}\beta$ index. Trager et al. (2000a) and TMB03 have chosen to use the stellar atmosphere computations of Tripicco & Bell (1995), who modeled the response of the Lick/IDS index system to changes in individual elements, in a differential sense, and ignore the α-enhanced isochrones (Maraston & Thomas 2003).

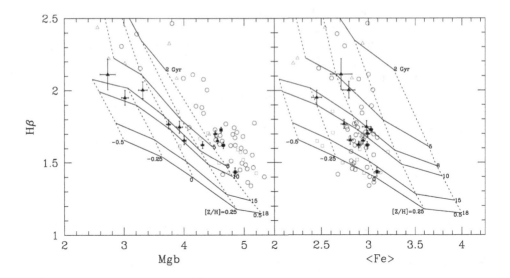

Fig. 25.2. The line strengths of local early-type galaxies compared with stellar population models from Worthey (1994). Solid lines: constant age. Dashed lines: constant metallicity. Solid squares and triangles: elliptical and S0 galaxies in Coma (Trager, Faber, & Dressler 2004). Open squares and triangles: elliptical and S0 galaxies in Fornax (Kuntschner 2000). Open circles: elliptical galaxies in Virgo, groups, and the field (González 1993). *Left*: Hβ as a function of Mgb. *Right*: Hβ as a function of $\langle$Fe$\rangle$. Note the significant difference in derived age and metallicity for most galaxies from each pair of indices; this is indicative of [α/Fe] $\neq$ 0 for these galaxies.

While more flexible for analyzing populations that may not resemble local stars, calibration of these theoretical corrections is required.

25.2.3 *Calibrations*

I show the results of α-enhanced models (TMB03) in Figure 25.3. It is clear that if these models are indeed properly calibrated, then it is possible to determine uniquely age, metallicity, and [α/Fe] for a single-burst population. Galaxies are unlikely to be SSPs, but globular clusters are (except maybe ω Cen: Smith 2004). I note here that the "metallicities" measured by the Trager et al. (2000a) technique are total metallicities [Z/H], not [Fe/H]. However, [Z/H] can be converted to [Fe/H] on the Zinn & West (1984) scale with the following (model-dependent) scaling: [Fe/H] = [Z/H] -0.94[α/Fe] (Tantalo, Chiosi, & Bressan 1998; Trager et al. 2000a; TMB03).

I will now exploit this fact and turn to a model of an individual cluster: 47 Tuc. Gibson et al. (1999) pointed out that the stellar population models of Jones & Worthey (1995) predicted an age in excess of 20 Gyr from Hγ and the Ca4227 or C$_2$4668 indices for 47 Tuc, a cluster with a CMD-based age of 11–14 Gyr (depending on the isochrone set used; Schiavon et al. 2002b). Vazdekis et al. (2001) brought the spectroscopic age down by using α-enhanced isochrones with He diffusion (from the Salaris & Weiss 1998 set). Schiavon et al. (2002a,b) used a two-pronged approach to solve this problem. First, they synthesized the spectrum directly from the CMD (Schiavon et al. 2002a), which was meant to reveal any

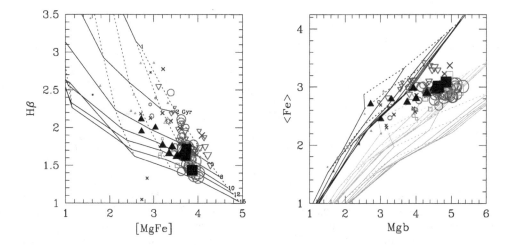

Fig. 25.3. The line strengths of the nuclei of local early-type galaxies compared with stellar population models from TMB03. Points and line types as in Fig. 25.2, except inverted triangles (S0s in groups and Virgo; Fisher, Franx, & Illingworth 1996) and crosses (bulges of S0/a–Sbc spirals in various environments; Proctor & Sansom 2002). Point sizes for all samples are proportional to velocity dispersion. *Left*: Ages and metallicities. Note that the metallicity (left to right: [Fe/H] = −1.35, −0.33, 0, 0.35, 0.67 dex) and age scales have changed. Note that field ellipticals (circles), nearly all S0s, and bulges span a large range in $H\beta$ strength and thus in SSP age, while cluster ellipticals (squares) are quite old and vary more in metallicity. *Right*: [α/Fe] (from left to right: [α/Fe] = 0, +0.3, +0.5 dex). In this plot, note that ellipticals tend to have [α/Fe] ≈ +0.3—in fact, this varies with velocity dispersion, with small ellipticals being more "solar" and large ones being more enhanced (Trager et al. 2000b)—while S0s and bulges tend to be more "solar."

problems with the stellar libraries. Schiavon et al. uncovered several: (a) it is necessary to have a strictly *homogeneous* set of atmospheric parameters for the stars in the library; (b) the metallicities of the library stars used in the synthesis must be that of the cluster to less than 0.05 dex, and must be on the same metallicity scale; and (c) the line strengths need to account for the CN bimodality of the cluster (which extends to at least the MSTO: Hesser 1978), in which roughly half the stars are CN-strong. Now that the necessary corrections to the line strengths are understood, the second step is to synthesize the line strengths from the best-fitting isochrone (Schiavon et al. 2002b). A significant problem arises in this step: the observed luminosity function does not match that of the model, due to a lack of AGB stars (in the Salaris isochrones) and a strong deficit of RGB stars (in both the Salaris and Padova isochrones). Once all of these problems are corrected for, nearly all of the synthesized spectral indices match the observations at the correct age (11–13 or 12–14 Gyr using the Salaris and Padova isochrones, respectively) and metallicity ([Fe/H] = −0.75 dex).

Now that we have confidence in our stellar population models to predict the age and metallicity of a single globular cluster (finally; Searle 1986), we turn to the calibration of the corrections for [α/Fe] ≠ 0. Maraston et al. (2003) have done this by comparing the Lick/IDS globular cluster data and the Galactic bulge cluster data of Puzia et al. (2002) with the TMB03 stellar population models, index by index. I show Puzia et al.'s data for a few

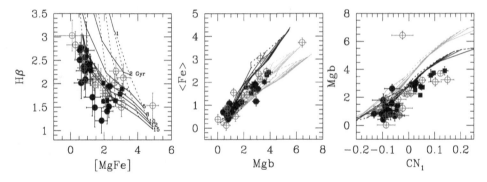

Fig. 25.4. The stellar populations of Milky Way (solid circles: halo clusters from Trager et al. 1998; solid squares: bulge clusters from Puzia et al. 2002) and M31 (Trager et al. 1998; halo clusters only) globular clusters. Left: ages and metallicities. Grids as in Fig. 25.3, although here extended down to [Fe/H] = −2.25 dex. The two Hβ-strong M31 clusters (V87, V116) are likely to be intermediate-aged, but the slightly "younger" Milky Way clusters are the anomalous blue horizontal branch clusters NGC 6388 and NGC 6441 (Rose & Tripicco 1986; Rich et al. 1997), which are as old as the other Galactic clusters. Center: [α/Fe] ratios; grids as in Fig. 25.3. Right: The "CN anomaly" (Burstein et al. 1984) in the Galactic and M31 clusters; grids are [α/Fe] = 0 (lower) and [α/Fe] = +0.5 (upper) from TMB03. Clearly, the Galactic and M31 clusters do not differ significantly in Mgb/CN$_1$, but as a whole are too strong in CN$_1$ at fixed Mgb given their typical [α/Fe] (middle panel; Maraston et al. 2003; TMB03). The highly aberrant M31 "cluster" V204 is likely to be one of the brightest stars in M31 (Berkhuijsen et al. 1988) rather than a globular cluster.

indices below (Fig. 25.4) but comment here that several indices do not match the models: CN$_1$ and CN$_2$ (Burstein et al. 1984); Ca4455 (this difficult index probably should be dropped from the Lick/IDS system); strikingly, C$_2$4668, an index with a large dynamic range at fixed temperature but varying metallicity (Worthey et al 1994; Worthey 1994); Na D; and the TiO indices (although this is not surprising, as Tripicco & Bell 1995 did not model TiO properly). TMB03 have solved the discrepancies in the CN indices and C$_2$4668 by allowing [N/α] = +0.5 (cf. Brodie & Huchra 1991; Smith et al. 1997), but Na D remains a mystery (Spinrad 1962a,b; Cohen, Blakeslee, & Côté 2003).

25.3   Globular Cluster Systems

Now that the required tools are in hand, I turn to what has been learned about the abundances and ages of the stellar populations of globular cluster systems of the Milky Way and M31 (§25.3.1), dwarf galaxies in the Local Group (§25.3.2), and early-type galaxies (§25.3.3).

25.3.1   *The Globular Cluster Systems of the Milky Way and M31*

Van den Bergh's (1969) spectroscopy of globular clusters in M31, the Fornax dwarf spheroidal galaxy, and the Milky Way opened the field of extragalactic globular cluster systems. He found that the metallicities of M31 globular clusters were on average more metal rich than those in the Milky Way with a similar or slightly larger metallicity spread, but the Fornax clusters were more metal poor and had a much smaller metallicity spread. Interestingly, van den Bergh also found that the Balmer line strengths of globular clusters may differ at fixed metallicity, even in the most metal-rich clusters (see the discussion in Rose 1985); this suggests that age may play a role in the stellar populations of some globular cluster systems or that blue horizontal branches may exist at much higher metallicities than seen in the Milky Way globular clusters.

In Figure 25.4 I plot the line strengths of globular clusters in the Milky Way and M31. Three primary results can be read from this figure.

(1) The mean metallicity and spread in metallicity of the globular cluster systems in the two dominant spirals of the Local Group are nearly identical when the bulge clusters of the Milky Way are included.[*]

(2) The M31 clusters are slightly more "solar" in their [α/Fe] ratios at the highest metallicities.

(3) The CN strengths at fixed Mgb are very similar in both systems but are significantly higher than the strength predicted by the models at the [α/Fe] ratios derived from the Mgb–$\langle$Fe$\rangle$ diagram.

Point (1) implies that metallicities of M31 clusters from integrated-light spectra interpreted using the TMB03 models can be assumed to be well calibrated onto the Zinn & West (1984) scale as long as [α/Fe] can be measured. Point (2) supports the identification of the two "intermediate-aged" (Hβ-strong) M31 clusters (V87 and V116) as being somewhat younger than the dominant cluster populations in the two systems (cf. Brodie & Huchra 1990). However, the Hβ-strong, metal-rich Galactic globular clusters are the bulge clusters NGC 6388 and NGC 6441, which have well-populated blue horizontal branches that do not exist in other metal-rich clusters (Rose & Tripicco 1986; Rich et al. 1997). Point (3) disagrees with the conclusions of Burstein et al. (1984), who found that M31 clusters have stronger CN strengths than Galactic globular clusters, solely because the Lick/IDS database does not contain the metal-rich Galactic bulge clusters, which span the high-CN strengths of the metal-rich M31 clusters. As discussed above, "enhanced" CN at fixed Mgb appears to be a result of [N/α] = +0.5 in these clusters.

I note, however, that many of the globular clusters in Figure 25.4 fall below the Hβ strengths of the TMB03 models. As discussed in §25.2.3 above, this is due to a number of factors. TMB03 use the Worthey et al. (1994) fitting functions, taken from the Lick/IDS stellar library of absorption-line strengths, in which the atmospheric parameters are somewhat inhomogeneous, and which are constrained by very few stars at metallicities below [Fe/H] < −1.25; this makes direct comparison with halo globular clusters problematic. Finally, it is possible that the TMB03 models may have problems with the RGB/AGB luminosity functions, although the Maraston (1998) synthesis method on which TMB03 is based uses the fuel-consumption theorem (Renzini & Buzzoni 1986), which should properly populate the luminosity functions.

25.3.2 The Globular Cluster Systems of Dwarf Galaxies in the Local Group

In Figure 25.5 I show the stellar populations for the globular cluster systems of three dwarf galaxies in the Local Group: the Large Magellanic Cloud (Beasley, Hoyle, & Sharples 2002), the Fornax dSph (Strader, Brodie, & Huchra 2003b), and the dIrr NGC 6822 (Cohen & Blakeslee 1998; see also Chandar, Bianchi, & Ford 2000; Strader et al. 2003b).

---

[*] Figure 25.4 ignores the inner bulge clusters of M31, for which spectral indices are unavailable. It may be that those clusters have high metallicity (as the Milky Way bulge globulars have higher metallicity than the halo clusters), which would weaken this conclusion.

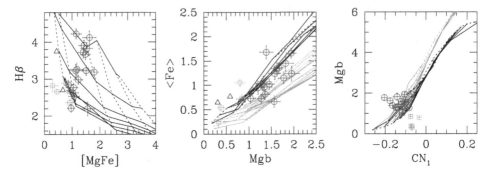

Fig. 25.5. The stellar populations of globular cluster systems of Local Group dwarf galaxies. Circles: LMC clusters from Beasley et al. (2002; for SWB classes IVB–VII only). Squares: Fornax dSph clusters from Strader et al. (2003a). Triangles: NGC 6822 clusters from Cohen & Blakeslee (1998; errors were not published). Panels and models as in Fig. 25.4. Note the CN anomaly in the Fornax dSph globular cluster system.

As shown previously by Rabin (1982), Beasley et al. (2002) find that the SWB sequence is indeed an age sequence, with the younger clusters being more metal rich. In fact, the LMC clusters represent the best training set we currently have for calibrating the stellar populations of intermediate-aged populations (Searle 1986), but the multivalued nature of the TMB03 model grids at very low metallicity (because of the blue horizontal branches) makes the old, metal-poor end difficult to calibrate without *a priori* knowledge of the horizontal branch morphology (Beasley et al. 2002). The $[\alpha/\mathrm{Fe}]$ ratios of the LMC clusters—which are only slightly supersolar—do not seem to decrease appreciably with decreasing age, so possibly age and $[\alpha/\mathrm{Fe}]$ are not tightly coupled (as is also seen for giant ellipticals, Trager et al. 2000b; Fig. 25.3). Moreover, the LMC clusters appear not to have a significant CN anomaly. The reasons for this are unclear, but may be related to the σ–$[\alpha/\mathrm{Fe}]$ relation discussed below for early-type galaxies.

The two smaller dwarfs considered in Figure 25.5 have very metal-poor globular cluster populations. The clusters in Fornax are quite old and even show a CN anomaly of the sort present in the Galactic and M31 clusters—that is, the CN strengths are too high at fixed Mgb strength. The NGC 6822 clusters have a range of ages, with at least one old cluster (H VII) and a number of younger clusters (Cohen & Blakeslee 1998; Chandar et al. 2000; Strader et al. 2003b). Da Costa & Mould (1988) employed a method similar to Rabin's (1982) study to probe the stellar populations of the globular cluster systems of NGC 147, NGC 185, and NGC 205, comparing the equivalent widths of the Balmer lines with Ca II K. They found that the majority of the clusters were old, although both NGC 205 and NGC 185 have intermediate-aged clusters and NGC 205 has at least two metal-rich clusters similar to those in the Galactic bulge (see their Fig. 3). However, line strengths on the Lick/IDS system do not exist for these clusters, and direct comparison with the stellar populations derived from Figure 25.5 is not possible.

25.3.3   The Globular Cluster Systems of Nearby Early-type Galaxies
Although heroic early efforts were made using 4–5 m class telescopes (Racine, Oke, & Searle 1978; Hanes & Brodie 1986; Mould, Oke, & Nemec 1987; Mould et al. 1990;

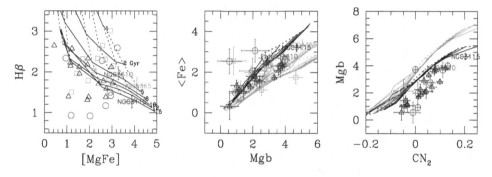

Fig. 25.6. The stellar populations of the globular cluster systems of nearby early-type galaxies. Panels and models as in Fig. 25.4. Triangles: NGC 3115 (Kuntschner et al. 2002). Circles: NGC 3610 (Strader et al. 2003c). Squares: NGC 4365 (Larsen et al. 2003; no CN data available). Host galaxies are labeled with their ID; data taken from Fisher et al. (1996; NGC 3115) or Trager et al. (1998; NGC 3610, NGC 4365). Cluster line-strength errors in the leftmost panel have been removed for clarity; typical errors are $\sigma_{H\beta} \approx \sigma_{[MgFe]} \approx 0.5$ Å (slightly smaller for NGC 3115). Note that the cluster systems of NGC 3115 and NGC 3610 apparently have a CN anomaly.

Brodie & Huchra 1991; Perelmuter, Brodie, & Huchra 1995; Jablonka et al. 1996; Minniti et al. 1996; Bridges et al. 1997), the era of measuring accurate stellar populations of the globular cluster systems of early-type galaxies began in earnest with the availability of multislit spectrographs on 8–10 m class telescopes. Kissler-Patig et al. (1998) and Cohen, Blakeslee, & Rhyzov (1998) studied NGC 1399 and M87, respectively, using the LRIS multislit spectrograph (Oke et al. 1995) on the Keck Telescopes. Cohen et al. (1998) studied the largest sample of extragalactic globular clusters around a single galaxy to date (150). Although the typical signal-to-noise ratio of these spectra is not as good as the most recent studies (e.g., Puzia et al. 2004), given the size of the data set, statistically significant conclusions about the stellar populations of the globular cluster system of an early-type galaxy could be drawn for the first time. In particular, $\sim 95\%$ of the clusters around M87 are old and can be classified into metal-poor ([Fe/H] ≈ -1.3) blue clusters and metal-rich ([Fe/H] ≈ -0.7) red clusters in a ratio of 2:3, thus resolving most of the debate about the cause of the bimodality in colors of globular cluster systems of early-type galaxies (e.g., Kundu & Whitmore 2001; Larsen et al. 2001). However, a small ($\sim 5\%$) population of intermediate-aged clusters can also be found in M87.

In Figure 25.6 I show recent results for the globular cluster systems of three early-type galaxies: NGC 3115 (Kuntschner et al. 2002), NGC 3610 (Strader et al. 2003c), and NGC 4365 (Larson et al. 2003), together with line strengths of their host galaxies (see figure caption for line-strength references). Although the error bars are large, the basic results of Cohen et al. (1998) are confirmed: there are generally three populations of globular clusters in early-type galaxies: (1) metal-poor, old, blue clusters, (2) metal-rich, old, red clusters, and (3) often small frostings of metal-rich, intermediate-aged, red(der) clusters (see also Puzia et al. 2004). This small frosting of intermediate-aged clusters may be a key to understanding the formation history of their host galaxies. A particularly interesting case is the globular cluster system of NGC 4365, as NGC 4365 has no evidence of having an intermediate-aged stellar population (Davies et al. 2001; Larsen et al. 2003; Fig. 25.6). It may be hoped that the multiple-population degeneracy of the Lick/IDS index system (Trager et al. 2000b) can be broken using globular clusters as tracers of individual star formation events in their host galaxies; such work is in progress (e.g., Larsen et al. 2003). Note also that these globular cluster systems also show a CN anomaly.

25.4 Early-type Galaxies and Bulges

The determination of the abundances and ages of galaxies is far more complex than for globular clusters, which are nearly all SSPs. Galaxies are composite stellar populations, and thus the interpretation of their spectra suffer from (currently) unavoidable difficulties. In particular, the absorption-line strength indices of composite populations are the luminosity-weighted sums of their SSP components. This means that young populations, which have low mass-to-light ratios, unduly bias the analysis of composite populations, if present (see, e.g., Trager et al. 2000b). However, in the absence of other knowledge (like the ages of the globular clusters possessed by the galaxy), line-strength indices only give the *SSP-equivalent* stellar population parameters for any individual galaxy. Therefore, I will proceed with a discussion of the abundances (and ages) obtained from modeling the integrated spectra of early-type galaxies and bulges as SSPs.

In Figure 25.3 I show the stellar populations of the nuclei of early-type galaxies (data from González 1993; Fisher et al. 1996; Kuntschner 2000; Trager et al. 2004) and spiral bulges (S0/a–Sbc; data from Proctor & Sansom 2002). Several results can be read directly from this figure.

(1) The nuclei of early-type galaxies and early-type spiral bulges have metallicities that range from just below solar to nearly 5 times the solar value.

(2) Field ellipticals, S0s in most environments, and early-type spiral bulges span a large range in $H\beta$ over a reasonably small range in [MgFe], and therefore span a large range in SSP-equivalent age.

(3) Cluster *ellipticals* are typically coeval to a factor of ~ 2 in age and vary mostly in metallicity (Bower, Lucey, & Ellis 1992; Kuntschner 2000).

(4) Ellipticals typically have $[\alpha/Fe] > 0$; $[\alpha/Fe]$ is proportional to velocity dispersion but apparently not to age (Trager et al. 2000b).

(5) S0s are slightly more "solar" in $[\alpha/Fe]$ than ellipticals. They also show a variation in $[\alpha/Fe]$ with velocity dispersion (Fisher et al. 1996), although not as strong as that in ellipticals.

(6) Bulges of early-type spirals (S0/a–Sbc) are similar to field ellipticals in the distribution of their ages, metallicities, and $[\alpha/Fe]$ ratios.

By analyzing correlations between the derived stellar population parameters, Trager et al. (2000b) found that the nuclei of elliptical galaxies (a) obey a $\log\sigma$–$[\alpha/Fe]$ relation (point 3 above, and suggested earlier by Worthey et al. 1992) that may (Thomas, Maraston, & Bender 2002) or may not (Trager et al. 2000b) depend on age, and (b) occupy only a thin plane in $\log t$–$[Z/H]$–$\log\sigma$ space (the "Z-plane"), such that age and metallicity are anti-correlated with a slope $\partial\log t/\partial\log Z = -3/2$ at fixed velocity dispersion and metallicity increases with velocity dispersion at fixed age.

The $\log\sigma$–$[\alpha/Fe]$ relation can be explained (Trager et al. 2000b) as either a varying IMF with galaxy mass—larger galaxies have "flatter" IMFs and therefore more high mass stars and therefore more Type II supernovae and therefore higher $[\alpha/Fe]$ and, at fixed age, higher $[Z/H]$ (Worthey et al. 1992; Matteucci 1994)—or that smaller galaxies have more efficient, metal-enriched winds (Vader 1986, 1987), such that Type II supernovae products are lost more efficiently in these objects and so have lower [α/Fe] at fixed Type Ia supernova yield, resulting in lower [Z/H] at fixed age. In the latter case, a high [α/Fe] is still required for massive ellipticals, so a very short star formation time scale (< 0.5 Gyr; Matteucci 1994) or a flattened IMF would be required for all elliptical galaxies.

A simple physical explanation for the Z-plane is more difficult to find. Multiple-burst population models seem to fit the data best, with small (1%–10%) bursts of recent (1–3 Gyr) star formation providing the enhanced Hβ strengths and suitably increased metallicities over a dominant (by mass) old population, but the modeling is not unique (Trager et al. 2000b). It is not yet clear if hierarchical merging models can reproduce this plane (Kauffmann & Charlot 1998). This plane has definite observational consequences. Because age and metallicity of elliptical galaxy nuclei are correlated at fixed velocity dispersion with a slope of 3/2, which preserves constant broad-band optical color and metal-line indices (§25.2.1), the Mg–σ relations (e.g., Mg$_2$–σ) are actually edge-on projections of this plane and therefore are *not* good indicators of a physically significant mass-metallicity relation (Faber 1977; Colless et al. 1999). In fact, *there is no mass-metallicity relation* in the SSP-equivalent populations of *field* elliptical galaxies; however, there *is* such a relation for the roughly coeval and presently old *cluster* elliptical galaxy population.

Although multiple ages in a population are hard to recover, the *light-weighted mean metallicity* is measured quite well from an integrated spectrum. An appendix in Trager et al. (2000b) demonstrates the calculation based on the CMD of the top of the RGB of M32 at $r \approx 1.5r_e$ (Grillmair et al. 1996) and extrapolated absorption-line strengths from $r \approx r_e$ (González 1993). The light-weighted metallicity from the CMD assuming a population between 8 and 15 Gyr old is [Z/H] = [Fe/H] = -0.25 dex; from the line strengths, a Worthey (1994) model (which uses the same isochrones) gives [Z/H] = [Fe/H] = -0.32 dex. A similar (though purely theoretical) calculation for [α/Fe] suggests that light-weighted abundance ratios are also measured quite reliably from integrated spectra.

Finally, by examining line strengths other than Mgb and $\langle$Fe$\rangle$, other elemental abundances beside the rather generic [α/Fe] are coming under scrutiny. Saglia et al. (2002), Cenarro et al. (2003), and Falcón-Barroso et al. (2003) have studied the Ca II triplet at 8600 Å in early-type galaxies and bulges and found that stellar population models (from either TMB03 or Vazdekis et al. 2003) overpredict the strength of the CaT* index by as much as 1.5 Å if [α/Fe] is set to the level suggested by Hβ, Mgb, and $\langle$Fe$\rangle$, and calcium is included with the α-elements. This suggests that either the IMF is significantly *steeper* than previously, assumed to include more dwarf stars (cf. the discussion about Na D earlier), in conflict with the mass-to-light ratios of the galaxies (Saglia et al. 2002), or that calcium does not act as an α-element in early-type galaxies and bulges (Worthey 1998; Trager et al. 2000a). This latter suggestion has recently been reinforced by Thomas, Maraston, & Bender (2003b), who examined the Ca4227 index (centered on the Ca II λ4227 line) of ellipticals from the Lick/IDS sample (Trager et al. 1998) and found that this index also required Ca to track Fe. More exactly, Thomas et al. required [α/Ca] = $+0.2$ for giant ellipticals, decreasing as velocity dispersion decreased. Therefore, giant galaxies are Ca underabundant. In fact, the Ca underabundance is nearly that of the Fe underabundance, [Fe/α] = $-$[α/Fe] = -0.3, so [Ca/Fe] $< +0.1$ in elliptical galaxies.

25.5   Summary

I conclude this review with an overview of the state of the art in determining the abundances and ages of globular clusters, early-type galaxies, and the bulges of spirals from integrated-light spectra. I then discuss the present difficulties we have with stellar population models and their application. I finish with a hint of what is to come in the immediate future.

25.5.1   The Current State of Play

It is now possible to determine accurately and precisely the age, metallicity, and abundance ratio $[\alpha/Fe]$ of a simple stellar population—one that has experienced only a single burst of star formation—from its integrated spectrum. We can therefore probe the stellar populations of globular clusters, including those around other, distant galaxies, with confidence. In Local Group galaxies, various researchers have used this technique to determine that the mean metallicity of globular cluster systems correlates with the mass of the host galaxy. The stellar content of many clusters too distant to resolve without adaptive optics or a large space telescope have been probed and the rough chemical compositions characterized. Overly strong CN lines, apparently indicative of N enhancements, have been detected in the Galactic, M31, and Fornax globular cluster systems, but not in the LMC. The stellar populations of the globular cluster systems of early-type galaxies can now be studied in detail, and the bimodality of globular cluster colors is now resolved into *three* populations: blue, old, metal poor and red, old, metal rich, which together dominate the cluster populations, and a frosting of red, intermediate-aged, metal-rich clusters. The CN anomaly is also present in these systems (Fig. 25.6), and so seems an *almost* generic property of globular cluster stellar populations, except in the LMC.

For more complex star formation histories, we can at present only parameterize this history as three (or five) numbers without additional information: the SSP-equivalent age, metallicity, and $[\alpha/Fe]$ (and now $[\alpha/Ca]$ and $[N/\alpha]$; TMB03). However, that parameterization is a useful one, and much progress has been made using this simplistic approach. Field ellipticals, S0s in many environments (except possibly in the richest clusters), and spiral bulges span a large range in SSP-equivalent age. Cluster ellipticals, on the other hand, are typically coeval to within a factor of 2 in age. SSP-equivalent metallicities of the nuclei of early-type galaxies and spiral bulges range from just below solar to nearly 5 times solar. At fixed velocity dispersion (mass), age and metallicity are correlated in such a way as to preserve constant broad-band optical color and metal-line strength; this makes the interpretation of the optical color-magnitude and Mg–σ relations of local early-type galaxies problematic. There *is* a mass-metallicity relation for early-type galaxies *at fixed age*, which manifests itself primarily in cluster galaxies. I note here that line-strength gradients (González 1993; Fisher et al. 1996) suggest that early-type galaxies globally follow the same relations, with lower metallicity and older ages (cf. Trager et al. 2000a). Early-type galaxies and spiral bulges typically have Fe-deficient ("α-enhanced") stellar populations. The value of this deficiency, $[\alpha/Fe]$, is well correlated with velocity dispersion. S0 galaxies are slightly more "solar" than giant ellipticals, where $\langle[\alpha/Fe]\rangle = +0.2$; bulges seem to follow the elliptical track, although the sample of bulges available is still small. Finally, work is progressing on tracking the abundances of other elements, including Ca, which seems to track Fe rather than the α-elements in elliptical galaxies.

25.5.2   Continuing Annoyances

Of course, although we have come a long way toward precision stellar populations, there are several significant issues still outstanding. Here are three.

Oxygen abundance   The oxygen abundance controls the temperature of the MSTO (e.g., Salaris & Weiss 1998), but no direct measure of [O/Fe] is available in the optical. Might the OH bands in the near-infrared be exploited for this purpose?

Blue horizontal branches   The presence or not of blue horizontal branch stars have been a continuing thorn in the side of those who analyze metal-rich stellar populations (e.g, Trager et al. 2000a). They apparently exist in dense, metal-rich clusters (Rose & Tripicco 1986; Rich et al. 1997), but the evolutionary path by which they appear in those clusters is unknown and therefore cannot yet be included in stellar population models. There is also the question of the importance of very metal-poor populations underlying the metal-rich populations that dominate the light (Rose 1985; Maraston & Thomas 2000). However, Rose (1985) showed that the strong Balmer lines in elliptical galaxies must be dominated by the light from dwarf stars, not giants. This, however, is not true in globular clusters, as seen in Figure 25.4.

Blue straggler stars   Given their colors and luminosities, these stars might be another significant worry, confusing age determinations (Rose 1985; Trager et al. 2000a). Again, an evolutionary path for these stars is not yet clear, and therefore they have not appeared in evolutionary syntheses. However, Rose (1985) found little evidence of such stars in his study of the blue line strengths of local elliptical galaxies.

25.5.3   Present and Future Directions

The future for this subject seems bright. Large surveys, such as the Sloan Digital Sky Survey, are beginning to use this method for studying the typical stellar populations of 10^5 galaxies (Bernardi et al. 2003; Eisenstein et al. 2003), although calibrations onto a well-calibrated and well-modeled system are not yet available. Two-dimensional spectroscopy using instruments like SAURON (Davies et al. 2001; de Zeeuw et al. 2002) are now examining the spatial distribution of the stellar populations of early-type galaxies and bulges and combining this information with kinematics. Spectral synthesis of galaxies (beginning with Vazdekis 1999) and principal-component analyses of spectra (e.g., Heavens, Jimenez, & Lahav 2000; Reichardt, Jimenez, & Heavens 2001; Eisenstein et al. 2003) are now coming over the horizon. Finally, several groups are pushing these techniques out to $z \approx 1$ (Kelson et al. 2001; Trager, Dressler, & Faber 2004), which will allow for the direct detection of the evolution of old stellar populations.

Acknowledgements.   It is a pleasure to thank my collaborators, G. Worthey, S. Faber, A. Dressler, M. Houdashelt, J. Dalcanton, D. Burstein, and J. J. González. I also gratefully acknowledge very helpful conversations with J. Brodie, A. Cole, J. van Gorkom, J. Johnson, D. Kelson, M. Kissler-Patig, H. Kuntschner, C. Maraston, B. Poggianti, T. Puzia, R. Schiavon, D. Thomas, and E. Tolstoy during the preparation of this review. An anonymous referee is thanked for a careful reading of the manuscript and especially for pointing out the lack of data on the inner bulge clusters of M31. I would like to thank the organizers, A. McWilliam and M. Rauch, for an enjoyable meeting. The stellar population community owes a significant debt of gratitude to L. Robinson and J. Wampler for the development of the Image Dissector Scanner (IDS), which revolutionized the study of integrated spectra of galaxies. I also would like to thank J. Nelson and CARA for the vision to develop and build the Keck Telescopes and J. B. Oke and J. Cohen for building the LRIS spectrograph, which together have pushed our study of stellar populations toward the high-redshift Universe. This research has been supported at various times by a Flintridge Foundation Fellowship, by a Carnegie Starr Fellowship, by NASA through Hubble Fellowship grant HF-01125.01-99A awarded by the Space Telescope Science Institute, which is operated by the Association of Universities for Research in Astronomy, Inc., for NASA under contract NAS 5-26555, and by the Kapteyn Astronomical Institute.

References

Beasley, M. A., Hoyle, F., & Sharples, R. M. 2002, MNRAS, 336, 168

Berkhuijsen, E. M., Humphreys, R. M., Ghigo, F. D., & Zumach, W. 1988, A&AS, 76, 65

Bernardi, M., et al. 2003, AJ, 125, 1882

Bono, G., Caputo, F., Cassisi, S., Castellani, V., Marconi, M. 1997, ApJ, 489, 822

Bower, R. G., Lucey, J. R., & Ellis, R. S. 1992, MNRAS, 254, 601

Bressan, A., Chiosi, C., & Fagotto, F. 1994, ApJS, 94, 63

Bridges, T. J., Ashman, K. M., Zepf, S. E., Carter, D., Hanes, D. A., Sharples, R. M., & Kavelaars, J. J. 1997, MNRAS, 284, 376

Brodie, J. P., & Huchra, J. P. 1990, ApJ, 362, 503

——. 1991, ApJ, 379, 157

Bruzual A., G. 1983, ApJ, 273, 105

Bruzual A., G., & Charlot, S. 1993, ApJ, 405, 538

Burstein, D., Faber, S. M., Gaskell, C. M., & Krumm, N. 1984, ApJ, 287, 586

Buzzoni, A. 1989, ApJS, 71, 817

Cassisi, S., Castellani, M., & Castellani, V. 1997, A&A, 317, 10

Cenarro, A. J., Gorgas, J., Vazdekis, A., Cardiel, N., & Peletier, R. F. 2003, MNRAS, 339, L12

Chandar, R., Bianchi, L., & Ford, H. C. 2000, AJ, 120, 3088

Charlot, S., & Bruzual A., G. 1991, ApJ, 126

Cohen, J. G., & Blakeslee, J. P. 1998, AJ, 115, 2356

Cohen, J. G., Blakeslee, J. P., & Côté, P. 2003, ApJ, 592, 866

Cohen, J. G., Blakeslee, J. P., & Rhyzov, A. 1998, ApJ, 496, 808

Colless, M., Burstein, D., Davies, R. L., McMahan, R. K., Saglia, R. P., & Wegner, G. 1999, MNRAS, 303, 813

Da Costa, G. S., & Mould, J. R. 1988, ApJ, 334, 159

Davies, R. L., et al. 2001, ApJ, 548, L33

de Vaucouleurs, G., & de Vaucouleurs, A. 1959, PASP, 71, 83

de Zeeuw, P. T., et al. 2002, MNRAS, 329, 513

Dorman, B., O'Connell, R. W., & Rood, R. T. 2003, ApJ, 591, 878

Eisenstein, D. J., et al. 2003, ApJ, 585, 649

Faber, S. M. 1972, A&A, 20, 361

——. 1977, in The Evolution of Galaxies and Stellar Populations, ed. B. M. Tinsley & R. Larson (New Haven: Yale Univ. Observatory), 157

Faber, S. M., & Jackson, R. E. 1976, 204, 668

Falcón-Barroso, J., Peletier, R. F., Vazdekis, A., & Balcells, M. 2003, ApJ, 588, L17

Fisher, D., Franx, M., & Illingworth, G. D. 1996, ApJ, 459, 110

Garnett, D. R. 2004, in Carnegie Observatories Astrophysics Series, Vol. 4: Origin and Evolution of the Elements, ed. A. McWilliam & M. Rauch (Cambridge: Cambridge Univ. Press), in press

Gibson, B. K., Madgwick, D. S., Jones, L. A., Da Costa, G. S., & Norris, J. E. 1999, AJ, 118, 1268

Girardi, L., Bressan, A., Bertelli, G., & Chiosi, C. 2000, A&AS, 141, 371

González, J. J. 1993, Ph.D. Thesis, Univ. of California, Santa Cruz

Greggio, L. 1997, MNRAS, 285, 151

Grillmair, C. J., et al. 1996, AJ, 112, 1975

Gunn, J. E., Stryker, L. L., & Tinsley, B. M. 1981, ApJ, 249, 48

Hanes, D. A., & Brodie, J. P. 1986, ApJ, 300, 279

Heavens, A. F., Jimenez, R., & Lahav, O. 2000, MNRAS, 317, 965

Hesser, J. E. 1978, ApJ, 223, L117

Jablonka, P., Bica, E., Pelat, D., & Alloin, D. 1996, A&A, 307, 385

Jones, L. A. 1998, Ph.D. Thesis, Univ. of North Carolina

Jones, L. A., & Worthey, G. 1995, ApJ, 446, L31

Kauffmann, G., & Charlot, S. 1998, MNRAS, 297, L23

Kelson, D. D., Illingworth, G. D., Franx, M., & van Dokkum, P. G. 2001, ApJ, 552, L17

Kissler-Patig, M., Brodie, J. P., Schroder, L. L., Forbes, D. A., Grillmair, C. J., & Huchra, J. P. 1998, AJ, 115, 105

Kundu, A., & Whitmore, B. C. 2001, AJ, 121, 2950

Kuntschner, H. 2000, MNRAS, 315, 184

Kuntschner, H., Ziegler, B. L., Sharples, R. M., Worthey, G., & Fricke, K. J. 2002, A&A, 395, 761

Kurucz, R. L. 1993, http://kurucz.harvard.edu/

Larsen, S. S., Brodie, J. P., Beasley, M. A., Forbes, D. A., Kissler-Patig, M., Kuntschner, H., & Puzia, T. H. 2003, ApJ, 585, 767

Larsen, S. S., Brodie, J. P., Huchra, J. P., Forbes, D. A., & Grillmair, C. J. 2001, AJ, 121, 2974

Lasker, B. M. 1970, AJ, 75, 21

Le Borgne, J.-F., et al. 2003, A&A, 402, 433

Lejeune, T., Cuisinier, F., & Buser, R. 1997, A&AS, 125, 229

——. 1998, A&AS, 130, 65

Maraston, C. 1998, MNRAS, 300, 872

Maraston, C., Greggio, L., Renzini, A., Ortolani, S., Saglia, R. P., Puzia, T. H., & Kissler-Patig, M. 2003, A&A, 400, 823

Maraston, C. & Thomas, D. 2000, ApJ, 541, 126

——. 2003, A&A, 401, 429

Matteucci, F. 1994, A&A, 288, 57

Minniti, D., Alonso, M. V., Goudfrooij, P., Jablonka, P., & Meylan, G. 1996, ApJ, 467, 221

Mould, J. R., Oke, J. B., de Zeeuw, P. T., & Nemec, J. M. 1990, AJ, 99, 1823

Mould, J. R., Oke, J. B., & Nemec, J. M. 1987, AJ, 92, 53

O'Connell, R. W. 1976, ApJ, 206, 370

——. 1986, in Stellar Populations, ed. C. A. Norman, A. Renzini, & M. Tosi (Cambridge: Cambridge Univ. Press), 167

——. 1999, ARA&A, 37, 603

Oke, J. B., et al. 1995, PASP, 107, 375

Peletier, R. F. 1989, Ph.D. Thesis, Rijksuniversiteit Groningen

Perelmuter, J.-M., Brodie, J. P., & Huchra, J. P. 1995, AJ, 110, 620

Peterson, R. C. 1976, ApJ, 210, L123

Peterson, R. C., Carney, B. W., Dorman, B., Green, E. M., Landsman, W., Liebert, J., O'Connell, R. W., & Rood, R. T. 2003, ApJ, 588, 299

——. 2004, in Carnegie Observatories Astrophysics Series, Vol. 4: Origin and Evolution of the Elements, ed. A. McWilliam & M. Rauch (Pasadena: Carnegie Observatories, http://www.ociw.edu/symposia/series/symposium4/proceedings.html)

Proctor, R. N., & Sansom, A. E. 2002, MNRAS, 333, 517

Puzia, T. H., et al. 2004, A&A, submitted

Puzia, T. H., Saglia, R. P., Kissler-Patig, M., Maraston, C., Greggio, L., Renzini, A., & Ortolani, S. 2002, A&A, 395, 45

Rabin, D. 1980, Ph.D. Thesis, California Institute of Technology

——. 1982, ApJ, 261, 85

Racine, R., Oke, J. B., & Searle, L. 1978, ApJ, 223, 82

Reichardt, C., Jimenez, R., & Heavens, A. F. 2001, MNRAS, 327, 849

Renzini, A., & Buzzoni, A. 1986, in Spectral Evolution of Galaxies, ed. C. Chiosi & A. Renzini (Dordrecht: Reidel), 135

Rich, R. M. 1988, AJ, 95, 828

Rich, R. M., et al. 1997, ApJ, 484, L25

Rose, J. A. 1985, AJ, 90, 1927

Rose, J. A., & Tripicco, M. J. 1986, AJ, 92, 610

Saglia, R. P., Maraston, C., Thomas, D., Bender, R., & Colless, M. 2002, ApJ, 579, L13

Salaris, M., & Weiss, A. 1998, A&A, 335, 943

Salasnich, B., Girardi, L., Weiss, A., & Chiosi, C. 2000, A&A, 361, 1023

Salpeter, E. E. 1955, ApJ, 121, 161

Schiavon, R. P., Faber, S. M., Castilho, B. V., & Rose, J. A. 2002a, ApJ, 580, 850
Schiavon, R. P., Faber, S. M., Rose, J. A., & Castilho, B. V. 2002b, ApJ, 580, 873
Searle, L. 1986, in Stellar Populations, ed. C. A. Norman, A. Renzini, & M. Tosi (Cambridge: Cambridge Univ. Press), 3
Searle, L., Wilkinson, A., & Bagnuolo, W. G. 1980, ApJ, 239, 803 (SWB)
Smith, G. H., Shetrone, M. D., Briley, M. M., Churchill, C. W., & Bell, R. A. 1997, PASP, 109, 236
Smith, V. V. 2004, in Carnegie Observatories Astrophysics Series, Vol. 4: Origin and Evolution of the Elements, ed. A. McWilliam & M. Rauch (Cambridge: Cambridge Univ. Press), in press
Spinrad, H. 1962a, PASP, 74, 146
——. 1962b, ApJ, 135, 715
Spinrad, H., & Taylor, B. J. 1969, ApJ, 157, 1279
——. 1971, ApJS, 22, 445
Strader, J., Brodie, J. P., Forbes, D. A., Beasley, M. A., & Huchra, J. P. 2003a, AJ, 125, 1291
Strader, J., Brodie, J. P., & Huchra, J. P. 2003b, MNRAS, 339, 707
Strader, J., Brodie, J. P., Schweizer, F., Larsen, S. S., & Seitzer, P. 2003c, AJ, 125, 626
Tantalo, R., Chiosi, C., & Bressan, A. 1998, A&A, 333, 419
Thomas, D., Maraston, C., & Bender, R. 2002, Ap&SS, 281, 371
——. 2003a, MNRAS, 339, 897 (TMB03)
——. 2003b, MNRAS, 343, 279
Tinsley, B. M. 1968, ApJ, 151, 547
——. 1972, A&A, 20, 383
Tinsley, B. M., & Gunn, J. E. 1976, ApJ, 203, 52
Trager, S. C., Dressler, A., & Faber, S. M. 2004, in preparation
Trager, S. C., Faber, S. M., & Dressler, A. 2004, in preparation
Trager, S. C., Faber, S. M., Worthey, G., & González, J. J. 2000a, AJ, 119, 1645
——. 2000b, AJ, 120, 165
Trager, S. C., Worthey, G., Faber, S. M., Burstein, D., & González, J. J. 1998, ApJS, 116, 1
Tripicco, M. J., & Bell, R. A. 1995, AJ, 110, 3035
Vader, J. P. 1986, ApJ, 305, 669
——. 1987, ApJ, 317, 128
van den Bergh, S. 1969, ApJS, 19, 145
Vazdekis, A. 1999, ApJ, 513, 224
Vazdekis, A., Casuso, E., Peletier, R. F., & Beckman, J. E. 1996, ApJS, 106, 307
Vazdekis, A., Cenarro, A. J., Gorgas, J., Cardiel, N., & Peletier, R. F. 2003, MNRAS, 340, 1317
Vazdekis, A., Salaris, M., Arimoto, N., & Rose, J. A. 2001, ApJ, 549, 274
Weiss, A., Peletier, R. F., & Matteucci, F. 1995, A&A, 296, 73
Westera, P., Lejeune, T., Buser, R., Cuisinier, F., & Bruzual, G. 2002, A&A, 381, 524
Whipple, F. L. 1935, Harvard Coll. Obs. Circ., 404, 1
Williams, T. B. 1976, ApJ, 209, 716
Worthey, G. 1994, ApJS, 95, 107
——. 1998, PASP, 110, 888
Worthey, G., Faber, S. M., & González, J. J. 1992, ApJ, 398, 69
Worthey, G., Faber, S. M., González, J. J., & Burstein, D. 1994, ApJS, 94, 687
Worthey, G., & Ottaviani, D. L. 1997, ApJS, 111, 377
Zinn, R., & West, M. J. 1984, ApJ, 55, 45

26

Abundances in spiral and irregular galaxies

DONALD R. GARNETT
Steward Observatory, University of Arizona

Abstract
This paper reviews the current data on gas-phase abundances in spiral and irregular galaxies. I discuss the spatial distribution of abundances in galaxies and what we know about the variation of abundance ratios for elements commonly observed in ionized nebulae. Brief comments are made on the astrophysical significance of various observed abundance phenomena.

26.1   Introduction

It is certainly appropriate to review gas-phase abundances in galaxies for this symposium, as the study of abundances in galaxies began in large part in Pasadena, with the work of Leonard Searle and Wallace Sargent. Searle (1971) demonstrated, through a relatively simple ionization analysis, that the radial gradients of [O III] emission-line strengths observed in spiral galaxies were most likely explained by radial gradients in the composition of the interstellar gas. At about the same time, Searle & Sargent (1972), as well as Peimbert & Spinrad (1970), showed that oxygen abundances in dwarf irregular galaxies were much lower than in the solar neighborhood, while helium abundances were close to the solar neighborhood values. These results implied that dwarf galaxies were not as evolved as the Milky Way, and that most of the helium in the Universe came not from stars, but rather from nucleosynthesis in the Big Bang.

Since then abundance data have accumulated for some 50 spiral galaxies and numerous dwarf galaxies. Here I present an overview of the patterns of metallicity and element abundance ratios observed in spiral and irregular galaxies. I will discuss the results for both types of galaxies rather than separately; many aspects can be discussed for the combined groups, although there are a number of differences that could constitute the topic of an entire conference alone. The observational data to be discussed represent a highly selected sample of abundances, gas masses, and stellar photometry from sources too numerous to mention individually. I will employ abundances derived almost exclusively from H II region spectra, since they contribute the largest set of abundance data for spirals and irregulars in the local Universe.

26.2   Spatial Abundance Profiles

Most of the data on abundances in external galaxies come from spectroscopy of H II regions. Accurate abundances depend on high signal-to-noise ratio measurements of electron temperature. While this has been relatively easy for metal-poor galaxies, few mea-

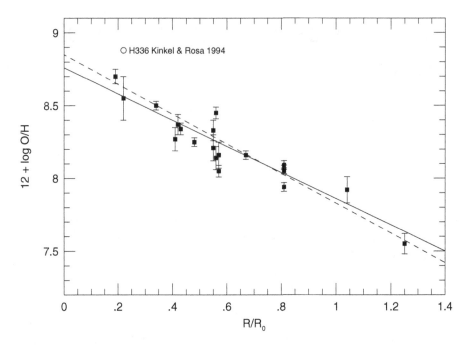

Fig. 26.1. The gradient in O/H across the disk of the spiral galaxy M101 vs. galactocentric radius. The open circle is the measurement for Searle 5 by Kinkel & Rosa (1994). (Adapted from Kennicutt et al. 2003.)

surements were available for metal-rich regions in spirals. New studies are beginning to crack this barrier, as T_e measurements are being made for solar-metallicity H II regions (Castellanos, Díaz, & Terlevich 2002; Kennicutt, Bresolin, & Garnett 2003). Another happy development is that abundances are being measured for luminous stars in nearby galaxies (e.g., Monteverde et al. 1997; Korn et al. 2002). The stellar abundances appear to be in good general agreement with those derived from H II regions.

The spatial distribution of abundances in galaxies depends coarsely on the Hubble type. Unbarred or weakly barred spiral galaxies typically have a strong radial gradient in metallicity (Fig. 26.1), as determined from O/H (Vila-Costas & Edmunds 1992; Zaritsky, Kennicutt, & Huchra 1994; Ferguson, Gallagher, & Wyse 1998; van Zee et al. 1998). The derived O/H can drop by a factor of 10–50 from the nucleus of a galaxy to the outer disk. In the best-studied spirals, such as M101, the O/H gradients are closely exponential over many disk scale lengths. Irregular galaxies, by contrast, show little spatial variation in abundances, to high levels of precision (Kobulnicky & Skillman 1996), indicating a well-mixed interstellar medium (ISM). Strongly barred spiral galaxies show evidence that their O/H gradients are shallower than in unbarred spirals.

26.2.1   *Metallicity versus Galaxy Luminosity/Mass*

Detailed examination of these data shows that there are significant correlations between abundances and abundance gradients in spirals and irregulars with galaxy structural properties. One well-established correlation is the relation between metallicity and galaxy

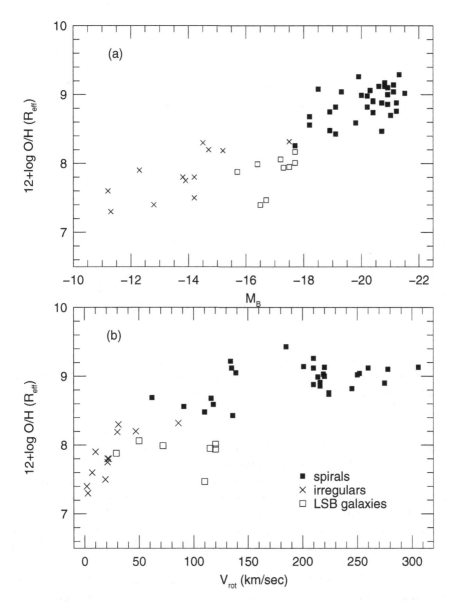

Fig. 26.2. The correlation of spiral galaxy abundance (O/H) at the disk half-light radius from the galaxy nucleus vs. galaxy blue luminosity (*a*) maximum rotational speed V_{rot} (*b*).

luminosity (Lequeux et al. 1979; Garnett & Shields 1987; Skillman, Kennicutt, & Hodge 1989). This is shown in the top panel of Figure 26.2, where I plot O/H determined at the half-light radius of the disk (R_{eff}) versus *B*-band absolute magnitude M_B. Zaritsky et al. (1994) noted the remarkable uniformity of this correlation over 11 magnitudes in galaxy luminosity, for ellipticals and star-forming spiral and irregular galaxies. The correlation be-

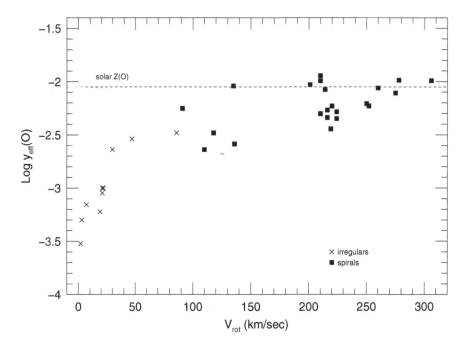

Fig. 26.3. Effective yields y_{eff} for nearby spiral and irregular galaxies versus rotation speed V_{rot}. Filled squares represent the data for spirals, while the crosses show the data for irregulars. (Adapted from Garnett 2002.)

tween O/H and rotation speed V_{rot} (bottom panel of Fig. 26.2) looks more interesting, as it suggests that the average metallicity approaches a constant for the most massive galaxies.

To the extent that blue luminosity reflects the mass of a system, the metallicity-luminosity correlation suggests a common mechanism regulating the global metallicity of galaxies. The most commonly invoked mechanism is selective loss of heavy elements in galactic winds (e.g., Marconi, Matteucci, & Tosi 1994). However, a metallicity-luminosity correlation could also occur if there is a systematic variation in gas fraction across the luminosity sequence. This question is of particular interest for understanding the evolution of dwarf galaxies (e.g., Lee et al. 2003a).

The question of loss of metals from galaxies is profound because of the existence of metals in low-column density Lyα forest systems (Ellison et al. 2000; Songaila 2001), which are probably gas clouds residing outside of galaxies. It is therefore a useful exercise to investigate what kinds of galaxies are potential candidates for ejecting heavy elements into the intergalactic medium.

This question can be examined further by studying metallicity as a function of gas fraction. In the context of the simple, closed-box, chemical evolution, Edmunds (1990) showed that outflows of gas and inflows of metal-poor gas cause galactic systems to deviate from the closed-box model in similar ways. Specifically, defining the effective yield

$$y_{\text{eff}} = \frac{Z}{\ln(\mu^{-1})}$$

(26.1)

where Z is metallicity and μ is gas fraction, outflows of any kind and inflows of metal-poor gas tend to make the effective yield smaller than the true yield of the closed-box model. Thus, comparing effective yields for a sample of galaxies can provide information on the relative importance of gas flows from one galaxy to another. Such a comparison is presented in Figure 26.3 (Garnett 2002), which plots the effective yields derived for 22 spiral and 10 irregular galaxies versus V_{rot} using Equation (26.1) and published data on abundances, atomic and molecular gas, and photometry. The data show a factor of 10–20 systematic increase in y_{eff} from the least massive irregulars to the most massive spirals.

This result is striking verification that the yields derived for dwarf irregulars are significantly lower than in spiral galaxies. The trend toward small y_{eff} in the least massive galaxies suggests that it is the loss of metals in galactic winds that drives the correlation, although other factors such as tidal stripping may also be important. The plot suggests that galaxies with $V_{rot} < 100$–150 km s^{-1} lose significant quantities of metals, while more massive ones retain essentially all their metals; this transition is consistent with estimates of Martin (1999).

26.2.2 Abundance Gradient Variations

Figure 26.4 shows that the steepness of abundance gradients (expressed in dex/kpc) decreases with galaxy luminosity. However, more luminous galaxies have larger disk scale lengths, and so if one looks at the gradient per disk scale length (Fig. 26.4, bottom panel), the correlation goes away. Considering the errors in the computed gradients (typically 25%), then the scatter in the slopes of gradients per scale length may be consistent with purely observational scatter. This is related to the exponential abundance gradients noted previously. It has been suggested (Combes 1998) that purely exponential gradients may be explained by "viscous disk" models (Lin & Pringle 1987; Clarke 1989) for disk evolution; if the time scale for viscous transport of angular momentum is comparable to the star formation time scale (with the two time scales perhaps connected through the gravitational instability), such models naturally produce an exponential stellar disk. Another interesting possibility (Prantzos & Boissier 2000) is that the variation in gradients is a result of evolution of disks that follow the scaling relations for galaxies in the cold dark matter model for galaxy formation (Mo, Mao, & White 1998). In the chemical evolution models of Prantzos & Boissier (2000), average metallicity correlates with galaxy mass, while the angular momentum parameter determines the gradient slope; galaxies with larger angular momentum have longer scale lengths and shallower abundance gradients. This would be remarkable since it is not obvious that the present-day composition of galaxies should have any memory of the initial conditions for galaxy formation.

26.2.3 Barred Spirals

Bars are interesting because the gravitational potential of a bar is expected to induce a large-scale radial gas flow, through radiative shocking and subsequent loss of angular momentum as the gas passes through the bar (Barnes 1991). The radial flow could significantly alter the metallicity distribution by mixing in gas from outer radii, thus weakening composition gradients. The evidence so far accumulated indicates that barred spirals generally have shallower composition gradients than weakly barred or nonbarred spirals (Martin & Roy 1994; Zaritsky et al. 1994). Martin & Roy (1994) have argued that the slope of the composition gradient correlates with both bar length and bar strength (the ratio of bar

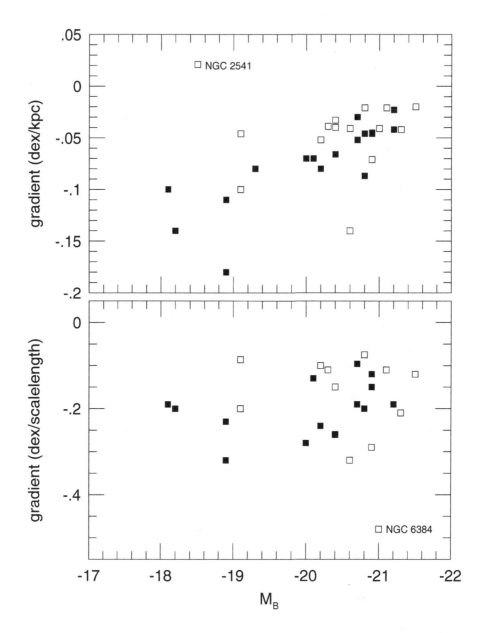

Fig. 26.4. The correlation of abundance gradient vs. M_B. The upper panel shows abundance gradients per kpc, while the lower panel shows gradients per unit disk scale length. (Adapted from Garnett et al. 1997.)

length to width). This is illustrated in Figure 26.5, which shows the O/H gradient per kpc versus bar length a relative to the photometric radius (*top panel*) and versus bar strength E_B = $10(1-b/a)$, where b is the bar width (*bottom panel*). The plots show that (1) the nonbarred spirals show a wide range of values for gradient slopes, although on average they are steeper

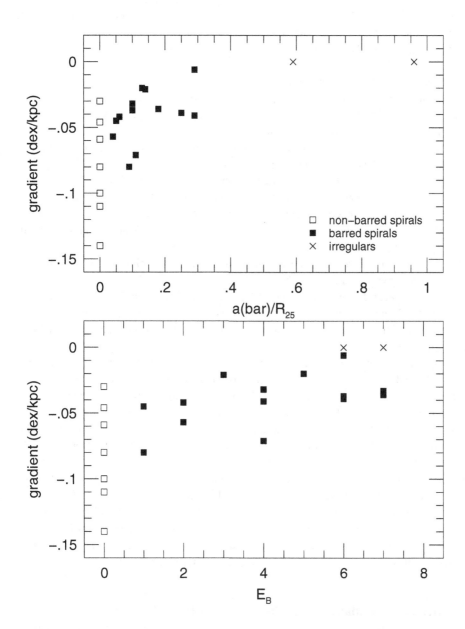

Fig. 26.5. *Top*: O/H gradient per kpc vs. bar length a relative to the photometric radius R_{25}. Filled squares are barred spirals, unfilled squares are nonbarred spirals, and crosses are irregular galaxies. *Bottom*: O/H gradient vs. bar strength $E_B = 10(1-b/a)$, where b/a is the ratio of bar minor and major axes. (Adapted from Martin & Roy 1994.)

than those for barred spirals, and (2) the O/H gradients for the barred spirals tend to get shallower with increasing bar length and bar strength.

Curiously, some barred spirals (e.g., NGC 1365; Roy & Walsh 1997) show an O/H gradient within the bar, with flattening only outside the bar. One way to explain this is if strong star formation occurs in the bar, building up abundances faster than the radial flow can homogenize them (Friedli, Benz, & Kennicutt 1994).

26.2.4   *Cluster Spirals and Environment*

Environment and interactions appear to play a significant role in the evolution of galaxies, particularly in dense environments. Interactions with satellites may be responsible for the significant fraction of lopsided spiral galaxies (Rudnick & Rix 1998). Disk asymmetry may affect the inferred spatial distribution of metals in the interstellar gas. For example, Kennicutt & Garnett (1996) noted an asymmetry in O/H between the NW and SE sides of M101, which may be related to the asymmetry in the structure of the disk. Zaritsky (1995) found a possible correlation between disk $B - V$ color and the slope of the O/H gradient, such that bluer galaxies tended to have steeper gradients. He suggested that accretion of metal-poor, gas-rich dwarf galaxies in the outer disk could steepen abundance gradients and make the colors bluer through increased star formation.

Rich clusters offer a variety of galaxy-galaxy and galaxy-intracluster medium interactions. The cluster environment affects the morphology of galaxies (Dressler 1980). It is also known that spirals near the center of rich clusters (e.g., the Virgo cluster) show evidence for stripping of H I (Warmels 1988; Cayatte et al. 1994). Such stripping is inferred to result from interaction of the galaxy ISM with the hot X-ray intracluster gas. The degree of H I stripping correlates with projected distance from the cluster core.

If field galaxies evolve through continuing infall of gas (Gunn & Gott 1972), then the truncation of H I disks in cluster spirals should have an effect on chemical evolution. Infall of metal-poor gas reduces the metallicity of the gas at a given gas fraction. Truncation of such infall should then cause the chemical evolution of cluster spirals to behave more like the simple closed-box model, and thus should have higher metallicities than comparable field spirals. This idea has led to several studies of abundances in Virgo spirals (Henry, Pagel, & Chincarini 1994; Henry et al. 1996; Skillman et al. 1996; note also Lee, McCall, & Richer 2003b for dwarf irregulars). The results indicate that cluster spirals with the largest H I deficiencies have higher O/H abundances than field spirals with comparable M_B, rotation speeds, and Hubble types, while spirals with only modest or little H I stripping have abundances comparable to those of similar field spirals. The samples studied so far have been small, and one must worry about possible systematic errors in abundances. This is an area that could benefit from further study with larger galaxy samples and more secure abundance measurements.

26.3   Element Abundance Ratios in Spiral and Irregular Galaxies

The abundance ratios of heavy elements are sensitive to the initial mass function (IMF), the star formation history, and variations in stellar nucleosynthesis with, for example, metallicity. In particular, comparison of abundances of elements produced in stars with relatively long lifetimes (such as C, N, Fe, and the s-process elements) with those produced in short-lived stars (such as O) probe the star formation history. Below, I review the accumulated data on C, N, S, and Ar abundances (relative to O) in spiral and irregular galaxies, covering 2 orders of magnitude in metallicity (as measured by O/H). The data are taken from a variety of sources in the literature on abundances for H II regions.

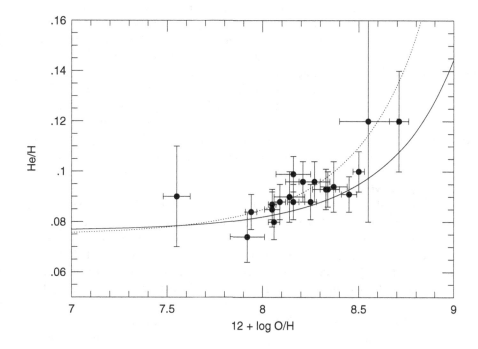

Fig. 26.6. He/H vs. O/H for H II regions in M101 (Kennicutt et al. 2003). The curves show two fits to the $\Delta Y/\Delta(O/H)$ relation for dwarf galaxies from Olive et al. (1997).

26.3.1 Helium

Helium, the second most abundant element, has significance for cosmology and stellar structure. Most ^{4}He was produced in the Big Bang, and the primordial mass fraction Y_p is a constraint on the photon/baryon ratio and thus on the cosmological model. The He mass fraction also affects stellar structure. Helium abundances are difficult to determine in any star but OB stars, and in these stars the surface abundance of He may be affected by evolution. On the other hand, He I recombination lines are relatively easy to measure in ionized nebulae, and so a large amount of data is available on He/H in the ISM.

A great deal of effort has been spent discussing the determination of helium abundances and Y_p in metal-poor dwarf galaxies (e.g., Steigman 2003), so I will be brief on this aspect. Peimbert & Torres-Peimbert (1974) made early estimates of Y_p by assuming that the He mass fraction varies linearly with metallicity (or O/H), and thus one could extrapolate He abundances in nebulae with a wide range of O/H to the pre-galactic He abundance at O/H = 0. Today there are very high signal-to-noise ratio data for He in approximately 40 metal-poor dwarf irregulars, so the extrapolation to O/H = 0 can be estimated to high statistical precision.

The good news is that the best current estimates of Y_p agree to within 5%. This is amazing agreement for measurements derived from spectroscopy of distant galaxies. Nevertheless, the differences in Y_p estimates are still significant for Big Bang nucleosynthesis theory, as the two largest studies obtain values for Y_p that disagree at the 4–5 σ significance level (cf. Olive, Skillmanm & Steigman 1997; Izotov & Thuan 1998; Peimbert, Peimbert, &

Torres-Peimbert 2000; Peimbert 2003). Depending on which estimate is considered most reliable, Y_p either agrees with the best current estimate of D/H, or it does not. At present the battleground for Y_p is focused on sources of systematic error, and these are likely to yield the greatest improvements in He measurements, rather than measuring more data points.

Helium abundances in spiral galaxies are less well determined, because of more uncertain electron temperatures. The He abundance does have an effect on ionization structure, so it is of interest to know how He/H varies with metallicity in the inner disks of spirals. Figure 26.6 shows He abundances in M101 H II regions, along with extrapolations of fits to the He/H – O/H correlation in metal-poor dwarf galaxies (Olive et al. 1997). The results suggest that $\Delta Y / \Delta (O/H)$ can be fit by a single linear relation up to at least solar abundances.

26.3.2   Carbon and Nitrogen

Carbon abundances in H II regions have been difficult to determine because the important ionization states (C^+, C^{+2}) have no strong forbidden lines in the optical spectrum. Only the UV spectrum shows collisionally excited lines from both C^+ and C^{+2}. A number of studies of carbon abundances in extragalactic H II regions were made with *IUE* (e.g., Dufour, Shields, & Talbot 1982; Peimbert, Pe na, & Torres-Peimbert 1986; Dufour, Garnett, & Shields 1988), but for the most part the *IUE* observations suffered from low signal-to-noise ratio.

The higher UV sensitivity of *HST* offered greatly improved measurements of UV emission lines from [C II] and C III]. Figure 26.7 shows data for C/O as a function of O/H in dwarf irregular and spiral galaxies from *HST* measurements (Garnett et al. 1995a, 1999; Kobulnicky & Skillman 1998). Note that some C (and O) is expected to be depleted onto interstellar dust grains. It is likely that the C/O values should all be increased by 0.1–0.2 dex, but a systematic variation in the fractional depletions with metallicity is not expected.

Figure 26.7 shows a trend of steeply increasing C/O for log O/H > -4, while C/O is roughly constant for lower O/H. This is in agreement with observations of C/O in disk stars in the Galaxy (Gustafsson et al. 1999). The notable trend in Figure 26.7 is the apparent "secondary" behavior of C with respect to O, despite the fact that C (i.e., ^{12}C) is primary. Tinsley (1979) demonstrated that such variations can be understood as the result of finite stellar lifetimes and delays in the ejection of elements from low- and intermediate-mass stars. If C is produced mainly in intermediate-mass stars, then the enrichment of C in the ISM is delayed with respect to O, which is produced in high-mass stars. On the other hand, C may also have a metallicity-dependent yield in massive stars that results from the effects of stellar mass loss (Maeder 1992). It is not yet certain which effect dominates to produce the rise in C/O. The biggest uncertainties revolve around the theoretical yields, primarily due to the unresolved uncertainties in the $^{12}C(\alpha, \gamma)^{16}O$ reaction rate (Hale 1998), and the effects of convective mixing during helium burning (Arnett 1996).

26.3.3   Nitrogen

The lower panel of Figure 26.7 shows the variation of N/O with O/H in spiral and irregular galaxies. The trend is very similar to that for C/O; N/O is roughly constant at log N/O ≈ -1.5 for log O/H < -3.7, and increases proportionally with O/H in more metal-rich galaxies. The second component is produced via the classical CNO cycle during hydrogen burning in stars and requires the presence of C and O in the star from birth ("secondary" N), while the first component is postulated to come from the CN cycle on freshly synthesized C

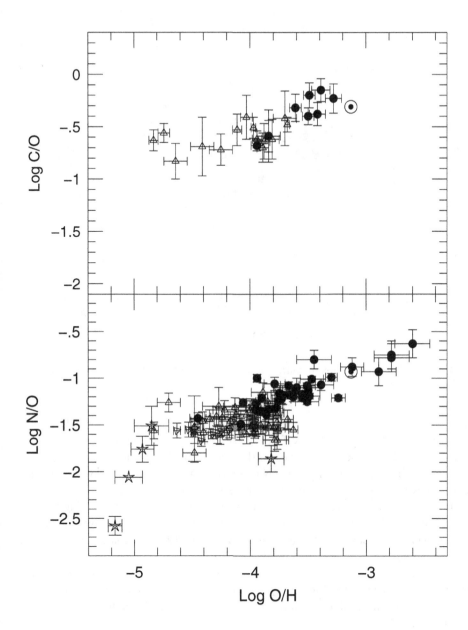

Fig. 26.7. *Top:* C/O abundance ratios from spectroscopy of H II regions in spiral and irregular galaxies. *Open symbols*: irregular galaxies; *filled symbols*: spiral galaxy H II regions. *Bottom:* N/O vs. O/H in spiral and irregular galaxies. *Open circles*: Garnett (1990); *open diamonds*: Thuan, Izotov, & Lipovetsky (1995); *filled circles*: Díaz et al. (1991), Garnett et al. (1997), and Kennicutt et al. (2003); *stars*: high-redshift absorption-line systems.

(from He burning) that has been convectively "dredged-up" into a hot H-burning zone at the base of the convective envelope, and does not require an initial seed of C or O ("primary" N).

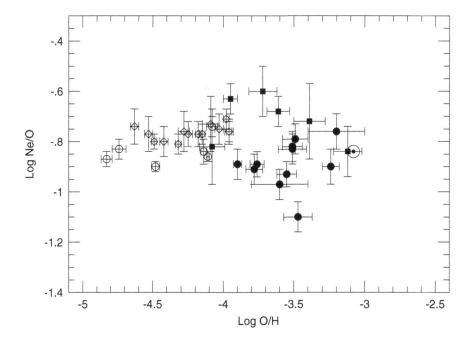

Fig. 26.8. Ne/O abundance ratios in spiral and irregular galaxies. Symbols are the same as in Fig. 26.7.

The latter process is most commonly thought to occur in the asymptotic giant branch (AGB) stage of intermediate-mass stars (Iben & Truran 1978), but has been found to occur in models of massive stars with increased convective overshooting or rotationally induced mixing (e.g., Langer et al. 1997). Some scatter is seen in N/O for dwarf galaxies (Kobulnicky & Skillman 1998). This may be the result of localized enrichment by Wolf-Rayet stars.

Recent studies of high-redshift Lyα absorption systems (plotted as stars in Fig. 26.7) have found objects with N/S, N/Si, and N/O ratios much lower than in the dwarf galaxies (Pettini, Lipman, & Hunstead 1995; Lu, Sargent, & Barlow 1998; Prochaska et al. 2002). A wide range in inferred N/O is seen in the damped Lyα systems, but the lowest values are as much as a factor of 10 smaller than the average for irregular galaxies. The results are consistent with the idea that the damped Lyα systems represent lines of sight through very young galaxies, with an age spread of a few hundred Myr, the time scale for enrichment of N from AGB stars.

26.3.4   *Neon, Sulfur, and Argon*

Neon, sulfur, and argon are products of the late stages of massive star evolution. ^{20}Ne results from carbon burning, while S and Ar are products of O burning. As they are all considered part of the α-element group, their abundances are expected to track O/H closely. Deviations from this may point to unsuspected variations in the stellar IMF or nucleosynthesis.

Neon abundances in extragalactic H II regions are derived primarily from optical mea-

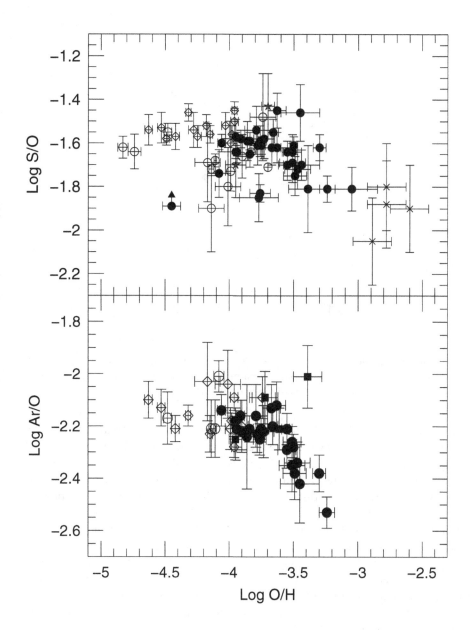

Fig. 26.9. *Top:* S/O abundance ratios in spiral and irregular galaxies. *Bottom:* Ar/O abundance ratios in spiral and irregular galaxies. Symbols are the same as in Fig. 26.7.

surements of [Ne III], although spacecraft measurements of the IR [Ne II] and [Ne III] fine-structure lines are becoming available. Few measurements of Ne are available for metal-rich objects. The data that are available show that Ne tracks O quite closely (Fig. 26.8), in agreement with results from planetary nebulae (Henry 1989).

Figure 26.9 shows data for sulfur and argon. For log O/H < −3.5, S/O and Ar/O are

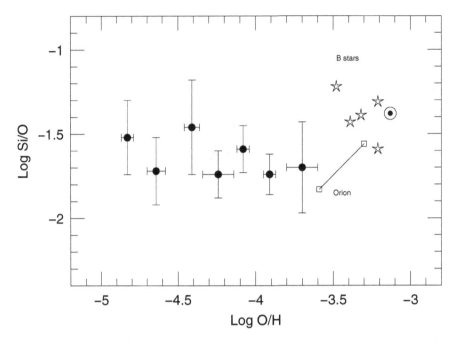

Fig. 26.10. Si/O abundance ratios in irregular galaxies (filled circles; from Garnett et al. 1995b). The open squares represent two different measurements for the Orion nebula, while the stars show averages for five samples of Galactic B stars.

essentially constant with O/H, and fall within the range predicted by Weaver & Woosley (1993). For log O/H > −3.5, however, there is evidence for declining S/O and Ar/O as O/H increases. The cause of the decline is not clear. It is possible that the ionization corrections for unseen S^{+3} have been underestimated in the more metal-rich H II regions. More observational study, especially IR spectroscopy, of metal-rich H II regions is needed to rule out ionization or excitation effects.

Because S and Ar are produced close to the stellar core, the yields of S and Ar may be sensitive to conditions immediately prior to and during the supernova explosion, such as explosive processing or fall-back onto the compact remnant (Weaver & Woosley 1993). If real, however, the declining S/O and Ar/O cannot be accounted for by simple variations in the stellar mass function and hydrostatic nucleosynthesis (Garnett 1989); some variation in massive star and/or supernova nucleosynthesis at high metallicities (perhaps due to strong stellar mass loss) may be needed.

26.3.5 Other Elements

Few other elements have been measured systematically in extragalactic H II regions over a wide range of O/H. Emission from [Fe II] and [Fe III] has been detected in a number of H II regions (Rodríguez 2002), but iron is severely depleted onto grains. Silicon can be measured in H II regions through the UV Si III] doublet at λλ1883, 1892 Å. Figure 26.10 shows Si/O measurements from Garnett et al. (1995b) for a small sample of metal-poor galaxies, along with data for several samples of B stars and the Orion nebula. Si/O appears to be roughly constant but smaller than the average for the Sun and the solar neighborhood B stars. Silicon is certainly depleted onto grains in the ISM, and the results in Figure 26.10 are consistent with a Si depletion of about −0.2 to −0.4 dex in the H II region sample. This is probably appropriate for low-density ionized gas (Sofia, Cardelli, & Savage 1994).

Iron has a variety of emission lines from [Fe II] and [Fe III] in the optical spectrum. The [Fe III] λ4658 Å is often observed in extragalactic H II regions. Izotov & Thuan (1999) measured [Fe III] in several metal-poor emission-line galaxies and derived Fe abundances. They obtained Fe/O ratios that were similar to the values found for metal-poor stars in the Galactic halo and used this to argue that the their emission-line galaxies were very young. However, Fe is greatly depleted onto grains in the ISM, so it is improbable that O/Fe ratios in H II regions can be used to interpret the enrichment history of the ISM without better understanding of the depletion factors.

26.4   Open Questions and Concluding Remarks

In conclusion, I would like to enumerate a few questions regarding the chemical evolution of galaxies that seem to need further investigation.

- What are the primary mechanisms determining the shape and slope of abundance gradients in spiral galaxies? We have seen that chemical evolution models tend to predict that composition gradients should get shallower in the inner disks spirals, but this has not been observed in real galaxies. Does viscous evolution play a role in maintaining an exponential gradient, by transferring new gas into the inner disk? This is essentially a hydrodynamical problem.
- How homogeneous are abundances in galaxies at a given place and time? Conflicting studies have argued for significant ($\pm$0.2–0.3 dex) variations in abundances on small scales, or for a very homogeneous composition ($<$ 0.1 dex variations). The question is relevant to the time scales for cooling and mixing of stellar ejecta with the ambient ISM. Most chemical evolution models assume instantaneous mixing, but is this a good approximation?
- Is infall of gas presently occuring in galaxies, and how does the rate evolve with time? This is a big unknown, since most chemical evolution models use relatively slow, ongoing infall of metal-poor gas to suppress the fraction of metal-poor G dwarfs. Observational evidence for classical infall is sketchy. The high-velocity clouds seen at high latitudes may represent infall, or may be part of a Galactic fountain flow, or may be associated with interaction between the Galaxy and its satellite galaxies.
- What galaxies may be losing metals to the intergalactic medium, and how much do they contribute? Is there a threshold mass above which galaxies retain metals?
- Does galaxy environment influence composition? The Virgo cluster studies need to be followed up by larger samples in a wider variety of cluster environments. The spiral-rich Ursa Major cluster and the Coma cluster are two obvious choices for continued study.
- Is there zero metallicity gas (or stars for that matter) in galaxies? Pre-enrichment by an initial stellar Population III also provides a solution to the G-dwarf problem and to the origin of metals seen in Lyα forest clouds. What is the composition of the huge gas reservoirs in the outer parts of spirals and irregulars?

We are seeing great improvements in the study of abundances in nearby galaxies, particularly with the new space-based UV and IR observatories, which are greatly improving the data for elements besides oxygen. With these new data, we are in a better position to connect the present-day abundance patterns in galaxies with those observed in high-redshift gas clouds. Eventually, emission-line spectroscopy of distant galaxies should greatly expand our information on heavy element abundances at early times, and allow us to trace the evolution of metallicity in the universe in greater detail. This will be an important complement to the absorption-line work on damped Lyα systems, as the connection between the emission-line gas and the stellar component is much clearer, and the disk component can be sampled more completely than with absorption studies.

Many of the questions mentioned reference hydrodynamics. Star formation, the evolution of galaxies, and stellar nucleosynthesis all involve hydrodynamics at fundamental levels, so improved hydrodynamical modeling of all of these phenomena should ultimately lead to a better understanding of galaxy evolution. Determining the mechanisms that connect abundances to galaxy structure should provide a continuing challenge to galaxy evolution theorists.

References

Arnett, D. 1996, Supernovae and Nucleosynthesis (Princeton: Princeton Univ. Press)

Barnes, J. 1991, in IAU Symp. 126, Dynamics of Galaxies and Their Molecular Cloud Distributions, ed. F. Combes & F. Casoli (Dordrecht: Kluwer), 363

Castellanos, M., Díaz, A. I., & Terlevich, E. 2002, MNRAS, 329, 315

Cayatte, V., Kotanyi, C., Balkowski, C., & van Gorkom, J. H. 1994, AJ, 107, 1003

Clarke, C. J. 1989, MNRAS, 238, 283

Combes, F. 1998, in Abundance Profiles: Diagnostic Tools for Galaxy History, ed. D. Friedli et al. (San Francisco: ASP), 300

Díaz, A. I., Terlevich, E., Vílchez, J. M., Pagel, B. E. J., & Edmunds, M. G. 1991, MNRAS, 253, 245

Dressler, A. 1980, ApJ, 236, 351

Dufour, R. J., Garnett, D. R., & Shields, G. A. 1988, ApJ, 332, 752

Dufour, R. J., Shields, G. A., & Talbot, R. J., Jr. 1982, ApJ, 252, 461

Edmunds, M. G. 1990, MNRAS, 246, 678

Ellison, S. L., Songaila, S., Schayes, J., & Pettini, M. 2000, AJ, 120, 1175

Ferguson, A. M. N., Gallagher, J. S., & Wyse, R. F. G. 1998, AJ, 116, 673

Friedli, D., Benz, W., & Kennicutt, R. C., Jr. 1994, ApJ, 430, L105

Garnett, D. R. 1989, ApJ, 345, 282

——. 1990, ApJ, 363, 142

——. 2002, ApJ, 581, 1019

Garnett, D. R., Dufour, R. J., Peimbert, M., Torres-Peimbert, S., Shields, G. A., Skillman, E. D., Terlevich, E., & Terlevich, R. J. 1995b, ApJ, 449, L77

Garnett, D. R., & Shields, G. A. 1987, ApJ, 317, 82

Garnett, D. R., Shields, G. A., Peimbert, M., Torres-Peimbert, S., Skillman, E. D., Dufour, R. J., Terlevich, E., & Terlevich, R. J. 1999, ApJ, 513, 168

Garnett, D. R., Shields, G. A., Skillman, E. D., Sagan, S. P., & Dufour, R. J. 1997, ApJ, 489, 63

Garnett, D. R., Skillman, E. D., Dufour, R. J., Peimbert, M., Torres-Peimbert, S., Terlevich, E., Terlevich, R. J., & Shields, G. A. 1995a, ApJ, 443, 142

Gunn, J. E., & Gott, J. R. 1972, ApJ, 176, 1

Gustafsson, B., Karlsson, T., Olsson, E., Edvardsson, B., & Ryde, N. 1999, A&A, 342, 426

Hale, G. M. 1998, in Stellar Evolution, Stellar Explosions, and Galactic Chemical Evolution, ed. A. Mezzacappa (Bristol: Institute of Physics), 17

Henry, R. B. C. 1989, MNRAS, 241, 453

Henry, R. B. C., Balkowski, C., Cayatte, V., Edmunds, M. G., & Pagel, B. E. J. 1996, MNRAS, 293, 635

Henry, R. B. C., Pagel, B. E. J., & Chincarini, G. L. 1994, MNRAS, 266, 421

Iben, I., Jr., & Truran, J. W., Jr. 1978, ApJ, 220, 980

Izotov, Y. I., & Thuan, T. X. 1998, ApJ, 500, 188

——. 1999, ApJ, 511, 639

Kennicutt, R. C., Jr., Bresolin, F., & Garnett, D. R. 2003, ApJ, 591, 801
Kennicutt, R. C., Jr., & Garnett, D. R. 1996, ApJ, 456, 504
Kinkel, U., & Rosa, M. 1994, A&A, 282, 37
Kobulnicky, H. A., & Skillman, E. D. 1996, ApJ, 471, 211
———. 1998, ApJ, 497, 601
Korn, A. J., Keller, S. C., Kaufer, A., Langer, N., Przybilla, N., Stahl, O., & Wolf, B. 2002, A&A, 385, 143
Langer, N., Fliegner, J., Heger, A., & Woosley, S. E. 1997, Nucl. Phys. A, 621, 457
Lee, H., McCall, M. L., Kingsburgh, R. L., Ross, R., & Stevenson, C. C. 2003a, AJ, 125, 146
Lee, H., McCall, M. L., & Richer, M. 2003b, AJ, 125, 2975
Lequeux, J., Rayo, J. F., Serrano, A., Peimbert, M., & Torres-Peimbert, S. 1979, A&A, 80, 155
Lin, D. N. C., & Pringle, J. E. 1987, ApJ, 320, L87
Lu, L., Sargent, W. L. W., & Barlow, T. A. 1998, AJ, 115, 55
Maeder, A. 1992, A&A, 264, 105
Marconi, G., Matteucci, F., & Tosi, M. 1994, MNRAS, 270, 35
Martin, C. L. 1999, ApJ, 513, 156
Martin, P., & Roy, J.-R. 1994, ApJ, 424, 599
Mo, H., Mao, S., & White, S. D. M. 1998, MNRAS, 295, 319
Monteverde, M. I., Herrero, A., Lennon, D. J., & Kudritzki, R.-P. 1997, ApJ, 474, 107
Olive, K. A., Skillman, E. D., & Steigman, G. 1997, ApJ, 483, 788
Peimbert, A. 2003, ApJ, 584, 735
Peimbert, M., Peimbert, A., & Torres-Peimbert, S. 2000, ApJ, 541, 688
Peimbert, M., Pe na, M. & Torres-Peimbert, S. 1986, PASP, 98, 1032
Peimbert, M., & Spinrad, H. 1970, ApJ, 159, 809
Peimbert, M., & Torres-Peimbert, S. 1974, ApJ, 193, 327
Prochaska, J. X., Henry, R. B. C., O'Meara, J. M., Tytler, D., Wolfe, A. M., Kirkman, D., Lubin, D., & Suzuki, N. 2002, PASP, 114, 933
Pettini, M., Lipman, K., & Hunstead, R. W. 1995, ApJ, 451, 100
Prantzos, N., & Boissier, S. 2000, MNRAS, 313, 338
Rodríguez, M. 2002, A&A, 389, 556
Roy, J.-R.., & Walsh, J. R. 1997, MNRAS, 288, 715
Rudnick, G., & Rix, H.-W. 1998, AJ, 116, 1163
Searle, L. 1971, ApJ, 168, 327
Searle, L., & Sargent, W. L. W. 1972, ApJ, 173, 25
Skillman, E. D., Kennicutt, R. C., Jr., & Hodge, P. W. 1989, ApJ, 347, 875
Skillman, E. D., Kennicutt, R. C., Jr., Shields, G. A., & Zaritsky, D. 1996, ApJ, 462, 147
Sofia, U. J., Cardelli, J. A, & Savage, B. D. 1994, ApJ 430, 650
Songaila, A. 2001, ApJ, 561, L153
Steigman, G. 2003, in Cosmochemistry: The Melting Pot of the Elements (XIIIth Canary Islands Winter School on Astrophysics), ed. C. Esteban, R. Garcia, & A. Herrero (Cambridge: Cambridge Univ. Press), in press
Thuan, T. X., Izotov, Y. I., & Lipovetsky, V. A. 1995, ApJ, 445, 108
Tinsley, B. M. 1979, ApJ, 229, 1046
van Zee, L., Salzer, J. J., Haynes, M. P., O'Donoghue, A. A. & Balonek, T. J. 1998, AJ, 116, 2805
Vila-Costas, M. B., & Edmunds, M. G. 1992, MNRAS, 259, 121
Warmels, R. H. 1988, A&AS, 72, 427
Weaver, T. A., & Woosley, S. E. 1993, Phys. Rep., 227, 65
Zaritsky, D. 1995, ApJ, 448, L17
Zaritsky, D., Kennicutt, R. C., Jr., & Huchra, J. P. 1994, ApJ, 420, 87

27

Chemical composition of the intracluster medium

MICHAEL LOEWENSTEIN
NASA/Goddard Space Flight Center and University of Maryland

Abstract
Clusters of galaxies are massive enough to be considered representative samples of the Universe, and to retain all of the heavy elements synthesized in their constituent stars. Since most of these metals reside in hot plasma, X-ray spectroscopy of clusters provides a unique and fundamental tool for studying chemical evolution. I review the current observational status of X-ray measurements of the chemical composition of the intracluster medium, and its interpretation in the context of the nature and history of star and galaxy formation processes in the Universe. I provide brief historical and cosmological contexts, an overview of results from the mature *ASCA* observatory database, and new results from the *Chandra* and *XMM-Newton* X-ray observatories. I conclude with a summary of important points and promising future directions in this rapidly growing field.

27.1 Introduction

27.1.1 Rich Clusters of Galaxies and their Cosmological Setting

Rich clusters of galaxies, characterized optically as concentrations of hundreds (or even thousands) of galaxies within a region spanning several Mpc, are among the brightest sources of X-rays in the sky, with luminosities of up to several 10^{45} erg s^{-1}. The hot intracluster medium (ICM) filling the space between galaxies has an average particle density $\langle n_{ICM} \rangle \approx 10^{-3}$ cm^{-3} and an electron temperature ranging from 20 to > 100 million K. Interpreted as virial temperatures, these correspond to masses of $\sim (1-20) \times 10^{14}\ M_\odot$; rich clusters are believed to be the largest gravitationally bound structures in the Universe. They are also notable for their high fraction (typically $\sim 75\%$) of member galaxies of early-type morphology, as compared to galaxy groups or the field. Within the framework of large-scale structure theory, rich clusters arise from the largest fluctuations in the initial random field of density perturbations, and their demographics are sensitive diagnostics of the cosmological world model and the origin of structure (e.g., Schuecker et al. 2003). Rich galaxy clusters are rare and (including their dark matter content) account for less than 2% of the critical density, ρ_{crit}, characterizing a flat Universe.

Embedded in the ICM of rich clusters—where 70%–80% of cluster metals reside— lies a uniquely accessible fossil record of heavy element creation. To the extent that the cluster galaxy stars where these metals were synthesized are representative, measurement of ICM chemical abundances provides constraints on nucleosynthesis—and, by extension, the epoch, duration, rate, efficiency, and initial mass function (IMF) of star formation—in the

Table 27.1. *Mass-to-Light Ratios and Mass Fractions*

Parameter	Universe	Clusters
$\langle M_{\text{total}}/L_B \rangle$	270	300
$\langle M_{\text{stars}}/L_B \rangle$	3.5	4
$\langle M_{\text{gas}}/L_B \rangle$	41	35
f_{baryon}	0.17	0.13
f_{stars}	0.013	0.013
f_{gas}	0.15	0.12
stars/gas	1/12	1/9

Universe. From this perspective it is useful to take an inventory of clusters, and compare with the Universe as a whole.

Consideration of recent results from the *Wilkinson Microwave Anisotropy Probe* supports a standard cosmological model wherein, to high precision, the average matter density totals $0.27\rho_{\text{crit}}$ in a flat Universe, and baryons amount to $0.044\rho_{\text{crit}}$ (Spergel et al. 2003). The estimate of Fukugita, Hogan, & Peebles (1998) of the total density in stars, $\sim 0.0035\rho_{\text{crit}}$, is corroborated by recent constraints based on the extragalactic background light (Madau & Pozzetti 2000), and the Two Micron All-Sky Survey and Sloan Digital Sky Survey (Bell et al. 2003). Since critical density corresponds to a mass-to-light ratio (*B* band) $M/L_B \approx 1000$, the cluster matter inventory compares to the Universe as a whole as indicated in Table 27.1.

Deviations from the typical rich cluster values displayed in Table 27.1 are found for both the total mass-to-light ratio and the baryonic contributions (Ettori & Fabian 1999; Mohr, Mathiesen, & Evrard 1999; Bahcall & Comerford 2002; Girardi et al. 2002; Lin, Mohr, & Stanford 2003)—indicative of differences and uncertainties in assumptions, method of calculation, and in extrapolation to the virial radius, as well as possible cosmic variance. However, it is clear that, at least to first order, observations are consistent with the theoretical expectation (e.g., Evrard 1997) that these largest virialized structures are "fair samples" of the Universe in terms of their mix of stars, gas, and dark matter. (A corollary of this is that bulge populations generally dominate the stellar mass budget in the field, as well as in clusters.) While clusters were the first systems identified with baryon budgets dominated by a reservoir of hot gas (White et al. 1993), this is now believed to apply to the Universe as a whole at past and present epochs (Davé et al. 2001; Finoguenov, Burkert, & Bohringer 2003).

An important *caveat* is that, since rich clusters do represent regions of largest initial over-density, star formation may initiate at higher redshift, proceed with enhanced efficiency, or be characterized by an IMF skewed toward higher mass stars when compared to more typical regions. If so, there is an opportunity to search for possible variations in star formation with epoch or environment, given a suitable class of objects for comparison.

The intergalactic medium in *groups* of galaxies may comprise one such sample. Groups generally include $\sim 2 - 50$ member galaxies, emit at X-ray luminosities $< 10^{44}$ erg s^{-1}, and have electron temperatures < 20 million K, corresponding to mass scales up to $\sim 10^{14}$ $M_{\odot}$ (Mulchaey 2000). Groups present their own unique interpretive challenges. They may not behave as closed boxes and evidently display a spread in their mass inventories, metallicities, and morphological mix of galaxies that reflect the theoretically expected cosmic variance in formation epoch and evolution (Davis, Mulchaey, & Mushotzky 1999; Hwang et al. 1999). However, since extending consideration to the poorest groups encompasses most of the galaxies (and stars) in the Universe, it is crucial to study the chemical composition of the intragroup, as well as the intracluster, medium.

27.1.2 Advantages of X-ray Wavelengths for Abundance Studies

From both scientific and practical perspectives, X-ray spectroscopy is uniquely well suited for studying the chemical composition of the Universe. Most of the metals in the Universe are believed to reside in the intergalactic medium; this is demonstrably true for rich galaxy clusters (e.g., Finoguenov et al. 2003). The concordance in mass breakdown between clusters and the Universe discussed above implies that, because of their deep potential wells, clusters, unlike most, if not all, galaxies, are "closed boxes" in the chemical evolution sense. Thus, modelers are provided with an unbiased and complete set of abundances that enables extraction of robust constraints on the stellar population responsible for metal enrichment.

For high temperatures, such as those found in the ICM, the shape of the thermal continuum emission yields a direct measurement of the electron temperature, and thus the ionization state. Complications arising from depletion onto dust grains, optical depth effects, and uncertain ionization corrections are minimal or absent. The energies of K-shell ($\rightarrow n = 1$) and/or L-shell ($\rightarrow n = 2$) transitions for *all* of the abundant elements synthesized after the era of Big Bang nucleosynthesis lie at wavelengths accessible to modern X-ray astronomical telescopes and instruments. The strongest ICM emission lines arise from well-understood H- and He-like ions, and line strengths are immediately converted into elemental abundances via spectral fitting. Of course, the quality and usefulness of X-ray spectra are limited by the available sensitivity and spectral resolution. In the following sections, I detail how the rapid progression in the capabilities of X-ray spectroscopy drives the evolution of our understanding of intracluster enrichment and its ultimate origin in primordial star formation in cluster galaxies. New puzzles are revealed with every insight emerging from subsequent generations of X-ray observatories, a situation that will surely continue with future missions.

27.2   Earliest Results on Cluster Enrichment

The 6.7 keV He-like Fe Kα emission line is the most easily detectable feature in ICM X-ray spectra, because it arises from a high-oscillator strength transition in an abundant ion and lies in a relatively isolated wavelength region. It was first detected in the *Ariel V* spectrum of the Perseus cluster (Mitchell et al. 1976), and *OSO-8* observations of the Perseus, Coma, and Virgo Clusters (Serlemitsos et al. 1977). Spectroscopic analysis of ~ 30 clusters (many observed with the *HEAO-1 A-2* satellite) revealed the ubiquity of this feature at a strength generally indicating an ICM Fe abundance of one-third to one-half solar (Mushotzky 1984). These early results demonstrated the origin of cluster X-ray emission as a thermal primordial plasma enriched by material processed in stars and ejected in galactic winds. Although not immediately realized, this provided the first indications of the prodigious magnitude of galactic outflows, since the measured amount of ICM Fe was of the same order as the sum of the Fe in all of the stars in all of the cluster galaxies.

Prior to 1993, Fe was the only element accurately measured in a large number of clusters, although some pioneering measurements of O, Mg, Si, and S features were obtained with the Solid State Spectrometer and Focal Plane Crystal Spectrometer aboard the *Einstein* X-ray Observatory (see Sarazin 1988 and references therein for these results, as well as a complete survey of the field prior to the the launch of the *ROSAT* and *Advanced Satellite for Cosmology and Astrophysics (ASCA)* X-ray observatories). Given the ambiguous origins of Fe, known to be synthesized in comparably large quantities by both Type Ia and Type II supernovae (henceforth, SNe Ia and SNe II) integrated over the history of the Milky Way, this represented a serious impediment to interpreting the data in terms of star formation history. Fortunately, the launch of *ASCA* provided the means to utilize the ICM as a repository of information on historical star formation and element creation.

27.3 ICM Abundances from the *ASCA* Database

With the combination of modest imaging capability, good spectral resolution, and large collecting area over a broad X-ray bandpass realized by its four telescope/detector pairs, the *ASCA* X-ray Observatory ushered in a new era of X-ray spectroscopy of astrophysical plasmas and revolutionized cluster elemental abundance studies (Arnaud et al. 1994; Bautz et al. 1994; Fukazawa et al. 1994). The clusters with accurate Fe abundance determinations cover a range of redshifts out to $z \approx 0.5$. This enabled the first investigation of the evolution of [Fe/H], as needed to understand the epoch and mechanism of enrichment (Mushotzky & Loewenstein 1997). The first results on gradients in ICM metallicity and abundance ratios were obtained (see §27.5), and the first significant sample of α-element abundances were measured with *ASCA*. Well over 400 observations of galaxy clusters and groups were made over the 1993–2000 duration of the *ASCA* mission. Because the *ASCA* data archive is now complete, and a comparable database based on observations with the current generation of X-ray observatories is many years away, an assessment of *ASCA* ICM enrichment results is timely.

The *ASCA* Cluster Catalog (ACC) project (Horner et al. 2004), initiated by K. Gendreau (NASA/GSFC), is the subject of the University of Maryland Ph.D. theses of D. Horner and W. Baumgartner, and includes contributions from R. Mushotzky, C. Scharf, and the author. Its overarching goal is to perform uniform and semi-automated spectral analysis on the integrated X-ray emission from every target in the *ASCA* cluster observational database, utilizing the most current processing and calibration. *ASCA* observed 434 galaxy clusters and groups in 564 pointings. Among these, 273 clusters are deemed suitable for spectroscopy, superseding previous samples (e.g., White 2000) by more than a factor of 2.

27.3.1 *Fe Abundance Trends*

The Fe abundance shows a remarkable uniformity in rich clusters, with a distribution that sharply peaks at ~ 0.4 solar (where solar Fe abundance is defined such that log (Fe/H) + 12 = 7.5; Grevesse & Sauval 1998). The Fe abundance tends to be higher in clusters with the highest central densities and shortest cooling times (see also Allen & Fabian 1998). It is unclear from these data alone whether this is an artifact of the presence of Fe gradients: since the X-ray emissivity is proportional to the gas density squared, X-ray spectra of more concentrated ICM are disproportionately weighted toward the more abundant central regions. There are also systematic trends with electron temperature, with Fe tending to be overabundant with respect to the average defined by the hottest clusters, for clusters with ICM temperatures $2 \text{ keV} < kT < 4 \text{ keV}$, and underabundant for $kT < 2 \text{ keV}$.

ASCA data revealed evidence for a *lack* of evolution in Fe abundance out to redshift 0.4 (Mushotzky & Loewenstein 1997; Rizza et al. 1998; Matsumoto et al. 2000, 2001) and no

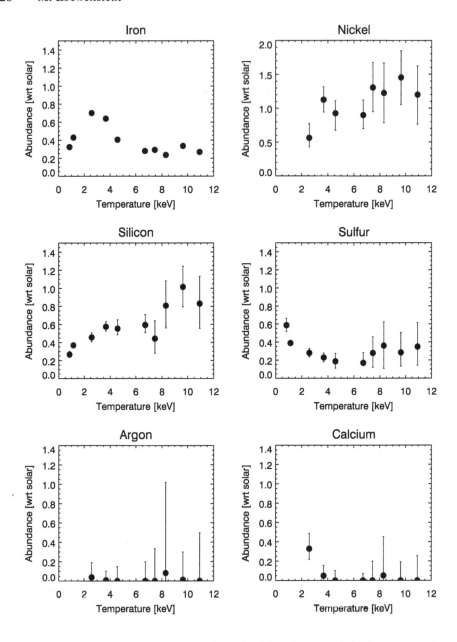

Fig. 27.1. Average abundances with respect to the solar abundances of Grevesse & Sauval (1998) as a function of temperature. 90% confidence errors are shown. (From Baumgartner et al. 2004.)

evidence of a decline at higher redshift (Donahue et al. 1999). *XMM-Newton* and *Chandra* measurements are further expanding the redshift range of precise cluster Fe abundance measurements (Jeltema et al. 2001; Arnaud et al. 2002; Maughan et al. 2003; Tozzi et al. 2003). On average, rich cluster Fe abundances are invariant out to $z \approx 0.8$, and show no sign of decline out to $z \approx 1.2$. The epoch of cluster enrichment is yet to be identified. Most of the star formation in the Universe occurred at $z > 1$ (Madau et al. 1996; Lanzetta et al. 2002; Dickinson et al. 2003), and fundamental plane considerations imply that most stars in clusters formed at $z > 2$ (van Dokkum et al. 1998; Jorgensen et al. 1999). Therefore, one expects *synthesis* of most cluster metals prior to the redshifts where they can presently be observed. However, it does not necessarily follow that the metals are in place in the ICM at such early epochs. As observations push back the enrichment era, scenarios where ICM metal injection from galaxies rapidly follows their synthesis during the early period of active star formation are favored over those with delayed metal release from galaxies, for example, via ongoing SN Ia-driven winds or stripping of enriched galaxy gas halo.

27.3.2 Metallicities, Abundance Patterns, and their Interpretation

Although often blended, emission features from O, Ne, Mg, Si, S, Ar, Ca, Fe, and Ni were measured in *ASCA* spectra of individual clusters. However, most of these are not detectable for the vast majority of the 273 observed systems. In order to obtain abundances with small statistical uncertainties, and smooth over both instrumental and astrophysical systematic effects and biases, Baumgartner et al. (2004) undertook joint analyses of multiple observations grouped into "stacks" of 13–47 clusters (Baumgartner et al. 2004) according to temperature (1 1 keV-wide bins centered on temperatures from 0.5–10.5 keV) and metallicity (high- and low-abundance bins). Ultimately, the two abundance stacks were merged for each temperature, and mean Si, S, Ar, Ca, Fe, and Ni abundances proved robust and reliable.

The results of the stacking analysis are summarized as follows (Fig. 27.1). For rich clusters ($kT > 4$ keV), the ratio of Si to Fe is $1.5 - 3$ times solar and displays an increasing trend with temperature. However, in contrast, the S-to-Fe ratio is solar or less (i.e. Si/S ≈ 3 times solar), and Ar and Ca are markedly subsolar with respect to Fe. The Ni-to-Fe ratio is $3 - 4$ times solar in these systems. This confirms and generalizes previous results (Mushotzky et al. 1996; Fukazawa et al. 1998; Dupke & White 2000a; Dupke & Arnaud 2001). Intermediate-to-poor clusters (2 keV $< kT <$ 4 keV) have subsolar ratios of both Si and S, with respect to Fe; however, the Si-to-S ratio remains high at about twice solar. This trend with temperature was previously identified by Fukazawa et al. (1998) and by Finoguenov, David, & Ponman (2000). The trend of higher α/Fe ratios in the hotter clusters with lower Fe/H (Fig. 27.2) is reminiscent of trends in Galactic stars and suggests the existence of ICM enrichment by a stellar population that is itself promptly enriched (i.e. by SNe II), as inferred for stars in bulges and elliptical galaxies.

Baumgartner et al. (2004) compare these ratios with those in various other systems, including the Galactic thin (Fig. 27.3) and thick disks, damped Lyα systems, Lyman-break galaxies, and the Lyα forest. In general, the high Si/Fe is not out of line with other systems of comparable Fe/H inferred to undergo rapid and efficient conversion of gas into stars. However, the low ratios of Ca to Fe and S to Si, and the high ratio of Ni to Fe evidently are unique to the ICM: a clear analog among observed classes of systems to the population responsible for enriching the ICM is not evident.

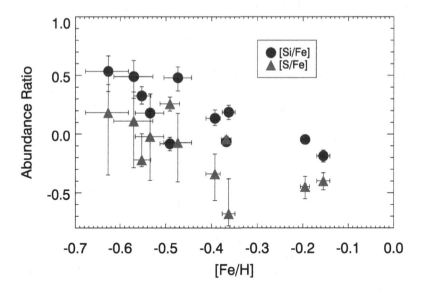

Fig. 27.2. Si/Fe and S/Fe ratios vs. Fe/H ratio, expressed as the logarithm with respect to solar. (From Baumgartner et al. 2004.)

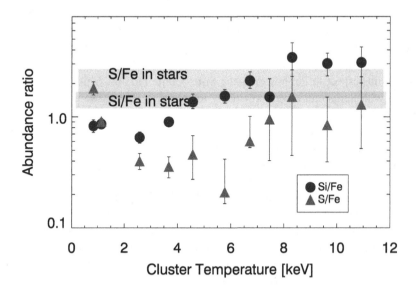

Fig. 27.3. Si/Fe and S/Fe ICM abundances vs. *kT* compared to those found in the Milky Way from Timmes, Woosley, & Weaver (1995). The range of Galactic Si/Fe is illustrated with the lower shaded rectangle, S/Fe with the upper, overlapping rectangle (Baumgartner et al. 2004).

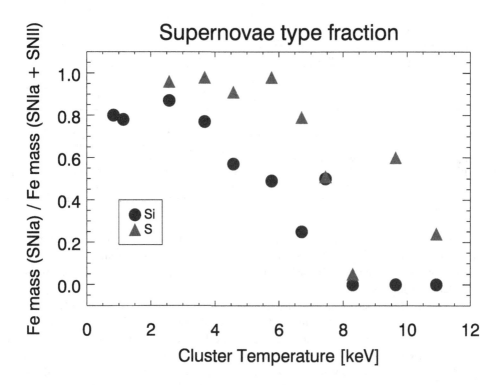

Fig. 27.4. Decomposition into SNe Ia and SNe II, using standard yields and the observed Si/Fe and S/Fe ratios, expressed in terms of the fraction of Fe from SNe Ia. (From Baumgartner et al. 2004.)

27.4 On the ICM Abundance Anomalies

Since bulges account for $\sim 90\%$ of the stellar mass in clusters, it is straightforward to estimate ICM enrichment under standard assumptions. Consider a simple stellar population with a local IMF (e.g., Kroupa 2001), assume that stars more massive than 8 $M_\odot$ explode as SNe II, that the rate of SN Ia explosions is as determined in nearby elliptical galaxies (0.16 SNU, where 1 SNU $\equiv$ 1 SN per 10^{10} solar blue luminosities per century; Cappellaro, Evans, & Turatto 1999) and injected for a duration of 10^{10} yr, and utilize SNe II and SNe Ia ("W7" model) yields from Nomoto et al. (1997a, b) with the former averaged over a standard IMF extending to 40 $M_\odot$ as in Thielemann, Nomoto, & Hashimoto (1996). The fractions of Fe originating in SNe Ia separately inferred from the Si/Fe and S/Fe ratios are shown in Figure 27.4, illustrating that *no* combination of standard SNe Ia and SNe II can explain the ICM abundance pattern. Baumgartner et al. (2004) demonstrate that this holds regardless of which published SN II yields are considered (see also Gibson, Loewenstein, & Mushotzky 1997; Loewenstein 2001).

In addition to ratios, it is instructive to consider the ICM mass in metals, normalized to the total starlight in cluster galaxies. Table 27.2 compares the mass-to-light ratios of the elements addressed in the stacking analysis, computed using the standard assumptions detailed above, with the observed averages in three ICM temperature regimes. The predicted SN II

Table 27.2. *Metal Mass-to-Light Ratios*

	Predicted				Observed		
	SNe II	SNe Ia W7	SNe Ia WDD1	SNe II W7	$kT > 8$ (keV)	$4 < kT < 8$ (keV)	$2 < kT < 4$ (keV)
Si	9.3×10^{-3}	2.4×10^{-4}	5.6×10^{-4}	9.3×10^{-3}	2.3×10^{-2}	1.3×10^{-2}	7.5×10^{-3}
S	3.1×10^{-3}	1.4×10^{-4}	3.4×10^{-4}	3.3×10^{-3}	4.5×10^{-3}	2.5×10^{-3}	2.3×10^{-3}
Ar	6.1×10^{-4}	2.6×10^{-4}	6.9×10^{-5}	6.3×10^{-4}	$< 9.6 \times 10^{-4}$	$< 5.5 \times 10^{-4}$	...
Ca	4.5×10^{-4}	2.0×10^{-4}	6.5×10^{-5}	4.7×10^{-4}	$< 6.6 \times 10^{-4}$	$< 3.8 \times 10^{-4}$	...
Fe	6.9×10^{-3}	1.2×10^{-3}	8.7×10^{-4}	8.1×10^{-3}	1.4×10^{-2}	1.5×10^{-2}	1.7×10^{-2}
Ni	4.5×10^{-4}	2.3×10^{-4}	5.7×10^{-5}	6.8×10^{-4}	3.1×10^{-3}	2.7×10^{-3}	...

and SN Ia contributions are shown separately, as well as their sum. For comparison, the predictions for a delayed detonation SN Ia model ("WDD1," Nomoto et al. 1997b) are shown along with those using the standard W7 model; use of the former generally exacerbates the problems detailed below.

There are a number of notable discrepancies. Both Fe and Ni are generally underpredicted, as is Si at high temperatures. One could achieve reconciliation for Fe and Si by roughly doubling the numbers of SNe Ia and SNe II, alterations that are not unreasonable since the SN Ia rate may be greater in the past and the IMF may be top-heavy (flat or bimodal; e.g., Loewenstein & Mushotzky 1996). One could then, perhaps, explain the Si/Fe trend with ICM temperature as selective mass loss of SN II products in systems with shallower gravitational potential wells. However, such a scenario overpredicts S, Ar, and Ca and underpredicts Ni.

The large observed amounts of Ni require a SN Ia rate, averaged over a Hubble time, of more than 10 times the estimated local rate, with a concomitant decrease in the number of SNe II to avoid overproducing Fe. However, this underpredicts Si. The problem is intractable using standard yields, unless a significant additional source of metals is considered.

What are some possible resolutions to this puzzle? The only explanation for the high Ni abundance is a high time-averaged SN Ia rate and/or a high SN Ia Ni yield—the rate must exceed $\sim 2(0.14/y_{\text{Ia,Ni}})$ SNU, where $y_{\text{Ia,Ni}}$ is the SN Ia Ni yield in solar masses. To account for those systems with high Si abundances, one could assume that SN II Si yields are twice those in the core collapse nucleosynthesis calculations currently in the literature. Alternatively, the Si overabundance could be explained if the number of massive SN II progenitor stars were approximately twice what a standard IMF predicts; but, in this case the S, Ar, Ca, and, perhaps, Fe yields require *ad hoc* reduction to avoid their overproduction.

If published yields are not grossly in error, one must appeal to an additional source of enrichment that preferentially synthesizes Si. These progenitors must be more plentiful, or their products more efficiently retained, in more massive clusters. An extensive search of the literature reveals several instances of SNe with the desired yield-pattern, including core collapse SN from very massive ($70\,M_\odot$) stars (Thielemann et al. 1996), massive ($\sim 30\,M_\odot$) metal-poor (~ 0.01 solar) stars (Woosley & Weaver 1995), or supermassive (with $70\,M_\odot$ He cores) metal-free stars that explode as hypernovae (Heger & Woosley 2002). While it is unlikely that a monolithic contribution from any of these objects truly explains the ICM

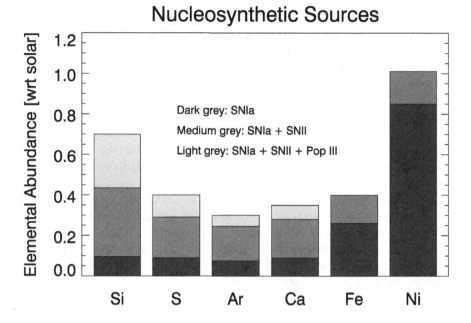

Fig. 27.5. Relative contributions to the enrichment of various elements in a typical clusters for a toy model including Population III hypernovae in addition to SNe Ia and SNe II in standard numbers with standard yields (Baumgartner et al. 2004).

abundance anomalies, their existence demonstrates that there are nucleosynthetic channels—likely associated with very massive and/or very metal-poor stars—that result in relative enhancements of Si. Figure 27.5 shows the respective contributions of SNe Ia, SNe II, and hypernovae (Heger & Woosley 2002) to each element in a toy model that reproduces a typical rich cluster ICM abundance pattern (Baumgartner et al. 2004).

There are now possible signatures of Population III and hypernovae in the level and pattern of enrichment in low-metallicity Milky Way stars (Umeda & Nomoto 2003), in the level and epoch of intergalactic medium enrichment as inferred from the Lyα forest (Songaila 2001), and in the reionization of the Universe (e.g., Wyithe & Loeb 2003). The number of hypernovae required in the ICM is reasonable in terms of the expected number of Population III progenitors (Loewenstein 2001).

27.5 *Chandra* and *XMM-Newton* **Results on Abundance Gradients**

ASCA data revealed clear evidence of central excesses of Fe in clusters with cooling flows and in clusters not involved in recent major mergers (Matsumoto et al. 1996; Ezawa et al. 1997; Ikebe et al. 1997, 1999; Xu et al. 1997; Kikuchi et al. 1999; Dupke & White 2000a, b), which are consistent with generally flat metallicity profiles outside of the central $\sim 2'$ (Tamura et al. 1996; Irwin, Bregman, & Evrard 1999; White 2000). Limitations in *ASCA* imaging precluded resolving gradients in Fe or drawing robust conclusions on the spatial distribution of α/Fe ratios, although there were indications that α elements may not display the central excess (Finoguenov et al. 2000; Fukazawa et al. 2000; Allen et al. 2001). *BeppoSAX* provided Fe profiles with better spatial resolution, and De Grandi & Molendi (2001) attributed the central Fe excess to injection by SNe Ia associated with the stellar population in the central cluster galaxy. If this is the case, milder or absent central excesses in SN II-synthesized α elements are expected, as claimed by Finoguenov et al. (2000).

These questions of gradients in abundance ratios are addressed through observations with the *Chandra* (Weisskopf et al. 2002) and *XMM-Newton* (Jansen et al. 2001) X-ray observatories. The former represents a huge leap in broad-bandpass imaging, with an angular resolution $\sim 0.''5$, the latter in collecting area (a total of ~ 5 times that of *ASCA*, with an order of magnitude better angular resolution). New results demonstrate that the central Fe excess is concentrated to within 100 kpc of the cluster center (Iwasawa et al. 2001; Kaastra et al. 2001; Lewis, Stocke, & Buote 2002; Smith et al. 2002), that there may be a central metallicity inversion (Johnstone et al. 2002; Sanders & Fabian 2002; Schmidt, Fabian, & Sanders 2002; Blanton, Sarazin, & McNamara 2003; Dupke & White 2003), and that abundance profiles are generally very flat from ~ 100 kpc out to a significant fraction of the virial radius (Schmidt, Allen, & Fabian 2001; Tamura et al. 2001). The apparent inversion may be an artifact of using an oversimplified model in spectral fitting (Molendi & Gastaldello 2001). The question of the presence of α/Fe gradients is not yet definitively settled (David et al. 2001), although early results indicate a mix of abundances in the central regions that is closer to the SN Ia pattern than is globally the case (Tamura et al. 2001; Ettori et al. 2002).

27.6   *XMM-Newton* **RGS Measurements of CNO**

The *XMM-Newton* Reflection Grating Spectrometer provides high-spectral resolution ($E/\Delta E$ ranges from 200 to 800 over the 0.35–2.5 keV RGS bandpass; den Herder et al. 2003) spectroscopy in the wavelength region containing the strongest X-ray features of the elements carbon through sulfur. Since individual lines are more distinctly resolved than in previous studies using CCD spectra (where $E/\Delta E \approx 50$), abundance determinations are much less sensitive to the assumed thermal emission model. C, N, and O lines are all widely accessible for the first time. These are unique probes of star formation and the era of ICM enrichment because of their distinctive nucleosynthetic origins. For an extended X-ray source such as a galaxy cluster, RGS spectra correspond to an emission-weighted average over the inner $\sim 1'$ (see Peterson et al. 2003)—within the central galaxy (if there is one) for a nearby cluster; thus, caution must be exercised in applying the results to the source as a whole.

27.6.1   *Oxygen*

The strong 0.65 keV OV III Lyα feature is measured with *XMM-Newton* in clusters out to redshifts as high as 0.3. The O/Fe ratio for the sample of cooling flow clusters in Peterson et al. (2003) is shown in Figure 27.6. [Note that Peterson et al. adopt the definition of solar abundances from Anders & Grevesse (1989), where log (Fe/H) + 12 = 7.67 and log (O/H) + 12 = 8.93, ~ 1.5 and ~ 1.7 times higher than currently more commonly used "cosmic abundances" of Fe and O, respectively.] O/Fe is generally subsolar, and overlaps with the range measured in Galactic stars of comparable Fe/H (Reddy et al. 2003). Low O abundances were previously reported in Abell 1795 (Tamura et al. 2001a), Abell 1835 (Peterson et al. 2001), Sèrsic 159-03 (Kaastra et al. 2001), Abell 496 (Tamura et al. 2001b),

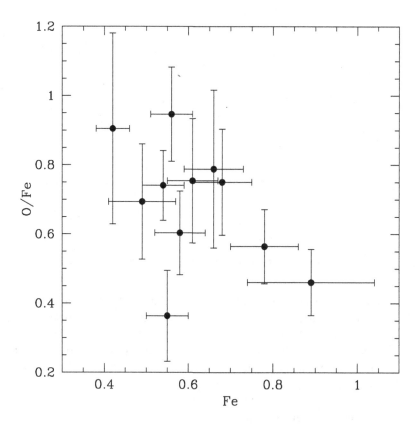

Fig. 27.6. O/Fe vs. Fe for cooling flow clusters. (Adapted from Peterson et al. 2003.)

NGC 5044, the central (elliptical) galaxy in the WP23 group (Tamura et al. 2003), and the elliptical galaxy NGC 4636 (Xu et al. 2002), which resembles a group in X-rays. The Galactic O/Fe-Fe/H trend is thought to reflect the delayed enrichment in Fe from SNe Ia relative to that of O from short-lived, massive SN II progenitors. In the intergalactic medium of clusters and groups, we may be seeing systematic variations in the relative stellar lock-up fractions of SN II and SN Ia products, relative retention of injected SN II and SN Ia material, or IMF. Low O/Si ratios in these systems were interpreted as a possible signature of enrichment by hypernovae associated with Population III (Loewenstein 2001). Intriguingly, as shown in Figure 27.7, Mg appears not be a good surrogate for O, as often assumed in studies of elliptical galaxies.

27.6.2 Carbon and Nitrogen

The Lyα lines of the H-like ions of C and N are weak compared to O, and fall in a less sensitive and less well-calibrated wavelength region of the RGS. As a result, there presently are few measurements of C and N: C and N in A 496 and M87 (Peterson et al. 2003) and N in the elliptical galaxies NGC 4636 (Xu et al. 2002) and NGC 5044 (Tamura et al. 2003). N is of particular interest, since its origin in intermediate-mass stars provides

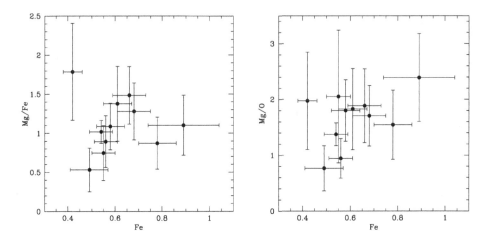

Fig. 27.7. Mg/Fe and Mg/O vs. Fe for cooling flow clusters. (From Peterson et al. 2003.)

crucial leverage in distinguishing the mass function of the enriching stellar population. In these objects C (when measured) and N are ~solar, and C/O and N/O are ~twice solar [solar as in Anders & Grevesse 1989: loG(C/H) + 12 = 8.56, log (N/H) + 12 = 8.05]. On a plot of N/O vs. O/H, these systems overlap with measurements in extragalactic H II regions (Pettini et al. 2002), where the secondary production of N dominates, but do extend to somewhat higher N/O at the same O/H.

27.7   *XMM-Newton* **Observations of M87**

Because of its brightness, M87 in the Virgo cluster is among the most extensively X-ray spectroscopically studied extended extragalactic objects. Its proximity enables one to extract detailed plasma conditions and composition, and to evaluate and correct for systematic effects. Previous work based on spectra from the *Einstein* (Canizares et al. 1982; Stewart et al. 1984; White et al. 1991; Tsai 1994a, b), *ROSAT* (Nulsen & Bohringer 1995; Buote 2001), and *ASCA* (Matsumoto et al. 1996; Hwang et al. 1997; Buote, Canizares, & Fabian 1999; Finoguenov & Jones 2000; Shibata et al. 2001) observatories is rapidly becoming superseded by *Chandra* and *XMM-Newton* data.

The RGS results for M87 show that, in the inner ~ 10 kpc, abundances of C, N, Ne, and Mg have best-fit values (relative to Anders & Grevesse 1989) of 0.7–1.0 solar, with 90% uncertainties of 0.2–0.3, while Fe is 0.77 ± 0.04 solar, and O 0.49 ± 0.04 solar (Sakelliou et al. 2002). With respect to the M87 stellar abundances (Milone, Barbuy, & Schiavon 2000), the α elements are apparently preferentially diluted with respect to Fe.

Gastaldello & Molendi (2002) derive detailed abundance distributions from analysis of *XMM-Newton* EPIC-MOS CCD data, obtaining O, Ne, Mg, Si, Ar, Ca, Fe, and Ni in 10 radial bins extending to $14'$. The MOS results are generally consistent with the RGS and indicate negative gradients on arcminute scales in all of the elements, with the possible exception of Ne. Gradients in α/Fe ratios are modest. Ratios with respect to Fe vary among the α elements: O/Fe, Ne/Fe, Mg/Fe, and S/Fe are subsolar; Si/Fe, Ar/Fe, and Ca/Fe are slightly supersolar. Ni/Fe increases from ~ 1.5 solar in the center to ~ 2.5 solar at 50 kpc.

Not surprisingly, Gastaldello & Molendi find that no unique combination of published SN Ia (either deflagration or delayed detonation models) and SN II yields is consistent with the full abundance pattern at any radius. Moreover, the combination of SN subtypes that provides the best fit varies with radius (i.e. delayed detonation models provide a better fit in the center, W7 further out where Ni/Fe is greater).

Finoguenov et al. (2002) utilize a cruder spatial division, thus obtaining higher precision for individual abundances (this is enhanced by their inclusion of *XMM-Newton* EPIC-PN CCD data). They propose explaining the observed abundance pattern and its radial variation with (1) a radially increasing SN II/SN Ia ratio, (2) high Si and S yields from SNe Ia (favoring delayed detonation models) and an *ad hoc* reduction in SN II S yields, and (3) a radial variation in SN Ia yields (corresponding to delayed detonation models with different deflagration-to-detonation transition densities). As they note, deflagration models must also play a role to account for the high Ni/Fe ratio found in other clusters. Whether such a complex model can coherently explain cluster abundance patterns in general is unclear. Matsushita, Finoguenov, & Bohringer (2003) follow this up with more sophisticated thermal modeling, and an analysis that includes individual line ratios, and they reach conclusions that are qualitatively similar in terms of the outwardly increasing importance of SNe II, the diversity of SNe Ia, and the yields of Si and S.

It is important to keep in mind that these new results are for the central ~ 70 kpc—still well within the influence of the M87 galaxy. The larger-scale cluster plasma (as studied in the ACC project discussed above) is not as profoundly affected by SNe Ia, and, indeed, there are indications of a transition to a more cluster-like abundance pattern at large radii in M87 (Finoguenov et al. 2002). In fact, the narrowing redshift window when the Fe in clusters must be synthesized and mixed into the ICM, as inferred from the lack of Fe abundance evolution, is becoming comparable to the time delay characteristic of many models of SN Ia explosions.

27.8   Concluding Remarks

The era of high-precision X-ray spectroscopy has arrived. Accurate and robust measurements can now be made of abundances and their variation in space and time, for a wide variety of elements in intracluster and intragroup media. This enables astronomers to utilize the ICM, where metals synthesized in cluster galaxy stars over the age of the Universe accumulated, as a laboratory for testing theories of the environmental dependence and impact of star formation.

27.8.1   Mean ICM Metallicities and Redshift Dependence

There are now hundreds of measurements, out to redshifts greater than 1, of Fe abundances and many tens of measurements (or significant upper limits) of Si, S, Ar, Ca, and Ni. The distribution function of Fe abundance outside of the central regions in rich clusters is sharply peaked around ~ 0.4 solar, and shows no evidence of a decline with increasing redshift out to $z \approx 1$. With the average Fe yield per current mass of stars ~ 3 times solar, a prodigious rapid enrichment is implied: a time average of $\sim 0.1\ M_\odot\ \mathrm{yr}^{-1}$ of Fe per (present-day) L_* galaxy integrated over 2 Gyr. This is equivalent to the nucleosynthetic yield of ~ 1 SN II yr^{-1} per L_* galaxy, corresponding to $\sim 100\ M_\odot\ \mathrm{yr}^{-1}$ per L_* galaxy forming stars with a Salpeter IMF, or ~ 20 SNU of SNe Ia! As a point of reference, note that the mean star formation rate in the Universe at high redshift, normalized to the cluster light, is ~ 0.02

$M_\odot$ yr^{-1} per L_* galaxy. So, despite a stellar fraction consistent with the universal average, several times the amount of metals that might be expected was synthesized in clusters. It is intriguing that Lyman-break galaxies were enriched to similar levels on a comparably brief time interval. In the extreme overdense regions that represent clusters of galaxies, the primordial IMF was skewed toward high masses, or an additional separate primordial population of massive stars was present. Clusters are thus strong candidates for the source of the missing metals at $z \approx 3$ (Finoguenov et al. 2003), even though only about 5% of the total universal baryon content resides in clusters.

27.8.2   *Abundance Gradients*

Centrally concentrated (< 100 kpc) excesses in Fe are common, and evidently exclusively occur in those clusters that can be characterized either as cooling flow clusters, as clusters with a massive central galaxy, or as clusters without recent major merging activity. An explanation of this phenomenon awaits additional *Chandra* and *XMM-Newton* results on the radial distribution of α elements. Early indications are that SNe Ia play a more important role at the location of central cluster galaxies, although other giant elliptical galaxies evidently did not retain such large masses of interstellar Fe. Metallicity gradients evidently are mild or absent beyond these central regions.

27.8.3   *Abundance Patterns*

There is more Si, Fe, and Ni than one might expect based on the observed stellar population. The relative abundances of elements measured in the ICM show clear departures from solar ratios and the abundance pattern in the Galactic disk and other well-studied galactic and protogalactic systems. The ICM abundance pattern shows systematic variations with ICM temperature (i.e. cluster gravitational potential well depth). These variations are apparent even in the relative abundances, which are not generally in solar ratios, among different α elements (O, Mg, Si, S, Ar, and Ca). Contribution from an additional, primordial source of metals may be required to finally explain these anomalies. The ICM Ni/Fe ratio is $2-4$ times solar, higher than in any other known class of object where this ratio is measured. Since both Fe and Ni are efficiently synthesized in SNe Ia, and given the importance of SNe Ia as fundamental probes of the cosmological world model, the origin of the Ni excess is clearly worthy of further investigation.

27.8.4   *The Future*

As the *Chandra* and *XMM-Newton* cluster databases mature, one can expect many more measurements of abundance gradients (of α elements, as well as Fe), tight constraints on the evolution of Si and O abundances out to $z > 0.4$ and Fe out to $z > 1$, and additional accurate measurements of C and N. *ASTRO-E2*, scheduled for launch in 2005, will provide true spatially resolved, high-resolution X-ray spectroscopy, which will yield cleaner measurements of N, O, Mg, and Ne, and their gradients.

Acknowledgements. I would like to acknowledge the organizers for their kind invitation to participate in this meeting, and for all of their efforts in assembling the conference and these proceedings. My gratitude extends to Richard Mushotzky and Wayne Baumgartner for their assistance in preparing this review.

References

Allen, S. W., & Fabian, A. C. 1998, MNRAS, 297, L63

Allen, S. W., Fabian, A. C., Johnstone, R. M., Arnaud, K. A., & Nulsen, P. E. J. 2001, MNRAS, 322, 589

Anders, E., & Grevesse, N. 1989, Geochim. Cosmochim. Acta, 53, 197

Arnaud, K. A. 1994, ApJ, 436, L67

Arnaud, M., et al. 2002, A&A, 390, 27

Bahcall, N. A., & Comerford, J. M. 2002, ApJ, 565, L5

Baumgartner, W. H., Loewenstein, M., Horner, D. J., & Mushotzky, R. F. 2004, ApJ, submitted

Bautz, M. W., Mushotzky, R., Fabian, A. C., Yamashita, K., Gendreau, K. C., Arnaud, K. A., Crew, G. B., & Tawara, Y. 1994, PASJ, 46, L131

Bell, E. F., McIntosh, D. H., Katz, N., & Weinberg, M. D. 2003, ApJ, 585, L117

Blanton, E. L., Sarazin, C. L., & McNamara, B. R. 2003, ApJ, 585, 227

Buote, D. A. 2001, ApJ, 548, 652

Buote, D. A., Canizares, C. R., & Fabian, A. C. 1999, MNRAS, 310, 483

Canizares, C. R., Clark, G. W., Jernigan, J. G., & Markert, T. H. 1982, ApJ, 262, 33

Cappellaro, E., Evans, R., & Turatto, M. 1999, A&A, 351, 459

Davé, R., et al. 2001, ApJ, 552, 473

David, L. P., Nulsen, P. E. J., McNamara, B. R., Forman, W., Jones, C., Ponman, T., Robertson, B., & Wise, M. 2001, ApJ, 557, 546

Davis, D. S., Mulchaey, J. S., & Mushotzky, R. F. 1999, ApJ, 511, 34

De Grandi, S., & Molendi, S. 2001, ApJ, 551, 153

den Herder, J.-W., et al. 2003, SPIE, 4851, 196

Dickinson, M., Papovich, C., Ferguson, H. C., & Budavàri, T. 2003, ApJ, 587, 25

Donahue, M., Voit, G. M., Scharf, C. A., Gioia, I. M., Mullis, C. R., Hughes, J. P., & Stocke, J. T. 1999, ApJ, 527, 525

Dupke, R. A., & Arnaud, K. A. 2001, ApJ, 548, 141

Dupke, R. A., & White, R. E. III 2000a, ApJ, 528, 139

——. 2000b, ApJ, 537, 123

——. 2003, ApJ, 583, L13

Ettori S., & Fabian A. C. 1999, MNRAS, 305, 834

Ettori S., Fabian A. C., Allen, S. W., & Johnstone, R. M. 2002, MNRAS, 305, 834

Evrard, A. E. 1997, MNRAS, 292, 289

Ezawa, H, Fukazawa, Y., Makishima, K., Ohashi, T., Takahara, F, Xu, H., & Yamasaki, N. Y. 1997, ApJ, 490, L33

Finoguenov, A., Burkert, A., & Bohringer, H. 2003, ApJ, 594, 136

Finoguenov, A., David, L. P., & Ponman, T. J. 2000, ApJ, 544, 188

Finoguenov, A., & Jones, C. 2000, ApJ, 539, 603

Finoguenov, A., Matsushita, K., Bohringer, H., Ikebe, Y., & Arnaud, M. 2002, A&A, 381, 21

Fukazawa, Y., et al. 1998, PASJ, 50, 187

Fukazawa, Y., Makishima, K., Tamura, T., Nakazawa, K., Ezawa, H., Ikebe, Y., Kikuchi, K., & Ohashi, T. 2000, MNRAS, 313, 21

Fukazawa, Y., Ohashi, T., Fabian, A. C., Canizares, C. R., Ikebe, Y., Makishima, K., Mushotzky, R. F., & Yamashita, K. 1994, PASJ, 46, L55

Fukugita, M., Hogan, C. J., & Peebles, P. J. E. 1998, ApJ, 312, 518

Gastaldello, F., & Molendi, S. 2002, ApJ, 572, 160

Gibson, B. K., Loewenstein, M., & Mushotzky, R. F. 1997, MNRAS, 290, 623

Girardi, M., Manzato, P., Mezzetti, M., Giuricin, G., & Limboz, F. 2002, ApJ, 569, 720

Grevesse, N., & Sauval, A. J. 1998, Space Science Reviews, 85, 161

Heger, A., & Woosley, S. E. 2002, ApJ, 567, 532

Horner, D. J., et al. 2004, ApJS, submitted

Hwang, U., Mushotzky, R. F., Burns, J. O., Fukazawa, Y., & White, R. A. 1999, ApJ, 516, 604

Hwang, U., Mushotzky, R. F., Loewenstein, M., Markert, T. H., Fukazawa, Y., & Matsumoto, H. 1997, ApJ, 476, 560

Ikebe, Y., et al. 1997, ApJ, 481, 660

Ikebe, Y., Makishima, K., Fukazawa, Y., Tamura, T., Xu, H., Ohashi, T., & Matsushita, K. 1999, ApJ, 525, 58

Irwin, J. A., Bregman, J. N., & Evrard, A. E. 1999, ApJ, 519, 518

Iwasawa, K., Fabian, A. C., Allen, S. W., & Ettori, S. 2001, MNRAS, 328, L5

Jansen, F., et al. 2001, A&A, 365, L1

Jeltema, T. E., Canizares, C. R., Bautz, M. W., Malm, M. R., Donahue, M., & Garmire, G. 2001, ApJ, 562, 124

Johnstone, R. M., Allen, S. W., Fabian, A. C., & Sanders, J. S. 2002, MNRAS, 336, 299

Jorgensen, I., Franx, M., Hjorth, J., & van Dokkum, P. G. 1999, MNRAS, 308, 833

Kaastra, J. S., Ferrigno, C., Tamura, T., Paerels, F. B. S., Peterson, J. R., & Mittaz, J. P. D. 2001, A&A, 365, L99

Kikuchi, K., Furusho, T., Ezawa, H., Yamasaki, N. Y., Ohashi, T., Fukazawa, Y., & Ikebe, Y. 1999, PASJ, 51, 301

Kroupa, P. 2001, MNRAS, 322, 231

Lanzetta, K. M., Yahata, N., Pascarelle, S., Chen, H.-W., & Fernández-Soto, A. 2002, ApJ, 570, 492

Lewis, A. D., Stocke, J. T., & Buote, D. A. 2002, ApJ, 573, L13

Lin, Y.-T., Mohr, J. J., & Stanford, S. A. 2003, ApJ, 591, 749

Loewenstein, M. 2001, ApJ, 557, 573

Loewenstein, M., & Mushotzky, R. F. 1996, ApJ, 466, 695

Madau, P., Ferguson, H. C., Dickinson, M. E., Giavalisco, M., Steidel, C. C., & Fruchter, A. 1996, MNRAS, 283, 1388

Madau, P., & Pozzetti, L. 2000, MNRAS, 312, L9

Matsumoto, H., Koyama, K., Awaki, H., Tomida, H., Tsuru, T., Mushotzky, R., & Hatsukade, I. 1996, PASJ, 48, 201

Matsumoto, H., Pierre, M., Tsuru, T. G., & Davis, D. 2001, A&A, 374, 28

Matsumoto, H., Tsuru, T. G., Fukazawa, Y., Hattori, M., & Davis, D. 2000, PASJ, 52, 153

Matsushita, K., Finoguenov, A., & Bohringer, H. 2003, A&A, 401, 443

Maughan, B. J., Jones, L. R., Ebeling, H., Perlman, E., Rosati, P., Frye, C., & Mullis, C. R. 2003, ApJ, 587, 589

Milone, A., Barbuy, B., & Schiavon, R. P. 2000, AJ, 120, 131

Mitchell, R. J., Culhane, J. L., Davison, P. J., & Ives, J. C. 1976, MNRAS, 175, 29p.

Mohr, J. J., Mathiesen, B., & Evrard, A. E. 1999, ApJ, 517, 627

Molendi, S., & Gastaldello, F. 2001, A&A, 375, L14

Mulchaey, J. S. 2000, ARA&A, 38, 289

Mushotzky, R. F. 1984, Phys. Scripta T7, 157

Mushotzky, R. F., & Loewenstein, M. 1997, ApJ, 481, L63

Mushotzky, R., Loewenstein, M., Arnaud, K. A., Tamura, T., Fukazawa, Y., Matsushita, K., Kikuchi, K., Hatsukade, I. 1996, ApJ, 466, 686

Nomoto, K., Hashimoto, M., Tsujimoto, T., Thielemann, F.-K., Kishimoto, N., Kubo, Y., & Nakasato, N. 1997a, Nuclear Physics A, 616, 79

Nomoto, K., Iwamoto, K., Nakasato, N., Thielemann, F.-K., Brachwitz, F., Tsujimoto, T., Kubo, Y., & Kishimoto, N. 1997b, Nuclear Physics A, 621, 467

Nulsen, P. E. J., & Bohringer, H. 1995, MNRAS, 274, 1093

Peterson, J. R., et al. 2001, A&A, 365, L104

Peterson, J. R., Kahn, S. M., Paerels, F. B. S., Kaastra, J. S., Tamura, T., Bleeker, J. A. M., Ferrigno, C., & Jernigan, J. G. 2003, ApJ, 590, 207

Pettini, M., Ellison, S. L., Bergeron, J., & Petitjean, P. 2002, A&A, 391, 21

Reddy, B. E., Tomkin, J., Lambert, D. L., & Allende Prieto, C. 2003, MNRAS, 340, 304

Rizza, E., Burns, J. O., Ledlow, M. J., Owen, F. N., Voges, W., & Bliton, M. 1998, MNRAS, 301, 328

Sakelliou, I., et al. 2002, A&A, 391, 903

Sanders, J. S., & Fabian, A. C. 2002, MNRAS, 331, 273

Sarazin, C. L. 1988, X-ray Emission from Clusters of Galaxies (Cambridge: Cambridge Univ. Press)

Schmidt, R. W., Allen, S. W., & Fabian, A. C. 2001, MNRAS, 327, 1057

Schmidt, R. W., Fabian, A. C., & Sanders, J. S. 2002, MNRAS, 337, 71

Schuecker, P., Bohringer, H., Collins, C. A., & Guzzo, L. 2003, A&A, 398, 867

Serlemitsos, P. J., Smith, B. W., Boldt, E. A.,Holt, S. S., & Swank, J. H. 1977, ApJ, 211, L63.

Shibata, R., Matsushita, K., Yamasaki, N. Y., Ohashi, T., Ishida, M., Kikuchi, K., Bohringer, H., & Matsumoto, H. 2001, ApJ, 549, 228

Smith, D. A., Wilson, A. S., Arnaud, K. A., Terashima, Y., & Young, A. J. 2002, ApJ, 565, 195

Songaila, A. 2001, ApJ, 561, L153 (erratum: 2002, 568, L139)

Spergel, D. N., et al. 2003, ApJS, 148, 175

Stewart, G. C., Fabian, A. C., Nulsen, P. E. J., & Canizares, C. R. 1984, ApJ, 278, 536

Tamura, T., et al. 1996, PASJ, 48, 671

——. 2001b, A&A, 365, L87

Tamura, T., Bleeker, J. A. M., Kaastra, J. S., Ferrigno, C., & Molendi, S. 2001a, A&A, 379, 107

Tamura, T., Kaastra, J. S., Makishima, K., & Takahashi, I. 2003, A&A, 399, 497

Thielemann, F.-K., Nomoto, K., & Hashimoto, M. 1996, ApJ, 460, 408

Timmes, F. X., Woosley, S. E., & Weaver, T. A. 1995, ApJS, 98, 617

Tozzi, P., Rosati, P., Ettori, S., Borgani, S., Mainieri, V., & Norman, C. 2003, ApJ, 593, 705

Tsai, J. C. 1994a, ApJ, 423, 143

——. 1994b, ApJ, 429, 119

Umeda, H., & Nomoto, K. 2003, Nature, 422, 871

van Dokkum, P. G., Franx, M., Kelson, D. D., & Illingworth, G. D. 1998, ApJ, 504, L17

Weisskopf, M. C., Brinkman, B., Canizares, C., Garmire, G., Murray, S., & Van Speybroeck, L. P. 2002, PASP, 114, 1

White, D. A. 2000, MNRAS, 312, 663

White, D. A., Fabian, A. C., Johnstone, R. M., Mushotzky, R. F., & Arnaud, K. A. 1991, MNRAS, 312, 663

White, S. D. M., Navarro, J. F., Evrard, A. E., & Frenk, C. S. 1993, Nature, 366, 429

Woosley, S. E., & Weaver, T. A. 1995, ApJS, 101, 181

Wyithe, S., & Loeb, A. 2003, ApJ, 588, L69

Xu, H., et al. 2002, ApJ, 579, 600

Xu, H., Ezawa, H., Fukazawa, Y., Kikuchi, K., Makishima, K., Ohashi, T., & Tamura, T. 1997, PASJ, 49, 9

28

Quasar elemental abundances and host galaxy evolution

FRED HAMANN[1], MATTHIAS DIETRICH[1,2], BASSEM M. SABRA[1,3],
and CRAIG WARNER[1]
(1) University of Florida, Gainesville, USA, (2) Georgia State University, Atlanta, USA,
(3) American University of Beirut, Lebanon

Abstract

High-redshift quasars mark the locations where massive galaxies are rapidly being assembled and forming stars. There is growing evidence that quasar environments are metal-rich out to redshifts of at least 5. The gas-phase metallicities are typically solar to several times solar, based on independent analyses of quasar broad emission lines and intrinsic narrow absorption lines. These results suggest that massive galaxies (e.g., galactic nuclei) experience substantial star formation before the central quasar becomes observable. The extent and epoch of this star formation (nominally at redshifts $z > 2$, but sometimes at $z > 5$) are consistent with observations of old metal-rich stars in present-day galactic nuclei/spheroids, and with standard models of galactic chemical evolution. There is further tantalizing (but very tentative) evidence, based on Fe II/Mg II broad emission-line ratios, that the star formation usually begins ≥ 0.3 Gyr before the onset of visible quasar activity. For the highest-redshift quasars, at $z \approx 4.5$ to ~6, this result suggests that the first major star formation began at redshifts ≥ 6 to ≥ 10, respectively.

28.1   Introduction

Quasars are no longer perceived merely as exotic high-redshift monsters. Rather, they are commonplace, in the sense that every massive galaxy today was at one time an active quasar host. Quasars are thus valuable probes of galaxy evolution. They light up the surrounding gas (in young galactic nuclei), and they provide a bright emission source for absorption-line studies, during a poorly understood, early phase of galaxy assembly. This review describes measurements of the gas-phase elemental abundances near quasars, and the implications for massive galaxy evolution. Please see Hamann & Ferland (1999) for a more comprehensive review of this topic.

28.1.1   *Quasars, Galaxies, and High-redshift Star Formation*

Recent studies show that quasars, or more generally active galactic nuclei (AGNs), are natural by-products of galaxy formation. The supermassive black holes (SMBHs) that power AGNs are not only common in the centers of galaxies, but the SMBH masses correlate directly with the mass of the surrounding galactic spheroid (Gebhardt et al. 2000; Merritt & Ferrarese 2001; Tremaine et al. 2002). Whatever processes lead to the formation of galactic spheroids (e.g., elliptical galaxies and the bulges of grand spirals) must also (somehow) create a central SMBH with commensurate mass. Most of the SMBHs in galaxies today are "dormant," or nearly so, because the mass accretion has declined or ceased. But all of the galaxies hosting SMBHs today must have been at one time "active," with the bright AGN phase corresponding to the final growth phase of the SMBH.

The bright AGN phase is expected to be brief. In the standard paradigm of AGN energy production, the mass accretion rate, $\dot{M}$, needed to maintain a given luminosity, L, is $L = \eta \dot{M} c^2$, where η is an efficiency factor believed to be of order 0.1. The luminosity can also be expressed as a fraction of the Eddington limit, $L = \gamma L_E \approx 1.5 \times 10^{46} \gamma M_8$ erg s^{-1}, where L_E is the Eddington luminosity, M_8 is the SMBH mass relative to 10^8 M$_\odot$, and γ is a constant typically between 0.1 and 1 for luminous quasars (Vestergaard 2004; Warner, Hamann, & Dietrich 2004). If the bright AGN phase corresponds to the final accretion stage where the SMBH roughly doubles its mass while accreting at $\sim$50% of the Eddington rate ($\gamma \approx 0.5$), then the lifetime of this phase should be $\sim 6 \times 10^7$ yr (based on the expressions above). This nominal lifetime fits well with the population demographics. In particular, the observed space density of present-day SMBHs, with masses from $< 10^6$ M$_\odot$ to $\sim 10^{10}$ M$_\odot$, can account for the high-redshift quasar population if every one of these SMBHs shined previously as an AGN for a few $\times 10^7$ yr (Ferrarese 2002).

Another important aspect of the AGN-galaxy relationship is the close link between AGNs and vigorous star formation. For example, high-redshift quasars are often strong sources of sub-mm dust emission, which is attributed to powerful starbursts in the surrounding galaxies (Omont et al. 2001; Cox et al. 2002). Several authors have noted that the growth in the quasar number density with increasing redshift matches well the increasing cosmic star formation rate (e.g., Franceschini et al. 1999). Quasars flourished at a time (corresponding to $z \approx 2$ to 3) when most massive galactic spheroids (e.g., elliptical galaxies and the bulges of grand spirals) were frantically forming most of their stars (Renzini 1998; Jimenez et al. 1999; Dunne, Eales, & Edmunds 2003). It seems likely that the same processes that dump matter into galactic nuclei to form an SMBH also induce substantial star formation.

We conclude that high-redshift quasars are bright beacons marking the locations where massive galactic nuclei are being assembled—vigorously making stars and building a central SMBH. However, we may still have something like the old "chicken versus egg" problem. Which came first, the galaxy or the quasar? The bulk of the stars or the central SMBH? How mature are the host galaxies when the central AGN finally becomes visible?

28.2   Quasar Abundance Studies

Quasar abundance studies seek to examine quantitatively the chicken versus egg problem. How chemically enriched are quasar environments and, by inference, how much star formation preceded the observed quasars? Does the degree of enrichment depend on the type of AGN or on properties (mass?) of the surrounding host galaxy? When does the star formation begin in quasar environments? Are the host environments of increasingly higher-redshift quasars less evolved and therefore less metal-rich? The answers to these questions will provide unique constraints on high-redshift star formation and early galaxy evolution.

Most of the effort in this field has been to measure the overall gas-phase metallicities near quasars. But there is still more we can learn from the relative metal abundances. The ratio of iron to α elements is of particular interest because the α elements, such as O, Ne, and Mg, derive exclusively from massive-star supernovae (Types II, Ib, and Ic), while Fe has a dominant contribution from intermediate-mass stars via Type Ia supernovae (Yoshii, Tsujimoto, & Nomoto 1996). The SN Ia contribution of Fe is delayed relative to the SN II+Ib+Ic products because of the longer lifetimes of SN Ia precursors (integrated over a stellar initial mass function). The amount of the delay is often assumed to be $\sim$1 Gyr, but Matteucci & Recchi (2001) showed that this number depends on environmental factors such as the star formation rate and initial mass function. They argue that $\sim$0.3 Gyr is more appropriate for young elliptical galaxies/spheroids (with high star formation rates). In any case, the delay can be used as an approximate "clock" to constrain the formation times of stellar populations. For example, observations of large gas-phase Fe/α ratios near metal-rich quasars would indicate that their surrounding stellar populations are already $\geq$0.3 Gyr old.

28.2.1 Spectral Diagnostics

The first step is to understand the abundance diagnostics in quasar spectra. The signature features are the broad emission lines (BELs), with profile widths greater than 1000–1500 km s^{-1} . Reverberation mapping studies show that quasar BELs form in photoionized gas within $\sim$1 pc of the central continuum source (Kaspi et al. 2000). Some of the first quasar studies noted that BELs identify the same variety of elements (hydrogen, carbon, nitrogen, oxygen, etc.) seen in much less exotic stellar and galactic environments (Burbidge & Burbidge 1967). In particular, there are no obvious abundance "anomalies." The first quantitative estimates of BEL region abundances (see Davidson & Netzer 1979 for an excellent review) confirmed the lack of abundance anomalies compared to solar element ratios, and showed further that the metallicities are probably within a factor of $\sim$10 of solar.

More recent abundance studies include absorption lines among the diagnostics. Quasar absorption lines are classified generically according to their full widths at half minimum (FWHMs). Roughly 10% to 15% of optically selected quasars have classic broad absorption lines (BALs), with FWHM $>$ 2000 to 3000 km s^{-1} . At the other extreme are the narrow absorption lines (NALs), which have typically FWHM $<$ a few hundred km s^{-1} . A useful working definition of the NALs is that the FWHM is not large enough to blend together important UV doublets, such as C IV $\lambda\lambda$1548, 1551, whose velocity separation is 500 km s^{-1} . Intermediate between the NALs and BALs are the so-called mini-BALs. The BALs and mini-BALs appear exclusively blueshifted with respect to the emission lines, with velocity shifts from near 0 km s^{-1} to $>$30,000 km s^{-1} in some cases. Their broad and smooth profiles at blueshifted velocities clearly identify outflows from the central quasar energy source. NALs are also frequently blueshifted, but they can appear at redshifted velocities up to $\sim$2000 km s^{-1} . NALs within $\pm$5000 km s^{-1} of the emission-line redshift are also called "associated" absorption lines (AALs). In general, NALs can form in a wide range of environments, from high-speed outflows like the BALs to cosmologically intervening clouds or galaxies having no relation to the quasar. One challenge in using NALs for abundance work is to understand the location of the absorbing gas.

The following sections outline the procedures and results related to each of these diagnostics. Please see Hamann & Ferland (1999) for a more complete discussion.

28.3   Intrinsic Narrow Absorption Lines

Figure 28.1 shows some typical examples of associated NALs in the spectrum of a bright quasar, PG 0935+417 with emission-line redshift z_{em} = 1.966. The AALs in this case have FWHMs from $\sim$12 to $\sim$133 km s^{-1} , and they are blueshifted by $\sim$1400 km s^{-1} to $\sim$3000 km s^{-1} with respect to z_{em}.

There are three critical steps involved in using NALs as abundance diagnostics. First, we must distinguish the *intrinsic* NALs from those that form in unrelated, cosmologically

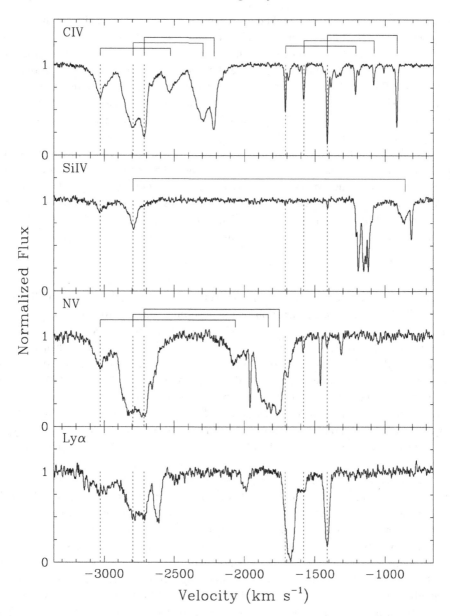

Fig. 28.1. Normalized Keck/HIRES spectra of "associated" NALs in the quasar PG 0935+417. Strong doublets are labeled above with open brackets: C IV $\lambda\lambda$1548, 1551, N V $\lambda\lambda$1238, 1242, and Si IV $\lambda\lambda$1393, 1403. The velocity scale is relative to $z_{em} = 1.966$ for the short wavelength component of each NAL doublet. The dotted vertical lines mark the velocities of the six strongest C IV systems.

intervening gas. We define "intrinsic" NALs loosely as forming in gas that is (or was) part of the overall AGN/host galaxy environment. Statistical studies suggest that the AALs are often intrinsic to the quasar/host galaxies (Foltz et al. 1986; Richards et al. 1999). Other observational criteria must be applied to determine which specific NALs are intrinsic, such as, (1) time-variable line strengths, (2) NAL profiles that are broad and smooth compared to the thermal line width, (3) excited-state absorption lines that require high gas densities, and (4) NAL multiplet ratios that reveal partial line-of-sight coverage of the background light source (see Barlow & Sargent 1997; Hamann et al. 1997b). Characteristics like variability, broad-ish line profiles, and partial covering often appear together, clearly indicating an intrinsic origin. Some NALs show none of these characteristics and the location of their absorbing gas remains ambiguous.

Second, we must derive column densities in a way that can account for an absorber that might be inhomogeneous and/or partially covering the background light source(s). In the most general case, the observed line "intensities" represent an average of the intensities transmitted over the projected area of an extended emission source(s). If the absorber happens to be homogeneous (has a constant line optical depth) across the face of a uniformly bright emission source, then the observed intensity is simply

$$\frac{I(\lambda)}{I_0(\lambda)} = 1 - C_f(\lambda) + C_f(\lambda) e^{-\tau_\lambda} \quad , \tag{28.1}$$

where $I_0(\lambda)$ is the intensity of the emission source at wavelength λ, τ_λ is the line optical depth, $C_f(\lambda)$ is the coverage fraction of the absorbing material $[0 \leq C_f(\lambda) \leq 1]$, and we ignore small contributions of line emission from the absorbing gas. If we measure two absorption lines having a known optical depth ratio (e.g., in a doublet), then we can use the measured strengths and ratios of those lines (providing two equations) to solve uniquely for both τ_λ and $C_f(\lambda)$ at each wavelength (Petitjean, Rauch, & Carswell 1994; Barlow & Sargent 1997; Hamann et al. 1997b). For resonance lines, the ionic column densities then follow simply from the optical depths, $N_{\text{ion}} \propto \int \tau_\lambda \, d\lambda$. The key is to obtain high-resolution spectra (so there are no unresolved line components) of NAL multiplets. Fortunately, there are many possibilities, such as the doublets shown in Figure 28.1, the H I Lyman series lines, and others.

If the absorbing medium is inhomogeneous, then more lines are needed to define the two-dimensional optical depth/column density distribution. However, a simple doublet analysis still provides a useful estimate of the optical depth spatial distribution (at each wavelength), subject to an assumed functional form for that distribution. We recently completed an extensive theoretical study of the effects of inhomogeneous absorption on observed absorption lines (Sabra & Hamann 2004), which builds upon the pioneering work of de Kool et al. (2001). We find that, for a wide range of inhomogeneous optical depth distributions, the spatially averaged value of the optical depth is very similar to the single value one would derive (from the same data) assuming homogeneous partial coverage (Eq. 28.1). Therefore, the most general situation effectively reduces to the simple homogeneous case for the purposes of an abundance analysis*.

The final step is to convert ratios of ionic column densities into abundance ratios using ionization corrections. The correction factors can be large because NALs are often highly ionized and we must compare a high ion, such as C IV, to H I Lyα to estimate the C/H abundance (metallicity). There can also be a range of ionizations present in the absorber but not enough measured lines to fully characterize this range. Nonetheless, Hamann (1997)

---

* There are second order effects that we will not go into here, but it boils down to making a reasonable assumption about the functional form of the column density/optical depth spatial distribution.

showed that we can always derive conservatively low estimates of the metal/hydrogen abundance ratios by assuming each metal line forms under conditions that most favor that ion. In particular, we can use the maximum possible value of the metal ion fractions, such as C IV/C, as indicated by photoionization calculations. Similarly, there are minimum ionization corrections for each metal ion that provide firm lower limits on the metal/hydrogen abundance (see also Bergeron & Stasińska 1986). These conservative estimates and firm lower limits provide useful constraints on the metallicity even if we have no knowledge of the degree or range of ionizations in the gas (see also Hamann & Ferland 1999).

28.3.1 *Results*

High-resolution spectra suitable for NAL abundance studies did not become available until the 1990s. The earliest results (Wampler, Bergeron, & Petitjean 1993; Petitjean et al. 1994, among others) were confirmed by later studies (Tripp, Lu, & Savage 1996; Hamann 1997; Hamann et al. 1997b; Savage, Tripp, & Lu 1998; Petitjean & Srianand 1999; Papovich et al. 2000). Quasar AALs often have metallicities $Z \geq Z_\odot$. Several of these studies also report N/C above solar, although in most cases the data are not adequate for this measurement. So far there are only two published reports (to our knowledge) of AAL abundances above redshift 4, and the results are slightly mixed. Wampler et al. (1996) estimated $Z \approx 2\, Z_\odot$ based on the tentative detection of O I $\lambda 1303$ in one AAL system at redshift 4.67, while Savaglio et al. (1997) reported $0.1 \leq Z \leq 1\, Z_\odot$ for an AAL complex at redshift 4.1. However, in both of these cases, and many others, the location of the absorber is not known. All of the confirmed intrinsic NALs (based on the criteria mentioned in §28.3) have metallicities $Z \geq Z_\odot$ (see references above). Petitjean et al. (1994) used a small sample of redshift ~ 2 quasars to show that there is a dramatic decline in absorber metallicity as one moves away from the quasar emission redshift, from $Z \geq Z_\odot$ in the AALs to $Z < 0.1\, Z_\odot$ at velocity shifts $\geq 10,000$ km s^{-1} (see their Fig. 14; also Carswell, Schaye, & Kim 2002). This result meets our expectation that AALs are often intrinsic to quasars, while most NALs at lower redshifts are unrelated (Rauch 1998).

There is a strong need now to improve on some of the early measurements and expand the database. We and collaborators have obtained high-resolution (7–9 km s^{-1}) spectra of 15 AAL quasars at $z_{em} \approx 2$–3 using the HIRES echelle spectrograph on the 10 m Keck I telescope. As a brief example consider the AALs of quasar PG 0935+417 shown in Figure 28.1. We find no evidence of line variability (above $\sim 10\%$) based on several observations that span ~ 2 years in the quasar rest frame. Nonetheless, the three systems of C IV doublets at 3000 to 2700 km s^{-1} are clearly intrinsic because of the large nonthermal line widths and doublet ratios that imply partial line-of-sight coverage of the background light source (see also Hamann et al. 1997a). The location of the three narrower absorption systems at 1700 to 1400 km s^{-1} remains unclear. However, a preliminary analysis of all of these lines (based on simple equivalent width measurements and assuming constant coverage fractions across the profiles) suggests that the metallicities in all cases are ~ 1 to $\sim 3\, Z_\odot$. A more detailed analysis of this data set is underway, but solar or higher metallicities appear to be common in the AALs, in agreement with earlier studies.

28.4 Broad Absorption Lines and Mini-BALs

Figure 28.2 shows the rest-frame UV spectrum of a fairly typical BAL quasar, PG 1254+047. The BALs in this case imply outflow velocities from roughly 15,000 to 27,000

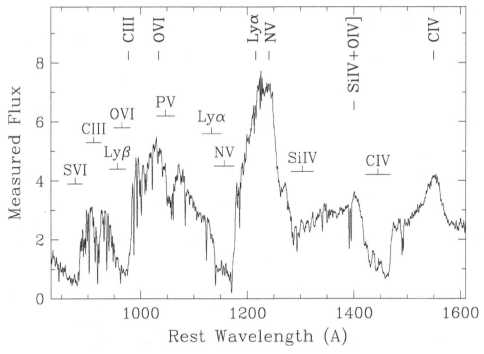

Fig. 28.2. *Hubble Space Telescope* spectrum of PG 1254+047 ($z_{\rm em}$ = 1.01) showing its strong BAL troughs. The BALs are labeled just above the spectrum, while the prominent broad emission lines are marked across the top. The measured flux has units 10^{-15} erg s^{-1} cm^{-2} Å^{-1}. (Adapted from Hamann 1998.)

km s^{-1}. The main obstacle to using BALs for abundance work is that the broad profiles blend together all the important doublets. Therefore, unlike the NALs, we cannot examine the doublet ratios for evidence of partial line-of-sight coverage. If we simply assume that the coverage fraction is unity (C_f = 1 in Eq. 28.1), then measurements of the column densities imply bizarre abundance ratios, such as, [Si/C] > 0.5 and [C/H] ≈ 1 to 2 (where $[x/y]$ = $\log(x/y) - \log(x/y)_\odot$). The surprising detections of P v λλ1118, 1128 BALs in at least half of the well-measured sources, including PG 1254+047 (Fig. 28.2), imply [P/C] ≥ 1.6 (Hamann 1998; Hamann et al. 2002b, and references therein; Junkkarinen et al. 2004).

The most likely explanation for these strange abundances is that they are incorrect. BALs are often much more optically thick than they appear because the absorber partially covers the background light source(s). Direct evidence for partial coverage has come from comparisons of widely spaced line pairs (similar to the doublet analysis but not involving doublets) in one well-measured BAL quasar (Arav et al. 2001). Doublet ratios in borderline NALs/mini-BALs, sometimes embedded within BAL profiles and believed to form in the same general outflow, also frequently indicate partial coverage (Barlow, Hamann, & Sargent 1997; Hamann et al. 1997a; Telfer et al. 1998; Srianand & Petitjean 2000). Less direct evidence comes from spectropolarimetry of scattered light in BAL troughs (Schmidt & Hines 1999), and from flat-bottomed BAL profiles that "look" optically thick even though they do not reach zero intensity.

Another clue comes from the surprisingly strong P V BALs. Hamann (1998) noted that the P V line should be nominally weak. Its ionization is very similar to C IV, but in the Sun phosphorus is ~1000 times less abundant than carbon. Therefore, if the P/C abundance is even close to solar in BAL regions, the P V line should not be present *unless* C IV and the other strong BALs of abundant elements are much more optically thick than they appear. In other words, there is partial coverage and unabsorbed light fills in the bottoms of BAL troughs. If we turn that argument around and assume solar P/C in PG 1254+047 (Fig. 28.2), we find that the true optical depth in its C IV line is ≥ 25 (Hamann 1998).

This interpretation of the P V may be disputed by Arav et al. (2001), who measured P V and many other BALs in one quasar spectrum with extraordinarily wide wavelength coverage. They estimate a metallicity near solar but $[P/C] \geq 1$. If that P/C result is correct, it would probably require a unique enrichment history (e.g., involving novae; Shields 1996). However, the uncertainties and challenges are substantial. We prefer to exclude BALs from quasar abundance studies.

28.5 Broad Emission Lines

Figure 28.3 shows part of the rest-frame UV spectrum of a quasar at redshift $z_{em} = 4.16$. The emission lines shown in this plot, plus C III] λ1909, have formed the basis for many BEL metallicity studies. The main advantage of the BELs is that they can be measured in any quasar using moderate resolution spectra. There is also no question that the lines form close to the quasar, nominally within ~1 pc of the central engine (§28.2.1).

One issue affecting the BEL analysis is that the emitting regions span simultaneously a wide range of densities and ionizations, with the higher ionizations occurring preferentially nearer the central continuum source (e.g., Ferland et al. 1992; Peterson 1993). Consequently, different lines can form in spatially distinct regions. Without a detailed model of the BEL environment, it is important for abundance studies to compare lines that form as much as possible in the same gas with similar excitation and radiative transfer dependencies. However, the range of physical conditions present in BEL regions provides a simplification: observed BEL spectra are flux-weighted averages over a diverse ensemble of "clouds." The tremendous advantage of this natural averaging is that we do not need to derive, or make specific assumptions about, the physical conditions in the different line-emitting regions (Hamann et al. 2002a). The formalism developed to simulate this situation (Baldwin et al. 1995) has been dubbed the "locally optimally emitting cloud" (LOC) model, because each line forms naturally where the conditions most favor its emission.

Another important consideration is that the combined emission in the metal lines is not sensitive to the overall metallicity. In particular, the strengths of prominent metal lines, such as C IV, relative to the hydrogen lines, such as Lyα, are not sensitive to the metal-to-hydrogen abundance ratio (Hamann & Ferland 1999). The main reason is that these BELs are the dominant coolants in the photoionized plasma where they form. Radiative equilibrium requires that the total energy emitted from any region in the plasma equals the total energy absorbed. Therefore, changing the metallicity by factors of several cannot produce a commensurate change in the overall line strengths without violating energy conservation.

Nonetheless, BELs are sensitive to the metal abundances in several ways (see Ferland et al. 1996; Hamann & Ferland 1999; Hamann et al. 2002a). First, departures from solar metallicity by orders of magnitude *will* change the total metal/H line emission ratio (as well as individual line ratios such as C IV/Lyα). For example, at very low metallicities

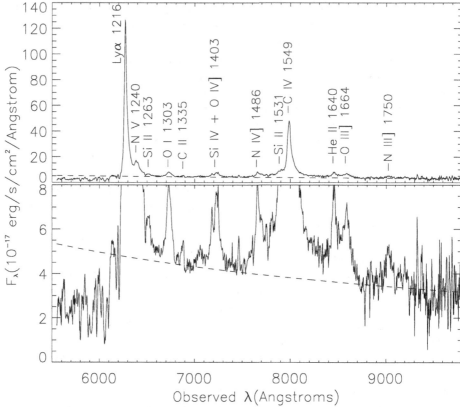

Fig. 28.3. Broad emission lines in the rest-frame UV spectrum of the redshift 4.16 quasar, BR 2248–1242. The lower panel is an expanded version of the upper plot. (From Warner et al. 2002.)

($\leq 0.03\ Z_\odot$) the metal lines no longer control the cooling and their emission strengths decline roughly commensurate with the metal/H abundance. This fundamental sensitivity to the extremes implies that typical BEL metallicities are conservatively within a factor of $\sim$30 of solar. Second, weak metal lines that are not important for the cooling do vary significantly with metallicity (while still preserving the overall energy balance). Abundance studies should, therefore, endeavor to include weaker lines, such as O III] λ1664 and N III] λ1750 (Fig. 28.3). Finally, the *relative* strengths of different metal lines can be sensitive to the metal/metal abundance ratios.

Taking advantage of this last point, Shields (1976) noted that in Galactic H II regions the N/O abundance scales roughly proportional to O/H (metallicity) because of a strong "secondary" contribution to the nitrogen enrichment (see also Pettini et al. 2002; Pilyugin, Thuan, & Vílchez 2003). The N/O $\propto$ O/H scaling dominates for metallicities above a few tenths solar. The N/O abundance therefore provides an indirect indicator of the overall metallicity. This technique has become the norm for BEL metallicity studies.

It should be noted that departures from the simple N/O $\propto$ O/H scaling can occur if there are metallicity-dependent stellar yields, or if the enrichment is dominated by star formation in discrete bursts (Kobulnicky & Skillman 1996; Henry, Edmunds, & Köppen 2000). The latter situation would lead to time-dependent fluctuations in N/O because of different delays in the stellar release of N and O. However, the overall trend for increasing N/O with O/H remains, and the best homogeneous data sets indicate that the scatter in the N/O $\propto$ O/H relation declines at high O/H (Pettini et al. 2002; Pilyugin et al. 2003). This is the regime of quasars. Moreover, there are no reports, to our knowledge, of high N/O ratios (solar or higher) in metal poor (significantly sub-solar) interstellar environments. Large N/O abundances are an indicator of high metallicities in any scenario that involves well-mixed interstellar gas. The situation with N/C can also be complicated because N and C arise from different stellar mass ranges, leading again to time-dependent effects (Henry et al. 2000; Chiappini, Romano, & Matteucci 2003). Nonetheless, in BEL analyses, we generally assume that N/C behaves approximately like N/O.

28.5.1 Results: Metallicity

Shields (1976) analyzed several ratios of UV intercombination (semi-forbidden) lines (see Fig. 28.3) and found that N/O and N/C are nominally solar to $\sim$10 times solar, consistent with solar to super-solar metallicities (see also Davidson & Netzer 1979; Uomoto 1984, and references therein). Hamann & Ferland (1993) and Ferland et al. (1996) later claimed that the metallicities are typically several times solar, based on a new analysis of the N V/C IV and N V/He II BEL ratios. More recently, Hamann et al. (2002a) used extensive photoionization calculations to quantify better the abundance sensitivities of all of these nitrogen line ratios. In particular, they explored the influence of non-abundance effects such as density, ionization, turbulence, and incident continuum shape, in the context of LOC calculations. They favored N III]/O III] and N V/(C IV + O VI λ1034) as the most robust indicators of the relative N abundance. N V/He II is also very useful, but it has a greater sensitivity to the ionizing continuum shape because it compares a collisionally excited line (N V) to a recombination line (He II; see also Ferland et al. 1996). Holding other parameters constant, the nitrogen line ratios scale almost linearly with the N/O and N/C abundances and should, therefore, be proportional to the metallicity if the enrichment follows N/O $\propto$ O/H.

Dietrich & Hamann (2004) used the Hamann et al. (2002a) calculations to estimate BEL metallicities in a sample of $\sim$700 quasars spanning redshifts $0 \leq z_{em} \leq 5$. They find typically several times $Z_{\odot}$ across the entire redshift range. Figure 28.4 shows results for a subsample of these quasars at $z_{em} \geq 3.5$, where the average metallicity is $\sim$5 $Z_{\odot}$ (from Dietrich et al. 2003). Notice that there is no evidence of a metallicity decrease at the highest redshifts. For the particular $z_{em} = 4.16$ quasar shown in Figure 28.3, where many lines are measurable owing to the narrow profiles and large equivalent widths, Warner et al. (2002) estimated $Z \approx 2 Z_{\odot}$. Dietrich et al. (2003) and Dietrich & Hamann (2004) noted that the metallicities derived from the N V lines ratios (most notably N V/He II) are typically $\sim$30% to a factor of $\sim$2 larger than estimates from the intercombination ratios (e.g., N III]/O III]). The reason for this discrepancy is not clear. It could arise from systematic measurement errors in the weaker lines (Warner et al. 2004), or, perhaps, from subtle excitation/radiative transfer effects in the emitting regions. In any case, a factor of $\leq$2 agreement among the different BEL diagnostics is quite good. Our current strategy is to average the results from the various nitrogen BEL ratios.

The absolute uncertainties are more difficult to quantify because they depend on the theoretical techniques and assumptions. Factors of a few can be expected, but the essential result seems secure. The metallicities in quasar BEL regions are minimally near solar and perhaps

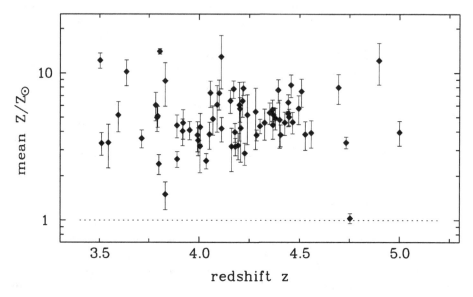

Fig. 28.4. Average metallicities derived from BEL ratios in a sample of high-redshift quasars. The error bars represent only the measurement uncertainties. (From Dietrich et al. 2003.)

typically several times above solar. This result is confirmed in general terms by the AAL data (§28.3.1; see also Kuraszkiewicz & Green 2004).

Another interesting result from the BELs is that more luminous quasars appear to be more metal-rich (Hamann & Ferland 1993, 1999; Osmer, Porter, & Green 1994; Dietrich & Hamann 2004). This trend is somewhat tentative because it is stronger in the N V/C IV and N V/He II ratios than in the intercombination lines, and it has not yet been tested in the AAL data. Nonetheless, it meets simple expectations. More luminous quasars are powered by more massive SMBHs, which reside in more massive galaxies (§28.1). There is a well-known relationship between mass and metallicity in normal galaxies (e.g., Trager et al. 2000), which is generally attributed to the effects of galactic winds. Massive galaxies reach higher metallicities because they retain their gas longer against the building thermal pressures from stellar mass loss and supernova explosions. Therefore, more luminous quasars in more massive galaxies could be more metal-rich. Warner et al. (2004) tried to test this idea by comparing BEL metallicities to estimates of the SMBH masses (a surrogate for the host galaxy masses) in the large quasar sample of Dietrich & Hamann (2004). The results are uncertain, but they do favor a positive correlation between SMBH mass and BEL metallicity, as expected. In fact, our best estimate for the slope of this trend agrees well with the mass-metallicity slopes observed in galaxies.

28.5.2   *Results: Fe/α*

BELs are the only diagnostics used so far to estimate Fe/α abundances as an age discriminator (§28.2). This work has relied on Fe II(UV)/Mg II λ2798, where "Fe II(UV)" represents a broad blend of many Fe II lines between roughly 2000 and 3000 Å. This blend is by far the most prominent Fe II feature in quasar spectra, and its UV wavelength makes it measurable at the highest redshifts in the near-infrared. The nearby Mg II doublet serves the α-element representative with an ionization similar to Fe II.

Measuring the flux in the broad Fe II blend presents a unique challenge (see Dietrich et al. 2004 for discussion). Spectra with wide wavelength coverage are essential to properly define the underlying continuum. It also helps to use a scaled template to fit the Fe II blend, based on either theoretical predictions (Wills, Netzer, & Wills 1985; Verner et al. 1999; Sigut & Pradhan 2003) or observations of a well-measured source (Vestergaard & Wilkes 2001).

The main difficulty, however, is understanding the theoretical relationship between the Fe II/Mg II emission ratio and the Fe/Mg abundance. The Fe II atom has hundreds of relevant energy levels and thousands of lines that can blend together and/or contribute to radiative pumping/fluorescence processes. Current interpretations of observed Fe II spectra are still based largely on the pioneering calculations by Wills et al. (1985). Their work suggests that the strong Fe II fluxes from typical quasars require a relative iron abundance (e.g., Fe/Mg) that is super-solar by a factor of roughly 3. More modern calculations (Verner et al. 1999; Sigut & Pradhan 2003; Baldwin et al. 2004; Verner 2004) use better atomic data and can, for the first time, incorporate a many-level Fe II atom into fully self-consistent treatments of the energy budget. So far these calculations have mostly just confirmed what Wills et al. knew already, that the uncertainties are large. However, work is now underway to understand these uncertainties. In particular, it will be important to quantify the sensitivities of the Fe II emission and the Fe II/Mg II ratio to various poorly constrained, non-abundance parameters in BEL regions. It might turn out that the Fe II(UV)/Mg II ratio is not the best Fe/α diagnostic (Verner 2004).

Putting these concerns aside for the moment, the observational results are becoming increasingly interesting. The best data indicate that the Fe II/Mg II emission ratio is the same on average at all redshifts (Thompson, Hill, & Elston 1999; Dietrich et al. 2002, 2004; Iwamuro et al. 2002). This includes the most recent observations of two quasars at redshifts $z_{em} \approx 6$ (Barth et al. 2003; Freudling, Corbin, & Korista 2003). If we accept the tentative Wills et al. (1985) result that Fe/Mg is nominally above solar in BELs, which they deduced from spectra of quasars with $z_{em} \leq 0.5$, then the newer data suggest that Fe/Mg is large at all redshifts. Therefore, SNe Ia played a role in the enrichment, and the star formation must have begun at least 0.3 Gyr prior to the observed quasar epochs. For the redshift ~ 6 quasars, this line of argument implies that the first major star formation occurred at redshift 10–20 (Freudling et al. 2003), consistent with the epoch of reionization deduced from recent *WMAP* data (Bennett et al. 2003).

28.6   Summary and Implications

Independent analysis of the BELs and intrinsic NALs indicates that quasar environments are metal-rich at all redshifts. Typical metallicities range from roughly solar to several times solar. There might also be a trend with luminosity in the sense that more luminous quasars, which reside in more massive host galaxies, are more metal rich. If the gas near quasars was enriched by a surrounding stellar population, then the high metallicities indicate that those populations are already largely in place by the time the quasars "turned on" or became observable. For example, simple "closed-box" chemical evolution with a normal stellar initial mass function will produce metallicities above solar only after $\geq 60\%$ of the original mass in gas is converted to stars. The quasar data imply that this degree of enrichment and evolution is common in massive galactic nuclei before redshift 2, and sometimes it occurs before redshift 5.

Unfortunately, the quasar data do not tell us the size (mass) of the stellar population responsible for the enrichment. However, the mass of gas in the BEL region sets a lower limit. The most recent estimates indicate that luminous quasars have BEL region masses that are conservatively $\sim 10^3$ to 10^4 $M_\odot$ (Baldwin et al. 2003). Normal galactic chemical evolution then suggests that the stellar population needed to enrich this gas to $Z \geq Z_\odot$ is 10 times more massive, or $\sim 10^4$ to 10^5 $M_\odot$. This mass may still be just the "tip of the iceberg" if one considers that quasar BEL regions are part of a much more massive reservoir that includes a $\geq 10^9$ $M_\odot$ SMBH and its surrounding accretion disk. BEL regions might, in fact, be constantly replenished by the flow of material through the accretion disk. If that flow is ~ 10 $M_\odot$ yr^{-1} of metal-rich gas (§28.1), then clearly a much more massive stellar population (perhaps with Galactic bulge-like proportions) would be needed for the enrichment (see Friaça & Terlevich 1998 for specific enrichment models). Observations of strong dust and sometimes CO emissions from high-redshift quasars (currently up to $z_{em} \approx 4.7$) indicate that there is often already $\geq 10^{10}$ $M_\odot$ of metal-enriched gas present (Omont et al. 2001; Cox et al. 2002). There must have been massive amounts of star formation in these quasar host galaxies before the observed quasar epoch. Efforts to date the stellar populations around quasars using Fe II/Mg II BEL ratios suggest that the star formation often begins in earnest ≥ 0.3 Gyr prior to the appearance of a visible quasar. These results are tentative because of theoretical uncertainties, but for the highest-redshift quasars yet studied ($z_{em} \approx 6$) they suggest that the first star formation occurred at $z \approx 10-20$.

In terms of the "chicken versus egg" problem (§28.1) the quasar abundance data are clear; a substantial stellar population is already in place by the time most quasars become visible. At low redshifts, direct imaging studies of quasars and their lower-luminosity cousins, the Seyfert 1 galaxies, show clearly that the host galaxies are already present with substantial, even moderately old, stellar populations on $>$ kpc scales (Nolan et al. 2001; Dunlop et al. 2003). At higher redshifts there is less direct imaging data, but at least some quasars still have substantial hosts (Kukula et al. 2001). It could be that the stars and the SMBH begin forming at the same time. But *visible* quasar activity might be delayed with respect to the surrounding star formation, even at the highest redshifts, because of the time needed to assemble the SMBH and/or blow out the dusty interstellar medium that obscures the youngest AGNs from our view (see also Romano et al. 2002; Kawakatu & Umemura 2003). This delay could explain observations showing that the quasar number density declines dramatically with increasing redshifts above $z \approx 3$, while the cosmic star formation rate appears to stay roughly constant out to at least $z \approx 4$ (Ivison et al. 2002).

Acknowledgements. We are grateful to our collaborators, especially Jack Baldwin, Tom Barlow, Gary Ferland, Vesa Junkkarinen, and Kirk Korista, who contributed insights and unpublished results to this review. We also acknowledge NSF grant AST99-84040.

References

Arav, N., et al. 2001, ApJ, 561, 118
Baldwin, J. A., et al. 2004, in preparation
Baldwin, J. A., Ferland, G. J., Korista, K. T., Hamann, F., & Dietrich, M. 2003, ApJ, 582, 590
Baldwin, J. A., Ferland, G. J., Korista, K. T., & Verner, D. 1995, ApJ, 455, L119

Barlow T. A., Hamann F., & Sargent W. L. W. 1997, in Mass Ejection from Active Galactic Nuclei, ed. N. Arav, I. Shlosman, & R. J. Weymann (San Francisco: ASP), 13

Barlow, T. A., & Sargent, W. L. W. 1997, AJ, 113, 136

Barth, A. J., Martini, P., Nelson, C. H., & Ho, L. C. 2003, ApJ, 594, L95

Bennett, C. L., et al. 2003, ApJS, 148, 1

Bergeron, J., & Stasińska, G. 1986, A&A, 169, 1

Burbidge, G., & Burbidge, E. M. 1967, Quasi-Stellar Objects (New York: Freeman)

Carswell, R., Schaye, J., & Kim, T.-S. 2002, ApJ, 578, 43

Chiappini, C., Romano, D., & Matteucci, F. 2003, MNRAS, 339, 63

Cox, P., et al. 2002, ApJ, 387, 406

Davidson, K., & Netzer H. 1979, Rev. Mod. Phys., 51, 715

de Kool, M., et al. 2001, ApJ, 548, 609

Dietrich, M., Appenzeller, I., Vestergaard, M., & Wagner, S. J. 2002, ApJ, 564, 581

Dietrich, M., & Hamann, F. 2004, in preparation

Dietrich M., Hamann, F., Appenzeller, I., Vestergaard, M. 2004, ApJ, submitted

Dietrich, M., Hamann, F., Shields, J. C., Constantin, A., Heidt, J., Jäger, K., Vestergaard, M., & Wagner, S. J. 2003, ApJ, 589, 722

Dunlop, J. S., McLure, R. J., Kukula, M. J., Baum, S. A., O'Dea, C. P., & Hughes, D. H. 2003, MNRAS, 340, 1095

Dunne, L., Eales, S. A., & Edmunds, M. G. 2003, MNRAS, 341, 589

Ferland, G. J., Baldwin, J. A., Korista, K. T., Hamann, F., Carswell, R. F., Phillips, M. M., Wilkes, B. J., & Williams, R. E. 1996, ApJ, 461, 683

Ferland, G. J., Peterson, B. M., Horne, K., Welsh, W. F., & Nahar, S. N. 1992, ApJ, 387, 95

Ferrarese, L. 2002, Current High-Energy Emission around Black Holes, ed. C.-H. Lee & H.-Y. Chang (Singapore: World Scientific), 3

Foltz, C. B., Weymann, R. J., Peterson, B. M., Sun, L., Malkan, M. A., & Chaffe, F. H., Jr. 1986, ApJ, 307, 504

Franceschini, A., Hasinger, G., Takamitsu, M., & Malquori, D. 1999, MNRAS, 310, L5

Freudling, W., Corbin, M. R., & Korista, K. T. 2003, ApJ, 587, L67

Friaça, A. C. S., & Terlevich, R. J. 1998, MNRAS, 298, 399

Gebhardt, K., et al. 2000, ApJ, 539, L13

Hamann, F. 1997, ApJS, 109, 279

——. 1998, ApJ, 500, 798

Hamann, F., & Ferland, G. J. 1993, ApJ, 418, 11

——. 1999, ARA&A, 37, 487

Hamann, F., Barlow, T. A., Cohen, R. D., Junkkarinen, V., & Burbidge, E.M. 1997a, in Mass Ejection from Active Galactic Nuclei, ed. N. Arav, I. Shlosman, & R. J. Weymann (San Francisco: ASP), 25

Hamann, F., Barlow, T. A., Junkkarinen, V., & Burbidge, E. M. 1997b, ApJ, 478, 80

Hamann, F., Korista, K. T., Ferland, G. J., Warner, C., & Baldwin, J. 2002a, ApJ, 564, 592

Hamann, F., Sabra, B., Junkkarinen, V., Cohen, R., & Shields, G. 2002b, in X-ray Spectroscopy of AGN with Chandra and XMM-Newton, ed. Th. Boller et al., MPE Rep. 279, 121 (astro-ph/0304564)

Henry, R. R. C., Edmunds, M. G., & Köppen, J. 2000, ApJ, 541, 660

Ivison, R. J., et al. 2002, MNRAS, 337, 1

Iwamuro, F., Motohara, K., Maihara, T., Kimura, M., Yoshii, Y., & Doi, M. 2002, ApJ, 565, 63

Jimenez, R., Friaça, A. C. S., Dunlop, J., Terlevich, R. J., Peacock, J., & Nolan, L. A. 1999, MNRAS, 305, L16

Junkkarinen, V., et al. 2004, in preparation

Kaspi, S., Smith, P. S., Netzer, H., Maoz, D., Jannuzi, B. T., & Giveon, U. 2000, ApJ, 533, 631

Kawakatu, N., & Umamura, M. 2003, ApJ, 583, 85

Kobulnicky, H. A., & Skillman, E. D. 1996, ApJ, 471, 211

Kukula, M. J., Dunlop, J. S., McClure, R. J., Miller, L., Percival, W. J., Baum, S. A., & O'Dea, C. P. 2001, MNRAS, 326, 1533

Kuraszkiewicz, J. K., & Green, P. J. 2004, ApJ, in press

Matteucci, F., & Recchi, S. 2001, ApJ, 558, 351

Merritt, D., & Ferrarese, L. 2001, ApJ, 547, 140

Nolan, L. A., Dunlop, J. S., Kukula, M. J., Hughes, D. H., Boroson, T., & Jimenez, R. 2001, MNRAS, 323, 308

Omont, A., Cox, P., Bertoldi, F., McMahon, R. G., Carilli, C., & Isaak, K. G. 2001, A&A, 374, 371

Osmer, P. S., Porter, A. C., & Green, R. F. 1994, ApJ, 436, 678

Papovich, C., Norman, C. A., Bowen, D. V., Heckman, T., Savaglio, S., Koekemoer, A. M., & Blades, J. C. 2000, ApJ, 531, 654

Peterson, B. M. 1993, PASP, 105, 247

Petitjean, P., Rauch, M., & Carswell, R. F. 1994, A&A, 291, 29

Petitjean, P., & Srianand, R. 1999, A&A, 345, 73

Pettini, M., Ellison, S. L., Bergeron, J., & Petitjean, P. 2002, A&A, 391, 21

Pilyugin, L. S., Thuan, T. X., & Vílchez, J. M. 2003, A&A, 397, 487

Rauch, M. 1998, ARA&A, 36, 267

Renzini A. 1998, in The Young Universe: Galaxy Formation and Evolution at Intermediate and High Redshift, ed. S. D'Odorico, A. Fontana, & E. Giallongo (San Francisco: ASP), 298

Richards, G. T., York, D. C., Yanny, B., Kollgaard, R. I., Laurent-Muehleisen, S. A., & Vanden Berk, D. E. 1999, ApJ, 513, 576

Romano, D., Silva, L., Matteucci, F., & Danese, L. 2002, MNRAS, 334, 444

Sabra, B. M., & Hamann, F. 2004, ApJ, submitted

Savage, B. D., Tripp, T. M., & Lu, L. 1998, AJ, 115, 436

Savaglio, S., Cristiani, S., D'Odorico, S., Fontana, A., Giallongo, E., & Molaro, P. 1997, A&A, 318, 347

Schmidt, G. D., & Hines, D. C. 1999, ApJ, 512, 125

Shields, G. A. 1976, ApJ, 204, 330

——. 1996, ApJ, 461, L9

Sigut, T. A. A., & Pradhan, A. K. 2003, ApJS, 145, 15

Srianand, R., & Petitjean, P. 2000, A&A, 357, 414

Telfer, R. C., Kriss, G. A., Zheng, W., & Davidsen, A. F. 1998, ApJ, 509, 132

Thompson, K. L., Hill, G. J, & Elston, R. 1999, ApJ, 515, 487

Trager, S. C., Faber, S. M., Worthey, G., & González, J. J. 2000, AJ, 119, 1645

Tremaine, S., et al. 2002, ApJ, 574, 740

Tripp, T. M., Lu, L., & Savage, B. D. 1996, ApJS, 102, 239

Uomoto, A. 1984, ApJ, 284, 497

Verner, E. M. 2004, in Carnegie Observatories Astrophysics Series, Vol. 4: Origin and Evolution of the Elements, ed. A. McWilliam & M. Rauch (Pasadena: Carnegie Observatories, http://www.ociw.edu/symposia/series/symposium4/proceedings.html)

Verner, E. M., Verner, D. A., Korista, K. T., Ferguson, J. W., Hamann, F., & Ferland, G. J. 1999, ApJS, 120, 101

Vestergaard, M. 2004, ApJ, in press

Vestergaard, M., & Wilkes, B. J. 2001, ApJS, 134, 1

Wampler, E. J., Williger, G. M., Baldwin, J. A., Carswell, R. F., Hazard, C., & McMahon, R. G. 1996, A&A, 316, 33

Wampler, E. J., Bergeron, J., & Petitjean, P. 1993, A&A, 273, 15

Warner, C., Hamann, F., & Dietrich, M. 2004, ApJ, submitted

Warner, C., Hamann, F., Shields, J. C., Constantin, A., Foltz, C. B., & Chaffee, F. H. 2002, ApJ, 567, 68

Wills, B. J., Netzer, H., & Wills, D. 1985, ApJ, 288, 94

Yoshii, Y., Tsujimoto, T., & Nomoto, K. 1996, ApJ, 462, 266

29

Chemical abundances in the damped Lyα systems

JASON X. PROCHASKA

UCO/Lick Observatory, University of California, Santa Cruz

Abstract

I introduce and review the data and analysis techniques used to measure abundances in the damped Lyα systems, quasar absorption-line systems associated with galaxies in the early Universe. The observations and issues associated with their abundance analysis are very similar to those of the Milky Way's interstellar medium. We measure gas-phase abundances and are therefore subject to the effects of differential depletion. I review the impact of dust depletion and then present a summary of current results on the age-metallicity relation derived from damped Lyα systems and new results impacting theories of nucleosynthesis in the early Universe.

29.1 Introduction

While high-resolution stellar spectroscopy provides the framework behind nearly all discussion of nucleosynthesis and chemical enrichment (see papers throughout this volume), these observational efforts are currently limited to the Milky Way and its nearest neighbors. Beyond the Local Group, it remains very difficult to precisely determine chemical abundances. Relative abundance measurements are limited to a few species (e.g., C, N, O, Ca, Mg) and a few dozen galaxies. At cosmological distances, the challenges related to galaxy spectroscopy are even more severe, and even the determination of a crude metallicity poses great difficulty (e.g., Kobulnicky & Koo 2000).

Within the Milky Way, absorption-line spectroscopy of the interstellar medium (ISM) provides abundance measurements for many elements in a range of physical environments. In terms of chemical abundance studies, this analysis is limited by two factors: (1) gas mixing occurs on short enough time scales that the majority of the ISM is chemically homogeneous (Meyer, Jura, & Cardelli 1998). Therefore, one cannot probe nucleosynthesis at a range of metallicity or age in the ISM; and (2) refractory elements like Si, Ti, Fe, and Ni are depleted from the gas phase. Their relative abundance patterns are principally reflective of differential depletion (see the review by Jenkins 2004). ISM absorption-line observations have also been carried out within the Magellanic Clouds (Welty et al. 1999, 2001). These observations reveal the metallicity of the LMC and SMC, but interpretations of the relative abundance patterns are complicated by dust depletion. Outside of the Milky Way and Magellanic Clouds, it is rarely possible to pursue ISM studies in other local galaxies. With current UV telescopes and instrumentation, there are too few bright UV sources behind nearby galaxies. The advent of the Cosmic Origins Spectrograph on the *Hubble Space Telescope* will improve the situation, but only to a modest degree.

Ironically, the laws of atomic physics, the expansion of the Universe, and the filtering of UV light by the Earth's atmosphere combine to make the early Universe the most efficient place for studying galactic elemental abundances. Quasar absorption-line spectroscopy provides a powerful, accurate means of studying nucleosynthesis and chemical enrichment in hundreds of galaxies over an epoch spanning several billion years at $z \approx 2-5$. These galaxies are called damped Lyα systems (DLAs). The name derives from the quantum mechanical damping of the Lyα profile owing to the large H I surface density that defines a DLA: $N(\text{H I}) \geq 2 \times 10^{20} \, \text{cm}^{-2}$. At high redshift, the plethora of UV resonance transitions that ISM researchers study in our Galaxy are conveniently redshifted to optical wavelengths where they can be examined using high-resolution spectrographs on 10 m-class telescopes. In this fashion, we are able to study the ISM of galaxies near the edge of the Universe using data that competes with the observations taken in the Galaxy. The analysis of the DLAs has been the focus of our research since the commissioning of the Keck telescopes, and we are now joined by several groups at observatories across the globe.

In this paper, I will introduce the techniques used in the analysis of the DLAs to the broad audience attending the fourth Carnegie Symposium. By reviewing this topic at a pedagogic level, my goal is to encourage greater communication between the stellar and damped Lyα communities. These fields of research offer complementary analysis into theories of nucleosynthesis and chemical enrichment. Although the fields suffer from unique systematic uncertainties, a synthesis of their results will ultimately lead to a deeper understanding on the origin of the elements.

29.2   The Data and Standard Analysis

The data that drive current elemental abundance research in the DLAs is very similar to the spectroscopy obtained for stars in the Milky Way. The largest data sets to date have been acquired with the HIRES echelle spectrograph (Vogt et al. 1994) on the Keck I telescope and the UVES-VLT spectrograph (Dekker et al. 2000). These spectrographs do provide resolution $R \geq 60,000$, yet the majority of damped Lyα research is conducted with $R \approx 40,000$ observations. While the "clouds" of gas that comprise the velocity profiles of the galaxies presumably have thermal widths below the resolution of the spectrograph, tests for "hidden" saturation (e.g., Prochaska & Wolfe 1996) have demonstrated that line saturation is not an important issue. Even the lower-resolution DLA studies by Pettini et al. (1994, 1997) on 4 m-class telescopes gave rather accurate column density measurements for the weak, unsaturated Zn II and Cr II profiles.

Figure 29.1 presents a sample of data showing several metal-line transitions for a typical damped Lyα system. The dashed vertical lines identify four transitions related to the $z = 1.776$ DLA, and we also identify a C IV doublet and Fe II multiplet related to two other absorption systems along the sightline. The line density described by this figure is typical of quasar spectra redward of the Lyα forest, i.e. $\lambda > (1 + z_{\text{QSO}}) \times 1215.67$ Å. Clearly, line blending is a rare phenomenon and spectral synthesis is generally unrequired. This is contrasted, of course, by analysis within the Lyα forest where contamination by coincident Lyα clouds is common. Another point to emphasize is that the majority of absorption systems along a quasar sightline tend to show absorption from C IV, Si IV, and Mg II doublets. These are trivially identified, and therefore line misidentifications are very unlikely in the analysis of the damped Lyα systems.

For many transitions, one approaches 1% statistical error in the column density measure-

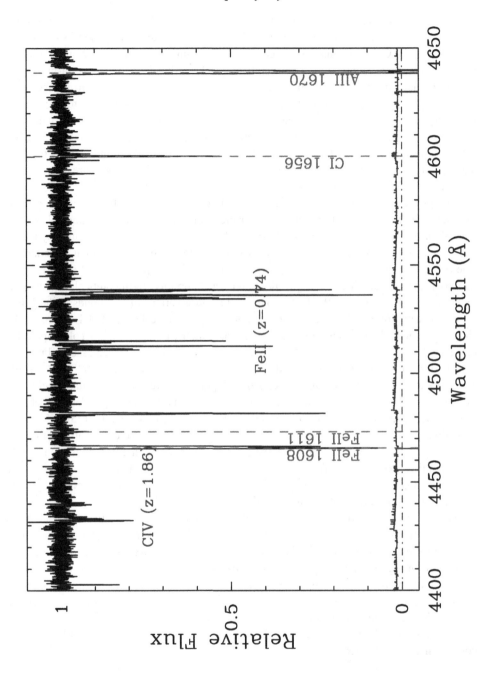

Fig. 29.1. A sample HIRES spectrum for the quasar Q1331+17, which exhibits a $z = 1.776$ DLA whose transitions are marked by vertical dashed lines. The S/N of these data is somewhat higher than most observations, while the resolution is typical (FWHM ≈ 7 km s^{-1}). In addition to the DLA transitions, there are several absorption lines arising from "metal-line systems" at $z = 0.74$ and $z = 1.86$.

ments with a signal-to-noise ratio (S/N) per pixel of only 30. Therefore, few (if any) of the damped Lyα systems have been observed at S/N > 100 per pixel or even the S/N of the data in Figure 29.1. Column densities are generally determined from either the summation of the observed optical depth (Savage & Sembach 1991) or through detailed Voigt-profile fitting. Oscillator strengths for the dominant transitions are almost exclusively from laboratory measurements and have typical errors of $< 20\%$. There are several important exceptions, however, notably the Fe II $\lambda 1611$ transition (one of the key Fe II transitions for metal-rich DLAs) whose best value is based on a theoretical calculation. Regarding the relative oscillator strengths, most of these transitions have been extensively analyzed in the Galactic ISM, and inaccuracies in the oscillator strengths have been corrected in the literature (e.g., Zsargó & Federman 1998; Howk et al. 2000).

Perhaps the most startling aspect of damped Lyα research for stellar spectroscopists is that the path from ionic column density measurements to elemental abundances is trivial. One often observes only one ion per element, the dominant species in a neutral hydrogen gas, and assumes no ionization corrections to compute gas-phase elemental abundances. The neglect of ionization corrections was unavoidable in the past (the first sets of observations provided few, if any, diagnostics) and was supported by theoretical expectations (Viegas 1994; Prochaska & Wolfe 1996) as well as observations of H I clouds in the Milky Way ISM with column densities comparable to the DLAs (see Jenkins 2004). Figure 29.2 shows a simple radiative transfer calculation for an ionizing flux incident on a plane-parallel, constant-density slab of hydrogen gas (Prochaska & Wolfe 1996). For an assumed number density of $n = 0.1 \, \text{cm}^{-3}$, we find that a damped Lyα system with $N(\text{H I}) > 10^{20.3} \, \text{cm}^{-2}$ would be $< 10\%$ ionized. More accurate and realistic calculations have come to similar conclusions (Vladilo et al. 2001), yet empirical confirmation remains an outstanding problem. Prochaska et al. (2002a) presented one of the few cases (the DLA at $z = 2.625$ toward Q1759+75) where transitions from multiple ionization states were unambiguously detected and argued that this

DLA with $N(\text{H I}) = 10^{20.65} \, \text{cm}^{-2}$ requires significant ionization corrections, contrary to the theoretical expectation. These authors also argued, however, that the conditions in the DLA toward Q1759+59 are probably unusual. Indeed, this expectation is supported by more recent studies (e.g., Prochaska et al. 2002b).

Without ionization corrections, the gas-phase abundances are trivially computed from the ionic column densities (e.g., Fe/H $= N(\text{Fe}^+)/N(\text{H}^0)$), and the precision of the elemental abundances match those of the column densities. Owing to systematic errors related to continuum fitting near the Lyα profile, the principal source of error tends to lie in $N(\text{H}^0)$, which generally limits the precision to ~ 0.1 dex. For relative abundance measurements the precision is often better than 0.05 dex at the 1σ level, surpassing all but the most accurate relative abundance measurements derived from stellar analyses. As we shall see in the next section, however, the precision achieved for these gas-phase abundances can be severely compromised by the effects of dust.

29.3   Dust

Since the pioneering studies by Strömgen (1948) and Spitzer (1954) on Ca$^+$ and Na^0 ions in the Milky Way ISM, astronomers have appreciated that refractory elements are depleted from the gas phase. These optical surveys gave the first convincing demonstration of depletion, and later UV spectroscopy revealed a more complete picture of dust. Similar observations of gas within the SMC and LMC have provided insight into depletion in other

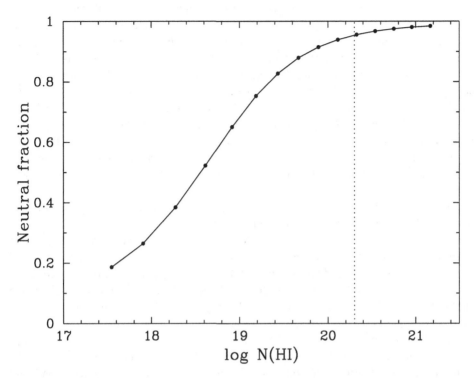

Fig. 29.2. Predicted neutral fractions for a plane-parallel slab of hydrogen gas with a range of H I surface densities $N(\text{H I})$. In this calculation, we assumed a hydrogen volume density $n_{\text{H}} = 0.1\,\text{cm}^{-3}$ and a standard extragalactic background radiation field. The results are based on a standard radiative transfer calculation (see Prochaska & Wolfe 1996 for more details).

galactic systems and have suggested depletion is universal with a generic pattern (see the Jenkins 2004 review).

Presently, the majority of the uncertainty, confusion, frustration, and pain associated with studying chemical abundances in the damped Lyα systems stems from dust (see the reviews by Draine 2004 and Jenkins 2004 for a complete discussion of dust). Dust plays two roles in the observations, one direct and one indirect. The direct effect is that refractory elements in the DLA (e.g., Fe, Ni, Cr) are depleted from the gas phase into and onto dust grains. One expects the processes are similar to those observed for the ISM of the Milky Way, although it is difficult to confirm this at high redshift. It is clear, however, that depletion levels in the DLA are significantly lower than typical sightlines through the Milky Way, instead often resembling warm gas in our Galactic halo or the gas in the LMC and SMC.

Dust would not pose such a difficult problem in a discussion of the DLA abundances if not for two points: (1) spectra of the "typical" DLA generally allow abundance measurements for only a few elements, primarily Si, Fe, Ni, Cr, and Zn; (2) there is an unfortunate degeneracy between the differential depletion patterns of these few elements (Zn exempted) and the nucleosynthetic pattern expected for Type II supernovae (e.g., Woosley & Weaver 1995). To wit, differential depletion implies enhancements of Si/Fe and roughly solar Fe, Cr, and Ni abundances, as does Type II supernova nucleosynthesis*. It is for this reason, above all others, that the non-refractory element Zn, an element with a speculative nucleosynthetic origin *at best*, has received such great prominence in DLA abundance research. Empirically, Zn roughly traces Fe in stars with metallicity [Fe/H] > −3 (see the review by Nissen 2004), and Zn is very nearly non-refractory. Therefore, the majority of the DLA community has adopted Zn as a proxy for Fe and have imposed dust corrections on the gas-phase abundances under the assumption that Zn/Fe should be solar in the DLAs (e.g., Vladilo 1998). These are sensible approaches, but the uncertainty in the nucleosynthetic origin of Zn gives me pause (as do issues relating to ionization corrections; see Jenkins 2004). If the Zn/Fe ratio is intrinsically ∼ +0.2 dex in the DLA, one may draw very different conclusions on the α/Fe ratios and ultimately the roles of various supernovae in the enrichment of these galaxies.

The other major issue related to dust in the DLAs is obscuration. The vast majority of DLAs have been identified toward bright quasars identified in optical or UV surveys. If the sightline to a given quasar penetrates a gas "cloud" with a large column of dust, it is possible the quasar will be removed from these magnitude-limited surveys. This was the concern of Ostriker & Heisler (1984), and Fall & Pei (1993) have developed a formalism to account for this selection bias. A full discussion of the likelihood that dust obscuration is influencing studies of damped Lyα abundances is somewhat beyond the scope of this paper. We will return to the topic in the next section, but we also refer the interested reader to the papers by Boissé et al. (1998), Ellison et al. (2001), and Prochaska & Wolfe (2002). My hope and current expectation is that the effects of dust obscuration are small at high redshift where the gas metallicities are lower and the observed sightlines show relatively low depletion levels and low molecular gas fractions (Ledoux, Petitjean, & Srianand 2004).

29.4   Chemical Evolution

The zeroth-order measure of a damped Lyα system (aside from its redshift) is the H I column density. By surveying the Lyα profile toward quasars at a range of redshifts, observers have traced the H I mass density of the Universe at a range of epochs (Wolfe et al. 1986; Lanzetta, Wolfe, & Turnshek 1995; Wolfe et al. 1995; Storrie-Lombardi & Wolfe 2000). The measurements for the damped systems provide a cosmic H I mass density because these galaxies dominate the neutral hydrogen gas density to at least $z = 4$ (see also Péroux et al. 2004).

The first-order measure of a DLA is its metallicity. This is determined from the gas-phase measurements of Fe$^+$, Si$^+$, Zn$^+$, Cr$^+$, and other ions. Because of dust depletion, one expects that the refractory elements (e.g., Fe, Ni, Cr) provide systematically lower metallicity values. When possible, therefore, observers have focused on non-refractory or mildly refractory elements, especially Zn. Indeed, this dictated the strategy of the first surveys by Pettini et al. (1994, 1997). These surveys provided the first ∼ 20 DLA metallicities, which showed the mean metallicity of the DLA (i.e. the neutral gas of the Universe) is ∼ 1/10 solar at $z = 2$.

In the 10 years since Pettini and collaborators initiated this field, the study of DLA abundances has evolved substantially, primarily owing to the birth of 10 m-class telescopes. Surveys with HIRES on the Keck Telescope (Lu et al. 1996; Prochaska & Wolfe 1999, 2000) have pushed the metallicity measurements to $z = 4$ and beyond and allowed the first exami-

---

* I suspect that this degeneracy between seemingly very different processes may not be a simple coincidence but may be the result of condensation temperatures correlating with even-numbered nuclei.

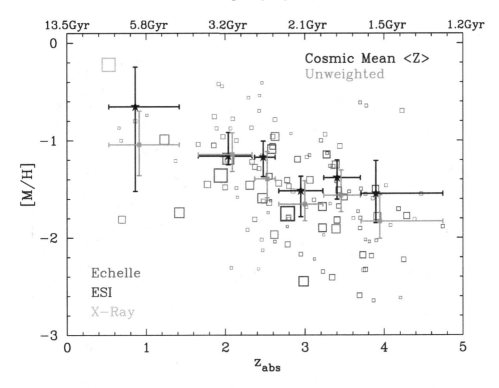

Fig. 29.3. Summary of the metallicity measurements vs. redshift for the 121 DLAs comprising the full, current sample. The area of the data points (squares) scales with the N(H I) values of the DLAs. The dark binned values with stars correspond to the cosmic mean metallicity $\langle Z \rangle$, which is the metallicity of the Universe in neutral gas.

nation of evolution in the mean metallicity. To extend the metallicity measurements above $z = 3$, these authors had to consider elements other than Zn because (1) it is difficult to measure its weak transitions along low-N(H I), low-metallicity sightlines and (2) the Zn II transitions are redshifted to observed wavelength $\lambda > 9000$ Å. The Prochaska & Wolfe (2000) survey was comprised of ~ 50 Fe measurements ranging from $z = 2$ to 4 and showed no statistically significant evolution in the mean metallicity. The advent of UVES on the VLT has led to an additional set of measurements (e.g., Molaro et al. 2000; Dessauges-Zavadsky et al. 2001), and the largest recent impact comes from new surveys using the Echellette Spectrograph and Imager (ESI) on the Keck II Telescope (Prochaska et al. 2003a, c). Owing to the high throughput of this moderate resolution spectrograph ($R \approx 10,000$) and an improved observing strategy, we have roughly doubled the sample of $z > 2$ metallicity measurements in $\sim 1/10$ the observing time. For the foreseeable future, instruments like ESI are going to lead this area of damped Lyα research.

Figure 29.3 presents a summary of the current set of damped Lyα metallicities, [M/H], as a function of redshift. A detailed discussion of these results is given in Prochaska et al. (2003b). In brief, both the unweighted and H I-weighted mean metallicities show evolution with redshift at 3σ significance with a slope $m \approx -0.25$ dex/Δz. The H I-weighted mean, often denoted $\langle Z \rangle$, is a true cosmic quantity; it represents the mean metallicity of the Uni-

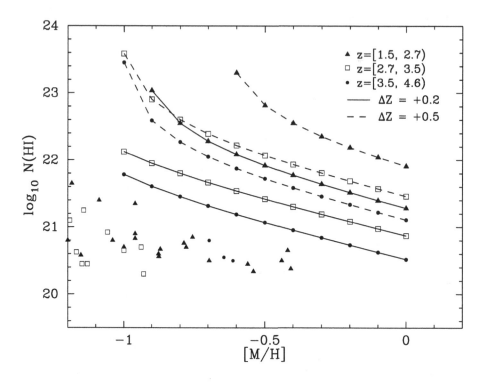

Fig. 29.4. This figure describes the robustness of the ⟨Z⟩ values presented in Fig. 29.3 to the presence of an outlier. We characterize the outlier by a range of H I column densities and [M/H] values. The curves are contours of constant ΔZ, the change in ⟨Z⟩ from including an outlier as a function of N(H I) and [M/H]. The point styles refer to three redshift intervals. The "free" points in the figure are observed DLA galaxies.

verse in neutral gas. Its determination allows direct comparisons with chemical enrichment and galaxy formation models in the early Universe (Pei, Fall, & Hauser 1999; Mathlin et al. 2000; Somerville, Primack, & Faber 2001). At present, there is a significant disagreement between the metallicities implied by the DLAs and the metal production inferred from a derivation of their star formation rates via observations of the C II* λ1335 transition. (Wolfe, Gawiser, & Prochaska 2003). This "missing metals" problem raises an important challenge for future observations and theoretical efforts related to the production of metals in the early Universe.

Prochaska et al. (2003b) noted that the set of 121 damped systems is the first sample with sufficient size to present a determination of ⟨Z⟩ robust to "reasonable" outliers. This point is emphasized in Figure 29.4, where we plot contours of ΔZ, the change in ⟨Z⟩, as a function of log$N$(H I) and [M/H] values for an assumed outlier. The point types correspond to various redshift bins and the line style indicates the magnitude change in ⟨Z⟩. Pairs of log$N$(H I), [M/H] values for the observed DLA are also shown in the figure as isolated points with point type according to their redshift. Consider the interval $1.5 < z < 2.7$. To impart a 0.2 dex change in ⟨Z⟩, one requires an outlier with 1/3 solar metallicity to have N(H I) $\approx 10^{22}$ cm^{-2}. An outlier with these characteristics would lie one magnitude off the observed

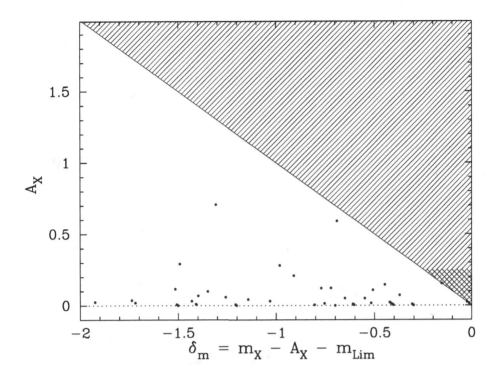

Fig. 29.5. Extinction corrections $A(X)$ derived from observed dust column densities against δ_m, the difference between the corrected brightness of the background quasar relative to the limiting magnitude of the damped Lyα survey (see Fig. 23 of Prochaska & Wolfe 2002). The shaded region denotes the area of parameter space that obscured quasars would occupy. The cross-hatched region designates the area of parameter space that we contend is populated by obscured quasars with foreground damped Lyα systems as inferred by the observed distribution of $A(X)$, δ_m values.

distribution of $N(\text{H I})$, [M/H] values. Even in the highest-redshift interval, which has the smallest sample size, it would take an outlier with a product of $N(\text{H I})$ and metallicity that is 3 times larger than any current observation. By definition, of course, an outlier lies off the main distribution of observed values. At present, however, to impose a large increase in $\langle Z \rangle$, one would have to introduce an outlier that lies far beyond the distribution of observed values. If such an outlier is identified, it would strongly suggest the existence of a currently unidentified population of DLAs with $\log N(\text{H I}) + [\text{M/H}] > 20.6$.

Indeed, Boissé et al. (1998) were the first to emphasize that the observed DLAs exhibit an upper limit to the sum $\log N(\text{H I}) + [\text{M/H}]$. They interpreted this upper limit in terms of dust obscuration; DLAs with a large product of $N(\text{H I})$ and metallicity may have larger dust optical depths and therefore may significantly obscure background quasars at UV wavelengths. We have argued that no evidence exists for significant obscuration at $z > 2$, as Ellison et al. (2001) and Prochaska & Wolfe (2002) discuss. The former authors have conducted a survey of DLAs toward radio-selected quasars and found no significant difference in the H I content of these DLAs, contrary to the expectation for dust obscuration. The latter authors have

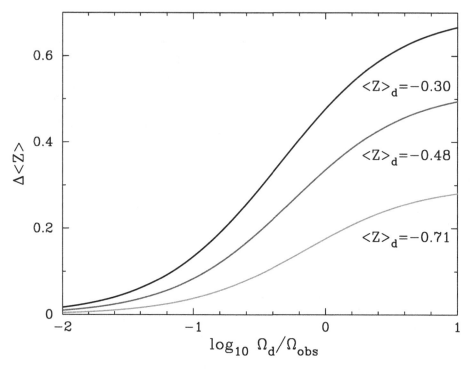

Fig. 29.6. Correction (ΔZ) to a $\langle Z \rangle$ value of 1/10 solar for an obscured gas component with mass density Ω_d and average metallicity $\langle Z \rangle_d$. The figure shows that if the observed gas mass density equals the observe quantity ($\Omega_d/\Omega_{obs} = 1$) and the observed gas has metallicity $\langle Z \rangle = 1/3$ solar, it would imply a factor of 2 correction to the observed mean metallicity.

argued that the observed distribution of inferred extinction values and apparent magnitudes indicate depletion plays a minor effect in the DLA analysis. In Figure 29.5, we present a figure similar to Figure 23 of Prochaska & Wolfe (2002), which includes the majority of the DLAs from our ESI survey. We refer the reader to the discussion in Prochaska & Wolfe (2002) for further details.

Independent of the above arguments, it is possible to assess the effects that dust obscuration will have on the measurements of $\langle Z \rangle$ for an assumed mass and metallicity of the obscured gas. Let $\langle Z \rangle_{tru} = \langle Z \rangle + \Delta Z$ and $\Omega_{tru} = \Omega_{obs} + \Omega_d$, where Ω_{obs} is the observed neutral gas density and Ω_d is the gas density that is obscured. Finally, assume that the obscured gas has mean metallicity $\langle Z \rangle_d$. Figure 29.6 shows the correction ΔZ to an observed mean metallicity of $\langle Z \rangle = -1$ dex for a range of Ω_d/Ω_{obs} values and three assumed $\langle Z \rangle_d$ values. We emphasize that the results in Figure 29.6 hold independently of the number of DLAs observed in the determination of $\langle Z \rangle$. The CORALS survey (Ellison et al. 2001) has argued that $\Omega_{tru} \approx \Omega_{obs}$, implying that Ω_d/Ω_{obs} is small. If Ω_d/Ω_{obs} is as large as 1, then one must worry about a significant correction to $\langle Z \rangle$ for $\langle Z \rangle_d$ values greater than 1/3 solar. If future surveys similar to CORALS demonstrate that Ω_d/Ω_{obs} is 1/10 or smaller, then it is unlikely dust obscuration will ever play a major role in the determination of $\langle Z \rangle$.

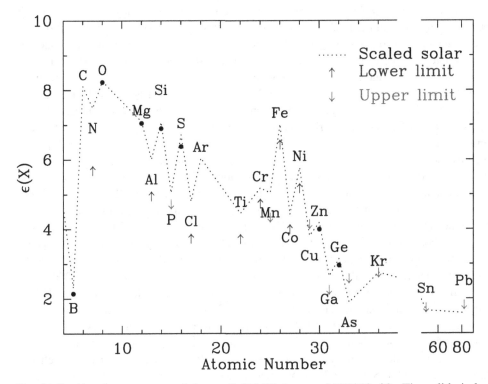

Fig. 29.7. Abundance pattern of the $z = 2.626$ DLA toward FJ0812+32. The solid circles and arrows represent detections and limits, respectively. The dotted line shows the solar abundances scaled to the oxygen abundance in the DLA.

29.5   Nucleosynthesis

While echelle observations on 10 m-class telescopes have led to a greater number of DLA metallicity measurements at a greater range of redshifts, the most significant advances from large telescopes have come through studies of the relative chemical abundances. These observations reveal the processes of nucleosynthesis during the first few billion years of the Universe and ultimately provide insight into the nature of the DLA galaxies and scenarios of galaxy formation (see Calura et al. 2004).

As discussed above, typical DLA observations have yielded relative abundance measurements for Si, Fe, Cr, Ni, and Zn with accuracies better than 10%. The principal difficulty in applying these observations to studies of nucleosynthesis is the effects of differential depletion. Specifically, the standard ISM depletion patterns (Savage & Sembach 1996) are very similar to the patterns expected for the yields from Type II supernovae (e.g., Woosely & Weaver 1995). This degeneracy led to arguments over the appropriate interpretation of Si/Fe enhancements and other ratios resulting from the first DLA surveys (Lu et al. 1996; Vladilo 1998; Pettini et al. 1999; Prochaska & Wolfe 1999). These arguments are still being discussed and may never be unambiguously resolved. Therefore, the community has turned its attention toward obtaining observations of elements that are not heavily depleted (e.g., N, O, Si, S, Zn, P) or whose differential depletion pattern runs contrary to expectations from nucleosynthesis (e.g., Mn/Fe, Ti/Fe; Dessauges-Zavadsky, Prochaska, & D'Odorico 2002).

One example of a nucleosynthetic diagnostic that is nearly free of depletion effects is the N/α ratio (where α in the DLA is generally given by Si and S). This abundance ratio is an excellent diagnostic of the time scales of star formation owing to the belief that N production is dominated by intermediate-mass stars (see Henry 2004). For this reason, among others, observations of N/α have played an important role in several contributions to this Symposium (see papers by, e.g., Garnett 2004 and Molaro et al. 2004). Regarding the DLAs over the past two years, observers have built a sample of ~ 30 measurements with metallicities ranging from [α/H] = -2 to nearly solar (Pettini et al. 2002; Prochaska et al. 2002b; Centurión et al. 2003). In general, these measurements track the N/α values observed in H II regions and stars in the local Universe. With our sample (Prochaska et al. 2002b), however, we speculated that the DLAs at low metallicity show a bimodal distribution of N/α values. In particular, we found that a small but significant fraction of these DLAs have very low N/α values, much lower than any value observed locally. If confirmed by future surveys, this bimodality may require revised yields of N in massive stars (e.g., Molaro et al. 2004), an initial (Population III?) epoch of star formation characterized by a top-heavy or truncated initial mass function (Prochaska et al. 2002b), or some other unappreciated physical mechanism. We defer additional discussion of this topic to the paper by Molaro et al. (2004).

At this Symposium, we reported the discovery of a DLA whose large N(H I) and [M/H] values will allow the detection and analysis of over 20 elements in a single galaxy (Prochaska, Howk, & Wolfe 2003). Figure 29.7 shows the abundance pattern of this galaxy and a comparison with the solar abundance pattern scaled to the galaxy's oxygen abundance ([O/H] $\approx 1/3$ solar). Aside from the analysis of stars in the Milky Way and its nearest neighbors (see, e.g., Hill 2004; Shetrone 2004; Venn et al. 2004), galaxies like this DLA will enable the most comprehensive nucleosynthetic analysis at any epoch in the Universe. The results presented in Figure 29.7 and future observations will have the following impacts on theories of nucleosynthesis.

(1) Observations of the B/O ratio test processes of cosmic ray and ν-wind spallation invoked to explain the production of the light elements B, Be, and Li (Fields & Olive 1999).

(2) Some of the galaxies will show measurements of all three CNO elements in the same DLA. These measurements yield clues to nucleosynthesis in intermediate-mass stars and place important time constraints on metal enrichment (Henry, Edmunds, & Köppen 2000).

(3) The relative abundances of the α-elements (e.g., O, Mg, Si, S) and the examination of odd-Z elements (e.g., P, Al, Ga, Mn) can be used to test predictions of explosive nucleosynthesis (e.g., Woosley & Weaver 1995). Furthermore, these abundances probe the initial mass function and mix of Type II vs. Type Ia supernovae in these protogalaxies. For the galaxy in Figure 29.7, the decline in relative abundance of the α-elements (e.g., [O/S] $\approx +0.3$) and the enhanced "odd-even effect" (e.g., [P/Si] < 0) suggest an enrichment history dominated by massive stars.

(4) Observations of Pb, Kr, Sn, and Ge will constrain theories of the s-process and r-process, and particularly AGB nucleosynthesis (e.g., Travaglio et al. 2001). It is important to emphasize that this scientific inquiry takes place in a relatively metal-rich gas (O/H $\approx 1/3$ solar) in a system with a strict upper limit to its age of 2.5 Gyr. The latter point is particularly relevant to theories on nucleosynthesis because this time scale limits the contribution from intermediate-mass stars.

While the results for the galaxy presented in Figure 29.7 will—on their own—place new constraints on theories of nucleosynthesis, the real excitement from its discovery is the promise of identifying many other galaxies with similar characteristics. We are currently pursuing several DLAs with N(H I) and [M/H] values similar to those of the $z = 2.626$ DLA toward FJ0812+32, whose observations should yield abundance measurements of ~ 20 elements in each DLA. These observations will reveal if the $z = 2.626$ DLA toward FJ0812+32 is a unique case or representative of the population of metal-rich DLAs. In addition to these efforts, we have begun a survey with ESI on Keck II to find an additional 10–50 of these DLAs. Several groups at this meeting are now involved with searches for extremely metal-poor stars at $z = 0$. Our complementary effort is to discover relatively metal-rich galaxies at very high z.

References

Boissé, P., Le Brun, V., Bergeron, J., & Deharveng, J.-M. 1998, A&A, 333, 841

Calura, F., Matteucci, F., Dessauges-Zavadsky, M., D'Odorico, S., Prochaska, J. X., & Vladilo, G. 2004, in Carnegie Observatories Astrophysics Series, Vol. 4: Origin and Evolution of the Elements, ed. A. McWilliam & M. Rauch (Pasadena: Carnegie Observatories, http://www.ociw.edu/symposia/series/symposium4/proceedings.html)

Centurión, M., Molaro, P., Vladilo, G., Péroux, C., Levshakov, S. A., D'Odorico, V. 2003, A&A, 403, 55

Dekker, H., D'Odorico, S., Kaufer, A., Delabre, B., & Kotzlowski, H. 2000, SPIE, 4008, 534

Dessauges-Zavadsky, M., D'Odorico, S., McMahon, R. G., Molaro, P., Ledoux, C., Péroux, C., & Storrie-Lombardi, L. J. 2001, A&A, 370, 426

Dessauges-Zavadsky, M., Prochaska, J. X., & D'Odorico, S. 2002, A&A, 391, 801

Draine, B. T. 2004, in Carnegie Observatories Astrophysics Series, Vol. 4: Origin and Evolution of the Elements, ed. A. McWilliam & M. Rauch (Cambridge: Cambridge Univ. Press), in press

Ellison, S. L., Yan, L., Hook, I. M., Pettini, M., Wall, J. V., & Shaver, P. 2001, A&A, 379, 393

Fall, S. M., & Pei, Y. C. 1993, ApJ, 402, 479

Fields, B. D., & Olive, K. A. 1999, ApJ, 516, 797

Garnett, D. R. 2004, in Carnegie Observatories Astrophysics Series, Vol. 4: Origin and Evolution of the Elements, ed. A. McWilliam & M. Rauch (Cambridge: Cambridge Univ. Press), in press

Henry, R. B. C. 2004, in Carnegie Observatories Astrophysics Series, Vol. 4: Origin and Evolution of the Elements, ed. A. McWilliam & M. Rauch (Cambridge: Cambridge Univ. Press), in press

Henry, R. B. C., Edmunds, M. G., & Köppen, J. 2000, ApJ, 541, 660

Hill, V. 2004, in Carnegie Observatories Astrophysics Series, Vol. 4: Origin and Evolution of the Elements, ed. A. McWilliam & M. Rauch (Cambridge: Cambridge Univ. Press), in press

Howk, J. C., Sembach, K. R., Roth, K. C., & Kruk, J. W. 2000, ApJ, 544, 867

Jenkins, E. B. 2004, in Carnegie Observatories Astrophysics Series, Vol. 4: Origin and Evolution of the Elements, ed. A. McWilliam & M. Rauch (Cambridge: Cambridge Univ. Press), in press

Kobulnicky, H. A., & Koo, D. C 2000, ApJ, 545, 712

Lanzetta, K. M., Wolfe, A. M., & Turnshek, D. A. 1995, ApJ, 440, 435

Ledoux, C., Petitjean, P., & Srianand, R. 2004, MNRAS, submitted (astro-ph/0302582)

Lu, L., Sargent, W. L. W., Barlow, T. A., Churchill, C. W., & Vogt, S. 1996, ApJS, 107, 475

Mathlin, G. P., Baker, A. C., Churches, D. K., & Edmunds, M. G. 2001, MNRAS, 321, 743

Meyer, D. M., Jura, M., & Cardelli, J. A. 1998, ApJ, 493, 222

Molaro, P., Bonifacio, P., Centurión, M., D'Odorico, S., Vladilo, G., Santin, P., & Di Marcantonio, P. 2000, ApJ, 541, 54

Molaro, P., Centurión, M., D'Odorico, S., & Péroux, C. 2004, in Carnegie Observatories Astrophysics Series, Vol. 4: Origin and Evolution of the Elements, ed. A. McWilliam & M. Rauch (Pasadena: Carnegie Observatories, http://www.ociw.edu/symposia/series/symposium4/proceedings.html)

Nissen, P. E. 2004, in Carnegie Observatories Astrophysics Series, Vol. 4: Origin and Evolution of the Elements, ed. A. McWilliam & M. Rauch (Cambridge: Cambridge Univ. Press), in press

Ostriker, J. P., & Heisler, J. 1984, ApJ, 278, 1

Pei, Y. C., Fall, S. M., & Hauser, M. G. 1999, ApJ, 522, 604

Péroux, C., Dessauges-Zavadsky, M., D'Odorico, S., Kim, T. S., & McMahon, R. G. 2004, in Carnegie Observatories Astrophysics Series, Vol. 4: Origin and Evolution of the Elements, ed. A. McWilliam & M. Rauch (Pasadena: Carnegie Observatories, http://www.ociw.edu/symposia/series/symposium4/proceedings.html)

Pettini, M., Ellison, S. L., Bergeron, J., & Petitjean, P. 2002, A&A, 391, 21

Pettini, M., Ellison, S. L., Steidel, C. C., & Bowen, D. V. 1999, ApJ, 510, 576

Pettini, M., Smith, L. J., Hunstead, R. W., & King, D. L. 1994, ApJ, 426, 79

Pettini, M., Smith, L. J., King, D. L., & Hunstead, R. W. 1997, ApJ, 486, 665

Prochaska, J. X., Castro, S., Djorgovski, S. G. 2003a, ApJS, 148, 317

Prochaska, J. X., Gawiser, E., Wolfe, A. M., Castro, S., & Djorgovski, S. G. 2003b, ApJS, 595, L9

Prochaska, J. X., Gawiser, E., Wolfe, A. M., Cooke, J., & Gelino, D. 2003c, ApJS, 147, 227

Prochaska, J. X., Henry, R. B. C., O'Meara, J. M., Tytler, D., Wolfe, A. M., Kirkman, D., Lubin, D., & Suzuki, N. 2002b, PASP, 114, 933

Prochaska, J. X., Howk, J. C., O'Meara, J. M., Tytler, D., Wolfe, A. M., Kirkman, D., Lubin, D., & Suzuki, N. 2002a, ApJ, 571, 693

Prochaska, J. X., Howk, J. C., & Wolfe, A. M. 2003, Nature, 427, 57

Prochaska, J. X. & Wolfe, A. M. 1996, ApJ, 470, 403

——. 1999, ApJS, 121, 369

——. 2000, ApJ, 533, L5

——. 2002, ApJ, 566, 68

Savage, B. D., & Sembach, K. R. 1991, ApJ, 379, 245

——. 1996, ARA&A, 34, 279

Shetrone, M. 2004, in Carnegie Observatories Astrophysics Series, Vol. 4: Origin and Evolution of the Elements, ed. A. McWilliam & M. Rauch (Cambridge: Cambridge Univ. Press), in press

Somerville, R. S., Primack, J. R., & Faber, S.M. 2001, MNRAS, 320, 504

Spitzer, L., Jr. 1954, ApJ, 120, 1

Strömgen, B. 1948, ApJ, 108, 242

Storrie-Lombardi, L. J., & Wolfe, A. M. 2000, ApJ, 543, 552

Travaglio, C., Gallion, R., Busso, M., & Gratton, R. 2001, ApJ, 549, 346

Venn, K. A., Tolstoy, E., Kaufer, A., & Kudritzki, R. P. 2004, in Carnegie Observatories Astrophysics Series, Vol. 4: Origin and Evolution of the Elements, ed. A. McWilliam & M. Rauch (Pasadena: Carnegie Observatories, http://www.ociw.edu/symposia/series/symposium4/proceedings.html)

Viegas, S. M. 1994, MNRAS, 276, 268

Vladilo, G. 1998, ApJ, 493, 583

Vladilo, G., Centurión, M., Bonifacio, P., & Howk, J. C. 2001, ApJ, 557, 1007

Vogt, S. S., et al. 1994, SPIE, 2198, 362

Welty, D. E., Frisch, P. C., Sonneborn, G., & York, D. G. 1999, ApJ, 512, 636

Welty, D. E., Lauroesch, J. T., Blades, J. C., Hobbs, L. M., & York, D. G. 2001, ApJ, 554, 75

Wolfe, A. M., Gawiser, E., & Prochaska, J. X. 2003, ApJ, 593, 235

Wolfe, A. M., Lanzetta, K. M., Foltz, C. B., & Chaffee, F. H. 1995, ApJ, 454, 698

Wolfe, A. M., Turnshek, D. A., Smith, H. E., & Cohen, R. D. 1986, ApJS, 61, 249

Woosley, S. E. & Weaver, T. A. 1995, ApJS, 101, 181

Zsargó, J., & Federman, S. R. 1998, ApJ, 498, 256

30

Intergalactic medium abundances

ROBERT F. CARSWELL
Institute of Astronomy, Cambridge, UK

Abstract
Galaxies at high redshifts will enrich the intergalactic medium with heavy elements through winds and supernova ejecta, and QSO absorption lines provide the most useful means of examining the evolution of the element abundances in this medium. Photoionization by background radiation at the densities appropriate for such a medium suggests that ions such as C IV and O VI will provide the strongest absorption lines for comparison with hydrogen. Direct detection of these lines provides heavy element abundance estimates of $\sim 10^{-1.5}$ to 10^{-3} solar in regions where the neutral hydrogen column densities are $\sim 10^{14.5}$ cm^{-2} or more. For lower-density regions, which are generally further from galaxies, average abundances may be inferred by stacking systems or pixel optical depths around the relevant wavelengths. Such methods have some shortcomings, and the results do not always agree very well, but suggest that heavy element abundances are $\leq 10^{-2.5}$ solar.

30.1 Introduction

The cold dark matter simulations describing the growth of structure in the Universe give a very good description of the nature of the intergalactic H I content, as confirmed by high-resolution spectroscopy of QSOs against which the corresponding H I absorption may be seen. What this shows is fairly complex structure with, as one would expect, a smooth transition between condensations, where self-gravity becomes important, and the surrounding medium. Under these circumstances what is truly intergalactic medium becomes a matter of where one wants to put the boundary between what is in galaxies and what is not.

There have been several papers describing the growth of structure in the Universe and the Lyα forest as a tracer of structure in the intergalactic medium (e.g., Miralda-Escudé et al. 1996; Jenkins et al. 1998; Croft et al. 2002). The models show an expected evolution in the gas component in the Universe, and the observational material shows similar behavior (Machacek et al. 2000). Using a range of methods, comparison between models and observations via QSO absorption in neutral hydrogen shows gratifying agreement (Rauch, Haehnelt, & Steinmetz 1997), though of course model selection is based in part on this agreement, and the data themselves for the Lyα forest are particularly uncertain below redshifts $z \approx 1.6$ or so, simply because at low redshifts Lyman forest studies require *HST*.

One of the topics being actively considered now comes under the general heading of "feedback," i.e. how the forming structures generate centers of radiation and heavy element production that then affect the surrounding intergalactic medium (Cen & Ostriker 1999).

HE2347−4342

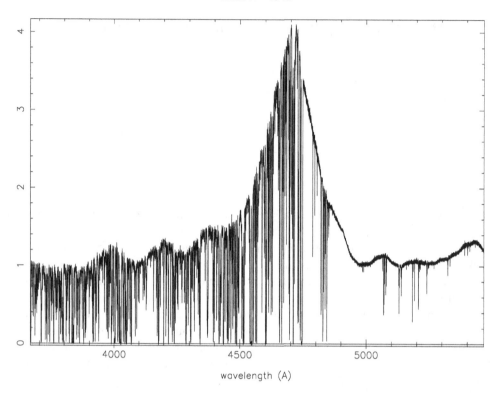

wavelength (A)

Fig. 30.1. The ESO VLT spectrum of the redshift $z = 2.88$ QSO HE2347−4342 showing the spectral region around the Lyα emission line. (Data from the ESO QSO absorption large program; PI = J. Bergeron.)

There have been a number of studies of the ionization of an inhomogeneous intergalactic medium by star-forming galaxies and quasars (Ciardi et al. 2000), the heating and ionization, particularly of helium (Theuns et al. 2002; Sokasian, Abel, & Hernquist 2003), the transport of heavy elements by galactic winds (e.g., Croft et al. 2002), and supernova-driven pregalactic outflows (Madau, Ferrara, & Rees 2001; Masao, Ferrara, & Madau 2002).

Here we concentrate on the observational constraints on the chemical enrichment of the intergalactic medium, particularly at redshifts $2 < z < 4$ where there is a wealth of material available from QSO absorption-line studies, especially from Keck HIRES and, more recently, UVES at the VLT. A sample spectrum showing the rich Lyα absorption at wavelengths shortward of the Lyα emission line, and the relative paucity of heavy element lines at longer wavelengths, is shown in Figure 30.1. It is the comparison of the H I line strengths and those from various ionization stages of heavier elements in such spectra that form the basis for intergalactic medium abundance studies.

Since the gas in the intergalactic medium is highly ionized, we consider briefly in the next section the background photoionizing flux. Following that we turn to the detection of heavy elements through the direct measurement of the ion transitions in high signal-to-noise ratio (S/N) spectra, and note that these generally correspond to the regions with higher densities

N(HI)

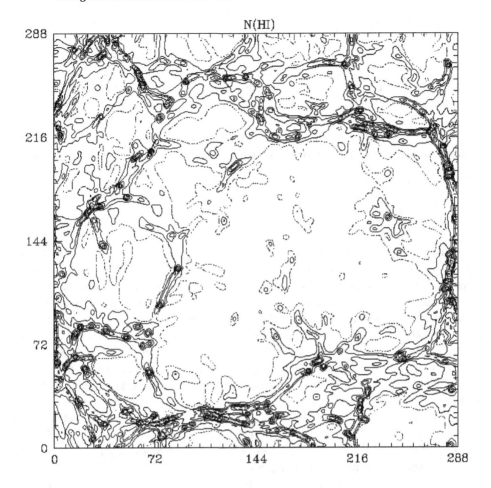

Fig. 30.2. A slice of the simulation by Miralda-Escudé et al. (1996) showing the neutral hydrogen column density at $z = 3$, with contours at $10^{12.5}$ cm^{-2} (dotted) and higher levels at intervals of 0.5 dex. (Reproduced from Miralda-Escudé et al. 1996.)

and higher H I column densities, which place them quite close to condensations that are associated with galaxies (see Fig. 30.2). We then describe some stacking techniques used to explore the abundances in lower-density regions, which represent the intergalacic medium far from galaxies, where the heavy element lines are too weak to be detected in individual cases.

30.2   Abundances and Ionization

The study of element abundances in the intergalactic medium became a matter of observation rather than speculation when Chaffee et al. (1985) drew attention to a $z = 3.32$ absorption system toward the redshift $z_{em} = 3.42$ QSO 0014+81, which has two components with H I column densities $\log N(\text{H I}) = 16.7$ and 16.3 cm^{-2}. The marginal presence of Si III at this redshift in their spectra leads to a metal-to-hydrogen ratio approximately $10^{-2.7}$ solar if the identification is correct. Other possible lines, notably C IV, C II, and O VI, were not detected, so the heavy element abundances in this system were clearly low. A subsequent investigation by Chaffee et al. (1986) concluded from a search for C III $\lambda977$ that the heavy element abundances in this system are likely to be $< 10^{-3.5}$ solar. Later work by Cowie et al. (1995) on the same object was based on higher-S/N Keck spectra, and they found both C II and Si IV at levels that turn out to be broadly consistent with a heavy element abundance $\sim 10^{-3}$ solar, but with relative ratios of the various heavy elements somewhat different from the solar values.

The range of values for the abundances in this case is not surprising. There are several assumptions that go into this general modelling and line strength prediction. The most important of these are:

- The clouds are predominantly photoionized and in ionization equilibrium, with a specified ionizing flux.
- the heavy elements are in fixed (usually solar) ratios.

The assumed ionizing flux changes not only with epoch, but, not suprisingly as our knowledge improves, with time of publication. Chaffee et al. (1986) assumed a Weymann-Malkan input spectrum (Bechtold et al. 1987) based on a QSO background with attenuation from the intervening galaxies. Later models by Haardt & Madau (1996) are also based on QSOs as the source, and the most recent models add a contribution from young galaxies (Haardt & Madau 2001). The neutral hydrogen fraction can be as little as 10^{-5} for typical cases, and this fraction depends on the ionizing flux in the region of the spectrum below the Lyman limit where it is not well known. Another major uncertainty in any such model is the nature of the UV photon flux from QSOs and the He II content of the intergalactic medium. Figure 30.3 shows the expected column densities for various ions as a function of total hydrogen number density per cm^3 assuming photoionization equlibrium in a uniform-density medium irradiated by the Haardt-Madau (1996, 2001) background flux estimates at redshift $z = 2$. It is evident from these curves that differences up to about 0.5 dex may be expected, even given best estimates for the photoionizing background fluxes.

The He II opacity is expected to be critically important for species such as O VI, since the ionization potential of He II is considerably less than that required to reach O VI. The O VI doublet at $\lambda\lambda1032, 1037$ Å is likely to be the strongest observable heavy element feature if the Universe is optically thin in the He II absorption continuum, but not otherwise. Thus, O VI searches are important for two reasons—metallicity studies and ionizing background considerations. However, there is a major complication in that the O VI doublet lies deep in the Lyα forest and so it is difficult to measure. The C IV $\lambda\lambda$ 1548, 1550 Å doublet is at longer wavelengths and does not suffer from such confusion, so, while the lines are expected to be weaker, it may be the most sensitive for heavy element abundance searches, especially at higher redshifts where the Lyα forest confusion is most severe.

30.3   C IV Searches

An early in-depth study of heavy elements associated with the lower H I column density systems, which are more typical of the Lyα forest, was that of Cowie et al. (1995). They found that roughly half of the Lyman forest lines with $\log N(\text{H I}) > 14.5$ cm^{-2} at redshifts $z \approx 3$ have measurable C IV with $\log N(\text{C IV}) > 12.0$; so inferred metallicities are $\sim 10^{-2}$ solar. Thus, most of these systems do not have primordial element abundances, as had been assumed at the time. The C IV lines are more usually in blended complexes, and if

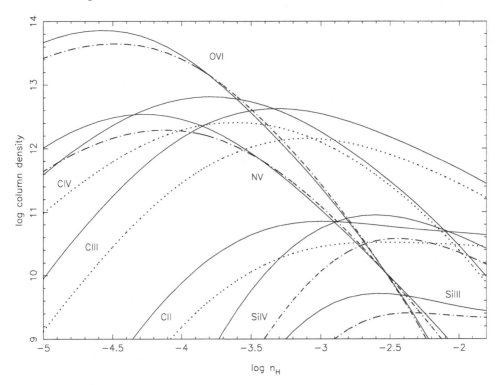

Fig. 30.3. The predicted column densities for ions in a system at redshift $z = 2.0$ for $\log N(\text{H I}) = 15.14$, heavy element abundances $10^{-2.65}$ solar, and a range of hydrogen densities n_H in the absorbing region. The solid lines are for the Haardt-Madau (2001) background, and the others are for the Haardt-Madau (1996) background flux. Carbon ions are shown as dotted lines to distinguish them from other species.

these are not transient (with a time scale $\sim 10^7$ yr), they must be parts of some gravitationally bound systems. Such velocity structure may be present in the corresponding H I lines, but in Lyα, particularly, saturation effects, and the fact that any thermal broadening of the lines is greater for hydrogen than for other species, usually make it extremely difficult to determine.

Studies by Songaila & Cowie (1996) and Songaila (1998) refined this further. They found C IV in about 75% of absorbers with $\log N(\text{HI}) > 14.5$ and more than 90% of those with $\log N(\text{HI}) > 15.2$. The latter group were found to have a narrow range of ionizations, covering less than a factor of 10 in number ratios of C IV/H I, C II/C IV, Si IV/C IV, and N V/C IV. The line widths imply photoionization rather than collisional ionization is dominant. The typical carbon abundance is $\sim 10^{-2}$ solar, and the Si/C ratio about 3 times solar. A typical example of such a system, with a total $\log N(\text{H I}) > 15.2$, is shown in Figure 30.4.

One very interesting finding was that Si IV/C IV decreases rapidly with redshift from high values (> 0.1) at $z > 3.1$. This could be the result of a change in the ionizing spectrum as the intergalactic medium becomes optically thin to He II ionizing photons. There have been indications of a change in Lyα forest line width at a similar redshift (Kim, Cristiani, & D'Odorico 2002; Theuns et al. 2002), which may arise for the same reason. However, later

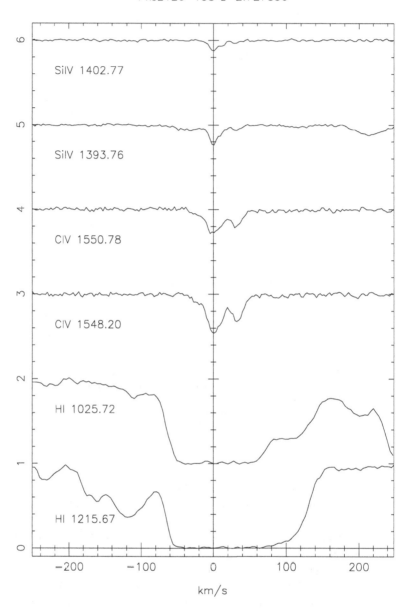

Fig. 30.4. The Lyα, Lyβ, C IV, and Si IV lines, on a common velocity scale, in the redshift system at $z = 2.727856$ in the spectrum of PKS 2126–158. Note the common velocity structure in the C IV and Si IV lines spreading over ~ 50 km s^{-1}, which cannot be seen in the saturated Lyman lines.

work shows that the trend in the Si IV/C IV ratio is less clear (Boksenberg 1997), so such an association may be spurious.

A feature of the systems in which C IV is generally found is that the H I column densities generally correspond to overdensities relative to the Universe mean of 5 or more, and these are likely to be close to galaxies, i.e. close to likely sources of chemical enrichment. However, the association is not as straightforward as the current simulations suggest, since the outflows from the galaxies themselves, necessary for the heavy element enrichment, disturb the local intergalactic medium. In detail there is a tendency for Lyα absorption to be weaker at the galaxy redshift, and stronger than the norm for a range of velocities a few hundred km s^{-1} from that redshift (see Adelberger et al. 2003). Despite this, the general trend for higher densities (and usually H I column densities) near to galaxies remains.

30.4 O VI Searches

As is evident from Figure 30.3 for the lower-density systems, which are the ones likely to be in the true intergalactic medium, O VI is the species that is most likely to be detected. The major difficulty in attempts to do this is the confusing lines from the Lyα forest, and higher-order Lyman lines. However, at redshifts ~ 2–3 the Lyman forest is not so dense as to preclude searches for individual O VI systems, and recently there have been some successes in this.

Simcoe, Sargent, & Rauch (2002) found several cases where the O VI doublet is present at redshifts $2.2 < z < 2.8$ from a search of Keck HIRES spectra. One of the search requirements was that the O VI lines be clearly present, and they found that these are generally associated with $\log N(\text{H I}) > 15.2$, along with C IV and lower-ionization species. The velocity structure is often complex, and for these systems there is frequently different velocity structure in different ions. They inferred oxygen abundances $> 10^{-1.5}$ solar on average for their sample. Figure 30.5 shows an example that illustrates this based on a UVES/VLT spectrum.

From a study of two QSOs, Carswell, Schaye, & Kim (2002) found that for $2.0 < z < 2.35$ O VI and C IV are generally present if $\log N(\text{H I}) > 14.5$, but usually without Si IV and the other lines. Their systems have lower H I column densities and often apparently simpler velocity structure. The heavy element lines are usually weak, and even at the lower redshifts they were dealing with the blending from Lyα and higher-order Lyman lines from the forest was sometimes considerable. In many cases, though these contaminants could be removed by using the profiles of other lines in the Lyman series, so the procedure is not as uncertain and arbitrary as it might appear. An example is shown in Figure 30.5, showing the fitted heavy element lines and the O VI contamination. Photoionization rather than collisional ionization must be the dominant mechanism in some cases, since the observed C IV lines are much narrower than they would have to be at the temperatures needed for collisional ionization. The heavy element abundances inferred are in the range [C/H]≈ -3 to -2, and the inferred hydrogen densities $\sim 10^{-4}$ cm^{-3} are very similar to the densities expected in the Universe models from the relation between $N(\text{H})$ and total hydrogen density given by Schaye (2002).

Bergeron et al. (2002) also undertook a study of O VI systems toward the $z = 2.45$ QSO Q0329–385, with generally similar results. They also investigated the oxygen-to-carbon ratio, and suggested that size constraints based on the Hubble flow require that this ratio be greater than the solar value. Overall they found that [O/C]≈ 0–1.3.

While the first reaction is to note the differences between the results described by Simcoe et al. (2002) and Carswell et al. (2002) particularly, they seem to be compatible. The Simcoe at al. sample is at higher H I column densities where complex velocity structure will, in any case, be more evident. It is not clear if the general velocity agreement found by Carswell

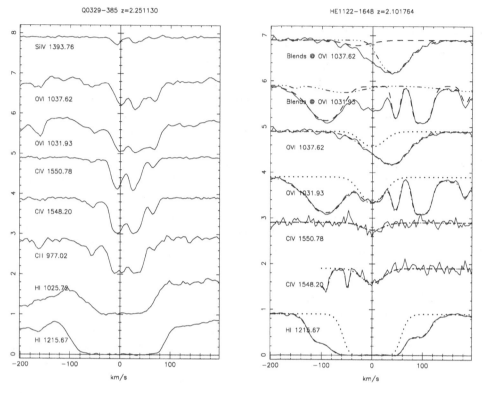

Fig. 30.5. *Left:* A complex absorption system with H I column density $N(\mathrm{H\ I}) \approx 10^{15.2}$ cm^{-2}. Note the differences in the velocity structure between the C III, C IV, O VI, and Si IV lines, and particularly that O VI shows significant offsets in this case. *Right:* An absorption system with $N(\mathrm{H\ I}) \approx 10^{14.9}$ cm^{-2} with simple velocity structure in the heavy elements. In this case Voigt profile fits to corresponding features are shown (dots), as are the blended lines that affect the observed O VI profiles (dash and dot-dash lines show individual blended components).

et al. is real or a result of having only two ions, C IV and O VI, to compare, particularly given the weakness of some of the lines and the uncertainties in allowing for Lyman lines from other systems blended with the O VI. It is interesting that the O VI systems with higher H I column densities tend to fall within the region of the O VI column density/line width diagram expected for a radiatively cooling flow of hot gas (Heckman et al. 2002, their Fig. 1), and so may be similar in nature. For the lower H I column densities, $14.5 < \log N(\mathrm{H\ I}) < 15$, the agreement is not as good, with about half of them having rather lower line widths for their column densities than other objects in that diagram. It is possible that these are the enriched intergalactic regions, which are now less affected by the galaxy that affected them.

The result of these studies has been to establish the presence of highly ionized enriched regions, with the expected densities given the H I column densities. However, the H I column densities correspond to similar overdensities as the higher-redshift C IV studies since the mean density of the Universe is lower in the redshift range where the O VI searches have been undertaken. So we wish to probe to lower H I column densities, and an effective method of doing this is finding some way of improving the S/N by averaging over many systems.

30.5   Stacked Spectra

One way to improve S/N and obtain an average result is by stacking spectra shifted to the Lyα rest position using the observed Lyα wavelengths and to look around at the wavelength regions covering C IV λλ1548, 1550 and anything else of interest. This has been attempted by several people, starting with an early attempt to detect O VI, in what would now be regarded as rather low-resolution spectra, by Norris, Peterson, & Hartwick (1983), though Williger et al. (1989) failed to confirm their tentative O VI detection in a different sample of objects. Lu (1991) obtained a tentative detection of C IV for systems with $\log N$(H I) $\approx$ 15.3 or more by this method, and suggested on this basis that the C/H ratio might be $\sim 10^{-3.2}$ solar.

Tytler & Fan (1994) used a higher-resolution spectrum to search for C IV by this stacking method and found none for H I column densities in the range $10^{13} - 10^{14}$ cm^{-2}. Keck spectra were analysed in a similar way by Lu et al. (2004), who found that for $13.5 < \log N$(H I) $<$ 14.0 C IV was not detectable at S/N>1800. From this they inferred that the abundance of carbon is at most $10^{-3.5}$ solar on average for these systems. Lu et al. (2004) removed a very few systems in which C IV was detectable, so their average value may be biased toward the low side, but not by very much.

However, there are potential problems with this method, as noted by Cowie & Songaila (1998) and investigated further by Ellison et al. (2000). For the high H I column density systems where C IV is seen in individual systems there is significant velocity structure, and the Lyα velocity centroid does not coincide with that of the C IV lines. This will be particularly serious for the earlier lower-resolution studies where blended Lyα features are not well resolved, but even for Keck HIRES spectra Ellison et al. find that there is a velocity offset with a σ = 17 km s^{-1} between the Lyα and C IV lines. Therefore, any stacking method using the Lyα centroids to determine the C IV position will have reduced sensitivity. How much the sensitivity is reduced depends on the clustering properties of the lines, and to what extent any velocity structure is determined and used for the H I. In Figure 30.4, for example, the Lyα shows a red component that is difficult to separate from the Lyα line alone, but at least that part may be removed if the Lyβ profile is examined as well. If the H I is chosen mainly from clean single systems, as in the Lu et al. (2004) sample, then one is genuinely determining the C IV/H I ratio for those particular systems. The Ellison et al. simulations show that the sensitivity difference is about 0.3 dex in the worst-case C IV/H I velocity spread they adopted, with a sensitivity limit going from C IV/H I = −3.1 to −2.8 for the 67 lines they used with $13.5 < \log N$(H I) $<$ 14.0.

For O VI there is the usual additional complication of Lyman forest blending with the lines of interest. There are two possible approaches to this. One is to treat the blended Lyman lines as an additional source of noise, and estimate this additional noise either by simulations or looking at the fluctuations in the summed spectrum near the O VI lines. Another is to best estimate the contamination from Lyman lines by fitting Voigt profiles and removing them before summing, as has been done by Simcoe, Sargent, & Rauch (2004). In any case, while the heavy element line strength will be affected by velocity offsets and uncertainties in the stacked spectrum the average line strength may be measured against a well-defined local continuum in that spectrum.

30.6   Pixel Optical Depths

A method that does not rely as much on line parameters, and in particular does not require stacking at estimated redshifts from Lyman lines, was introduced by Cowie & Songaila (1998). They correlate pixel optical depths of Lyman and heavy element species, and so avoid any aspect of profile fitting, and find that the mean Lyα to C IV optical depth ratio is roughly constant down to a detection limit corresponding to $N(\text{H I}) \approx 10^{13.5}$ cm^{-2}. This gives abundances about an order of magnitude higher than the stacking method, in an interesting contrast with the Lu et al. (2004) result. Ellison et al. (2000) comment that a velocity offset between the Lyα and C IV lines should, as we have seen, only partly account for the difference even in the worst case.

Cowie & Songaila (1998) did not detect O VI by this method in their sample, which was generally at redshifts $z > 3$, but subsequently Schaye et al. (2002) found a signal corresponding to O VI for redshifts $2 < z < 3$. A detailed description of the method and its effectiveness when used on cosmological simulations is given by Aguirre, Schaye, & Theuns (2002). They also give a recipe for how to implement the method, which in summary (for O VI to be specific) is:

- Measure residual flux relative to continuum in Lyα, and take as $\tau = -\log(\text{relative flux})$. For saturated lines, use $\min(\tau_{\text{Lyn}} f_\alpha \lambda_\alpha / f_n \lambda_n)$ for Lyman series, else τ is ill-determined.
- Determine $\tau(\text{O VI})$ at $\lambda\lambda 1032, 1038$ Å in corresponding pixels, and choose $\min[\tau(1032), 2 \times \tau(1038)]$ to minimize contamination.
- Correct for contamination by removing pixel optical depths appropriate for other lines, e.g. in the Lyman series.
- Bin into H I values (in log space), and take median $\tau(\text{Ly}\alpha)$ and $\tau(\text{O VI})$ in each bin.

Then, $\tau(\text{O VI})$ is O VI optical depth plus perhaps an offset that is independent of $\tau(\text{H I})$, which may arise particularly from errors in the continuum estimate. The continuum estimate is especially uncertain in the Lyα forest, where even in relatively uncrowded spectra at $z \approx 2$–2.5 errors of order 2% are not uncommon, and even for the C IV regions the presence of broad emission lines (not to mention small flux calibration errors causing small ripples and even jumps in the spectra) may result in continuum uncertainties of up to about 0.5%, even in spectra with S/N≈ 80 per pixel. Typical results for O VI are shown in Figure 30.6. Note the detections of O VI, down to apparent optical depths $\sim 10^{-1}$ for redshifts $z < 3.2$, and possibly up to $z \approx 3.5$, as revealed by the correlated O VI and H I optical depths.

This method shows the presence of O VI, but converting the measured quantities to abundances involves calibration via simulations. Even then, uncertainties in the continuum levels in the individual spectra can be important. A statistician might be concerned at taking a minumum optical depth from several lines and converting that to an optical depth at Lyα, or the strongest heavy element transition, when there is noise present, and of omitting those pixels where there is apparent emission rather than absorption. Comparison of the results from those from the same analysis applied to simple simulated spectra is all that is needed to make allowance for any such biases. Another possible source of bias is the difference in intrinsic line widths between the O VI and corresponding H I lines if thermal broadening is important, as it appears to be in those systems where O VI and C IV are strong enough to be measured directly. For pure thermal broadening the O VI lines have velocity widths $\frac{1}{4}$ that of H I, so in individual cases the H I optical depth may still be a significant fraction of the peak value, while for the corresponding O VI this is not true.

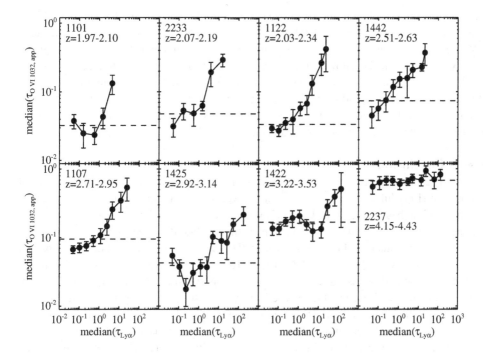

Fig. 30.6. Results from a pixel analysis search for O VI. H I-O VI pixel pairs were binned in H I optical depth, and the data points indicate the median optical depth in each bin. The dashed lines represent O VI detection limits. (Reproduced from Schaye et al. 2000.)

30.7 Anomalous Abundances

Much of the discussion above has been aimed at addressing the question of average abundances in the intergalactic medium, and seeing if there are trends with H I column density and the matter density in the regions. There will, of course, be a spread in the abundances for any simple parameterization such as this, but there is little information yet on how large this spread is. However, there are a few anomalously high-abundance regions that may be telling us something, even if it is not clear what.

The Carswell et al. (2002) compilation of O VI systems contained one that shows O VI and C IV, at redshift $z = 2.030096$ toward HE 1122−1648, which is unusual in two ways: the Lyα is very weak and, unlike the other systems studied, N V is clearly present. An examination of other spectra at similar redshifts suggests that, while such systems are by no means common, this particular example is not unique. The narrow heavy element lines again point to temperatures that are too low for collisional ionization to be effective, and if they are photoionized by the intergalactic background then the density is $\log n_H \approx -4$ to -3.5 cm^{-3} and the heavy element abundances ∼solar. In particular, the nitrogen-to-carbon ratio is ∼solar or even higher. Size estimates based on these models suggest a length scale ∼ 100 pc. There are no nearby high H I column density systems in this case, so if it is enriched material it is not close to a galaxy with a significant interstellar medium. This could be due to local ionization or dynamic effects, or geometry, so may not be a significant problem. However, the nitrogen overabundance does set such systems apart, and leads one to speculate that it is not normal galaxy ejecta but that from a QSO. This could be either the background object, though the velocity is then $\sim 30,000$ km s^{-1}, or from a foreground active galactic nucleus.

While such systems may only be of marginal importance in the overall scheme of things, their existence has to be allowed for in the various stacking techniques that are used to search for heavy elements at low H I column densities. One such contaminant in a sample of a few hundred systems could lead to an average abundance $\sim 10^{-2}$ solar while the "typical" abundance could be much less. The use of median, as opposed to average, values for pixel or stacked spectra, as is common, should avoid such problems.

30.8   Conclusions

The direct detection of highly ionized heavy elements corresponding to neutral hydrogen at H I column densities $> 10^{14}$ cm^{-2} reveals that parts of the higher-density intergalactic medium has been enriched by outflow of material from galaxies. The heavy element abundances, where they have been determined, suggest that the lower-density regions have lower element abundances. For systems with N(H I) $> 10^{15}$ cm^{-2} the abundance of oxygen is $\sim 10^{-1.5}$ solar, while for $10^{14.5} < N$(H I) $< 10^{15}$ cm^{-2} estimates are $\sim 10^{-2.5}$ solar. At lower column densities less direct techniques, such as stacking by redshift or pixel optical depth, are required. These techniques each suffer from some biases and uncertainties, and we are still at the stage where these have not been explored fully. Uncertainties in ion velocity relative to the H I position reduce the sensitivity of the redshift stacking method, and for pixel stacking the results should be compared with simulations. The two techniques can give raw abundance estimates that are almost an order of magnitude apart in the worst cases, and estimated corrections to both do not yet make them fully compatible. So, on the basis of what we know so far, for the low-density regions the true intergalactic medium is far from galaxies, and the heavy element abundances could be anywhere below a rather uncertain value of $10^{-2.5}$ solar. Any constraints on the large-scale feedback from galaxies into the intergalactic medium are not very tight.

References

Adelberger, K. L., Steidel, C. C., Shapley, A. E., & Pettini, M. 2003, ApJ, 584, 45
Aguirre, A., Schaye, J., & Theuns, T. 2002, ApJ, 576, 1
Bechtold, J., Weymann, R. J., Lin, Z., & Malkan, M. A. 1987, ApJ, 315, 180
Bergeron, J., Aracil, B., Petitjean, P., & Pichon, C. 2002, A&A, 396, L11
Boksenberg, A. 1997, in Structure and Evolution of the Intergalactic Medium from QSO Absorption Line
    Systems, ed. P. Petitjean & S. Charlot (Paris: Editions Frontières), 85
Carswell, R. F., Schaye, J., & Kim, T.-S. 2002, ApJ, 578, 667
Cen, R., & Ostriker, J. P. 1999, ApJ, 519, L109
Chaffee, F. H., Foltz, C. B., Bechtold, J., & Weymann, R. J. 1986, ApJ, 301, 116
Chaffee, F. H., Foltz, C. B., Weymann, R. J., Rošer, H.-J., & Latham, D. W. 1985, ApJ, 292, 362
Ciardi, B., Ferrara, A., Governato, F., & Jenkins, A. 2000, MNRAS, 314, 611
Cowie, L. L., & Songaila, A. 1998, Nature, 394, 44
Cowie, L. L., Songaila, A., Kim, T.-S., & Hu, E. 1995, AJ, 109, 1522
Croft, R. A. C., Hernquist, L., Springel, V., Westover, M., & White, M. 2002, ApJ, 580, 634
Ellison, S. L., Songaila, A., Schaye, J., & Pettini, M. 2000, AJ, 120, 1175
Haardt, F. & Madau, P. 1996, ApJ, 461, 20
——. 2001, in Clusters of Galaxies and the High Redshift Universe Observed in X-rays, Recent Results of
    XMM-Newton and Chandra, XXXVIth Rencontres de Moriond, ed. D. M. Neumann & J. T. T. Van,
    http://www-dapnia.cea.fr/Conferences/Morion_astro_2001/index.html (astro-ph/0106018)
Heckman, T. M., Norman, C. A., Strickland, D. K., & Sembach, K. R. 2002, ApJ, 577, 691
Jenkins, A., et al. 1998, ApJ, 499, 20

Kim, T.-S., Cristiani, S., & D'Odorico, S. 2002, A&A, 383, 747

Lu, L. 1991, ApJ, 379, 99

Lu, L., Sargent, W. L. W., Barlow, T. A., & Rauch, M. 2004, AJ, submitted (astro-ph/9802189)

Machacek, M. E., Bryan, G. L., Meiksin, A., Anninos, P., Thayer, D., Norman, M., & Zhang, Y. 2000, ApJ, 532, 118

Madau, P., Ferrara, A., & Rees, M. J. 2001, ApJ, 555, 92

Masao, M., Ferrara, A., & Madau, P. 2002, ApJ, 571, 40

Miralda-Escudé, J., Cen, R., Ostriker, J. P., & Rauch, M. 1996, ApJ, 471, 582

Norris, J., Peterson, B. A., & Hartwick, F. D. A. 1983, ApJ, 273, 450

Rauch, M. 1998, ARA&A, 36, 267

Rauch, M., Haehnelt, M. G., & Steinmetz, M. 1997, ApJ, 481, 601

Schaye, J. 2001, ApJ, 559, 507

Schaye, J., Rauch, M., Sargent, W. L. W., & Kim, T.-S. 2000, ApJ, 541, L1

Simcoe, R. A., Sargent, W. L. W., & Rauch, M. 2002, ApJ, 578, 737

——. 2004, in Carnegie Observatories Astrophysics Series, Vol. 4: Origin and Evolution of the Elements, ed. A. McWilliam & M. Rauch (Pasadena: Carnegie Observatories, http://www.ociw.edu/symposia/series/symposium4/proceedings.html)

Sokasian, A., Abel, T., & Hernquist, L. 2003, MNRAS, 340, 473

Songaila, A. 1998, AJ, 115, 2184

Songaila, A., & Cowie, L. L. 1996, AJ, 112, 335

Theuns, T., Schaye, J., Zaroubi, S., Kim, T.-S., Tzanavaris, P., & Carswell, B. 2002, ApJ, 567, L103

Tytler, D., & Fan X.-M. 1994, ApJ, 424, 87

Williger, G. M., Carswell, R. F., Webb, J. K., Bobsenberg, A., & Smith, M. G. 1989, MNRAS, 237, 635

31

Conference summary

BERNARD E. J. PAGEL
Astronomy Centre, University of Sussex, UK

Abstract
Highlights of the meeting include new insights into *r*- and *s*-processes and an explosion of interesting abundance data on stars in the Galactic halo and in dwarf spheroidal galaxies. I include in this summary a few suggestions on the chemical evolution of the thick and thin disks, and on the present and past distribution of baryons and metals in the Universe.

31.1 Introduction

It has been a great pleasure and honor for me to be able to attend this fascinating Symposium, and I should like to express my warm thanks to Andy McWilliam and Michael Rauch for inviting me. Clearly the Carnegie Observatories (or Mount Wilson as I still like to think of them) are carrying their illustrious traditions forward into their second century of existence.

31.2 History

George Preston recalled some highlights associated with the names of G. E. Hale, A. S. King, H. N. Russell, Horace Babcock, Lawrence Aller, Paul Merrill et al. I was intrigued to hear that after the discovery of stellar technetium Merrill began to pay much more attention to unconventional new ideas; I think he would have enjoyed this conference!

Margaret Burbidge described the steps leading up to current ideas on nucleosynthesis, beginning with Hoyle's pioneering work in 1946, through the peaks in the standard abundance distribution, and her work with Geoff on the barium star HD 46407, to B²FH and subsequent developments. I remember visiting Margaret and Geoff in their small apartment in Botolph Lane, Cambridge, all covered in tracings, in 1954.

31.3 The Overall Picture

George Wallerstein raised some of the questions that form a backdrop to much of our proceedings, notably the issue of galaxy formation by monolithic collapse versus a series of mergers, and the extent to which the Galactic halo could have been formed from dwarf spheroidals. Abundance characteristics like the α/Fe ratio preclude such a possibility for the classical globular clusters, but some field stars, especially with retrograde orbits, might originate in captured dwarfs.

31.4 Nucleosynthetic Yields

31.4.1 Massive Stars

Yields from massive stars were discussed by Dave Arnett, Claudia Travaglio, and Ken Nomoto. There are basically two kinds of process: hydrostatic burning (hydrogen, helium, carbon and neon burning, and *s*-process) and explosive synthesis (oxygen and silicon burning and the *r* and *rp*-processes). Laser plasma experiments have led to improved knowledge of opacities and equations of state, but the structure of supernova remnants (SNRs) like Cas A shows up the limitations of one-dimensional models; 2-D and 3-D models are under development and help to "educate" the 1-D models (e.g., by discovering instabilities), but the explosion mechanism is still uncertain.

The abundance peculiarities of extremely metal-deficient stars (notably Christlieb's star HE 0107–5240 with [Fe/H] = –5.3), include excesses (relative to iron) of C, N, O, α-elements, Zn, and Co, and deficiencies of Cr and Mn. Ken Nomoto was able to explain these abundance patterns from nucleosynthesis by Pop III supernovae (SNe) with masses anywhere between 20 and 130 $M_\odot$ forming massive black holes with mixing–fallback in the ejecta, combined with the SN-induced star formation model in which, for a given yield, Fe/H is inversely proportional to the explosion energy. Strong mixing and fall-back explains the cases with strongly enhanced C and O and corresponds to an observed class of faint SNe (IIp), while higher explosion energies (possibly assisted by rotation) produce hypernovae (some of classes Ic and IIn), in which complete Si burning leads to α-element excess and the peculiarities in the iron group.

The *r*-process, reviewed by John Cowan, still holds many mysteries, since its path involves highly unstable nuclei, and there is no certainty as to where it occurs; Grant Mathews favors the neutrino-heated hot SN bubble, but neutron star pairs are another candidate and, in any case, magnetic fields, jets, and neutrino oscillations may be involved. Some intriguing nuclear information has come from H-bomb tests (Stephen Becker).

Among the lowest-metallicity stars, there are huge variations in the relative abundance of *r*-process and iron-group elements, while among the *r*-process nuclides themselves there is sometimes excellent agreement with the solar system *r*-process distribution and sometimes not, especially between the $A \simeq 130$ Xe peak and the higher Pt peak. So, as previously suggested by Wasserburg, Busso, & Gallino (1996) on other grounds, there may be two or more distinct *r*-processes, analogous to the weak and main *s*-processes. This, in turn, relates to radioactive cosmochronology, to which we return later.

31.4.2 Low- and Intermediate-mass Stars

Intermediate-mass stars are significant sources of He, C, N, and the main *s*-process elements. We heard from Dick Henry that, compared with the old work of Renzini and Voli, widely used despite their own health warnings for so many years, more recent synthetic models like those of van den Hoek & Groenevegen predict higher mass loss rates, less helium production, and an increase in C and N production at lower metallicities. Hot-bottom burning sets in at around $3\,M_\odot$, and C and N production is further enhanced by rotation. There is fair agreement with abundances found in planetary nebulae. A somewhat contentious issue is raised by the bimodal distribution of N/α ratios in damped Lyα systems, the upper branch corresponding to N/O in H II galaxies and the lower branch nearly an order of magnitude lower; could this be an age effect, or is there something funny going on with the initial mass function? Mercedes Mollá described galactic chemical evolution models with a new set of yields for carbon and nitrogen, and Amanda Karakas discussed the production of aluminum (including ^{26}Al) and heavy magnesium isotopes in AGB stars.

Rotation also influences mixing processes following the first dredge-up (Corinne Charbonnel). At the bump in the luminosity function where the H-burning shell crosses the previous lowest boundary of the outer convection zone, the μ-barrier is partially lifted, enabling various extra-mixing processes to take place as a result of rotation. Consequences include lowering of the ^{12}C/^{13}C ratio, destruction of ^{3}He and the production of a lithium "flash" leading to enhanced mass loss attested by a dust shell after a brief super-lithium rich phase.

Significant, perhaps even dominant, contributions to Galactic enrichment in ^{7}Li, ^{13}C, ^{15}N, and ^{26}Al come from novae (Sumner Starrfield); UV spectra of some fast novae show enhanced abundances of N, O, Ne, Mg, and Al, but not C or Si.

Dramatic advances in the theory of the *s*-process have resulted from the efforts of the Torino group, presented by Maurizio Busso and Oscar Straniero. Evidence from branchings in the Kr–Rb region confirms that most of the main *s*-process with neutrons from ^{13}C takes place during shell H-burning phases between thermal pulses, with just a minor top-up from ^{22}Ne during the pulses themselves. This renders obsolete the old idea of an exponential distribution of exposures (Seeger, Fowler, & Clayton 1965), which I described in my book as arguably the most elegant result in the whole of nucleosynthesis theory! The steps at magic numbers are still there, however. A free parameter in the theory is the mass of the ^{13}C pocket that gets ingested into the intershell zone. This seems to be more or less constant, leading to increasing neutron exposures with lowering metallicity, and eventually to a "strong" *s*-process with lots of lead (confirmed by Judy Cohen).

31.5   Modeling Galactic Chemical Evolution

According to taste, there is a great variety in the degree of sophistication and complexity that one can put into galactic chemical evolution models, ranging from chemodynamical and cosmological SPH plus semi-analytic simulations (as described by Brad Gibson), to numerical models with parameterized star formation and gas flow laws (Francesca Matteucci), to highly simplified analytical toy models, which still have a useful contribution to make, as was shown at this meeting by Verne Smith and Matt Shetrone, for example. The wide range in *r*-process/Fe ratios in the most metal-poor stars suggests some kind of inhomogeneous model (Sally Oey), although the simplest version of this postulates a threshold that is not observed and predicts far too many metal-free stars.

31.6   Stellar Abundance Analysis

Substantial revisions in the "official" solar oxygen abundance in the last two years led to some doubts as to how good stellar abundances really are and the influence of bandwagon effects. Bengt Gustafsson described the activities of "sheep" (who follow the crowd) and "goats" (who like to put a spanner in the works, but not too big a one). I believe the comparison with emission lines from H II regions is a good check (see Table 31.1). To one place of decimals in the log (to quote two would be like a second marriage—a triumph of hope over experience!), the agreement is gratifyingly good, or rather was, until Kim Venn dropped her WLM bombshell at this meeting! The H II O abundance is typical for such a dwarf galaxy, whereas the stellar one is surprisingly high. We heard from Andreas Korn and

Table 31.1. *CNO Abundances in Local Group Galaxies*

Object	C	N	O	Reference
Local Galaxy:				
Sun	8.4	(7.8?)	8.7	Allende Prieto et al. 2001, 2002
Sun	8.6	7.9	8.7	Holweger 2001
Orion nebula	8.5	7.8	8.7	Esteban et al. 1998
diffuse ISM			8.6 to 8.7	Meyer, Jura, & Cardelli 1998
cepheids	8.0/8.6		8.3/8.9	Luck et al. 1998
M 31:				
H II regions		7.0/8.2	8.5/9.2	Dennefeld & Kunth 1981
SNR		7.4/8.0	8.3/8.7	Blair, Kirshner, & Chevalier 1982
4 AF supergiants	8.3		8.8	Venn et al. 2000
M 33:				
H II regions		7.9–.16R	9.0–.12R	Vílchez et al. 1988
SNR		7.8–.12R	8.8–.07R	Blair & Kirshner 1985
B,A supergiants			9.0–.16R	Monteverde et al. 1997
LMC:				
H II regions, SNR	7.9	6.9	8.4	Garnett 1999
cepheids	7.7/8.3		8.0/8.8	Luck et al. 1998
PS 34-16 (early B)	7.1	7.5	8.4	Rolleston et al. 1996
LH 104-24 (early B)	7.5	7.7	8.5	Rolleston et al. 1996
NGC 1818/D1	7.8	7.4	8.5	Korn et al. 2002
NGC2004 (4 early B)	8.1	7.0	8.4	Korn et al. 2002
4 F supergiants	8.1			Russell & Bessell 1989
SMC:				
H II regions, SNR	7.5	6.6	8.1	Kurt et al. 1999
cepheids	7.4/7.8		8.0/8.3	Luck et al. 1998
10 A supergiants	< 7.3/< 8.7	6.8/7.7	8.1	Venn 1999
3 F supergiants	7.7		8.1	Spite, Barbuy, & Spite 1989
2 F supergiants	7.8			Russell & Bessell 1989
NGC 6822:				
H II regions		6.6	8.3	Pagel, Edmunds, & Smith 1980
H II regions			8.4	Pilyugin 2001
2 A supergiants			8.4	Venn et al. 2001
WLM:				
H II regions		6.5	7.8	Skillman, Terlevich, & Melnick 1989
A supergiants		7.5	8.4	Venn 2004

Norbert Przybilla how important it still is to get better atomic data for both LTE and non-LTE abundance analyses, and from Mariagrazia Franchini about the possibilities of using the Lick indices to get α/Fe ratios.

31.7 Abundance Effects from Internal Stellar Evolution

Dave Lambert discussed the wide range of effects observed in AGB and post-AGB stars. Hot-bottom burning may break through into the Ne-Na and Mg-Al cycles, leading to peculiarities in ^{26}Mg/^{24}Mg ratios in NGC 6752 and some field dwarfs. Hot-bottom burning can also lead to ^{7}Li and fluorine production, and a final shell-flash seems to be responsible for the effects found in FG Sge and H-poor carbon stars like Sakurai's object and R Cr B. RV Tau stars, a class of post-AGB stars, are apparently metal-poor because refractories are locked on dust, resulting in an enhancement of *s*-process/Fe ratios, which in turn correlate with heavy/light *s*-process ratios (Maarten Reyniers). There is now direct evidence for *s*-process enhancements in planetary nebulae (Harriet Dinerstein).

Light elements are affected by mixing processes even in the main sequence stage. Anne Boesgaard has discovered a beryllium dip in open clusters like the Hyades, which shows interesting contrasts to the lithium dip. Within the dip, Be is down, though not as much as Li, which is reminiscent of effects of rotational mixing on $^6Li/^7Li$ (Pinsonneault et al. 1999), whereas on the cool side Be is undepleted for quite a long way, which corresponds more to a vertically stratified model with a deeper exclusion zone for Be than for Li.

31.8   Abundances in Stellar Populations

31.8.1   *The Milky Way Halo*

The study of the oldest stars in our Galaxy, some of which appear to have lower metallicity than anything seen at high redshifts up to now, was justly described by John Norris as "near-field cosmology." The records are held by G77–61 ([Fe/H] $\simeq$ −5.5) and Christlieb's star ([Fe/H] $\simeq$ −5.3), discovered in the Hamburg Objective Prism survey, but the numbers fall below expectations from the Simple Model for the halo below [Fe/H] = −4 (Norbert Christlieb searched 5 million stars to get 2000 extremely metal-poor candidates), a result explained by Tsujimoto, Shigeyama, & Yoshii (2000) in terms of SN-induced star formation. Further candidates are likely to be found from the huge database provided by the Sloan Digital Sky Survey with the aid of automated spectrum analysis (Carlos Allende Prieto). Lithium abundances on the Spite plateau are a bit uncomfortably low for Big Bang nucleosynthesis in the light of deuterium and cosmic microwave background fluctuation spectrum data, so there has to be quite a bit of depletion, by a factor of 3 or so, and the presence or absence of a metallicity dependence hinges on subtleties in the abundance analysis. Some globular cluster stars show higher lithium abundances and a scatter (Boesgaard).

Below [Fe/H] = −3, very large relative overabundances of the CNO elements are sometimes found (see Nomoto's and Judy Cohen's talks), and in a few cases this also applies to Mg, Si, and Eu, but for most stars α/Fe is quite flat down to [Fe/H] = −4, while Co/Fe rises and Mn, Cr/Fe fall with decreasing metallicity. Among neutron-capture elements, Sr varies wildly, in contrast to Ba, and the ratio of *r*-process to iron is highly variable.

New data from UVES on the ESO VLT exhibit a remarkably uniform and flat trend in α/Fe ratios between [Fe/H] = −2 and −4 (Monique Spite). I remarked that only analytical toy models could account for such a flat trend, but Francesca disagreed. According to Monique's data, oxygen shows only a modest overabundance down to [Fe/H] = −3.5, apart from the odd exception; this hot issue, also discussed by Suchitra Balachandran and in a poster by Garik Israelian, was delicately tiptoed around by speakers and audience.

With the discovery of uranium in CS 31082–001, radioactive cosmochronology has at last become respectable (Roger Cayrel). For a long time, the Th/Eu ratio has been used for this purpose, following a suggestion of mine (Pagel 1989), and at first I was a bit miffed at this not having been mentioned, but now that this ratio has been exposed as a quite unreliable one (see also Otsuki, Mathews, & Kajino 2002), I suppose I really should be grateful!

31.8.2   *The Thin and Thick Disks*

Thick-disk stars (essentially the same as Bengt Strömgren's "Intermediate Population II") resemble bulge stars with minor differences, i.e. they continue to display the

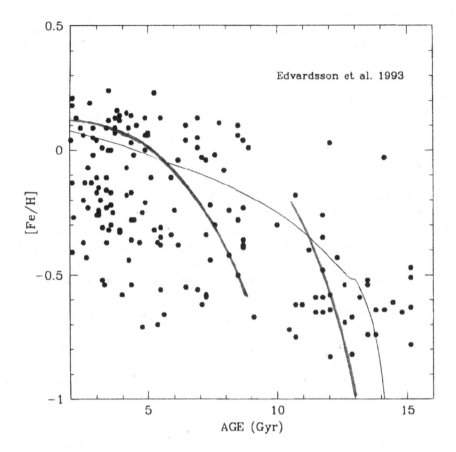

Fig. 31.1. Thin curves: age-metallicity relation from the "two-inflow" model of Chiappini, Matteucci, & Gratton (1997). Thick curves: sketch of suggested age-metallicity relations for the thick and thin disks. Data points from Edvardsson et al. (1993). Adapted from Chiappini et al. (1997); reproduced from Pagel (2001).

α-rich effect noted by George Wallerstein 40 years ago (Wallerstein 1962), up to quite high metallicities, symptomatic of an old get-rich-quick stellar population, while in the halo both α-rich and non-α-rich stars are found, the latter belonging to the outer halo and showing similarities in composition to dwarf spheroidals and irregulars (Poul Nissen). In the thick disk, however, as appears from recent work reported by Sophia Feltzing (as well as the Edvardsson et al. 1993 survey and other studies), the α/Fe ratio does actually come down at high metallicities, indicating a significant contribution from SNe Ia as well as possible effects of a star formation threshold (Cristina Chiappini). While the time needed for a significant contribution from SNe Ia is often assumed to be of the order of 1 Gyr, in reality there must be a spread, so that it is still hard to be quantitative about the age and formation time scale of the thick disk, but a sketch of how the chemical evolution of both disks might have gone is shown in Figure 31.1. A new version of the Edvardsson et al. (1993) survey is under way (Bacham Reddy).

31.8.3   The Galactic Bulge

Our host Andy McWilliam has been making heroic efforts over the years, in collaboration with Mike Rich, to gain accurate abundances for red giants in the bulge. Their latest results on K giants show trends in α-elements that are rather like those in the thick disk, with O, Mg, and Si high (as well as Al and Eu), but O/Fe coming down at the highest metallicities. Ca/Fe behaves like in the thin disk, which is interesting because Lambert et al. (1996) found a similar effect in the RR Lyrae stars, presumably representing the thick disk. What is this telling us about the origin of calcium? Near-infrared spectra enable the abundance analysis to be extended to M giants in globular clusters (Livia Origlia).

31.8.4   Globular Clusters

The globular cluster metallicity scale has been refined by using Fe II (Bob Kraft). Galactic globular clusters all have [Fe/H] ≥ -2.4, but in many respects show similar abundance patterns to field stars in the same metallicity range (Chris Sneden). The main differences are the Na,Al–O anticorrelation (although Al abundances are still a bit dodgy) and the persistence of CNO anomalies down to the main sequence, suggesting a multi-generational aspect to globular clusters. Ruprecht 106 is exceptional and may be a captured system, and the inner-halo–outer halo effects found in field stars have a counterpart in globular clusters, e.g. M4 versus M5 (Jon Fulbright). Still stronger anomalies are found in a few cases (Inese Ivans).

The most intriguing globular cluster (or galaxy remnant?) is, of course, ω Centauri, with metallicities ranging from [Fe/H] $= -2$ to zero, and high s-process abundances (Verne Smith; Elena Pancino). Lanthanum, a heavy s-process product, shows a stepwise increase with Fe/H at [Fe/H] $= -1.6$, while the light s-process element yttrium varies less. ω Cen displays both cluster-like (Na–O anticorrelation) and dwarf galaxy-like properties (low α/Fe, Eu/Fe). The metal-poor component shows rotation, whereas the metal-rich one does not, all of which suggests a merger of two extragalactic globular clusters and capture into the Milky Way. A model by Takuji Tsujimoto envisages three star formation episodes, the first terminated by a wind whereafter AGB stars enriched the remaining gas with heavy s-process elements; in the third phase SN Ia-induced star formation occurred, and the remaining gas was stripped by passage through the Milky Way disk. He ascribes the difference from other globular clusters to formation by colliding protocluster clouds; low-impact velocities favor SN-induced star formation and chemical evolution, whereas high velocities prevent it.

31.8.5   Nearby Galaxies

The LMC has stars of all ages, despite the gap in the cluster age distribution, and a fairly well-defined age-metallicity relation; it remains to be clarified how "bursty" the star formation history has been (Vanessa Hill). Carbon and nitrogen abundances are low, as in blue compact galaxies, suggesting youth, but α/Fe and O/Fe ratios are also low, suggesting a long star formation time scale. There are also substantial differences between clusters of similar age and metallicity in both Clouds (Jennifer Johnson).

Stellar wind analysis techniques (Fabio Bresolin) applied to A-supergiants in gas-rich dwarf galaxies NGC 6822, Sextans, and GR 8 give oxygen abundances in good agreement with H II regions in the same galaxies, and also metal abundances, which indicate [α/Fe] $\simeq 0$ down to [Fe/H] $\simeq -1.5$ (Kim Venn), but for WLM there is a discrepancy (see Table 31.1). The common finding of solar α/Fe is somewhat puzzling: is there something fundamental about it (e.g., as a consequence of star formation bursts), or does it imply largely common star formation histories? These differ considerably in detail, although age-metallicity relations are often similar within the uncertainties (Eva Grebel). In any case, there is not the large variety in α/Fe ratios suggested by Gilmore & Wyse (1991).

High-resolution spectroscopy with large telescopes has also led to detailed information about ages and compositions in dwarf spheroidal galaxies (Matt Shetrone). These divide into those with a simple star formation history dominated by an early burst (e.g., Dra, Sex, UMi, Scl) and those with a more complicated one (e.g., Sgr, For, Car, Leo I and II). The simple group provides new insights into nucleosynthesis: O,Mg/Fe start to go down when [Fe/H] ≥ -1.5, suggesting that star formation was interrupted, and Ca and Ti are less enhanced than O and Mg, due to explosive versus hydrostatic synthesis or a contribution from SNe Ia? The metallicity dependence of Mn and Cu is not due to the latter effect, as can be deduced also from their behavior in the Galactic halo. Y and Na are low, while Ba/Y is high, indicating *s*-process production in metal-poor AGB stars.

Complex star formation histories place constraints on hierarchical galaxy formation models (Tammy Smecker-Hane). Dwarf spheroidals mostly follow a metallicity-luminosity relation, $Z \propto L^{0.3}$, as do gas-rich dwarfs, though with a lower zero-point, but not for the reasons given by Dekel & Silk (1986), which require a simple history culminating in a terminal wind, and it seems that total luminosity is more important for metallicity than are the details of star formation history (Carmen Gallart). Furthermore, giant elliptical galaxies cannot have been made up from dwarf spheroidals. The main difference between gas-rich and gas-poor dwarf galaxies is that the latter have been robbed of their gas by the Milky Way and M 31, probably through ram pressure (Jay Gallagher).

The metallicity-luminosity relation extends all the way up to M 31, which has a disturbed, metal-rich halo with [Fe/H] $\simeq -0.6$ like 47 Tuc (Mike Rich) and many blue horizontal-branch stars indicating great age. The globular cluster system rotates and includes a globular cluster/captured dSph very like ω Cen.

31.9 The Interstellar Medium

X-ray satellites *ASCA* and *Chandra* now permit spatially resolved abundance determinations in SNRs (John Hughes). In SNe II these are lumpy and diverse, and the accuracy of relative abundances depends on the species compared being in the same place. Cas A has at least four components, of which one is a featureless nonthermal continuum while others represent O-burning (Si, S) and incomplete Si-burning (Fe). One can have "inside-out" configurations in which iron-group elements overtake the lighter ones. SN Ia remnants are smoother, but may contain lumps of hot iron!

Various solid pieces of stars are found as presolar grains in meteorites (Don Clayton) and provide constraints on the chemical evolution of the Galaxy. "Mainstream" SiC grains come from AGB stars (with relatively low ^{12}C/^{13}C and high ^{14}N/^{15}N), but there are also "X" grains from SNe where these ratios are reversed. The silicon isotopes present something of a mystery because the plot of ^{29}Si excess vs. ^{30}Si excess has a slope of 1.3 and passes above the (solar) origin, although the stellar sources must have been older and presumably less metal rich than the Sun. My suggestion is that there may be a bias toward high metallicity in the sample, e.g. if stellar winds are metallicity-dependent. Isotopic and elemental patterns in the grains suggest a well-mixed Galactic chemical evolution (Larry Nittler).

However, interstellar dust does not consist of unmodified solid ejecta from stars; these are subject to shocks, sputtering, etc. in the interstellar medium (ISM), and some return to the gas phase with a typical turnover time of 3×10^8 yr and there is grain growth in the ISM (Bruce Draine). Various clues suggest the composition and size distribution of the dust: 2200 Å absorption comes from sp^2-bonded carbon in sheets (graphite or PAH), diffuse interstellar bands from large molecules (?), 3.4 μm features from C–H stretches in linear chains and mid-infrared features from PAHs and Si–O stretches. The ratio of visual absorption to reddening increases with the size of the particles, which is mostly under 1μm.

UV spectra (from *Copernicus* to *FUSE*) reveal how much of the standard abundance distribution is depleted from the gas to the dust phase, but sometimes ionization corrections are needed (Ed Jenkins). The revised solar O and C abundances lead to a more consistent picture than one had before (cf. Table 31.1), and there is now some information about the composition of high-velocity clouds and the Magellanic Stream, which resembles that in dwarf irregular/blue compact galaxies. In the Galactic plane, the D/H ratio is uniform in the local bubble (100 pc), but shows anomalous variations further afield—only 7.5×10^{-6} on the sightlines to λ Sco and δ Ori, but 2.2×10^{-5} on that to γ^2 Vel (Jeff Linsky). Could deuterium be locked on grains? This would hardly account for the unusually large value, but a spatially and temporally varying infall of relatively unprocessed matter might. No deuterated molecules were detected in a cloud 28 kpc from the Galactic Center by Don Lubowich, but molecular features are rather weak there anyway.

Absorption lines of molecules such as CO, CN, HNC, and HCO^+ can now be studied by a new technique using interferometry (Tommy Wiklind). In both diffuse and dark clouds, CO and HCO^+ are much more abundant than expected from gas-phase reaction networks, while O_2 is expected but not seen. CO has been detected in emission from far-infrared-luminous galaxies/AGNs up to a redshift of 4.7. and many molecules have been found in absorption up to $z = 0.9$; relative abundances are similar to those found locally.

31.10   The Local Universe

Galactic winds resulting from SN activity in starburst galaxies can strongly influence chemical evolution, depending on the depth of the potential well and the structure of ambient gas (Crystal Martin). Hot SN ejecta are removed in the wind, but not much of the ISM. High-density winds appear in X-ray images (e.g., NGC 3077 and NGC 1569), whereas low-density winds are detected from blue-shifted absorption lines (e.g., in ultra-luminous infrared galaxies). Mass flow rates are comparable to star formation rates, but in big galaxies it is not clear that all the material involved escapes, nor is it clear whether large or small galaxies make the dominant contribution to the intergalactic medium.

Abundances in stellar populations can be studied on the basis of integrated light using the Lick indices (Scott Trager) calibrated on globular clusters. The age-metallicity degeneracy is broken using Hβ, provided there is no emission or extended blue horizontal branch, but one does need accurate and complete isochrones and a spectral library. Some results indicate nitrogen enhancements in red giants of M 31 and Fornax, but not in old clusters of the LMC, while conversely these have an α/Fe enhancement while Fornax does not. Ellipticals in the Coma cluster are older than field ellipticals, both with α enhancement, while S0s have a spread in age with less α enhancement, but the ages are less robust than the chemical results, which will be extended to less prominent elements in the near future. The calcium triplet in ellipticals is anomalously weak, either because of bad fitting functions or because of some anomaly in Ca/Mg (see Galactic bulge!).

Another approach to composition variations between and across galaxies comes from H II regions (Don Garnett). There are some problems, including temperature fluctuations and the calibration of R_{23} (which I sometimes think of like Macbeth as "Bloody instructions which, being taught, return to plague the inventor"!), but there is quite good agreement with young stars (Table 31.1). Oxygen abundance gradients in non-barred spirals such as M 101 are quite constant at −0.2 dex per scale length, a simple result that may need a complicated explanation. The metallicity-luminosity relation translates into an effective yield (abundance/ln gas fraction) that increases with rotational velocity (representing mass) up to a point and then levels off, presumably when SN-driven winds no longer escape. Iron abundances can sometimes be found from Fe II lines—prominent in AGNs and some peculiar stars like η Carinae— taking account of UV pumping from Lyα and other sources; for example, in Orion it is about 1/10 solar in the gas phase (Ekaterina Verner).

One of the benefits of the SDSS is the possibility to study the star formation rate density in the local Universe from a complete sample of Hα emission (Jarle Brinchmann). The major contribution comes from dusty, high-metallicity, high-surface brightness spirals and the star formation rate is about 1/4 of the past average, in agreement with the Madau plot and its various updatings (e.g., Steidel et al. 1999).

ASCA and *XMM-Newton* spectra of numerous clusters of galaxies have provided new details about the composition of the intracluster medium (Michael Loewenstein). There is no evolution with redshift up to $z = 0.8$, but some variation with temperature of the X-ray gas. CNO/Fe are as in the Galactic disk, with subsolar metallicity, but Si and Ni are relatively overabundant, and the composition cannot be represented by any combination of conventional SNe Ia and SNe II. Could there be a Population III contribution?

31.11 The High-redshift Universe

The composition of the broad emission-line gas in quasars bears witness to the effect of rapid star formation accompanying that of the central black hole (Fred Hamann). In other words, there exists a get-rich-quick population suitable for the black hole's future role as the core of a massive spheroidal system. Absolute abundances are model dependent, but the high relative abundance of N compared to C and especially O is suggestive of a high metallicity like twice solar, and the mass of the region is similar to that of a globular cluster. Intrinsic, narrow absorption lines confirm at least solar abundance of carbon.

Next in metallicity (and in ambient mass density) after quasars come the Lyman-break galaxies (Kurt Adelberger), about which much has been found out from the lensed object cB58. The abundance of oxygen is about 1/3 solar, while the interstellar lines indicate a lower abundance of iron-group elements in the gas phase. The P Cygni profile of C IV resembles that of stars in the LMC. Lyα has a violet-shifted component indicating outflows at 300–600 km s^{-1}, which perturb and enrich the intergalactic medium out to 0.5 Mpc, leading to a correlation between Lyman-break galaxies and intergalactic C IV. So we witness SN feedback in action.

Next in the scale come the damped Lyα systems which may be the raw material for disk galaxies today (Jason Prochaska, Paolo Molaro, Francesco Calura). Metallicity, measured by zinc, shows only a mild evolution with redshift from 2 to 6, and there are problems from dust: selection bias and differential depletion from the gas phase. As was already mentioned by Dick Henry, N/α is bimodal, the low branch being attributed by Jason to massive Pop III stars, but I prefer Paolo's explanation in terms of the yield from conventional massive stars,

Table 31.2. *Baryon and Metal Budgets (after Finoguenov et al. 2003)*

Component	$Z, 10^{-2}$	$\Omega_Z, 10^{-5}$		$\Omega_b, 10^{-2}$	
		$z = 0$			
Stars	1.2^a	$2.3 - 4.6$	— Most	$0.2 - 0.4$	
O VI absorbers	0.26	$1.8 - 5.2$	— metals	$0.7 - 2.0$	— Most
Lyα forest	0.01	0.1		1.2	— baryons
X-ray gas, clusters	0.7	1.4		0.2	
Total	0.28	$5.6 - 11.3^{\,b}$		$2.3 - 3.8$	
Predicted				3.9	
		$z = 2.5$			
Protocl. gas	0.7	1.4	— Most	0.2	
ISM, dust	0.8	$0.8 - 1.7$	— metals	$0.1 - 0.2$	
Lyα forest	0.01	$0.1 - 0.3$		$1 - 5$	— Most baryons
DLAs	0.1	0.1		0.1	
Total	0.1	$2.4 - 3.5$		$1.4 - 5.5$	
Predicted		$1.6 - 3.2$		3.9	

^{a} i.e. solar.
^{b} $\Rightarrow$ yield $\equiv \Omega_Z/\Omega_{\text{stars}} \simeq 0.028$

enhanced by rotation as described by Meynet & Maeder (2002). Sub-damped Lyα absorber systems, with $N(\text{HI}) \simeq 10^{19}$ cm^{-2}, contribute significantly to the gas and metal budgets and show stronger evolution with z (Céline Peroux).

Finally, we come to the intergalactic medium, aka the Lyα forest (Bob Carswell, Rob Simcoe), which is sparse at low redshift and best seen at $z = 2$ to 3. Usable lines are from C IV, N V, Si IV, and (buried in the Lyα forest) O VI. O VI/C IV gives electron densities agreeing with SPH simulations, and this, with ionizing flux estimated from Si IV/C IV, enables some abundance estimates to be made: [C/H] $\simeq -2$ in the neighborhood of galaxies. More sensitivity is obtained by stacking C IV spectra and computing pixel optical depths; some weird enriched regions have been found in this way. No chemically pristine regions have been found (Rob Simcoe). Typically [C/H] $\simeq -2.5$ whenever Lyα is seen, although it can go down as low as -3.5; the problem is that one is fighting against ever diminishing column densities!

31.12   Conclusions

Highlights of this conference have been in my view the new insights into the *s*- and *r*-processes and the explosive increase in details of stellar abundances in the Galactic halo and in dwarf spheroidals, together with their star formation history and age-metallicity relations.

I do not think a conference on the origin and evolution of the elements would be complete without a survey of where the baryons and metals are in the Universe as a whole, and where they were at a substantial redshift—questions that have been addressed in the last few years by Persic & Salucci (1992) and Fukugita, Hogan, & Peebles (1998), as far as baryons are concerned, and including metals by Pettini (1999), Pagel (2002), and Lilly, Carollo, & Stockton (2004). Table 31.2 is based on work by Finoguenov, Burkert, & Böhringer (2003), whose numbers I quote with their kind permission.

While it has long been known that stars only account for a tenth or so of the baryonic matter density deduced from primordial deuterium and the cosmic microwave background angular fluctuation spectrum, there has been uncertainty as to where the missing baryons reside at present, although at $z = 2.5$ they are likely to be in the ionized gas associated with the Lyα forest. There is probably less such gas around today, much of the remainder being associated with O VI gas, which also contains like half the metals, assuming 0.2 solar abundance (the metal density is somewhat more robust than the baryon density, according to Mathur, Weinberg, & Chen 2003); the other half is mainly in stars.

Pettini (1999) drew attention to the fact that, while a quarter of all stars had been born by a redshift of 2.5, it was not possible to account for the corresponding quarter of today's metals on the basis of damped Lyα absorbers, Lyman-break galaxies, or the Lyα forest. This has given rise to two bold, but possible hypotheses involving dust in SCUBA galaxies (Dunne, Eales, & Edmunds 2003) and the arguments of Finoguenov et al. based on non-evolution of intracluster gas and its metal content since $z = 3$; so together these two sites might account for the bulk of the missing metals. Finoguenov et al. argue for early enrichment of the intracluster/protocluster gas by a top-heavy initial mass function. I am not sure if this is necessary; the issue may be decidable from the sort of data presented by Michael Loewenstein. In any case, the numbers in the table indicate an average yield for the whole Universe of about twice solar, similar to what one can get from a conventional Salpeter mass function.

References

Allende Prieto, C., Lambert, D. L., & Asplund, M. 2001, ApJ, 556, L63
——. 2002, ApJ, 573, L137
Blair, W. P., & Kirshner, R. P. 1985, ApJ, 289, 582
Blair, W. P., Kirshner, R. P., & Chevalier, R. A. 1982, ApJ, 254, 50
Chiappini, C., Matteucci, F., & Gratton, R. 1997, ApJ, 477, 765
Dekel, A., & Silk, J. 1986, ApJ, 303, 39
Dennefeld, M., & Kunth, D. 1981, AJ, 86, 989
Dunne, L., Eales, S. A., & Edmunds, M. G. 2003, MNRAS, 341, 589
Edvardsson, B., Andersen, J., Gustafsson, B., Lambert, D. L., Nissen, P. E., & Tomkin, J. 1993, A&A, 275, 101
Esteban, C., Peimbert, M., Torres-Peimbert, S. & Escalante, V. 1998, MNRAS, 295, 401
Finoguenov, A., Burkert, A., & Böhringer, H. 2003, ApJ, 594, 136
Fukugita, M., Hogan, C. J., & Peebles, P. J. E. 1998, ApJ, 503, 518
Garnett, D. R. 1999, in IAU Symp. 190, New Views of the Magellanic Clouds, ed. Y.-H. Chu et al. (Dordrecht: Kluwer), 266
Gilmore, G., & Wyse, R. F. G. 1991, ApJ, 367, L55
Holweger, H. 2001, in Solar and Galactic Composition, ed. R. F. Wimmer-Schweingruber (New York: AIP), 23
Korn, A. J., Keller, S. C., Kaufer, A., Langer, N., Przybilla, N., Stahl, O., & Wolf, B. 2002, A&A, 385, 143
Kurt, C. M., Dufour, R. J., Garnett, D. R., Skillman, E. D., Mathis, J. S., Peimbert, M., Torres-Peimbert, S., & Ruiz, M.-T. 1999, ApJ, 518, 246
Lambert, D. L., Heath, J. E., Lemke, M., & Drake, J. 1996, ApJ, 103, 183
Lilly, S. J., Carollo, C. M., & Stockton, A. N. 2004, in Origins 2002: the Heavy Element Trail from Galaxies to Habitable Worlds, ed. C. E. Woodward & E. P. Smith (San Francisco: ASP), in press (astro-ph/0209243)
Luck, R. E., Moffett, T. J., Barnes, T. G., & Gieren, W. P. 1998, AJ, 115, 605
Mathur, S., Weinberg, D., & Chen, X. 2003, ApJ, 582, 82
Meyer, D. M., Jura, M., & Cardelli, J. A. 1998, ApJ, 493, 222

Meynet, G., & Maeder, A. 2002, A&A, 390, 561

Monteverde, M. I., Herrero, A., Lennon, D. J. & Kudritzki, R.-P. 1997, ApJ, 474, L107

Otsuki, K., Mathews, G. J., & Kajino, T. 2002, NewA, 8, 767

Pagel, B. E. J. 1989, in Evolutionary Phenomena in Galaxies, ed. J. E. Beckman & B. E. J. Pagel (Cambridge: Cambridge Univ. Press), 201

——. 2001, in Cosmic Evolution, ed. E. Vangioni-Flam, R. Ferlet, & M. Lemoine (New Jersey: World Scientific), 223

——. 2002, in Chemical Enrichment of Intra-cluster and Intergalactic Medium, ed. R. Fusco-Fermiano & F. Matteucci (San Francisco: ASP), 489

Pagel, B. E. J., Edmunds, M. G., & Smith, G. 1980, MNRAS, 193, 219

Persic, M., & Salucci, P. 1992, MNRAS, 258, 14P

Pettini, M. 1999, in Chemical Evolution from Zero to High Red-shift, ed. J. Walsh & M. Rosa (Berlin: Springer), 233

Pilyugin, L. S. 2001, A&A, 374, 412

Pinsonneault, M. H., Steigman, G., Walker, T. P., & Naranayan, V. K. 1999, ApJ, 527, 180

Rolleston, W. R. J., Brown, P. J. F., Dufton, P. L., & Howarth, I. D. 1996, A&A, 315, 95

Russell, S. C., & Bessell, M. S. 1989, ApJS, 70, 865

Seeger, P. A., Fowler, W. A., & Clayton, D. D. 1965, ApJS, 11, 121

Skillman, E. D., Terlevich, R., & Melnick, J. 1989, MNRAS, 240, 563

Spite, M., Barbuy, B., & Spite, F. 1989, A&A, 222, 35

Steidel, C. C., Adelberger, K. L., Giavalisco, M., Dickinson, M., & Pettini, M. 1999, ApJ, 519, 1

Tsujimoto, T., Shigeyama, T., & Yoshii, Y. 2000, ApJ, 531, L33

Venn, K. A. 1999, ApJ, 518, 405

——. 2004, in Carnegie Observatories Astrophysics Series, Vol. 4: Origin and Evolution of the Elements, ed. A. McWilliam & M. Rauch (Pasadena: Carnegie Observatories, http://www.ociw.edu/symposia/series/symposium4/proceedings.html)

Venn, K. A., et al. 2001, ApJ, 547, 765

Venn, K. A., McCarthy, J. K. Lennon, D. J., Przybilla, N., Kudritzki, R.-P., & Lemke, M. 2000, ApJ, 541, 610

Vílchez, J. M., Pagel, B. E. J., Díaz, A. I., Terlevich, E., & Edmunds, M. G. 1988, MNRAS, 235, 633

Wallerstein, G. 1962, ApJS, 6, 407

Wasserburg, G. J., Busso, M., & Gallino, R. 1996, ApJ, 466, L109

Credits

The following figures in this volume were reproduced with permission from the original author and publisher.

Figure 4.7: Reprinted with permission from Sneden, C., & Cowan, J. J. 2003, Science, 299, 70, "Genesis of the Heaviest Elements in the Milky Way Galaxy." Copyright 2003 AAAS.

Figure 17.4–17.5: Grebel, E. K., Gallagher, J. S., & Harbeck, D. 2003, AJ, 125, 1926, "The Progenitors of Dwarf Spheroidal Galaxies." Reproduced with permission from the American Astronomical Society.

Figure 18.1, 18.9: Ferguson, A. M. N., et al. 2002, AJ, 124, 1452, "Evidence for Stellar Substructure in the Halo and Outer Disk of M31." Reproduced with permission from the American Astronomical Society.

Figure 18.3: Bellazzini, M., et al. 2003, A&A, 405, 867, "The Andromeda Project. I. Deep HST-WFPC2 V, I Photometry of 16 Fields toward the Disk and the Halo of the M31 Galaxy. Probing the Stellar Content and Metallicity Distribution." Reproduced with permission from *Astronomy and Astrophysics*.

Figure 18.8: Rich, R. M., et al. 2004, AJ, in press (astro-ph/0309296), "Deep Photometry in a Remote M31 Major Axis Field near G1." Reitzel, D. B., et al. 2004, AJ, in press (astro-ph/0309295), "Keck Spectroscopy of Red Giant Stars in the Vicinity of M31's Massive Globular Cluster G1." Reproduced with permission from the American Astronomical Society.

Figure 18.10: Da Costa, G. S., Armandroff, T. E., & Caldwell, N. 2002, AJ, 124, 332, "The Dwarf Spheroidal Companions to M31: WFPC2 Observations of Andromeda III." Reproduced with permission from the American Astronomical Society.

Figure 18.13: Pritchet, C. J., & van den Bergh, S. 1994, AJ, 107, 1730, "Faint Surface Photometry of the Halo of M31." Reproduced with permission from the American Astronomical Society.

Figure 21.4: Draine, B. T. 2003, ApJ, 598, 1026, "Scattering by Interstellar Dust Grains. II. X-rays." Reproduced with permission from the American Astronomical Society. Reproduced with permission from the American Astronomical Society.

Figure 23.2: Pagani, L., et al. 2003, A&A, 402, L77, "Low Upper Limits on the O_2 Abundance from the Odin Satellite." Reproduced with permission from *Astronomy and Astrophysics*.

Figure 24.4: Martin, C. L., Kobulnicky, H. A., & Heckman, T. M. 2002, ApJ, 574, 663, "The Metal Content of Dwarf Starburst Winds: Results from Chandra Observations of NGC 1569." Reproduced with permission from the American Astronomical Society.

Figure 26.6: Kennicutt, R. C., Bresolin, F., & Garnett, D. R. 2003, ApJ, 591, 801, "The Composition Gradient in M101 Revisited. II. Electron Temperatures and Implications for the Nebular Abundance Scale." Reproduced with permission from the American Astronomical Society.

Figure 28.3: Warner, C., et al. 2002, ApJ, 567, 68, "The Metallicity of the Redshift 4.16 Quasar BR2248–1242." Reproduced with permission from the American Astronomical Society.

Figure 28.4: Dietrich, M., et al. 2003, ApJ, 589, 722, "Quasar Elemental Abundances at High Redshifts." Reproduced with permission from the American Astronomical Society.

Figure 30.2: Miralda-Escudé, J., et al. 1996, ApJ, 471, 582, "The Ly-α Forest from Gravitational Collapse in the Cold Dark Matter + Lambda Model." Reproduced with permission from the American Astronomical Society.

Figure 30.6: Schaye, J., et al. 2000, ApJ, 541, L1, "The Detection of Oxygen in the Low-density Intergalactic Medium." Reproduced with permission from the American Astronomical Society.

Figure 31.1: Pagel, B. E. J. 2001, in Cosmic Evolution, ed. E. Vangioni-Flam, R. Ferlet, & M. Lemoine (New Jersey: World Scientific), 223, "Chemical Evolution of the Galaxy: The G-dwarf Problem and Radioactive Chronology Revisited Taking Account of the Thick Disk." Reproduced with permission from World Scientific Publishing Co.